Novel Perspectives in MicroRNAs

Novel Perspectives in MicroRNAs

Editor: Daniel Fletcher

MURPHY & MOORE
www.murphy-moorepublishing.com

www.murphy-moorepublishing.com

MURPHY & MOORE

Cataloging-in-Publication Data

Novel perspectives in MicroRNAs / edited by Daniel Fletcher.
p. cm.
Includes bibliographical references and index.
ISBN 978-1-63987-722-5
1. MicroRNA. 2. Small interfering RNA. 3. RNA. I. Fletcher, Daniel.
QP623.5.S63 N68 2023
572.88--dc23

Murphy & Moore Publishing
1 Rockefeller Plaza,
New York City,
NY 10020, USA

ISBN 978-1-63987-722-5

Contents

Preface

In my initial years as a student, I used to run to the library at every possible instance to grab a book and learn something new. Books were my primary source of knowledge and I would not have come such a long way without all that I learnt from them. Thus, when I was approached to edit this book; I became understandably nostalgic. It was an absolute honor to be considered worthy of guiding the current generation as well as those to come. I put all my knowledge and hard work into making this book most beneficial for its readers.

MicroRNA (miRNA) refers to a small single-stranded non-coding RNA molecule. It can be found in animals, plants and some viruses. miRNAs function in post-transcriptional regulation of gene expression and RNA silencing. They work through base-pairing along with complementary sequences in mRNA molecules. Several mammalian cell types contain a high concentration of miRNAs as circulating in extracellular fluid, which are released into body fluids such as blood and cerebrospinal fluid, and could be used as biomarkers in a variety of disorders. MiRNAs play a role in gene regulation and can also produce histone modifications and DNA methylation at promoter sites, which can affect the target gene expression. MiRNAs are important in modulating complex enzymatic cascades, such as the hemostatic blood coagulation system. The topics included in this book on microRNAs are of utmost significance and bound to provide incredible insights to the readers. Researchers and students engaged in the study of this molecule will be assisted by it.

I wish to thank my publisher for supporting me at every step. I would also like to thank all the authors who have contributed their researches in this book. I hope this book will be a valuable contribution to the progress of the field.

Editor

1

Relationship between Altered miRNA Expression and DNA Methylation of the DLK1-DIO3 Region in Azacitidine-Treated Patients with Myelodysplastic Syndromes and Acute Myeloid Leukemia with Myelodysplasia-Related Changes

Michaela Dostalova Merkerova [1,*], Hana Remesova [1], Zdenek Krejcik [1], Nikoleta Loudova [1], Andrea Hrustincova [1], Katarina Szikszai [1], Jaroslav Cermak [1], Anna Jonasova [2] and Monika Belickova [1]

[1] Institute of Hematology and Blood Transfusion, U Nemocnice 1, 128 20 Prague 2, Czech Republic; hana.remesova@uhkt.cz (H.R.); zdenek.krejcik@uhkt.cz (Z.K.); nikoleta.loudova@uhkt.cz (N.L.); andrea.mrhalkova@uhkt.cz (A.H.); katarina.szikszai@uhkt.cz (K.S.); jaroslav.cermak@uhkt.cz (J.C.); monika.belickova@uhkt.cz (M.B.)

[2] General University Hospital, 128 08 Prague, Czech Republic; atjonas@hotmail.com

[*] Correspondence: Michaela.Merkerova@uhkt.cz;

Abstract: The *DLK1–DIO3* region contains a large miRNA cluster, the overexpression of which has previously been associated with myelodysplastic syndromes (MDS). To reveal whether this overexpression is epigenetically regulated, we performed an integrative analysis of miRNA/mRNA expression and DNA methylation of the regulatory sequences in the region (promoter of the *MEG3* gene) in CD34+ bone marrow cells from the patients with higher-risk MDS and acute myeloid leukemia with myelodysplasia-related changes (AML-MRC), before and during hypomethylating therapy with azacytidine (AZA). Before treatment, 50% of patients showed significant miRNA/mRNA overexpression in conjunction with a diagnosis of AML-MRC. Importantly, increased level of *MEG3* was associated with poor outcome. After AZA treatment, the expression levels were reduced and were closer to those seen in the healthy controls. In half of the patients, we observed significant hypermethylation in a region preceding the *MEG3* gene that negatively correlated with expression. Interestingly, this hypermethylation (when found before treatment) was associated with longer progression-free survival after therapy initiation. However, neither expression nor methylation status were associated with future responsiveness to AZA treatment. In conclusion, we correlated expression and methylation changes in the *DLK1–DIO3* region, and we propose a complex model for regulation of this region in myelodysplasia.

Keywords: microRNA; myelodysplastic syndromes; acute myeloid leukemia; azacitidine; 14q32; MEG3

---

1. Introduction

Myelodysplastic syndromes (MDS) represent a heterogeneous spectrum of hematopoietic stem cell (HSC) disorders characterized by inefficient hematopoiesis, peripheral blood cytopenia, and dysplasia in one or more myeloid cell lineages; they also have a tendency to evolve into acute myeloid leukemia with myelodysplasia-related changes (AML-MRC), which develops in approximately 30–40% of patients [1,2]. In recent years, the hypomethylating agent azacitidine (AZA) has become the standard therapy for higher-risk MDS and AML-MRC. AZA prolongs patient survival, improves clinical outcomes and life quality, and delays progression to AML. However, response to AZA therapy occurs in only approximately 40–50% of MDS patients [3–5]. AZA is a cytidine analog, which at low doses functions as a DNA methyltransferase inhibitor, causing DNA hypomethylation; at high doses, AZA shows direct cytotoxicity to abnormal bone marrow (BM) hematopoietic cells through its incorporation into DNA and RNA, resulting in cell death [6].

The *DLK1–DIO3* genomic region is located on chromosome 14q32 and contains three paternally expressed protein-coding genes (*DLK1, RTL1,* and *DIO3*), three maternally expressed noncoding genes (*MEG3, MEG8,* and antisense *RTL1*), many small nucleolar RNAs (snoRNAs), and one of the largest miRNA clusters (54 miRNAs) in the human genome [7] (Figure 1). Expression within the *DLK1–DIO3* region is under the control of three differentially methylated regions (DMRs): Intergenic DMR (IG-DMR), *DLK1*-DMR, and *MEG3*-DMR [8]. The *MEG3*-DMR overlaps with the promoter and the 5′-end of the *MEG3* gene and includes several binding sites for CCCTC-binding factor (CTCF). CTCF blocks the interaction between enhancers and promoters, causing transcriptional repression. It exerts its regulatory function by binding to unmethylated DNA, thus preventing the expression of target genes [9].

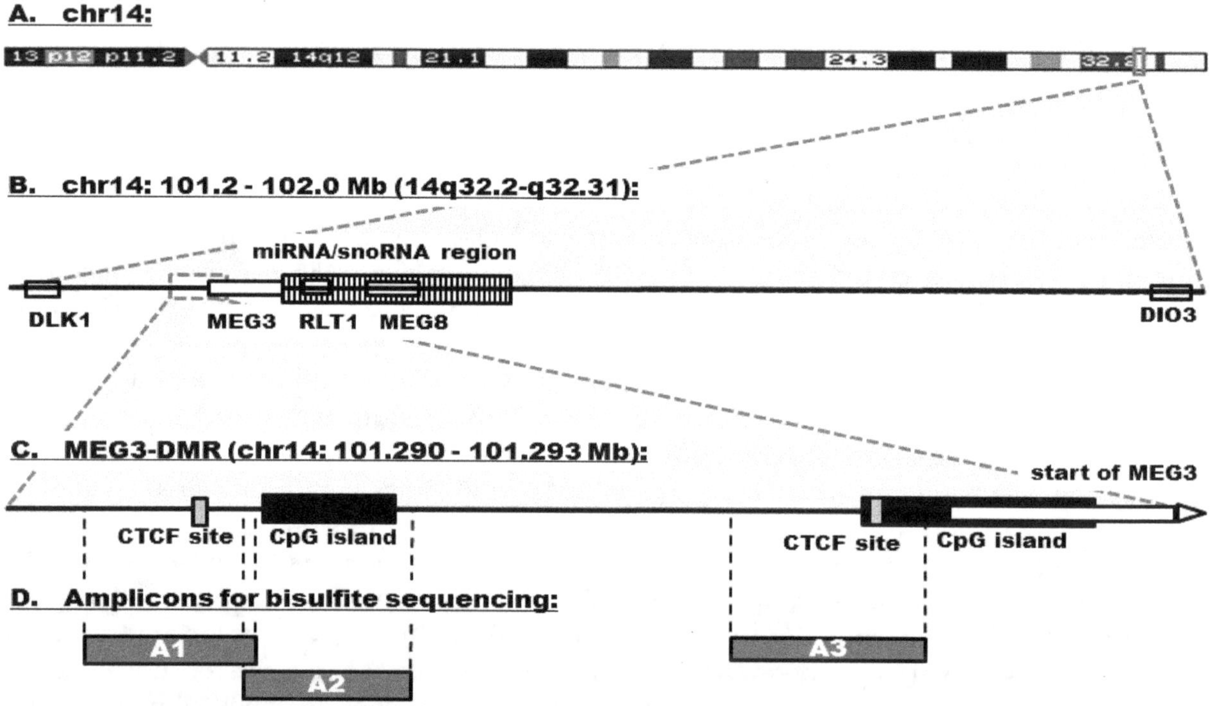

Figure 1. Genomic organization of the *DLK1–DIO3* domain. (**A**) The region is located on chromosome 14q32.2. (**B**) It contains three paternally expressed protein-coding genes (*DLK1, RTL1,* and *DIO3*), three maternally expressed noncoding genes (*MEG3, MEG8,* and antisense *RTL1*), many snoRNAs, and a large cluster of 54 miRNAs (striped bar). (**C**) The expression within the region is under the control of *MEG3*-DMR that overlaps with the promoter and the 5′-end of the *MEG3* gene. *MEG3*-DMR contains several CTFC binding sites (gray bars) and CpG islands (black bars). (**D**) The schema shows localization of amplicons (A1–A3) in the region that were used for methylation analyses.

The miRNAs in the *DLK1–DIO3* region possess oncogenic and tumor suppressor properties and are frequently deregulated in various cancers [7]. Their deregulation has been linked to abnormal induction of apoptosis and suppression of proliferation [10–12]. They are also involved in hematopoietic stem/progenitor cell (HSPC) differentiation [13–15]. The upregulation of miRNAs in the *DLK1–DIO3* region has been reported in acute promyelocytic leukemia (APL, a subtype of AML with t(15;17)) [16–19] and MDS [20–23]. In APL, a comparison of diagnostic and remission patient samples showed that strong upregulation of these miRNAs was correlated with hypermethylation at *MEG3*-DMR, including the CTCF binding site motifs [24]. Only one publication has studied the methylation of this region in MDS patients; Benetatos et al. (2010) showed that hypermethylation of the *MEG3* promoter occurred in 35% of MDS patients [25].

Using SNP array-based karyotyping in cytogenetically normal MDS patients, we previously demonstrated that upregulation of the miRNAs located in the *DLK1–DIO3* domain seen in MDS is probably not caused by any chromosomal aberration or uniparental disomy [22,26]. Due to the similarity between MDS and AML, it may be expected that overexpression of these miRNAs is associated with aberrant hypermethylation of the locus. In this study, we performed a thorough, integrative analysis investigating the expression of miRNAs and mRNAs encoded in the *DLK1–DIO3* region and compared these data with the methylation status of *MEG3*-DMR in CD34+ BM cells (i.e., undifferentiated hematopoietic stem/progenitor cells (HSPCs), including blasts) from higher-risk MDS/AML-MRC patients, particularly focusing on the effects of hypomethylating AZA therapy.

2. Materials and Methods

2.1. Patients

BM samples from 12 patients with higher-risk MDS (N = 7) or AML-MRC (N = 5) without previous malignancy, chemotherapy, or transplantation were obtained between the years 2010–2013 from the First Department of Medicine, Department of Hematology, General University Hospital, and from the Institute of Hematology and Blood Transfusion in Prague, Czech Rep. Paired samples (N = 24) before and during AZA therapy (collected at the time of the best response, between cycles 4 and 11) were collected from each patient. Patient age ranged from 63–82 years (average 69), and the female/male distribution was 6/6. AZA was administered at 75 mg/m^2/day for 7 consecutive days every 28 days. The hematological evaluation of the response to treatment was performed according to the International Working Group (IWG) criteria for MDS and AML [27,28]. Seven patients were considered as responders, and five patients were nonresponders. The detailed clinical characteristics of all patients are summarized in Table S1. All samples from this cohort were used for parallel investigation of miRNA/mRNA expression and DNA methylation. BM samples obtained from age-matched healthy donors with no adverse medical histories were used as controls. Due to limited amounts of sample, we had to use different controls for each type of analysis. Written informed consent was obtained from all test subjects in accordance with the approval of the Institutional Review Board (No. NS9634-4/2008).

2.2. Cell Separation and Nucleic Acid Extraction

Mononuclear cells (MNCs) were separated from BM aspirates by Ficoll-Paque density gradient centrifugation (GE Healthcare, Munich, Germany). CD34+ cells were isolated from MNCs by the Direct CD34 Progenitor Cell Isolation MACS Kit (Miltenyi Biotec, Bergisch Gladbach, Germany). Total RNA was extracted from separated CD34+ cells using the acid guanidinium thiocyanate-phenol-chloroform method, and the RNA samples were incubated with DNase I (Qiagen, Hilden, Germany) to prevent genomic DNA contamination. The salting-out method was used to isolate DNA from CD34+ cells. The concentrations of DNA and RNA were assessed using a NanoDrop and a Qubit 2.0 Fluorometer (both from Thermo Fisher Scientific, Waltham, MA, USA). The integrity of total RNA was evaluated by an Agilent Bioanalyzer 2100 (Agilent Technologies, Santa Clara, CA, USA).

2.3. miRNA Expression

The miRNA expression data were retrieved from our previous project, which studied miRNA expression changes on a genome-wide level before and during AZA therapy using Agilent Human miRNA Microarrays, Release 19.0, 8 × 60K (Agilent Technologies, Santa Clara, CA, USA) [29]. That study included data on miRNA expression from a complete cohort of 12 patients and four controls. For this publication, we have focused on only miRNAs located in the *DLK1–DIO3* region.

2.4. Gene Expression

Data (NCBI GEO accession number GSE77750) from our previous genome-wide analysis performed using Illumina microarrays (HumanHT-12 v4 Expression, Illumina, San Diego, CA, USA) were used [30]. For this project, the expression of only five genes in the *DLK1–DIO3* region (*DLK1, RTL1, DIO3, MEG3,* and *MEG8*) and of the *CTCF* was further considered within our patient cohort (the same 12 patients with MDS/AML-MRC) and 10 controls.

For validation, the expression of *MEG3* was measured in an independent cohort of samples using RT-qPCR. The cohort included 79 patients with primary MDS (12 MDS with del(5q), 8 MDS with single lineage dysplasia (SLD), 8 MDS with ring sideroblasts and SLD (RS-SLD), 12 MDS with multilineage dysplasia (MLD), 8 MDS-RS-MLD, 10 MDS with excess blasts 1 (EB1), 21 MDS-EB2), 12 AML-MRC, and 13 age-matched healthy controls. SuperScript IV VILO Master Mix and the TaqMan Gene Expression Assay with TaqMan Universal Master Mix (all from Thermo Fisher Scientific, Waltham, MA, USA) were used. The data were normalized to *HTFR* as the reference gene and processed with the ddCt method.

2.5. Bisulfite Sequencing

Genomic DNA isolated from CD34+ BM cells from the 24 samples (the same set of paired samples from 12 patients as in the expression analyses) and three healthy controls was used for analysis of the methylation status of *MEG3*-DMR. Bisulfite conversion was performed using EpiTect Bisulfite Kit (Qiagen, Hilden, Germany). Based on the publication by Manodoro et al. [24], three regions with significant methylation changes in APL were chosen, and previously published primers for these amplicons (A1, A2, and A3) were used (Table S2). Amplicons were generated using FastStart High Fidelity PCR System (Roche Applied Science, Mannheim, Germany) and purified with Agencourt AMPure beads (Beckman Coulter, Krefeld, Germany). The purified PCR products were sequenced using the Roche 454 GS Junior System. Image processing was performed the GS RunBrowser software (Roche Applied Science). Methylation status of CpG sites was quantified using QUMA web-based tool [31].

2.6. Statistical Analyses

Statistical analyses were performed using the GraphPad Prism v7.03 (GraphPad Prism Software, La Jolla, CA, USA). Unpaired and paired t-tests were used to compare miRNA/mRNA expression levels and clinical parameters between different groups of samples. The chi-square test was applied for comparison of categorical clinical variables. Hierarchical clustering analysis was done using average linkage and Euclidean distance. Correlation analyses were done by computing Pearson correlation coefficients (r). Overall survival (OS) and progression-free survival (PFS) curves were generated by the Kaplan-Meier method, and the differences between groups were assessed by the log-rank test. OS and PFS were defined as the time from the beginning of treatment until death from any cause (OS) or disease progression (PFS). Differences were considered statistically significant at $p < 0.05$.

3. Results

3.1. Expression of miRNAs Encoded within the DLK1–DIO3 Region

Within the microarray miRNA expression data, signals of 25 miRNAs located in the *DLK1–DIO3* region were detected. Their levels varied by three orders of magnitude in microarray signal intensity. Before treatment, increased expression of these 25 miRNAs was detected in 6 of 12 samples (50% of patients), and in the remaining 6 samples, the miRNA expression was similar to that detected in controls (Figure 2A).

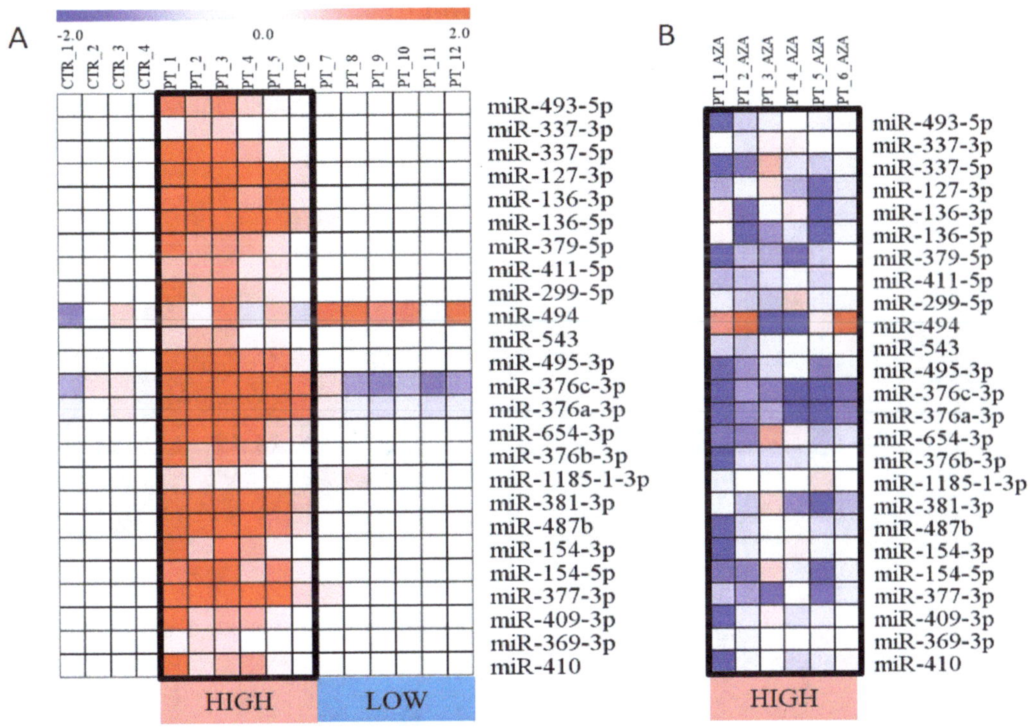

Figure 2. Expression of miRNAs encoded in the *DLK1–DIO3* region. (**A**) miRNA expression in untreated patients (PT) and healthy controls (CTR). The expression level is calculated as the binary logarithm of fold change (logFC) compared to the mean expression of controls. The miRNAs are aligned in the heatmap according to their location on chr14. (**B**) Reduction of miRNA expression following AZA treatment. The changes in miRNA levels are calculated as the logFC between paired samples (AZA treatment vs. pretreatment). Only those patients with miRNA overexpression before treatment are shown. Both heatmaps use a color gradient intensity scale to express visually the logFC values in a range of colors (blue—downregulation, red—upregulation, white—unchanged expression).

Furthermore, we compared miRNA expression as paired samples from pretreatment and during AZA therapy. miRNA expression was reduced following therapy specifically in the patients with overexpressed miRNAs before treatment, and their levels decreased to approximately half the pretreatment expression levels. In contrast, the levels of these miRNAs remained unchanged in the patients with baseline miRNA levels before AZA treatment (Figure 2B).

3.2. Expression of Genes Related to the DLK1–DIO3 Region

Expression of five genes encoded in the *DLK1–DIO3* region (*DLK1, RTL1, DIO3, MEG3,* and *MEG8*) and of the *CTCF* gene were measured using microarrays in the same cohort of patients as for the miRNA analysis. Signals of probes for the *RTL1* and *DIO3* genes were below the detection limit. Expression of *DLK1, MEG3,* and *MEG8* followed the same trend as that of the clustered miRNAs; their levels were increased in the untreated patients with elevated miRNA expression and subsequently reduced during AZA therapy. In the patients with baseline miRNA expression, the transcript levels remained unchanged compared with the levels seen in controls. Interestingly, expression of the *CTCF* gene was also moderately higher ($p = 0.185$) in the patients with increased miRNA expression in the *DLK1–DIO3* region and then reversed to baseline after therapy ($p = 0.026$) (Figure S1). Quantification of *MEG3* expression in a larger, independent cohort of patients showed significant upregulation ($p = 0.0049$) of *MEG3* in MDS/AML-MRC patients compared to healthy donors (Figure 3A). Interestingly, this upregulation occurred more frequently in patients with higher-risk diagnoses than in those with early subtypes of MDS (Figure 3B) and was associated with poor OS (hazard ratio [HR] = 2.6013, 95%

confidence interval [CI] = 1.209 to 5.648, $p = 0.0008$) and PFS (HR = 3.127, 95% CI = 1.411 to 6.933, $p < 0.0001$) in untreated patients (Figure 3C,D).

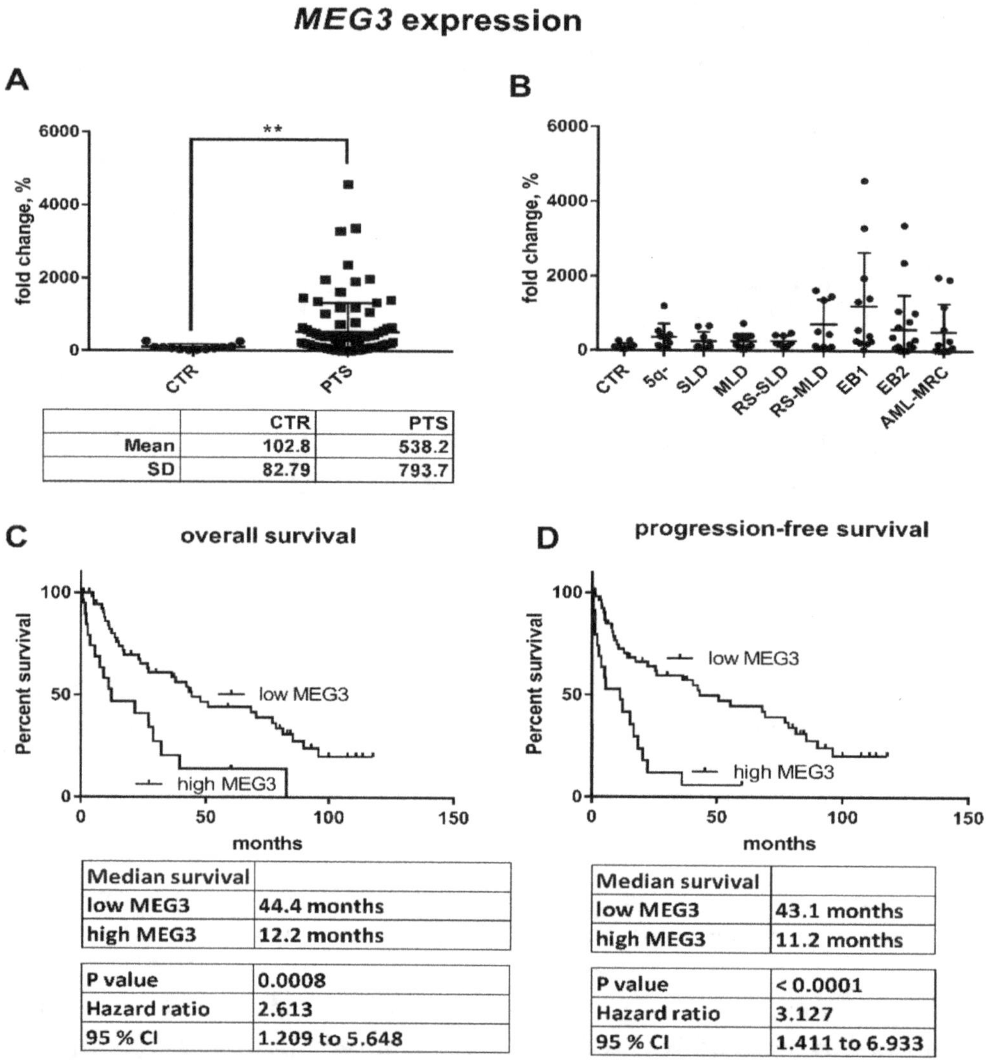

	CTR	PTS
Mean	102.8	538.2
SD	82.79	793.7

Median survival	
low MEG3	44.4 months
high MEG3	12.2 months

P value	0.0008
Hazard ratio	2.613
95 % CI	1.209 to 5.648

Median survival	
low MEG3	43.1 months
high MEG3	11.2 months

P value	< 0.0001
Hazard ratio	3.127
95 % CI	1.411 to 6.933

Figure 3. Relative *MEG3* expression in the validation cohort of MDS/AML-MRC patients. (**A**) The expression of *MEG3* was significantly different between healthy controls (CTR) and all patients (PTS) (** $p < 0.01$). (**B**) Further distribution of the patients according to their diagnosis revealed increased expression of *MEG3* particularly in the patients with advanced disease. (**C**) Overall survival and (**D**) progression-free survival of patients with low vs. high *MEG3* expression (the cut-off was set up to mean level of *MEG3*).

3.3. DNA Methylation in the MEG3-DMR

For DNA methylation analysis, amplicon libraries from three distinct regions overlapping the *MEG3*-DMR were prepared, and a total of 53,849 read sequences were generated using amplicon bisulfite sequencing. After a quality control check, 471 sequence reads on average were analyzed per amplicon per sample. To validate the reproducibility of the sequencing, we compared the methylation level of two overlapping CpGs independently sequenced in amplicons A1 and A2, which proved to have a high level of correlation (r = 0.901, $p < 0.001$).

Average methylation profiles examined throughout all three amplicons revealed several differentially methylated CpGs among various groups of samples (Figure 4A). The comparison between patients and controls showed significant hypermethylation of almost all CpGs in the A3 region in patients (overall hypermethylation >30% compared with the mean of the controls, which was observed in 50% of untreated patients) (Figure 4A,B). Following AZA therapy, a majority of CpGs from the A2 region (Figure 4C) and one CpG (#236) from the A1 region (Figure 4B) were significantly hypomethylated compared with the pretreatment status, particularly in the patients with increased miRNA levels. Finally, we examined methylation of CpGs in two CTCF binding sites located in our amplicons (CTCF_B: Amplicon A1 CpGs #311-319, and CTCF_D: Amplicon A3 CpGs #308-314) and detected significant hypermethylation of A3 CpG #314 in patient samples (Figure 4B).

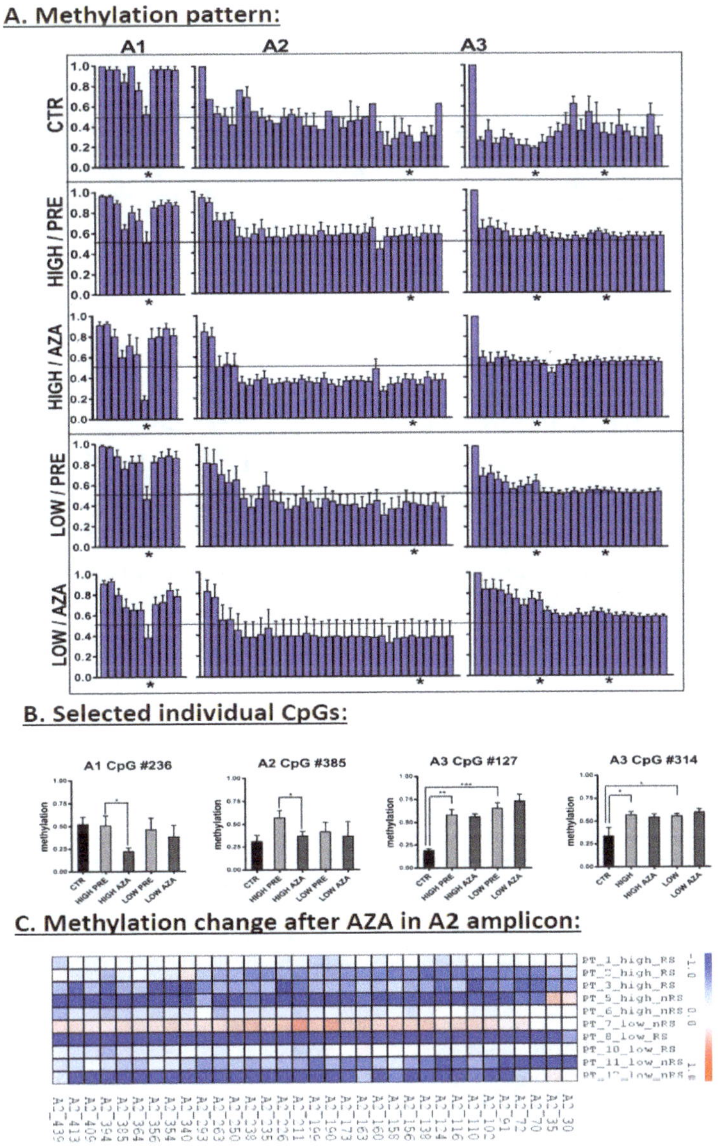

Figure 4. DNA methylation of the *MEG3*-DMR. (**A**) Average methylation profiles examined in the A1–A3 amplicons. Samples were separated into one control group (CTR) and four patient groups that were then divided according to the miRNA expression (HIGH or LOW groups) and AZA therapy (PRE—pretreatment, AZA—AZA-treatment). The horizontal line spanning the plots indicates 50% of methylation. (**B**) Methylation of individual CpGs that are marked in the first section by stars below the graphs. * $p < 0.05$, ** $p < 0.01$, *** $p < 0.001$. (**C**) Methylation change in the A2 amplicon after AZA treatment. Methylation change is expressed as the logFC in AZA-treated samples compared with paired pretreatment samples using a color gradient intensity scale (blue—hypomethylation, red—hypermethylation, white—unchanged methylation). RS—responders to AZA, nRS—nonresponders to AZA.

3.4. Correlation of Expression and Methylation Data

We performed a series of pairwise correlation analyses of the expression of selected miRNAs/mRNAs across all patient data (Table S3). Based on this analysis, we identified strong correlation of expression levels of majority of clustered miRNAs. Moreover, the level of *MEG3* positively correlated with the levels of *MEG8, DLK1,* and the miRNAs. However, *CTCF* expression was not significantly correlated with the levels of any of the tested genes/miRNAs.

Pairwise correlation analyses were performed also for DNA methylation data (Table S3). A significant positive correlation of the methylation levels of individual CpGs located nearby was detected. Four separate clusters regulated in common were identified: CpGs from the A1 amplicon, those from the A2 amplicon, and two separate CpG clusters within the A3 amplicon (A3_1—CpGs #54-171 and A3_2—CpGs #227-388).

Importantly, the correlation of methylation status of individual CpGs with miRNA/mRNA expression levels identified that methylation in a part of the A3 region (CpGs #54–231) is negatively correlated with the expression of the majority of the studied miRNAs/mRNAs, with the exception of miR-136-3p.

3.5. Clinical Characteristics with Relation to the Expression and Methylation Data

Because only a half of the patients showed overexpression of miRNAs/mRNAs from *DLK1–DIO3* region, we divided the cohort according to the expression activity in the locus and analyzed the two resulting groups of patients according to clinical features (age, sex, karyotype, BM blasts, diagnosis, risk category using criteria of the International Prognostic Scoring System (IPSS), and response to hypomethylating therapy) (Table 1). Of these characteristics, only the percentage of BM blasts (patients with low expression: 10.4 $\pm$ 2.8%, patients with high expression: 31.6 $\pm$ 9.0%; $p = 0.048$) and diagnosis (patients with low expression: 6 MDS patients, patients with high expression: 1 MDS and 5 AML-MRC patients; $p = 0.015$) were significantly associated with expression activity. Concerning cytogenetics, neither particular aberrations nor their classification based on IPSS showed a correlation with expression levels. The rest of the variables, including overall response rate (ORR) to AZA ($p = 0.558$), were not statistically significant.

Table 1. Clinical characteristics of patients and their stratification according to expression activity in the *DLK1–DIO3* locus.

Variable		Patients with Baseline miRNA Expression	Patients with Increased miRNA Expression	*p*-Value
Age (years)	Mean (range)	73 (65–88)	69 (63–81)	0.294
Sex, *n*	Male Female	4 2	2 4	0.248
Karyotype, *n*	Normal Abnormal	2 4	1 5	0.505
Marrow blasts	Mean (range)	10.4 (5.1–20.0)	31.6 (18.0–75.0)	0.048
Diagnosis, *n*	MDS AML-MRC	6 0	1 5	0.015
IPSS, *n*	Intermediate-1/2 High	3 3	0 6	0.182
Response to AZA therapy	ORR, *n* (%)	3 (50%)	4 (67%)	0.558

IPSS—International Prognostic Scoring System, ORR—overall response rate.

Further, we investigated the survival of patients after the initiation of AZA therapy in relation to miRNA expression and DNA methylation in pretreatment samples. Concerning miRNA levels, we observed no differences in either OS ($p = 0.901$) or PFS ($p = 0.611$) between the patient with baseline vs. increased expression activity (Figure S2). For analysis of DNA methylation, we compared survival of the patients divided according to their overall methylation status in the three examined amplicons. Importantly, we found that PFS (but not OS) significantly correlated with methylation status of the A3 locus. Log-rank test showed that patients with higher methylation of this locus had a significantly longer PFS than those with lower methylation (HR = 0.220, 95% CI = 0.056 – 0.865, $p = 0.030$) (Figure 5).

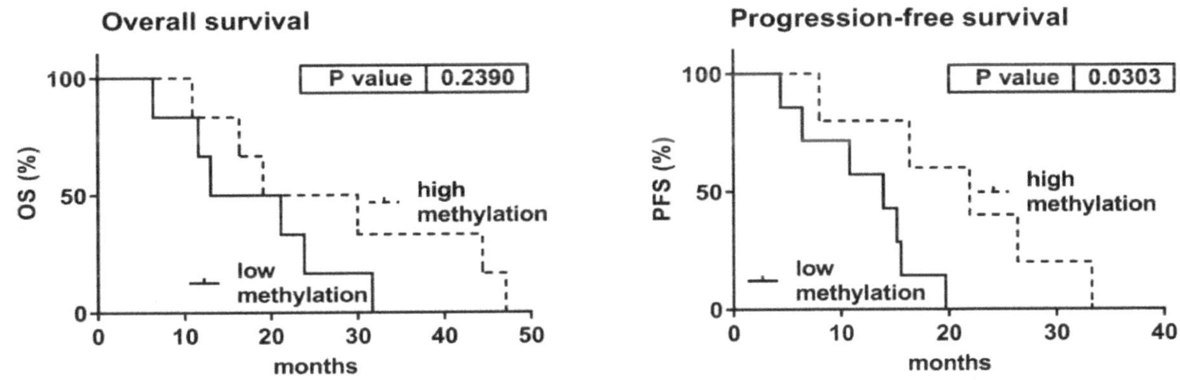

Figure 5. Patient survival after the initiation of AZA therapy according to DNA methylation of the A3 amplicon before treatment. DNA methylation in each sample was considered low or high according to the median value of overall methylation in the locus.

Finally, we investigated the potential application of the altered DNA methylation of the *MEG3*-DMR for prognostic purposes regarding the responsiveness to AZA treatment. However, we did not find any association between AZA response and methylation level in the analyzed amplicons. Although we observed a substantial decrease in DNA methylation in the A2 region following AZA therapy (Figure 4C), this change occurred in the majority of the patients regardless of their response status.

4. Discussion

In several recent publications, apparent overexpression of miRNAs located within the *DLK1–DIO3* region has been observed in MDS patients [20–23]. Following lenalidomide treatment, higher expression levels of these miRNAs diminished in the patients with low and intermediate-1 risk with a del(5q) aberration [22]. Overexpression of these miRNAs has been linked to hypermethylation of the *MEG3*-DMR in APL [24]. Therefore, we investigated the expression activity within the *DLK1–DIO3* region in the context of the methylation status of the *MEG3*-DMR in myelodysplasia. We focused on patients with higher-risk MDS/AML-MRC diagnoses treated with the hypomethylating agent AZA and examined gene expression levels and DNA methylation status in this chromosomal region following treatment. Although performed on a limited number of patients, our study provides a unique dataset focused particularly on the *DLK1–DIO3* region and includes information on miRNA and mRNA expression and methylation data. Furthermore, all the data were obtained from the same cohort of patients before and during hypomethylation therapy.

Initially, we studied expression within the *DLK1–DIO3* region. Our data confirmed that expression activity throughout the region was regulated uniformly because levels of the miRNAs and mRNAs significantly correlated within our patient cohort. In the untreated MDS/AML-MRC patients, 50% of samples showed significant overexpression of miRNAs and mRNAs encoded within the region. Other publications described increased expression of *DLK1–DIO3*-related miRNAs, even in somewhat higher proportions (approximately 90%) of MDS/AML-MRC patients [20–23].

Evaluation of expression data with clinical features showed that expression in the pretreatment samples was related to BM blast count, patient diagnosis, and outcome (OS and PFS). High expression was significantly associated with the diagnosis of AML-MRC and poor outcome, whereas low expression was related to MDS and favorable outcome. Because overexpression of these miRNAs has been linked to a specific cytogenetic aberration [t(15;17)] in primary AML (i.e., APL) [19], we inspected the chromosomes of our patients. However, no chromosomal abnormality or cytogenetic risk category was linked to the expression changes. Therefore, we conclude that the increased expression activity could be related to the process of leukemic transformation. Interestingly, *MEG3* overexpression has also been detected in CD34+ cells from patients with primary myelofibrosis (found in 65% of patients), and this upregulation was related to increased blast counts and higher percentages of circulating CD34+ cells [32].

As half of the patients in our cohort had increased expression activity in the *DLK1–DIO3* region, we examined whether this increase was affected by hypomethylating therapy. Indeed, we observed a reduction in expression to near normal levels after treatment with AZA. To distinguish whether this effect was directly caused by hypomethylation of DNA in the regulatory sequences of the region or if it was an indirect consequence of other changes caused by AZA exposure, we thoroughly examined the methylation pattern in the *MEG3*-DMR.

We investigated DNA methylation in the same regions (located in the *MEG3*-DMR) that were hypermethylated in APL so that we could compare methylation status in APL and myelodysplasia. In APL, the hypermethylation spread over all the three tested amplicons and diminished in remission, whereas we identified a distinct methylation pattern in our cohort of MDS/AML-MRC patients. In untreated patients, we revealed significant hypermethylation solely in the A3 amplicon (that closely precedes the *MEG3* start), whereas the remaining two amplicons had similar levels of methylation as the healthy controls. The hypermethylation in the A3 amplicon was seen in 50% of the MDS/AML-MRC untreated patients. For comparison, Benetatos et al. identified hypermethylation of the *MEG3* promoter in 35% of MDS and 48% of AML patients using conventional methylation-specific PCR [25]. In our patients, methylation of CpGs from the 5′ upstream part of the A3 locus negatively correlated with the expression of genes encoded in the *DLK1–DIO3* region. Interestingly, hypermethylation of the A3 locus before treatment had a positive impact on PFS after AZA therapy initiation.

Various methylation patterns were observed in the remaining amplicons. Of them, the hypomethylation seen in the whole A2 region following AZA therapy was remarkable. Although significant in only the patients with increased miRNA levels, this methylation change was apparent in a majority of patients regardless of expression activity in the *DLK1–DIO3* region. Because of its widespread detection, this hypomethylation might be attributed to the hypomethylating effect of AZA and could therefore influence the deregulated expression of the miRNA cluster seen in a portion of our patients.

In APL, Manodoro et al. documented an atypical pattern of hypermethylation in the *DLK1–DIO3* locus associated with higher expression of the clustered miRNAs that they attributed to the presence of CTCF binding sites [24]. A CTCF insulator binds to several target sites (to their unmethylated forms) in the *MEG3*-DMR, inhibiting transcriptional activity in the region [9]. Here, we detected significant hypermethylation at one of the CTCF binding sites in the A3 amplicon. This hypermethylation occurred in a majority of patients and was not affected by AZA treatment. Interestingly, expression of the *CTCF* gene was higher in the untreated patients with increased miRNA expression, which might interfere with the potential impact of hypermethylation of its binding site in the *MEG3*-DMR. Unfortunately, expression of the CTCF insulator was not considered in the publication studying APL [24]. In summary, aberrant hypermethylation at one of the CTCF binding sites in the region might affect the activity of CTCF during transcriptional regulation of the *DLK1–DIO3* region, but this effect is probably not as large as in the APL study.

Finally, we investigated the potential application of the deregulation observed in the *DLK1–DIO3* locus for prognostic purposes regarding AZA treatment. Although we identified changes in expression and methylation levels before AZA therapy, these changes were not associated with future responsiveness to treatment. Concerning survival analyses, only methylation status of the A3 amplicon could be useful for the prediction of PFS during treatment. Kaplan-Meier curves suggested that A3 hypermethylation before treatment might serve as a potential positive prognostic marker for PFS after AZA initiation. However, since we evaluated only a small number of patients and no mutation experiments in animal models were done, strict interpretation of these data must be done with caution.

Epigenetic changes have frequently been demonstrated in MDS, especially in the subtypes with higher blast counts. Aberrant methylation has been found to correlate with poor prognosis and survival even in early stage patients [7,33]. Expression in the *DLK1–DIO3* region is known to be epigenetically regulated, and both expression and methylation of this locus have been recognized as altered in myelodysplasia. Here, we correlated changes in expression activity with specific methylation changes in its regulatory sequences in higher-risk MDS and AML-MRC patients treated with AZA and thoroughly investigated a complex regulatory mechanism within this region. Based on our data, we assumed that epigenetic alterations affect expression activity in the locus. Some of these changes have also been documented in APL [24]; however, other observations have had distinct features. Although further investigation in larger datasets and mutation experiments in animal models are required, we conclude that expression of miRNAs located in the *DLK1–DIO3* region is regulated, at least in part, in a different manner than in APL. It seems unlikely that the increased expression in this region in myelodysplasia can be simply attributed to an individual alteration in DNA methylation because we observed several changes that may affect the resulting state. Other mechanisms besides hypermethylation of the *MEG3*-DMR, such as the influence of transcription factors, may also play yet unrecognized roles in the regulation of the *DLK1–DIO3* region in myelodysplasia.

Supplementary Materials:
Figure S1: Expression of genes within the DLK1–DIO3 region, Figure S2: Overall survival (A) and progression-free survival (B) of MDS/AML-MRC patients after the initiation of AZA therapy according to pretreatment levels of 14q32 miRNAs, Table S1: Patient characteristics, Table S2: List of primers used for the generation of amplicons for bisulfite sequencing, Table S3: Matrix showing pairwise correlations between expression of mRNAs/miRNAs and methylation of CpG sites.

Author Contributions: Conceptualization, M.D.M.; Formal analysis, M.D.M.; Funding acquisition, M.D.M.; Investigation, H.R., Z.K. and M.B.; Methodology, H.R., Z.K., N.L., A.H. and K.S.; Project administration, M.D.M. and M.B.; Supervision, J.C., A.J. and M.B.; Writing—original draft, M.D.M.; Writing—review & editing, H.R., Z.K., J.C., A.J. and M.B.

Acknowledgments: The authors thank Kyra Michalova and Zuzana Zemanova (Center of Oncocytogenetics, Faculty Hospital and First Faculty of Medicine, Charles University, Prague) for the cytogenetic data.

References

1. Aul, C.; Bowen, D.T.; Yoshida, Y. Pathogenesis, etiology and epidemiology of myelodysplastic syndromes. *Haematologica* **1998**, *83*, 71–86. [PubMed]
2. Garcia-Manero, G. Myelodysplastic syndromes: 2014 update on diagnosis, risk-stratification, and management. *Am. J. Hematol.* **2014**, *89*, 97–108. [CrossRef] [PubMed]
3. Fenaux, P.; Mufti, G.J.; Hellstrom-Lindberg, E.; Santini, V.; Finelli, C.; Giagounidis, A.; Schoch, R.; Gattermann, N.; Sanzet, G.; List, A.; et al. Efficacy of azacitidine compared with that of conventional care regimens in the treatment of higher-risk myelodysplastic syndromes: A randomised, open-label, phase III study. *Lancet Oncol.* **2009**, *10*, 223–232. [CrossRef]
4. Kornblith, A.B.; Herndon, J.E.; Silverman, L.R.; Demakos, E.P.; Odchimar-Reissig, R.; Holland, J.F.; Powell, B.L.; De Castro, C.; Ellerton, J.; Larson, R.A.; et al. Impact of azacytidine on the quality of life of patients with myelodysplastic syndrome treated in a randomized phase III trial: A Cancer and Leukemia Group B study. *J. Clin. Oncol.* **2002**, *20*, 2441–2452. [CrossRef] [PubMed]
5. Vardiman, J.; Reichard, K. Acute Myeloid Leukemia with Myelodysplasia-Related Changes. *Am. J. Clin. Pathol.* **2015**, *144*, 29–43. [CrossRef] [PubMed]
6. Loiseau, C.; Ali, A.; Itzykson, R. New therapeutic approaches in myelodysplastic syndromes: Hypomethylating agents and lenalidomide. *Exp. Hematol.* **2015**, *43*, 661–672. [CrossRef] [PubMed]

7. Benetatos, L.; Hatzimichael, E.; Londin, E.; Vartholomatos, G.; Loher, P.; Rigoutsos, I.; Briasoulis, E. The microRNAs within the DLK1-DIO3 genomic region: Involvement in disease pathogenesis. *Cell. Mol. Life Sci.* **2013**, *70*, 795–814. [CrossRef] [PubMed]
8. Kagami, M.; O'Sullivan, M.J.; Green, A.J.; Watabe, Y.; Arisaka, O.; Masawa, N.; Matsuoka, K.; Fukami, M.; Matsubara, K.; Kato, F.; et al. The IG-DMR and the MEG3-DMR at human chromosome 14q32.2: Hierarchical interaction and distinct functional properties as imprinting control centers. *PLoS Genet.* **2010**, *6*, e1000992. [CrossRef] [PubMed]
9. Rosa, A.L.; Wu, Y.; Kwabi-Addo, B.; Coveler, K.J.; Reid Sutton, V.; Shaffer, L.G. Allele-specific methylation of a functional CTCF binding site upstream of MEG3 in the human imprinted domain of 14q32. *Chromosome Res.* **2005**, *13*, 809–818. [CrossRef] [PubMed]
10. Formosa, A.; Markert, E.K.; Lena, A.M.; Italiano, D.; Finazzi-Agro', E.; Levine, A.J.; Bernardiniet, S.; Garabadgiual, A.V.; Melino, G.; Candi, E. MicroRNAs, miR-154, miR-299-5p, miR-376a, miR-376c, miR-377, miR-381, miR-487b, miR-485-3p, miR-495 and miR-654-3p, mapped to the 14q32.31 locus, regulate proliferation, apoptosis, migration and invasion in metastatic prostate cancer cells. *Oncogene* **2014**, *33*, 5173–5182. [CrossRef] [PubMed]
11. Niu, C.S.; Yang, Y.; Cheng, C. MiR-134 regulates the proliferation and invasion of glioblastoma cells by reducing Nanog expression. *Int. J. Oncol.* **2013**, *42*, 1533–1540. [CrossRef] [PubMed]
12. Jin, Y.; Peng, D.; Shen, Y.; Xu, M.; Liang, Y.; Xiao, B.; Lu, J. MicroRNA-376c inhibits cell proliferation and invasion in osteosarcoma by targeting to transforming growth factor-alpha. *DNA Cell Biol.* **2013**, *32*, 302–309. [CrossRef] [PubMed]
13. Choong, M.L.; Yang, H.H.; McNiece, I. MicroRNA expression profiling during human cord blood-derived CD34 cell erythropoiesis. *Exp. Hematol.* **2007**, *35*, 551–564. [CrossRef] [PubMed]
14. Tenedini, E.; Roncaglia, E.; Ferrari, F.; Orlandi, C.; Bianchi, E.; Bicciato, S.; Tagliafico, E.; Ferrari, S. Integrated analysis of microRNA and mRNA expression profiles in physiological myelopoiesis: Role of hsa-mir-299-5p in CD34+ progenitor cells commitment. *Cell Death Dis.* **2010**, *1*. [CrossRef] [PubMed]
15. Wang, F.; Yu, J.; Yang, G.; Wang, X.; Zhang, J. Regulation of erythroid differentiation by miR-376a and its targets. *Cell Res.* **2011**, *21*, 1196–1209. [CrossRef] [PubMed]
16. Jongen-Lavrencic, M.; Sun, S.M.; Dijkstra, M.K.; Valk, P.J.M.; Löwenberg, B. MicroRNA expression profiling in relation to the genetic heterogeneity of acute myeloid leukemia. *Blood* **2008**, *111*, 5078–5085. [CrossRef] [PubMed]
17. Valleron, W.; Laprevotte, E.; Gautier, E.; Quelen, C.; Demur, C.; Delabesse, E.; Agirre, X.; Prósper, F.; Kiss, T.; Brousset, P. Specific small nucleolar RNA expression profiles in acute leukemia. *Leukemia* **2012**, *26*, 2052–2060. [CrossRef] [PubMed]
18. Ley, T.J.; Miller, C.; Ding, L.; Raphael, B.J.; Mungall, A.J.; Robertson, A.G.; Hoadley, K.; Triche, T.J.; Laird, P.W.; Baty, J.D.; et al. Genomic and epigenomic landscapes of adult de novo acute myeloid leukemia. *N. Engl. J. Med.* **2013**, *368*, 2059–2074. [PubMed]
19. Dixon-McIver, A.; East, P.; Mein, C.A.; Cazier, J.; Molloy, G.; Chaplin, T.; Lister, T.A.; Young, B.D.; Debernardi, S. Distinctive patterns of microRNA expression associated with karyotype in acute myeloid leukaemia. *PLoS ONE* **2008**, *3*, e2141. [CrossRef] [PubMed]
20. Votavova, H.; Grmanova, M.; Merkerova, M.D.; Belickova, M.; Vasikova, A.; Neuwirtova, R.; Cermak, J. Differential expression of microRNAs in CD34+ cells of 5q- syndrome. *J. Hematol. Oncol.* **2011**. [CrossRef] [PubMed]
21. Merkerova, M.D.; Krejcik, Z.; Votavova, H.; Belickova, M.; Vasikova, A.; Cermak, J. Distinctive microRNA expression profiles in CD34+ bone marrow cells from patients with myelodysplastic syndrome. *Eur. J. Hum. Genet.* **2011**, *19*, 313–319. [CrossRef] [PubMed]
22. Krejčík, Z.; Beličková, M.; Hruštincová, A.; Kléma, J.; Zemanová, Z.; Michalová, K.; Čermák, J.; Jonášová, A.; Merkerová, M.D. Aberrant expression of the microRNA cluster in 14q32 is associated with del(5q) myelodysplastic syndrome and lenalidomide treatment. *Cancer Genet.* **2015**, *208*, 156–161. [CrossRef] [PubMed]
23. Merkerova, M.D.; Krejcik, Z.; Belickova, M.; Hrustincova, A.; Klema, J.; Stara, E.; Zemanova, Z.; Michalova, K.; Cermak, J.; Jonasova, A. Genome-wide miRNA profiling in myelodysplastic syndrome with del(5q) treated with lenalidomide. *Eur. J. Haematol.* **2015**, *95*, 35–43. [CrossRef] [PubMed]
24. Manodoro, F.; Marzec, J.; Chaplin, T.; Miraki-Moud, F.; Moravcsik, E.; Jovanovic, J.V.; Wang, J.; Iqbal, S.;

Taussig, D.; Grimwade, D.; et al. Loss of imprinting at the 14q32 domain is associated with microRNA overexpression in acute promyelocytic leukemia. *Blood* **2014**, *123*, 2066–2074. [CrossRef] [PubMed]

25. Benetatos, L.; Hatzimichael, E.; Dasoula, A.; Dranitsaris, G.; Tsiara, S.; Syrrou, M.; Georgiou, I.; Bourantas, K.L. CpG methylation analysis of the MEG3 and SNRPN imprinted genes in acute myeloid leukemia and myelodysplastic syndromes. *Leuk. Res.* **2010**, *34*, 148–153. [CrossRef] [PubMed]
26. Merkerova, M.D.; Bystricka, D.; Belickova, M.; Krejcik, Z.; Zemanova, Z.; Polak, J.; Hajkova, H.; Brezinova, J.; Michalova, K.; Cermak, J. From cryptic chromosomal lesions to pathologically relevant genes: Integration of SNP-array with gene expression profiling in myelodysplastic syndrome with normal karyotype. *Genes Chromosomes Cancer* **2012**, *51*, 419–428. [CrossRef] [PubMed]
27. Cheson, B.D.; Greenberg, P.L.; Bennett, J.M.; Lowenberg, B.; Wijermans, P.W.; Nimer, S.D.; Pinto, A.; Beran, M.; de Witte, T.M.; Stone, R.M.; et al. Clinical application and proposal for modification of the International Working Group (IWG) response criteria in myelodysplasia. *Blood* **2006**, *108*, 419–425. [CrossRef] [PubMed]
28. Cheson, B.D.; Bennett, J.M.; Kopecky, K.J.; Büchner, T.; Willman, C.L.; Estey, E.H.; Schiffer, C.A.; Doehner, H.; Tallman, M.S.; Lister, T.A.; et al. Revised recommendations of the International Working Group for Diagnosis, Standardization of Response Criteria, Treatment Outcomes, and Reporting Standards for Therapeutic Trials in Acute Myeloid Leukemia. *J. Clin. Oncol.* **2003**, *21*, 4642–4649. [CrossRef] [PubMed]
29. Krejcik, Z.; Belickova, M.; Hrustincova, A.; Votavova, H.; Jonasova, A.; Cermak, J.; Dyr, J.E.; Merkerova, M.D. MicroRNA profiles as predictive markers of response to azacitidine therapy in myelodysplastic syndromes and acute myeloid leukemia. *Cancer Biomark.* **2018**, *22*, 101–110. [CrossRef] [PubMed]
30. Belickova, M.; Merkerova, M.D.; Votavova, H.; Valka, J.; Vesela, J.; Pejsova, B.; Hajkova, H.; Klemaet, J.; Cermak, J.; Jonasova, A. Up-regulation of ribosomal genes is associated with a poor response to azacitidine in myelodysplasia and related neoplasms. *Int. J. Hematol.* **2016**, *104*, 566–573. [CrossRef] [PubMed]
31. Kumaki, Y.; Oda, M.; Okano, M. QUMA: Quantification tool for methylation analysis. *Nucleic Acids Res.* **2008**, *36*, W170–W175. [CrossRef] [PubMed]
32. Pennucci, V.; Zini, R.; Norfo, R.; Guglielmelli, P.; Bianchi, E.; Salati, S.; Sacchi, G.; Prudente, Z.; Tenedini, E.; Ruberti, S.; et al. Abnormal expression patterns of WT1-as, MEG3 and ANRIL long non-coding RNAs in CD34+ cells from patients with primary myelofibrosis and their clinical correlations. *Leuk. Lymphoma* **2015**, *56*, 492–496. [CrossRef] [PubMed]
33. Boultwood, J.; Wainscoat, J.S. Gene silencing by DNA methylation in haematological malignancies. *Br. J. Haematol.* **2007**, *138*, 3–11. [CrossRef] [PubMed]

2

MicroRNAs as Diagnostic and Prognostic Biomarkers in Ischemic Stroke

Ceren Eyileten [1,†], Zofia Wicik [2,†], Salvatore De Rosa [3], Dagmara Mirowska-Guzel [1], Aleksandra Soplinska [1], Ciro Indolfi [3,4], Iwona Jastrzebska-Kurkowska [5], Anna Czlonkowska [5] and Marek Postula [1,*]

1 Department of Experimental and Clinical Pharmacology, Medical University of Warsaw, Center for Preclinical Research and Technology CEPT, 02-097 Warsaw, Poland; ceren.eyileten-postula@wum.edu.pl (C.E.); dmirowska@wum.edu.pl (D.M.-G.); ola@soplinska.pl (A.S.)
2 Rheumatology Division, Hospital das Clinicas HCFMUSP, Universidade de Sao Paulo, Sao Paulo, SP 01246-903, Brazil; zofiawicik@gmail.com
3 Division of Cardiology, Department of Medical and Surgical Sciences, "Magna Graecia" University, 88100 Catanzaro, Italy; saderosa@unicz.it (S.D.R.); indolfi@unicz.it (C.I.)
4 URT-CNR, Department of Medicine, Consiglio Nazionale delle Ricerche of IFC, Viale Europa S/N, 88100 Catanzaro, Italy
5 2nd Department of Neurology, Institute of Psychiatry and Neurology, 02-957 Warsaw, Poland; ikurkowska@ipin.edu.pl (I.J.-K.); czlonkow@ipin.edu.pl (A.C.)
* Correspondence: mpostula@wum.edu.pl;
† These authors contributed equally to this work.

Abstract: Stroke is the second-most common cause of death worldwide. The pathophysiology of ischemic stroke (IS) is related to inflammation, atherosclerosis, blood coagulation, and platelet activation. MicroRNAs (miRNAs) play important roles in physiological and pathological processes of neurodegenerative diseases and progression of certain neurological diseases, such as IS. Several different miRNAs, and their target genes, are recognized to be involved in the pathophysiology of IS. The capacity of miRNAs to simultaneously regulate several target genes underlies their unique value as diagnostic and prognostic markers in IS. In this review, we focus on the role of miRNAs as diagnostic and prognostic biomarkers in IS. We discuss the most common and reliable detection methods available and promising tests currently under development. We also present original results from bioinformatic analyses of published results, identifying the ten most significant genes (HMGB1, YWHAZ, PIK3R1, STAT3, MAPK1, CBX5, CAPZB, THBS1, TNFRSF10B, RCOR1) associated with inflammation, blood coagulation, and platelet activation and targeted by miRNAs in IS. Additionally, we created miRNA-gene target interaction networks based on Gene Ontology (GO) information derived from publicly available databases. Among our most interesting findings, miR-19a-3p is the most widely modulated miRNA across all selected ontologies and might be proposed as novel biomarker in IS to be tested in future studies.

Keywords: miRNA; bioinformatic analysis; ischemic stroke; miRNA-gene target interaction; network; biomarker; diagnosis; prognosis

---

1. Introduction

Stroke is the second-most common cause of death worldwide, accounting for almost 6.5 million stroke deaths each year. Approximately 795,000 strokes occur in the United States each year. On average, every 40 s, someone in the United States has a stroke, and on average, every 4 min, someone dies of a stroke [1,2]. In Europe, strokes account for 405,000 deaths (9%) in men and 583,000 (13%) deaths in women each year [3]. While hemorrhagic strokes accounts for 15% and a further 5% are due to unknown etiology, approximately 80% of all acute strokes are ischemic, mainly from large vessel occlusion due to either artery-to-artery embolism or cardiac embolism. [4,5]. Ischemic stroke (IS) is characterized by the sudden loss of blood circulation to an area of the brain, resulting in irreversible brain injury and subsequent neurologic deficits occurring already few minutes after the onset of ischemia [6].

MicroRNAs (miRNA) are small, endogenous, single-stranded, noncoding RNA molecules ranging in length from 18–25 nt that are found in eukaryotic cells. They regulate approximately 60% of the mammalian protein coding genes, primarily through the interaction with mRNAs. This effect is exerted by binding to complementary regions of messenger transcripts to repress their translation or less frequently inducing their degradation. Up to now, more than 5000 human miRNAs have been identified, and about 1000 are estimated to exist [7]. Several studies showed that miRNAs can be useful for the treatment of particular diseases. One advantage of miRNA-based therapies is that synthetic mimics and inhibitors can be used to alter endogenous miRNA levels in clinic. This approach has been utilized for cancer and hepatitis C virus therapy [8,9]. Apart from their treatment value, it was shown that expression of miRNAs plays important roles in many physiological and pathological processes, including epigenetics and neurodegenerative diseases and progression of specific neurological diseases [10,11]. It has been shown that some pathological processes of the central nervous system, and that Alzheimer's disease, multiple sclerosis, and stroke are associated to specific alteration of circulating miRNAs [12]. This suggests that circulating miRNAs could be used as clinical biomarkers, to provide a sort of "liquid biopsy" taken from peripheral blood but providing information on pathophysiological processes going on in the brain and underlying stroke [13–15].

Potential advantages of this approach include the early diagnosis of acute cerebrovascular events, as already proven before with coronary artery disease, peripheral arterial disease and ischemia, acute myocardial infarction, myocarditis, heart failure, takotsubo cardiomyopathy in-stent restenosis, diabetes, or platelet dysfunction [7,16–29].

Several conventional methods were used by researchers in order to detect miRNAs such as, northern blotting, quantitative polymerase chain reaction (qPCR), and microarrays. However, each of these methods has its own individual limitations. Northern analysis was a widely used method for miRNA analyses, as it is easy to perform, specificity is relatively high, sequences with even partial homology can be used as hybridization probes, and the cost of running many gels is low once the equipment is set up [30]. However, northern blot technology also has disadvantages. For example, contamination due to radiolabeling, low detection efficiency, and poor sensitivity for oligonucleotide probes may complicate the measurements [31]. RT-PCR and microarrays allow reliable and effective detection for miRNAs. Techniques like northern blot and RT-PCR allow testing for only a few miRNAs per experiment. On the contrary, using miRNA microarray, it is possible to identify more than several hundred differentially regulated miRNAs in a particular disease or condition in one single run [7]. Several new technical developments have been brought about in the field of RT-PCR in order to further improve their efficiency. Among the others, the coupling with microfluidic cards should be mentioned, together with the digital droplet PCR, which allows very high analytical standards, minimizing reaction volumes [32]. Additionally, next-generation-sequencing technology has quickly emerged as the preferred platform for studying and discovering novel circulating miRNAs. It is possible to sequence multiple samples in one time by pooling with next-generation-sequencing and also possible to construct comprehensive expression profiles for every assessed sample, on the other hand potentially large investment in bioinformatics analysis, time and personnel are required [33,34]. Beside these techniques, there are new promising methodologies have been developed to detect miRNAs, for example, sensitive miRNA biosensors such as silicon nanowires, gold nanoparticles, silver nanoclusters, or conducting polymer/carbon nanotube hybrids [35–38].

Since circulating miRNAs can be detected by several techniques in plasma, serum, or whole blood, they have become the focus of interest as potential new diagnostic and prognostic biomarkers and tools for understanding their role in IS. Therefore, we systematically reviewed all available studies assessing the clinical usefulness of circulating miRNAs as diagnostic or prognostic markers in IS. Thus, in this article, we present an overview of the current knowledge on diagnostic and prognostic value of miRNAs related to IS based on human studies and report the results of a quantitative bioinformatic analysis highlighting the most promising miRNAs for clinical application in IS.

2. Circulating miRNAs and Stroke

Acute myocardial ischemia and IS share several common pathophysiological aspects. Previous studies implied that the circulating miRNAs could be exploited as diagnostic and/or prognostic indicators in several cardiovascular diseases (CVD) [27]. However, limited number of researches on role of miRNAs in IS and its relation to clinical outcomes were published.

2.1. PCR-Based Analysis for miRNA Expression

Quantitative reverse transcription PCR (RT-qPCR) technology is the gold standard for gene expression measurement. Several companies have developed qPCR-based assays for the detection of miRNA expression [39]. RT-qPCR technique allowed us to achieve high sensitivity, high specificity, and quantitative data of miRNA expression analysis. However, RT-qPCR is low-throughput profiling for a genome-wide miRNA-profiling assay [40].

In one of the first study, alterations in plasma miRNA levels in patients with IS were compared to healthy controls. The circulating levels of miR-30a, miR-126, and let-7b were analyzed in plasma samples of patients with IS obtained at different time-points both in acute phase (24 h from the onset of symptoms) and recovery phase (at 1 week, 4 weeks, 24 weeks, and 48 weeks after the episode). Circulating miR-30a and miR-126 expression were down regulated in IS patients regardless of etiology in all time points, except the last measurement at 48 weeks when it increase to baseline value. Interestingly, the expression pattern of circulating let-7b in IS patients with large-vessel atherosclerosis (LA) was increased in contrast to what observed in patients with other stroke subtypes (i.e., small-vessel disease (SV), cardioembolic stroke (CE) and undetermined cause (UDN) groups). Follow-up evaluations revealed that all circulating miRNAs levels returned to the baseline 48 weeks after the episode. Moreover, based on the level of three miRNAs authors defined a specific clinical score system that was further correlated with functional status as evaluated with the modified Rankin Scale (mRS), suggesting that miR-30a, miR-126, and let-7b could be used as a reliable marker for the diagnosis of IS (See Tables 1 and 2) [41]. It is particularly interesting that a modulation of miR-30 has also been reported in patients with acute myocardial infarction [42]. If on one hand this strengthens the hypothesis that miRNAs are "smart" biomarkers capable of reflecting the pathophysiological processes underlying CVS, on the other hand it undermines its usefulness as a disease-specific biomarker.

In another study, two miRNAs selected based on literature search, namely let-7e and miR-338, were analyzed in serum from IS patients in three phases of the ischemic event (i.e., acute- 1–7 days, subacute- 8–14 days, recovered- over 15 days) and miRNA levels were compared with 51 healthy volunteers. In this study only let-7e expression levels were significantly higher at all time-points with significantly higher levels in IS patients at the acute stage. Moreover, let-7e level was correlated with serum high sensitivity C-reactive protein (hs-CRP) level ($r = 0.67$, $p = 0.033$). A prognostic role of both miRNAs failed to be proven as there was no correlation between miRNAs expression level and National Institutes of Health Stroke Scale (NIHSS) scores, which is a widely used tool that determine the severity of a stroke. However, serum let-7e showed a specificity up to 73.4% and a sensitivity of 82.8% in IS patients at the acute stage, whereas serum miR-338 in IS patients showed a specificity up to 53.2% and a sensitivity of 71.9% in the acute stage. Thus, authors suggested that let-7e expression in serum may serve as a useful noninvasive circulating biomarker for the acute stage of IS [43].

Huang et al. sought to evaluate another let-7 family member, let-7e-5p, in two independent case-control IS populations. The results showed that the expression level of let-7e-5p was significantly higher in IS patients than in control subjects. Logistic regression analysis revealed that let-7e-5p expression was associated with an increased risk of IS (adjusted OR, 1.89; 95% CI, 1.61~2.21; $p < 0.001$). Also, diagnostic accuracy of acute phase specific miRNAs were tested through calculating the area under curve (AUC) of receiver operating characteristic (ROC) curves. The addition let-7e-5p to the traditional risk factor model improved the diagnostic potential to an AUC of 0.82 (95% CI, 0.78~0.85). In order to find targets of let-7e both bioinformatics and target gene expression analysis were performed. It showed that let-7e-5p expression was negatively correlated with several genes (ATF2, CASP3, FGFR2, NLK, PTPN7, RASGRP1, and TGFBR1). Therefore, the study showed that let-7e-5p expression is significantly higher in IS patients and is associated with the occurrence of IS. Moreover, authors suggested that let-7e-5p may be involved in the pathogenesis of IS by regulating CASP3 and NLK expression, as two genes enriched in the MAPK signaling pathway [44].

Gong et al. performed study in order to determine prognostic value of several let-7 family members (let-7a, let-7b, let-7c, let-7d, let-7e, let-7f, let-7g, miR-98) and relation to massive cerebral infarction (MCI) within first 48 h from the onset of symptoms. In their report, the expression of let-7f was down regulated in IS with MCI in comparison to healthy controls and IS without MCI, and up regulated in group without MCI at baseline, i.e., 48 h. When comparing relative expression of let-7f between groups with and without hemorrhagic transformation (HT), authors found up-regulation of let-7f in the MCI without HT after two weeks from the baseline. However, when compared MCI with and without HT significant up-regulation of let-7f was found in the first group after two weeks. It is worth to mention that the level of hs-CRP was negatively correlated with the relative expression of let-7f in the MCI group. Another important finding of the study showed that the expression level of let-7f in the MCI without HT is positively correlated with patients status based on Glasgow Coma Scale score and negatively with hs-CRP concentration ($r = -0.88$, $p < 0.0001$). Also in this study target gene expression analysis was performed and the relative expression of let-7f was negatively correlated with interleukin-6 (IL-6) expression in the MCI without HT (48 h and 2 weeks) ($r = -0.40$, $p < 0.001$), but not in the MCI with HT group, which may suggest that the downregulation of let-7f expression in patients with MCI without HT may induce inflammation [45]. The large involvement of let-7 with cerebrovascular disease, as proven by multiple evidences from independent studies is particular interesting. In fact, let-7 expression and availability is influenced by a circular noncoding RNA named CircPVT1, involved in the modulation of cell senescence [46].

Leung et al. investigated and compared plasma concentrations of miR-124-3p and miR-16 as diagnostic markers in acute stroke. Ninety-three patients with IS, 19 patients with hemorrhagic stroke, and 23 healthy controls enrolled in the study. Plasma concentrations of miRNAs were determined by RT- PCR. Median plasma 124-3p concentrations taken within 24 h of symptom onset were higher in hemorrhagic stroke patients than that in IS patients, while median miR-16 concentration in IS patients were higher than that in hemorrhagic stroke patients. Authors concluded that both miR-124-3p and miR-16 are diagnostic markers to discriminate hemorrhagic stroke and IS [47]. Another study aimed to investigate circulating miRNAs namely miR-15a, miR-16, and miR-17-5p [48]. The selection of specific miRNA was based on assumption that miR-15a and miR-16 are increased in the serum of patients with ischemic events and miR-17-5p may be a critical factor in post-stroke adult neurogenesis [49–52]. In this study, Wu et al. evaluated the utility of these three miRNAs in peripheral blood of 106 IS patients and 120 healthy controls as IS serum biomarkers for diagnostic value. Serum levels of miR-15a, miR-16, and miR-17-5p were significantly higher in IS patients compared to control subjects. Serum miR-15a levels showed a significant positive correlation with age. Besides, there was a strong negative correlation between serum miR-16 levels and high-density lipoprotein (HDL) and apolipoprotein A1 (ApoA1). Moreover, in multivariate logistic regression model, it was found that serum miR-17-5p level was a significant and independent predictor for IS (OR 3.968; CI 95% 1.001–14.29, $p = 0.035$). Also, ROC analysis showed that selected miRNAs were useful IS biomarkers (See Table 2) [48]. In line with these findings, miR-16 was recently associated to peripheral ischemia and was suggested as a potential mediator of remote vascular remodeling [19]. Hence, it could reflect systemic vascular dysfunction, rather than specific and acute localization of ischemic disease.

Jin and Xing evaluated 28 miRNAs selected based on previous publications that were described to possess pro-angiogenic or anti-angiogenic properties. Selected miRNAs were evaluated in plasma samples of IS patients and controls taken within 24 h after ischemic events. In the exploring stage performed in 10 patients and 10 controls, 11 differentially expressed miRNAs (DEM) were identified and included into the validating stage. In the second stage, in order to validate the significantly expressed miRNAs, 106 IS patients and 110 controls were analyzed. The expression of miR-126, miR-130a, and miR-378 in plasma decreased in IS patients; however, miR-222, miR-218, and miR-185 plasma levels were increased. At logistic regression analysis they found that miR-126 (OR 0.840; CI 95% 0.766–0.9220), miR-130a (OR 0.885; CI 95% 0.827–0.948), miR-222 (OR 1.064; CI 95% 1.004–1.126), miR-218 (OR 1.138; CI 95% 1.036–1.250), and miR-185 (OR 1.099; CI 95% 1.003–1.205) were independent predictor factors for IS. Moreover, the combined analysis of these five DEMs demonstrated a good diagnostic performance for IS with sensitivity of 87.7%, and specificity of 54.5%. Additionally, miR-126, miR-378, and miR-101 were negatively associated, and miR-222, while miR-218, miR-206 positively associated with NIHSS score what could be used for disease severity management of IS [53].

So far, several studies showed the importance of miR-145 in stroke. In one of the first study, circulating miR-145 was evaluated in 32 IS patients and compared to 14 control participants. The results showed significant up-regulation of circulating miR-145 expression and non-significant downregulation after one month in IS patients ($N = 11$). Authors suggested that miR-145 might be a desirable biomarker in IS [54].

In the next study nine previously reported stroke associated miRNAs (miR-21, miR-23a, miR-29b, miR-124, miR-145, miR-210, miR-221, miR-223 and miR-483-5p) were screened both in 146 IS group and 96 control subjects. In this validation process only miR-145 was significantly up-regulated, but miR-23a and miR-221 were significantly downregulated. Serum miR-23a and miR-221 were moderatly negatively correlated with plasma hs-CRP. Moreover, a strong positive correlation existed between serum miR-145 and hs-CRP (r = 0.6713), and a moderate correlation existed between serum miR-145 and IL-6 (r = 0.5896). Interestingly, miR-145 level was positively correlated with infarct volume and NIHSS scores (r = 0.6249 and r = 0.6288, respectively). Finally, the prediction value using both hs-CRP and miR-145 was significantly higher than for hs-CRP alone (See Table 2). In 49 IS patients long term follow-up lasting 2 years was performed and showed that the expression of miR-145 was highest within the acute phase (1 to 7 days) of stroke, but decreased within the recovery phase (1 month, 6 months, and 2 years) in the serum of patients [55]. The abovementioned role of miR-145 was confirmed through bioinformatics analyses (Gene Ontology and Kyoto Encyclopedia of Genes and Genomes enrichment) that included two mRNA and 1 miRNA microarray expression profile data from the Gene Expression Omnibus database. Based on this analysis two miRNAs, namely miR-145 and miR-122, may represent potential biomarkers in IS. Also, three novel miRNAs (miR-99b, miR-542-3p, and miR-455-5p) were deregulated what may suggest their roles in the pathological processes of IS [56].

Contrary to recent studies, Tsai et al. did not find significant modulation of miR-145 in IS. However, they found that miR-21 level was significantly higher, but miR-221 was significantly lower compared to healthy controls. MiR-145 was excluded from further analysis, as there were no significant differences between groups. At multivariable logistic regression analysis both increased miR-21 level (OR 6.16; 95% CI 2.82, 14.64) and decreased miR-221 level (OR 10.38; 95%CI 4.52, 26.45) were associated with increased risk of stroke when added to classical risk factors like age, sex, diabetes, hypertension, smoking, and hyperlipidemia. Also ROC analysis showed that adding both miR-21 and miR-221 into the model substantially improved AUC value for IS prediction. Hence, the authors suggested that miR-21 and miR-221 can be novel biomarkers for stroke, but not miR-145 [57].

Also, Zhou and Zhang identified decreased plasma miR-21 and miR-24 levels in 68 IS patients compared with 21 healthy controls. Moreover, miR-21 expressions correlated with miR-24 level, and both miRNAs negatively correlated with early outcome of IS based on NIHHS score ($r = -0.703$, $p < 0.05$ for miR-21; $r = -0.694$, $p < 0.05$ for miR-24) [58].

Previous study analyzed the changes of miR-223 levels in 75 healthy control samples and compared to 79 IS patients within 72 h from the onset of symptoms. It was found that miR-223 expression correlates with stroke subtype (i.e., LA and SA subtype) and negatively correlates with NIHSS scores, but the correlations with infarct volume and insulin-like growth factor 1 (IGF-1) levels were low [59]. Beside miRNA studies in circulating blood miR-223 was detect also in exosomes. Exosomes are 30–100 nm vesicles that cells secrete into extracellular space when multivesicular bodies fuse with cell membrane. MiR-223 was shown to be one of the most highly expressed miRNAs in plasma exosomes of healthy human and results about its function in ischemia injury are inconsistent [60]. Chen et al. found that the level of miR-223 in circulating exosomes was elevated after onset of IS, and exosomal miR-223 expression was positively correlated to NIHSS score. Moreover, stroke patients with poor outcomes based on mRS inclined to have a greater exosomal miR-223 expression. Therefore, increased exosomal miRNA-223 possess both diagnostic and prognostic value for IS [61].

In another study serum exosomal miRNAs, namely miR-9 and miR-124, were evaluated in 65 IS patients and compared with 66 non-stroke volunteers in order to explore their diagnostic and prognostic value. The team found that both miR-9 and miR-124 levels were increased in IS patients. Moreover, both miRNAs expression were positively correlated with NIHSS score ($r = 0.7126$ and 0.6825 respectively, $p < 0.01$), infarct volume evaluated by magnetic resonance imaging (MRI) ($r = 0.6768$ and 0.6312, respectively, $p < 0.01$) and serum IL-6 concentration ($r = 0.6980$ and 0.6550, respectively, $p < 0.01$). Finally, diagnostic value for IS of exosomal miR-9 and miR-124 was confirmed in ROC analysis [62]. Interestingly, another small study done by Liu et al. showed that serum expression of miR-124 was downregulated within the first 24 h after the ischemic event, but there were no differences in miR-9 and miR-219 levels between 31 IS and 11 healthy controls. Also, only miR-124 and miR-9 were negatively correlated with infarct lesion volume ($r = -0.423$, $p = 0.022$; and $r = -0.608$, $p < 0.001$), but not with NIHSS score. Besides, a significant negative correlation existed between plasma hs-CRP levels and serum miR-124 and miR-9 levels ($r = -0.421$, $p = 0.023$; and $r = -0.511$, $p = 0.004$) [63].

In the latest study, 17 previously reported stroke-associated miRNAs were measured in serum using RT-qPCR. Researchers first evaluated miRNA levels in randomly selected 30 IS patients compared with 30 control participants. MiR-21, miR-145, miR-29b, and miR-146b were significantly increased but miR-23a and miR-221 levels were decreased within 24 h after stroke onset compared with the control group. Further verification was done for miR-21, miR-23a, miR-29b, miR-145, miR-146b, and miR-221 with a larger number of patients. MiR-146b was significantly upregulated in the serum of 128 IS patients compared with 102 control participants. In addition to this, positive correlation was found between upregulated serum miR-146b level and plasma hs-CRP, infarct volume and NIHSS score, and serum IL-6 of patients. Importantly, the combination of plasma hs-CRP and serum miR-146b gained a better sensitivity/specificity for prediction of IS (AUC from 0.782 to 0.863). Besides, it was found that decreased miR-221 level negatively correlated with plasma hs-CRP level but not serum IL-6 level in IS patients. Authors suggested that upregulated serum miR-146b might be a potential diagnostic biomarker for IS evaluation [64].

2.2. MiRNA Profiling and RNA Sequencing Strategy

MiRNA microarrays are commonly used for miRNA expression profiling analysis. They have advantages and disadvantages comparison to RT-qPCR technologies. First of all, they allow us a powerful high-throughput profiling for a genome-wide miRNA profiling assay and the protocol can be easily standardized. Despite, there are some major disadvantages associated with miRNA microarrays. For example, generally a large amount of high-quality RNA samples are needed for the microarray experiments, which is often a major challenge for miRNA expression analysis of clinical samples. Moreover, lower miRNA detection sensitivity and specificity is the main limitation of microarray analysis compared to RT-qPCR analysis [65]. Beside miRNA microarrays, the next-generation-sequencing technology has quickly emerged as the preferred platform for studying circulating miRNA profiling. Comparison to microarray technology RNA-sequencing technology has higher sensitivity and specificity, however high cost is a practical consideration for researchers [66].

In one of the first study using microarray strategy Jickling et al. identified candidate miRNAs that could serve as a diagnostic tool in 24 IS patients. In a group with IS six miRNAs, expression was increased (i.e., miR-122, miR-148a, let-7i, miR-19a, miR-320d, miR-4429), and two miRNAs, expression was decreased (miR-363 and miR-487b) in comparison to healthy controls. By using gene target database, they also evaluated potential gene targets for selected miRNAs and they found two main pathways which are regulated by these miRNAs, i.e., the nuclear factor-κB (NF-κB) and Toll-like receptor signaling pathways, the pathways, which are involved in immune activation, leukocyte extravasation and thrombosis [67].

Using screening technique Li et al. discovered 115 miRNAs that were differentially expressed in IS patients compared to controls. For the further analysis they selected based on well-defined criteria and literature search 13 miRNAs. In order to validate the initial results they performed RT-qPCR of selected miRNAs and found that expression of 8 miRNAs (miR-32-3p, miR-106b-5p, miR-423-5p, miR-451a, miR-1246, miR-1299, miR-3149, and miR-4739) significantly increased, and the expression of 5 miRNAs (miR-224-3p, miR-377-5p, miR-518b, miR-532-5p, and miR-1913) significantly decreased in the serum of IS patients. According to functional assays only upregulated miR-32-3p, miR-106b-5p, and miR-1246, and downregulated miR-532-5p in IS serum might play a vital role in the pathogenesis of IS. Moreover, through bioinformatic analysis, they found that stroke-related genes vascular endothelial growth factor-A (VEGFA), myeloid cell leukemia-1 (Mcl-1), and superoxide dismutase 2 (SOD2) might be the targets of miR-106b, therefore authors suggested that miR-106b may affect multiple pathways such as apoptosis, oxidation, angiogenesis, and neurogenesis in IS [68–71].

MiRNA profiling strategy in the whole blood samples was proposed to analyze the utility of miRNA for the disease progression and stroke subtype evaluation in young patients with IS. This strategy showed that 157 miRNAs were differentially regulated across stroke samples. In total, 138 miRNAs were upregulated, and 19 miRNAs were downregulated. Interestingly, different miRNAs expression was observed between stroke subtypes groups, i.e., 8 miRNAs were downregulated (hsa-let-7f, miR-126, -1259, -142-3p, -15b, -186, -519e, and -768-5p) and 17 miRNAs were upregulated (hsa-let-7e, miR-1184, -1246, -1261, -1275, -1285, -1290, -181a, -25*, -513a-5p, -550, -602, -665, -891a, -933, -939, and -923) across subtypes of stroke. Further analysis showed also the utility of miRNA profiling in prognosis [72].

Profile of circulating miRNA was also evaluated by Sepramaniam et al. In three different cohorts that contained 169 patients with IS. In total 314 miRNAs were detected upon profiling. In the initial step of the analysis they found 105 different miRNAs that were deregulated in stroke cases. Among the 105 miRNAs, 58 were downregulated while 47 were upregulated. Moreover, further analysis help to significantly distinguish the stroke etiology i.e., LA, CE and SV based on 32 miRNAs (let-7a, let-7d*, let-7g, let-7i, miR-126, -130a, -187*, -18a*, -20a, -22*, -26b, -30b, -30c, -30e*, -320b, -320d, -324-5p, -331-3p, -340, -342-3p, -361-5p, -363, -422a, -423-3p, -501-5p, -502-3p, -505*, -574-3p, -675,-886-5p, -92a, and -93*), however detailed analysis was not provided in the study. Finally, the patients were segregated into acute and recovery phase and 26 miRNAs (let-7d*, miR-125b-2*, -1261, -1299, -130a, -1321, -208a, -22*, -23a, -27a*, -320b, -320d, -30c, -340, -422a, -423-3p, -488, -502-5p, -549a, -574-3p, -574-5p, -617, -627, -886-5p, -92a, and -93*) were unique for acute stroke and 16 miRNAs (let-7a, let-7g, miR-129-5p, -192-5p, -196a*, -26b, -30b, -30e*, -370, -381, -493*, -525-5p, -652, -920, -933, and -96) were unique for stroke patients at 7 days after the event ("recovered patients"). Diagnostic accuracy for acute phase of IS tested through calculating the AUC of ROC curves showed that 5 different miRNAs could serve as a potential biomarkers i.e., miR-125b-2*, -27a*, -422a, -488 and -627 [73] (See Figure 1).

Table 1. Human studies evaluating miRNAs as diagnostic/prognostic biomarkers in stroke.

Ref	miRNAs	Sampling/Sampling Time Point	Number of Stroke pts/Controls	Inclusion Criteria	Exclusion Criteria	Stroke Subtype	Prognostic or Diagnostic Value	Regulation of miRNAs	Correlation
Long et al. [41]	miR-30a, miR-126, Let-7b,	Plasma/24 h, 1 w, 4 w, 24 w, 48 w	197/50	First-ever stroke patients with cerebral infarction	Exclusion criteria included TIA, subarachnoid hemorrhage, embolic brain infarction, brain tumors, and cerebrovascular malformation, pulmonary fibrosis, endocrine and metabolic diseases (except type 2 diabetes), inflammatory and autoimmune diseases, and serious chronic diseases, for example, hepatic cirrhosis and renal failure. Cardioembolic stroke and documented atrial fibrillationwere also excluded from the study.	LA 51, SA 48, CE 50, UDN 48	Diagnostic value	MiR-30a and miR-126 were downregulated in 24 h, 1 w, 4 w, and 24 w. After 48 w miRNA levels increased to the baseline	No correlation was found between HDL, LDL, triglyceride, systolic and diastolic blood pressures, diabetes and smoking status and miRNAs.
Peng et al. [43]	miR-338, Let-7e	Serum and CSF/1–7 days (acute phase), 8–14 days (subacute phase), over 15 days (recovery)	72/51	Diagnosis of an initial episode of cerebral infarction based on clinical history and MRI results, ages ranging from 55 to 75 years, patient arrival at the hospital after 4.5 h but within 24 h after the event, NIHSS score of 4 to 15, and without hemorrhagic transformation.	Exclusion criteria for all enrolled patients included recurrent stroke, tumors, abnormal renal or liver function, infectious diseases, immune diseases, blood disorders, and psychiatric illness including depression and schizophrenia.	NA	Let-7e may be a biomarker in acute phase/had no prognostic value	Let-7e significantly higher at all time-points in serum. MiR-338 and Let-7e in CSF was upregulated in 8–14 days (subacute).	No correlation was found between NIHSS and Let-7e, correlation was found between CRP levels and Let-7e.
Huang et al. [44]	Let-7e-5p	Whole blood/24 h	Two groups: 44/44, 302/02	First-ever IS patients	Patients with a history of stroke, peripheral arterial occlusive disease or cancer were excluded from this study.	NA	Diagnostic value for Let-7e-5p	Let-7e-5p significantly higher with IS patients than controls	Negatively correlated with ATF2, CASP3, FGFR2, NLK, PTPN7, RASGRP1 and TGFBR1 genes.

Table 1. *Cont.*

Ref	miRNAs	Sampling/Sampling Time Point	Number of Stroke pts/Controls	Inclusion Criteria	Exclusion Criteria	Stroke Subtype	Prognostic or Diagnostic Value	Regulation of miRNAs	Correlation
Gong et al. [45]	Let-7f	Serum/after 48 h and 2 w	88/130	Selection criteria included age >18 years, within 48 h after stroke attack, based on CT or MRI, the patient had an infarct of at least 67% of the middle cerebral artery territory, with or without the additional infarction of the anterior or posterior cerebral artery on the same side.	Exclusion criteria included unconsciousness due to metabolic disturbances or medication, any sedation or surgery, a pre-stroke score on the mRS of more than 2; and the presence of a concurrent serious illness which may affect the patient's outcome, such as severe cardiopulmonary complications.	NA	Prognostic value	Let-7f was downregulated in IS with MCI and upregulated in IS without MCI at 48 h.	Let-7f was negatively correlated with hs-CRP in IS MCI patients, also negatively correlated with IL-6 in MCI without HT
Leung et al. [47]	miR-124-3p and miR-16	Plasma/(≤6 h), (6–24 h)	93 IS + 19 HS/23	Patients aged 18 years old and above were included to the study, HS or IS confirmed by CT scan and/or MRI, who presented within 24 h of symptom onset.	NA	NA	Diagnostic value	MiR-124-3p levels were markedly higher in patients with HS patients compared to IS patients only in cases presenting early (≤6 h), increased miR-16 were found in patients with IS compared to those with HS in patients presenting late after symptom onset (6–24 h)	Plasma concentrations of miR-124-3p, but not miR-16, positively correlated with lesion volume on CT in HS patients; however, both plasma miR-124-3p and miR-16 did not correlate with lesion volume on MRI in IS patients.
Wu et al. [48]	miR-15a, miR-16, miR-17-5p	Serum/before treatment	106/120	The cohort included 55 men and 51 women with a mean age of 64.8 years (range, 39–88 years).	Symptoms indicative of subarachnoid hemorrhage, even if no imaging findings of hemorrhage were found on CT or MRI, intracranial hemorrhage, acute myocardial infarction, critical limb ischemia.	NA	Diagnostic value	MiR-15a, miR-16, and miR-17-5p were significantly higher in IS patients compared to control subjects	MiR-15a was significantly correlated with age, strong negative correlation between miR-16 levels and HDL and ApoA1 was found.

Table 1. *Cont.*

Ref	miRNAs	Sampling/Sampling Time Point	Number of Stroke pts/Controls	Inclusion Criteria	Exclusion Criteria	Stroke Subtype	Prognostic or Diagnostic Value	Regulation of miRNAs	Correlation
Jin F. and Xing J. [53]	miR-126, miR-17-5p, miR-17-3p, miR-18a, miR-19a, miR-20a, miR-19b-1, miR-92a, Let-7b, Let-7f, miR-130a, miR-210, miR-378, miR-296, miR-101, miR-221, miR-222, miR-328, miR-15b, miR-16, miR-26b, miR-27b, miR-218, miR-206, miR-338-3p, miR-497, miR-195a-3p, miR-185	Plasma/24h	106/110	Within 24 h post the onset of symptom, diagnosed with IS according to patient history, laboratory and neurological examination, CT scan, MRI, and/or MRA.	Patients with infection, renal or hepatic failure, hematological malignancies, solid tumors, immunosuppressive therapy, or treatment with thrombolytic therapy were excluded from the study.	NA	Diagnostic value, disease severity management	MiR-126 and miR-130a decreased in the IS patients while miR-222, miR-218, and miR-185 increased in the IS patients.	MiR-126, miR-378, miR-101 negatively, miR-222, miR-218, miR-206 were positively correlated with NIHSS
Gan et al. [54]	miR-145	Whole blood/NA/after one month second sampling (N = 11)	32/14	IS patients between the ages of 18 and 49 years were included. IS was confirmed by either MRI or CT imaging of the brain.	Excluded from the study were subjects with hemorrhage stroke.	NA	No diagnostic and prognostic value was found	Upregulation of miR-145	No correlation was found
Jia et al. [55]	miR-21, miR-23a, miR-29b, miR-124, miR-145, miR-210, miR-221, miR-223, miR-483-5p	Serum/24 h	146/96	Patients with IS within 24 h after symptom onset were included.	Exclusion criteria included under 18 years old, being on thrombolytic or anticoagulant therapies, intracerebral hemorrhage or hemorrhagic transformation, other complicating neurological or neuropsychological diseases, cancer, comorbidity with proinflammatory conditions and clinical signs of infection at any time during the study.	NA	Diagnostic value	MiR-145 was upregulated, miR-23a and miR-221 were significantly downregulated	Positive correlation between miR-145 and hs-CRP, IL-6, infarct volume and NIHSS scores was found, serum miR-23a and miR-221 were moderate negatively correlated with plasma hs-CRP

Table 1. *Cont.*

Ref	miRNAs	Sampling/Sampling Time Point	Number of Stroke pts/Controls	Inclusion Criteria	Exclusion Criteria	Stroke Subtype	Prognostic or Diagnostic Value	Regulation of miRNAs	Correlation
Tsai et al. [57]	miR-145, miR-21, miR-221	Serum/7 days	167/157	Patients with IS based on the World Health Organization criteria. The blood samples from the patients were taken within 7 days of the onset of stroke. Demographic data and histories of hypertension, diabetes, hypercholesterolemia and cigarette smoking were obtained from each study subject.	NA	NA	Diagnostic value	MiR-21 was downregulated, miR-221 was upregulated, no significance for miR-145 was found	MiR-21 expressions was correlated with miR-221 levels.
Zhou et al. [58]	miR-21, miR-24	Plasma/24 h	68/21	ACI participated included in the study. The diagnosis of ACI was conducted based on patient history, lab examination, neurological deficit, MRI and MRA results.	Patients with a history of tumor, immune disease, blood disease, acute infectious disease, renal or liver failure were excluded.	NA	Diagnostic value	MiR-21 and miR-24 were downregulated	MiR-21 expressions were positively correlated with miR-24 level, and negatively correlated with NIHHS score
Chen et al. [61]	miR-223	Exosomes/plasma/less than 72 h	50/33	Stroke patients included in the study. Demography feature, related previous history including hypertension, diabetes mellitus, hyperlipidemia, cardiopathy, associated laboratory test, and imaging information including blood glucose, blood lipid, electrocardiogram, cardiac ultrasonography, carotid artery ultrasonography, MRI, and MRA were also collected for analysis.	Exclusive criteria included recurrent stroke or stroke onset longer than 72 h, renal or liver failure, acute infectious disease, tumor, hematologic disease, and patients who are unable to cooperate with physical examination.	LA 25, SA 17, CE 8	Diagnostic and prognostic value	MiR-223 was upregulated	MiR-223 was positively correlated with NIHSS score

Table 1. *Cont.*

Ref	miRNAs	Sampling/Sampling Time Point	Number of Stroke pts/Controls	Inclusion Criteria	Exclusion Criteria	Stroke Subtype	Prognostic or Diagnostic Value	Regulation of miRNAs	Correlation
Wang et al. [59]	miR-223	Whole blood/less than 72 h	79/75	IS patients included in the study.	The exclusion criteria included recurrent stroke, intracranial tumor, multiple trauma, hematological system diseases, renal or liver failure, acute infectious diseases and other diseases affecting the hemogram. If the time from the onset of stroke symptoms to blood sample collection was longer than 72 h, the patient was excluded.	LA 37, SA 9, CE 5, UN 28	Diagnostic value	MiR-223 in IS patients were greatly increased compared to the control	MiRNA-223 was negatively correlated with NIHSS scores, plasma level of IGF-1 was positively correlated with that of miRNA-223
Ji et al. [62]	miR-9, miR-124	Exosomes/plasma/the mean time of enrollment blood draw was 16.5 h.	65/66	IS patients were recruited after either MRI or CT imaging of the brain.	Patients with intracerebral hemorrhage or unknown etiology were excluded.	NA	Diagnostic and prognostic value	MiR-9 and miR-124 in IS patients were increased compared to the control	The levels of both miR-9 and miR-124 were positively correlated with NIHSS scores, infarct volumes and serum concentrations of IL-6.
Liu et al. [63]	miR-9, miR-124, miR-219	Serum/24 h	31/11	Patients with IS 24 h after symptom onset were enrolled to the study.	Exclusion criteria were being under 18 years old, being on thrombolytic or anticoagulant therapies, intracerebral hemorrhage or hemorrhagic transformation, other complicating neurological or neuropsychological diseases, cancer, comorbidity with proinflammatory conditions, and clinical signs of infection at any time during the study.	NA	Inflammatory value	MiR-124 was downregulated	Both serum miR-124 and miR-9 levels within 24 h were negatively correlated with infarct volume and plasma hs-CRP levels. All three miRNAs were negatively correlated with MMP-9 levels.
Jickling et al. [67]	miR-122, miR-148a, Let-7i, miR-19a, miR-320d, miR-4429, miR-363, miR-487b	Whole blood/NA	24/24	Patients with IS were enrolled to the study. Stroke diagnosis restricted diffusion on brain MRI (positive DWI-MRI).	Patients with infection (current or within 2 weeks of stroke), immunosuppressive therapy, lymphoma, leukemia, or treatment with thrombolytic therapy were excluded from study.	LA 8, SA 8, CE 8	Diagnostic value	In patients with IS, miR-122, miR-148a, let-7i, miR-19a, miR-320d, miR-4429 were decreased and miR-363, miR-487b were increased compared to vascular risk factor controls.	MiRNAs may regulates NF-κB and toll-like receptor signaling pathways, which are involved in immune activation, leukocyte extravasation and thrombosis.

Table 1. *Cont.*

Ref	miRNAs	Sampling/Sampling Time Point	Number of Stroke pts/Controls	Inclusion Criteria	Exclusion Criteria	Stroke Subtype	Prognostic or Diagnostic Value	Regulation of miRNAs	Correlation
Li et al. [68]	In total 115 miRNAs were screened miR-32-3p, miR-106b-5p, miR-423-5p, miR-451a, miR-1246, miR-1299, miR-3149, miR-4739, miR-224-3p, miR-377-5p, miR-518b, miR-532-5p, miR-1913	Serum/24 h	117/82	IS patients (aged >45) within 24 h after stroke onset were enrolled to the study.	Exclusion criteria included other types of stroke (TIA, subarachnoid hemorrhage, brain tumors, and cerebrovascular malformation); severe systemic diseases, i.e., pulmonary fibrosis, endocrine, and metabolic diseases (except type 2 diabetes); inflammatory and autoimmune diseases; and serious chronic diseases, for example, hepatic cirrhosis and renal failure.	NA	Diagnostic value of upregulated miR-32-3p, miR-106b-5p, miR-1246, and downregulated miR-532-5p	MiR-32-3p, miR-106b-5p, miR-423-5p, miR-451a, miR-1246, miR-1299, miR-3149, miR-4739 were upregulated	MiR-106b may affect multiple pathways such as apoptosis, oxidation, angiogenesis, and neurogenesis in IS.
Tan et al. [72]	in total 157 miRNAs were screened hsa-let-7f, miR-126, -1259, -142-3p, -15b,-186, -519e, -768-5p hsa-let-7e, miR-1184, -1246, -1261, -1275, -1285, -1290, -181a, -25*, -513a-5p, -550, -602, -665, -891a, -933, -939, -923	Whole blood/within 6–18 months	19/5	Asian stroke patients between the ages of 18 to 49 were enrolled to the study. Blood samples collected from stroke patients within 6–18 months in time scale from the index stroke. IS was confirmed either with CT or MRI of the brain.	NA	LA 8, SA 3, CE 5, UN 3	Diagnostic value	In total 138 miRNAs were upregulated and in total 19 miRNAs were downregulated. hsa-let-7f, miR-126, -1259, -142-3p, -15b,-186, -519e, -768-5p were downregulated, hsa-let-7e, miR-1184, -1246, -1261, -1275, -1285, -1290, -181a, -25*, -513a-5p, -550, -602, -665, -891a, -933, -939, -923 were upregulated.	Among the upregulated miRNAs, the expression of miR-101, -106b, -130a, -144, -18a, -18b, -19a, -19b, -194, -22, -22, -29b, -29c and -363 were the highest for LA (mRS = 3) stroke sample and positively correlated to the profile observed for LA mRS > 2

Table 1. *Cont.*

Ref	miRNAs	Sampling/Sampling Time Point	Number of Stroke pts/Controls	Inclusion Criteria	Exclusion Criteria	Stroke Subtype	Prognostic or Diagnostic Value	Regulation of miRNAs	Correlation
Sepramaniam et al. [73]	In total 314 miRNAs were screened. hsa-let-7e, hsa-miR-125b-2*, hsa-miR-1261, hsa-miR-129-5p, hsa-miR-1321, hsa-miR-135b, hsa-miR-145, hsa-miR-184, hsa-miR-187*, hsa-miR-196a*, hsa-miR-198, hsa-miR-200b*, hsa-miR-210, hsa-miR-214, hsa-miR-220c, hsa-miR-25*, hsa-miR-602, hsa-miR-611, hsa-miR-617, hsa-miR-623, hsa-miR-627, hsa-miR-637, hsa-miR-638, hsa-miR-659, hsa-miR-668, hsa-miR-671-5p, hsa-miR-675, hsa-miR-920, hsa-miR-933, hsa-miR-943, hsa-miR-99a, hsa-let-7a, hsa-let-7b*, hsa-let-7c, hsa-let-7d*, hsa-let-7f, hsa-let-7g, hsa-let-7i,	Whole blood/24 h, 48 h, 7 days, from 2 months to 2 years from stroke onset	169/118	Patients with IS were enrolled to the study. IS was confirmed through either MRI or CT imaging of the brain.	NA	NA	Diagnostic value, potential diagnostic biomarkers; miR-125b-2*, -27a*, -422a, -488 and -627	Among the significant 105 miRNAs, 58 were downregulated while 47 were upregulated. Upregulated miRNAs: hsa-let-7e, hsa-miR-125b-2*, hsa-miR-1261, hsa-miR-129-5p, hsa-miR-1321, hsa-miR-135b, hsa-miR-145, hsa-miR-184, hsa-miR-187*, hsa-miR-196a*, hsa-miR-198, hsa-miR-200b*, hsa-miR-210, hsa-miR-214, hsa-miR-220c, hsa-miR-25*, hsa-miR-602, hsa-miR-611, hsa-miR-617, hsa-miR-623, hsa-miR-627, hsa-miR-637, hsa-miR-638, hsa-miR-659, hsa-miR-668, hsa-miR-671-5p, hsa-miR-675, hsa-miR-920, and downregulated miRNAs: hsa-let-7a, hsa-let-7b*, hsa-let-7c, hsa-let-7d*,	

Table 1. *Cont.*

Ref	miRNAs	Sampling/Sampling Time Point	Number of Stroke pts/Controls	Inclusion Criteria	Exclusion Criteria	Stroke Subtype	Prognostic or Diagnostic Value	Regulation of miRNAs	Correlation
	hsa-miR-106b*, hsa-miR-126, hsa-miR-1299, hsa-miR-130a, hsa-miR-151-5p, hsa-miR-18a*, hsa-miR-182, hsa-miR-183, hsa-miR-186, hsa-miR-192, hsa-miR-20a, hsa-miR-208a, hsa-miR-22*, hsa-miR-500, hsa-miR-500*, hsa-miR-501-5p, hsa-miR-502-5p, hsa-miR-502-3p, hsa-miR-505*, hsa-miR-532-5p, hsa-miR-574-5p, hsa-miR-574-3p, hsa-miR-576-5p, hsa-miR-625, hsa-miR-629, hsa-miR-652, hsa-miR-7, hsa-miR-886-5p, hsa-miR-92a, hsa-miR-93*, hsa-miR-96							hsa-let-7f, hsa-let-7g, hsa-let-7i, hsa-miR-106b*, hsa-miR-126, hsa-miR-1299, hsa-miR-130a, hsa-miR-151-5p, hsa-miR-18a*, hsa-miR-182, hsa-miR-183, hsa-miR-186, hsa-miR-192, hsa-miR-20a, hsa-miR-208a, hsa-miR-22*, hsa-miR-500, hsa-miR-500*, hsa-miR-501-5p, hsa-miR-502-5p, hsa-miR-502-3p, hsa-miR-505*, hsa-miR-532-5p, hsa-miR-574-5p, hsa-miR-574-3p, hsa-miR-576-5p, hsa-miR-625, hsa-miR-629, hsa-miR-652, hsa-miR-7, hsa-miR-886-5p, hsa-miR-92a, hsa-miR-93*, hsa-miR-96	

Table 1. *Cont.*

Ref	miRNAs	Sampling/Sampling Time Point	Number of Stroke pts/Controls	Inclusion Criteria	Exclusion Criteria	Stroke Subtype	Prognostic or Diagnostic Value	Regulation of miRNAs	Correlation
Wang et al. [74]	Microarray revealed 17 upregulated miRNAs and 103 downregulated miRNAs in MRI(−) acute stroke patients compared with MRI(+) acute stroke patients, 33 upregulated miRNAs and 36 downregulated miRNAs in MRI(+) acute stroke hsa-miR-106b-5P, hsa-miR-4306, hsa-miR-320e hsa-miR-320d	Plasma/0–3 h 23 patients, 3–6 h 37 patients, 6–12 h 31 patients, 12–24 h 45 patients.	136/116	The inclusion criteria consisted of having IS or TIA and having no history of coronary artery disease.	Patients were excluded if they had received intravenous thrombolytic or anticoagulant therapy before the initial blood samples were collected.	LA 60, SA 51, CE 23, UDN 2	Diagnostic value of hsa-miR-106b-5P, hsa-miR-4306, hsa-miR-320e, and hsa-miR-320d	hsa-miR-106b-5p hsa-miR-4306 increased, hsa-miR-320e and hsa-miR-320d decreased which are associated with diagnostic value	NA
Tian et al. [75]	hsa-mir-4454, hsa-mir-140-3p, hsa-mir-106b-5p, hsa-mir-25-3p, hsa-mir-16-5p, hsa-mir-223-3p, hsa-mir-484, hsa-mir-130b-3p, hsa-mir-151a-3p, hsa-mir-130a-3p, hsa-mir-93-5p, hsa-mir-107	Plasma/6 h	40/30	Inclusion criteria: time duration from stroke onset to admission was less than 6 h.	Patients with immune disease, trauma, coronary heart disease, organ failure, tumor, and infection were excluded from the study.	LA 10, SA 8, CE 14, UN 7, ODE 1	Diagnostic and prognostic value of miR-16	MiR-140, miR-106b, miR-130a, miR-16, miR-223, miR-93, miR-484, miR-25, miR-130b, miR-107, and miR-151 were upregulated and miR-4454 was downregulated	Plasma concentrations of miR-16 were related to TOAST criteria, OCSP criteria, and the prognosis of HACI patients

Table 1. *Cont.*

Ref	miRNAs	Sampling/Sampling Time Point	Number of Stroke pts/Controls	Inclusion Criteria	Exclusion Criteria	Stroke Subtype	Prognostic or Diagnostic Value	Regulation of miRNAs	Correlation
Tiedt et al. [76]	hsa-let-7b-3p, hsa-let-7d-3p, hsa-let-7f-5p, hsa-let-7i-5p, hsa-miR-1, hsa-miR-16-2-3p, hsa-miR-17-3p, hsa-miR-17-5p, hsa-miR-18a-3p, hsa-miR-18a-5p, hsa-miR-20a-5p, hsa-miR-26b-5p, hsa-miR-92a-3p, hsa-miR-99b-5p, hsa-miR-101-3p, hsa-miR-125a-5p, hsa-miR-125b-5p, hsa-miR-126-5p, hsa-miR-130a-3p, hsa-miR-140-3p, hsa-miR-143-3p, hsa-miR-181a-5p, hsa-miR-193a-5p, hsa-miR-378a-3p, hsa-miR-423-3p, hsa-miR-532-5p, hsa-miR-660-5p, hsa-miR-3158-3p, hsa-miR-3158-5p, hsa-miR-3184-5p hsa-miR-3688-3p, hsa-miR-3688-5p	Plasma/24 h, 48 h, 72 h, 90th day.	260/160	IS and TIA patients were enrolled to the study.	Patients with active malignant disease, inflammatory or infectious diseases, surgery within the last three months and prior medication with low-molecular or unfractionated heparin within the last month were excluded. Further, for the discovery sample patients with prior use of antiplatelet medication within the last month, a history of myocardial infarction, stroke, or TIA, or signs for silent CNS infarction on neuroimaging were also excluded. For the replication sample, patients with prior medication with low-molecular or unfractionated heparin within the last month were excluded.	LA 61, SA 18, CE 79, UN 96	Diagnostic value of miR-125a-5p, miR-125b-5p, miR-143-3p	MiR-125a-5p, miR-125b-5p, and miR-143-3p were upregulated.	The transformed infarct volumes of IS patients (N = 188) were correlated with expression levels of miR-125a-5p, miR-125b-5p, and miR-143-3p.

Table 1. *Cont.*

Ref	miRNAs	Sampling/Sampling Time Point	Number of Stroke pts/Controls	Inclusion Criteria	Exclusion Criteria	Stroke Subtype	Prognostic or Diagnostic Value	Regulation of miRNAs	Correlation
Mick et al. [77]	ex-RNAs by RNASeq (331 miRNAs, 97 piRNAs, and 43 snoRNAs) for RT-qPCR analysis: miR-877-5p, miR-124-3p, miR-320d, snoRNA SNO1402, hsa-miR-656-3p, hsa-miR-941	Plasma/NA	2763 participants included from (Framingham Heart Study; Offspring Cohort Exam 8), unbiased next-generation sequencing conducted using plasma from 40 participants from the cohort.	Subjects were diagnosed with stroke based on review of medical records, including relevant hospitalizations, and clinic reported events by at least 2 neurologists agreeing on one of the following manifestations: definite cerebrovascular accident, atherothrombotic infarction of the brain, cerebral embolism, intracerebral hemorrhage, or subarachnoid hemorrhage.	NA	NA	NA	Observational study	miR-877-5p, miR-124-3p, and miR-320d) and one snoRNA (SNO1402) were independently associated with prevalent stroke, hsa-miR-656-3p and hsa-miR-941 were significantly associated with incident stroke
Chen et al. [64]	Let-7b, miR-23a, miR-126, miR-15a, miR-16, miR-17-5p, miR-19b, miR-29b, miR-339-5p, miR-21, miR-221, miR-32-3p, miR-106-5p, miR-532-5p, miR-145, miR-146b, miR-210	Serum/24 h	128/102	Patients with IS within 24 h after symptoms were enrolled to the study.	Exclusion criteria were being under 18 years old, being on thrombolytic or anticoagulant therapies, intracerebral hemorrhage or hemorrhagic transformation, other complicating neurological or neuropsychological diseases, cancer, comorbidity with proinflammatory conditions, and clinical signs of infection at any time during the study.	NA	Diagnostic value for miR-146b. There is no significance of the expression of serum miRNAs among 5 IS groups in this study. It was suggested that miR-146b may be a biomarker for IS but not for separating subtypes of IS.	MiR-21, miR-145, miR-29b and miR-146b were significantly upregulated and miR-23a and miR-221 levels were significantly downregulated	Positive significant correlation was found between serum miR-146b level and plasma hs-CRP, infarct volume and NIHSS score, and serum IL-6 of patients.

Table 1. *Cont.*

Ref	miRNAs	Sampling/Sampling Time Point	Number of Stroke pts/Controls	Inclusion Criteria	Exclusion Criteria	Stroke Subtype	Prognostic or Diagnostic Value	Regulation of miRNAs	Correlation
Wu et al. [78]	754 miRNAs were screened, 71 miRNAs were upregulated and 49 miRNAs were downregulated in IS patients	Serum/NA	50/50 then it was confirmed with a larger cohort 177 IS, 81 TIA patients and 42 controls	IS and TIA patients were enrolled to the study. The diagnosis of IS was based on the acute occurrence of focal neurological deficit lasting for more than 24 h and was confirmed by the positive findings of brain CT and MRI.	Patients with a history of hemorrhagic infarction, peripheral arterial occlusive diseases, chronic liver/kidney diseases, primary/metastatic neoplasms or other malignant diseases were excluded.	NA	Prognostic value for miR-23b-3p and miR-29b-3p	MiR-23b-3p, miR-29b-3p, miR-181a-5p and miR-21–5p were markedly increased in IS patients. MiR-23b-3p, miR-29b-3p and miR-181a-5p were also significantly elevated in TIA patients	MiR-23b-3p levels in IS patients were positively related with discharge mRS scores, while miR-23b-3p and miR-29b-3p levels in IS patients were negatively related with discharge BI scores.

Abbreviations: CE, cardioembolic stroke; LA, large artery stroke; SA, small artery stroke; ST, stroke types; TP, time point of blood sampling; UDN, stroke due to undetermined cause, ODE, other determined etiologies; miR, microRNA; hs-CRP, high sensitivity C-reactive protein; HDL, high density lipoprotein; LDL, low density lipoprotein; NIHSS, National Institutes of Health Stroke Scale; IL-6, interleukin 6; TOAST, Trial of Org 10,172 in Acute Stroke Treatment; OCSP, Oxfordshire Community Stroke Project; HACI, hyperacute cerebral infarction; mRS, Modified Rankin Scale; BI, Barthel index; CSF, cerebrospinal fluid; CNS, central nervous system; NA, no data; IS, ischemic stroke; ACI, acute cerebral infraction; MCI, massive cerebral infarction; HS, hemorrhagic stroke; MRI, Magnetic resonance imaging; MRA, magnetic resonance angiography; CT, computed tomography; ApoA1, apolipoprotein A1; MMP-9, matrix metallopeptidase 9; NF-κB, Nuclear factor-κB; TIA, transient ischemic attack; IGF-1, Insulin-like growth factor 1; h, hour; w, week.

Table 2. The results of area under the receiver operating characteristic curves from human studies.

Ref	MicroRNA	AUC		Specificity	Sensitivity
Long et al. [41]	miR-30a	24 h	0.91	80%	94%
		1 week	0.91	84%	93%
		4 weeks	0.92	84%	90%
		24 weeks	0.93	84%	92%
	miR-126	24 h	0.92	84%	92%
		1 week	0.94	86%	90%
		4 weeks	0.93	84%	92%
		24 weeks	0.92	82%	92%
	Let-7b	24 h	0.93	84%	92%
		1 week	0.92	84%	90%
		4 weeks	0.92	86%	92%
		24 weeks	0.91	80%	89%
Wu et al. [48]	miR-15a,	0.698		NA	NA
	miR-16,	0.820			
	miR-17-5p	0.784			
	miR-15a + miR-16 + miR-17-5p	0.845			
Peng et al. [43]	Let-7e	0.86		73.4%	82.8%
	miR-338	0.63		53.2%	71.9%
Huang et al. [44]	Let-7e-5p	0.82		NA	NA
Jin F. and Xing J. [53]	miR-126	0.654		NA	NA
	miR-130a	0.642			
	miR-222	0.584			
	miR-218	0.624			
	miR-185	0.601			
	miR-126 + miR-130a + miR-222 + miR-218 + miR-185	0.767			
Chen et al. [61]	miR-223	0.859		78.8%	84%
Sepramaniam et al. [73]		Cohort 1	Cohort 2	NA	NA
	miR-125b-2*	0.95	0.85		
	miR-27a*	0.89	0.86		
	miR-422a	0.92	0.86		
	miR-488	0.87	0.86		
	miR-627	0.84	0.76		
Tian et al. [75]	miR-16 (overall patients)	0.775		87%	69.7%
	miR-16 (LAA patients)	0.952		91.3%	100%
Wang et al. [74]		Total	MRI(+)	NA	NA
	hsa-miR-106b-5p	0.999	0.962		
	hsa-miR-4306	0.877	0.952		
	hsa-miR-320e	0.953	0.981		
	hsa-miR-320d	0.977	0.987		

Table 2. *Cont.*

Ref	MicroRNA	AUC	Specificity	Sensitivity
Jia et al. [55]	CRP	0.794	NA	NA
	CRP+ miR-145	0.896		
	CRP + miR-23a	0.816		
	CRP + miR-221	0.819		
Tsai et al. [57]	miR-21/miR-221 (traditional risk factors)	0.93	NA	NA
Ji et al. [62]	miR-9	0.8026	NA	NA
	miR-124	0.6976		
Tiedt et al. [76]	miR-125a-5p + miR-125b-5p + miR-143-3p	0.663 (IS vs. TIA)		
	miR-125a-5p + miR-125b-5p + miR-143-3p + IL6 + NSE	0.661 (IS vs. TIA)		
Chen et al. [64]	miR146b	0.776	NA	NA
	CRP	0.782		
	IL-6	0.684		
	CRP+miR146b	0.863		
	IL-6+miR146b	0.819		
	CRP+IL-6+miR146b	0.866		

Abbreviations; MiRNA, microRNA; AUC, area under curve; NA, no data; CRP, C-reactive protein; LAA, large artery atherosclerosis; IL-6, interleukin 6; NSE, neuron specific enolase; MRI, magnetic resonance imaging; IS, ischemic stroke; TIA, transient ischemic attack.

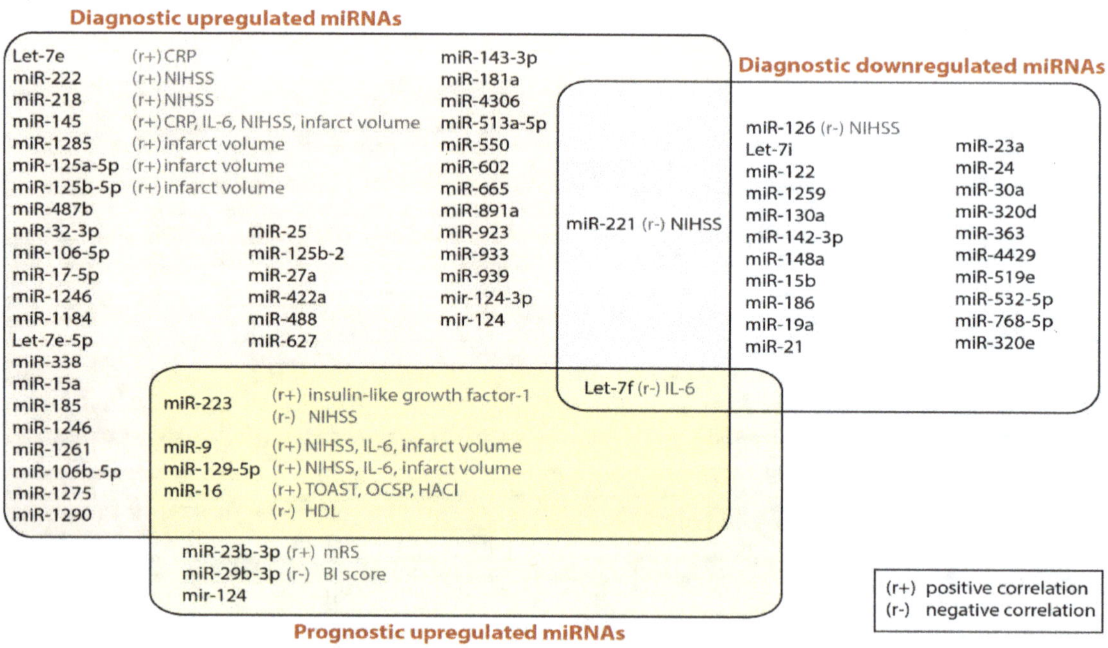

Figure 1. Regulation of diagnostic and prognostic miRNAs serving as biomarkers in ischemic stroke, based on human studies. Abbreviation: miRNA, microRNA; CRP, C-reactive protein; HDL, high density lipoprotein; NIHSS, National Institutes of Health Stroke Scale; IL-6, interleukin 6; TOAST, Trial of Org 10,172 in Acute Stroke Treatment; OCSP, Oxfordshire Community Stroke Project; HACI, hyperacute cerebral infarction; mRS, Modified Rankin Scale; BI, Barthel index. Refs: [41,43–45,47,48,53,55,57,58,61,62,67,68,72–76,78].

Profiling of miRNAs expression in plasma samples taken from IS patients was performed in 136 individuals with MRI evaluation on admission. In this population 76 patients had already ischemic changes detected with MRI (MRI+). Initial analysis revealed 120 miRNAs in MRI (−) group differentially expressed than in MRI (+) group and 69 miRNAs in MRI (+) in comparison to control group. Seventeen miRNAs were significantly changed (i.e., 10-fold) in plasma samples from both MRI (−) and (+) IS patients compared with control subjects and were further evaluated. However, only two miRNAs (hsa-miR-106b-5p and hsa-miR-4306) showed a gradient of increase, and only two miRNAs (hsa-miR-320e and hsa-miR-320d) showed a gradient of decrease from control patients to MRI(−) and MRI(+) IS patients. Also, ROC analysis performed on data from all IS patients using 4 miRNAs profile allowed to discriminate patients with IS from healthy controls. Thus, the authors showed that hsa-miR-106b-5p, hsa-miR-4306, hsa-miR-320e, and hsa-miR-320d might facilitate the diagnosis and clinical management of IS [74].

Another study aimed to identify specific circulating miRNAs that would facilitate the diagnosis of hyperacute cerebral infarction less than 6 h from the acute event, and validate their usefulness in follow-up up to 3 months after the onset of cerebral infarction. Forty patients with hyperacute cerebral infarction and 30 age-matched healthy volunteers were recruited in this study. Seven hyperacute cerebral infarction and four age-matched healthy volunteers were selected randomly for microarray analysis. Thirty-three patients and 23 controls were selected for RT-qPCR validation. In discovery phase microarray analysis showed 11 miRNAs that were upregulated (miR-140, miR-106b, miR-130a, miR-16, miR-223, miR-93, miR-484, miR-25, miR-130b, miR-107, and miR-151) and 1 miRNA that was downregulated (miR-4454), however in validation phase, only miR-16 was significantly different between patients and the control group. In multivariate logistic regression analysis only miR-16 (OR 1.669, 95% CI 1.071 ± 2.602, $p = 0.024$) was predictive for hyperacute cerebral infarction stroke. Moreover, miR-16 expression was significantly higher in the poor prognosis group than in the good prognosis group during follow-up and the diagnostic accuracy of miR-16 as a biomarker for was confirmed in ROC analysis [75].

In previously published research that consisted from three stages: discovery stage (20 IS patients and 20 healthy control subjects), validation stage (40 IS patients and 40 controls), and replication stage (200 IS patients and 100 controls). RNA sequencing was performed in order to study expression changes of circulating miRNAs. Out of 32 miRNAs, three miRNAs were upregulated in IS patients compared to both controls (i.e., miR-125a-5p, miR-125b-5p, and miR-143-3p) and transient ischemic attacks (TIA) patients. In long-term follow-up lasting up to 90 days both miR-125b-5p and miR-143-3p levels decreased, starting at day two, with no significant difference compared to controls. Expression levels of miR-125a-5p subsequently increased and stayed constantly elevated in comparison to control group. It is worth mentioning that in the studied population, different stroke subtypes were included, and miR-125a-5p, miR-125b-5p, and miR-143-3p were similar across etiological subgroups, i.e., patients with LA, CE, and stroke of undetermined etiology. Moreover, miR-125a-5p, miR-125b-5p, and miR-143-3p differentiated between controls and IS patients with an area AUC of 0.90 (sensitivity: 85.6%; specificity: 76.3%). Importantly, they found the expression levels of miR-125a-5p, miR-125b-5p, and miR-143-3p to be independent of infarct volume and stroke etiology. This finding emphasizes potential utility of these miRNAs as a broadly applicable diagnostic marker for IS. Based on another step of the experiment it was shown that the most important source of these miRNAs are platelets. Moreover, significantly higher concentration of miR-143-3p were found in extracellular microvesicles isolated from IS patients in comparison to healthy controls, but not for miR-125a-5p or miR-125b-5p [76].

One of the largest studies was done by Mick et al. They identify and validate ex-RNAs (extracellular noncoding RNAs) from plasma in participants of the Framingham Heart Study (FHS). The results demonstrate that when studied in a large observational cohort, miRNAs are significantly associated with stroke. In comparison to previously published studies there are some differences making this study highly valuable. First of all it was the largest, unbiased, community-based report of association between plasma circulating ex-RNAs (miRNAs, piRNAs, and snoRNAs) and stroke.

Secondly as a method for ex-RNA discovery RNA sequencing was used that allowed to search not only for miRNAs, but also some other type of non-coding RNA like piRNAs, and snoRNAs. They selected the most abundantly expressed ex-RNAs by RNASeq (331 miRNAs, 97 piRNAs, and 43 snoRNAs) for RT-qPCR analysis in the entire FHS Offspring Cohort. Based on proposed strategy 3 miRNAs (miR-877-5p, miR-124-3p, and miR-320d) and one snoRNA (SNO1402) were independently associated with prevalent stroke. Moreover, two other miRNAs i.e., miR-656-3p and miR-941 were significantly associated with incident stroke risk adjusting for each other and potentially confounding clinical variables. However, such a strategy did not allow analysis of the relation between IS subtypes and miRNAs [77].

Wu et al. evaluated the expression patterns of specific miRNAs in TIA patients. In their study, 754 miRNAs were initially screened by the TaqMan Low Density Array (TLDA) in two pooled serum samples from 50 IS patients and 50 controls. After miRNA profiling, significantly changed miRNAs validated by RT-qPCR in the same cohort and further confirmed in another larger cohort including 177 IS, 81 TIA patients, and 42 controls. Consequently, TLDA screening showed that 71 miRNAs were upregulated, and 49 miRNAs were downregulated in IS patients. RT-qPCR validation confirmed that serum levels of miR-23b-3p, miR-29b-3p, miR-181a-5p, and miR-21–5p were markedly increased in IS patients. Strikingly, miR-23b-3p, miR-29b-3p, and miR-181a-5p were also significantly elevated in TIA patients. Logistic regression and ROC curve analyses showed that these changed miRNAs may function as predictive and discriminative biomarkers for IS and TIA. Thus, authors suggested that, their distinctive expression signatures may contribute to assessing neurological deficit severity of IS and subsequent stroke risk after TIA [78].

Summing up the above reported results, accumulating evidence indicates that circulating miRNAs might be used as an innovative diagnostic and prognostic biomarker and potential novel therapeutic target through its potential roles in inflammatory processes. However, to better evaluate the diagnostic and prognostic role of miRNAs, more studies are required to investigate the intricate interactions between the miRNAs and their target genes in IS. Results of the studies presented in this review should be interpreted with consciousness. As we discussed above studies found controversy findings. The discrepancy of the results might be the reason of demographic differences between populations, heterogeneity of populations, various cohort sizes and study designs. It is necessary to conduct further studies to validate the current hypotheses and closely determine the association between various miRNAs and their contribution to cardiovascular diseases development.

3. Future Perspectives for Using miRNAs in Diagnosis and Prognosis in Ischemic Stroke

The use of circulating miRNAs as disease biomarkers in stroke is a very attractive and promising concept. In fact, they might offer several advantages over traditional disease biomarkers: (i) they can reflect specific cellular pathophysiological alterations; (ii) experimental evidence suggests that they could potentially indicate the specific etiology of stroke; (iii) since specific miRNAs showed significant modulation well before the development of the acute and irreversible cerebral damage, they could potentially allow early stroke diagnosis and/or the identification of subjects at risk before they develop an acute stroke.

Nevertheless, a number of challenges still need to be addressed before circulating miRNAs could enter the clinical arena: (a) only few studies have tested their clinical usefulness, most of which are quite small and present methodological flaws; (b) the large variation observed for miRNAs within the same patients group could make the identification of diagnostic cutoffs particularly challenging; (c) most of the available studies that evaluated the clinical usefulness of circulating miRNAs in stroke were performed in Asian populations and their results could therefore not completely apply to other ethnicities; (d) most studies did not assess the impact of other clinical variables on the prognostic potential, which could be additive in some cases and thus potentially helpful if used to correct/adjust biomarker values; (e) finally, the current standard method for the reliable measurement of circulating miRNAs is RT-PCR, which presents some limitations for the use as clinical biomarker in the setting of an acute disease. Although several alternative methodological approaches are being tested, a reliable, fast and cheaper analytical methodology has not been established, yet.

4. Bioinformatics Analysis

We performed bioinformatics analysis in order to identify the most commonly regulated circulating miRNA among the selected studies that are involved in inflammation, blood coagulation, and platelet activation. The miRNAs list was narrowed down by searching human studies literature. R programming was used to build a miRNA-Gene-Network based on the interactions of the miRNAs and target genes and their functions. TargetScan and MirTarBase databases were used to predict the target genes of the miRNAs [79,80]. To identify the genes associated with inflammation process; platelet activation; blood coagulation we performed a screening of the Gene Ontology (GO) terms for the presence of the key words using the biomaRt package in R [81]. Key words used for screening the GO terms are given in the Supplementary Materials (See Table S1). Electronic database Pubmed and Scopus was searched between December 2017 and March 2018, and original studies were reviewed to evaluate the potential diagnostic or/and prognostic role of circulating miRNAs associated with IS. Review articles and meta-analyses were also investigated, and their secondary references were examined for possible inclusion. Our search was limited to human studies only and did not exclude studies on the basis of ethnicity of study participants. Search terms comprised of the following search syntax: "Search ("micrornas" [MeSH Terms] OR "mir" [MeSH Terms] OR "mirna" [MeSH Terms] OR "circulating miRNA" [MeSH Terms] OR "circulating microRNA" [MeSH Terms]) AND ("ischemic stroke" [MeSH Terms] OR "stroke" [All Fields]) AND ("diagnostic" [MeSH Terms] OR "diagnosis" [All Fields]) OR ("prognostic" [MeSH Terms] OR "prognosis" [All Fields])" Filters: Humans. A total of 52 records were identified after duplicates removal. Titles and abstracts were screened by two independent operators, with exclusion of 27 records for any of the following reasons: (a) they were not related to the specific research question (n = 11); (b) they did not present original data (n = 14); or they were not human studies (n = 2). Finally, 25 articles were selected to be used in this review.

MiRNA-target interaction network was constructed using R programming and visualized using Cytoscape software v 3.6.1 [82]. miRNA-target interactions used for constructing the network were obtained from TargetScan and MirTarBase databases. In the next step we selected miRNAs targeting at least 5 genes associated with specific GO process (inflammation process; platelet activation; blood coagulation), and target genes regulated by at least 5 miRNAs (See Figure S1b). In the third step we subsetted miRNAs regulating at least 5 target genes from the network created in the first step. Final networks were created by using circular layout and direct force layout (See Figure S1c). Circular networks were sorted by the degree of connections between miRNAs and their targeted genes. It enabled to retrieve top miRNAs and top shared targets involved in analyzed GO process (See Figures 2a, 3a, and 4a). To retrieve gene–gene interactions between our top targets we used interactome datasets from String app v 1.2.2 for Cytoscape [83].

The pathophysiology of IS is related with inflammation, blood coagulation process and platelet activation. In course of thromboembolic stroke, activated platelets orchestrate a thrombo-inflammatory cascade by promoting thrombus formation and growth, activating leukocytes, and potentiating cerebral endothelium injury. In line with this aim, we focused on miRNAs involved in inflammation, blood coagulation, and platelet activation in our bioinformatics analysis in order to identify to account for all potential pathophysiological mechanisms underlying the development of IS. In fact, the study that were selected included different subtypes of IS, such as thromboembolic, LA, SA, and UDN. We adopted such broad criteria is the intention to identify a pattern of possibly few miRNAs that could help to recognize the different underlying etiologies and/or main pathophysiological mechanisms. Using a similar approach, we were recently able to identify as few as two circulating miRNAs to be used for the differential diagnosis of heart failure of ischemic origin from other forms of non-ischemic origin [22].

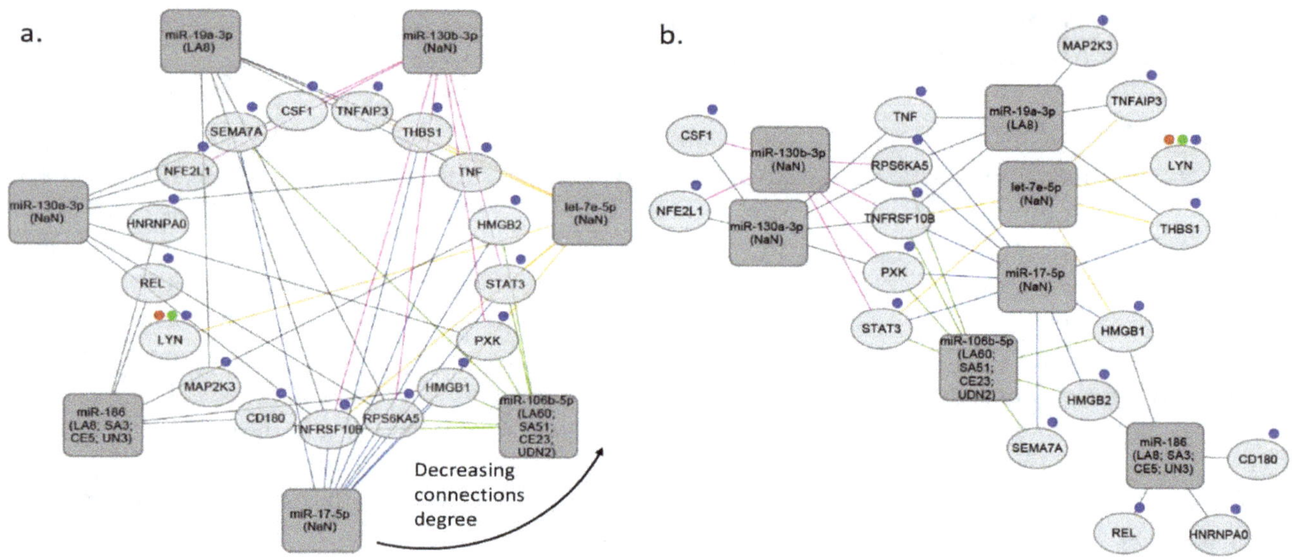

Figure 2. miRNA-target gene networks based on inflammatory response. (**a**) Inflammatory response-network sorted by the degree of connections, (**b**) Inflammatory response-interaction network. The rectangles indicate the stroke type miRNAs, the ellipses indicate target genes. Red, green and blue marks represent specific GO process-blood coagulation, platelet activation, and inflammation process, respectively. LA, large artery stroke; SA, small artery stroke; UDN, stroke due to undetermined cause; CE, cardioembolic stroke; NaN, no data.

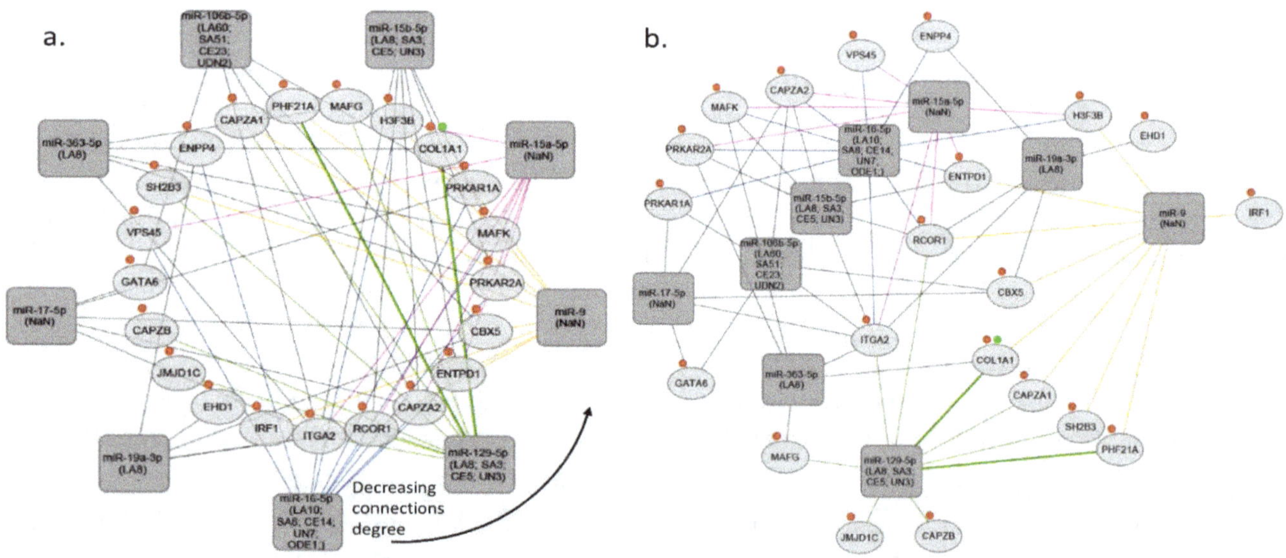

Figure 3. miRNA-target gene networks based on blood coagulation. (**a**) Blood coagulation- network sorted by the degree of connections, (**b**) Blood coagulation-interaction network. The rectangles indicate the stroke type miRNAs, the ellipses indicate target genes. Red and green marks represent specific GO process-blood coagulation and platelet activation, respectively. LA, large artery stroke; SA, small artery stroke; UDN, stroke due to undetermined cause; CE, cardioembolic stroke; NaN, no data; ODE, other determined etiology; UN, determined cause.

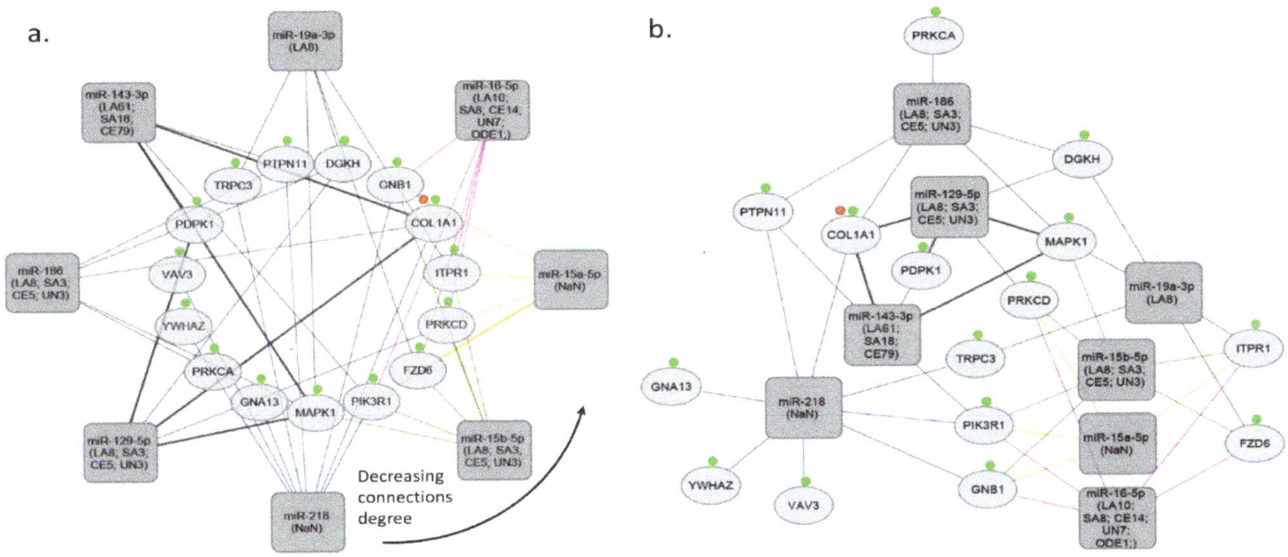

Figure 4. miRNA-target gene networks based on platelet activation. (**a**) Platelet activation-network sorted by the degree of connections, (**b**) Platelet activation-interaction network. The rectangles indicate the stroke type miRNAs, the ellipses indicate target genes. Red and green marks represent specific GO process-blood coagulation and platelet activation, respectively. LA, large artery stroke; SA, small artery stroke; UDN, stroke due to undetermined cause; CE, cardioembolic stroke; NaN, no data; ODE, other determined etiology; UN, determined cause.

Our analysis showed that 2 common miRNAs namely, miR-17-5p and miR-106b-5p may regulate both inflammatory response and blood coagulation. 1 common miRNA, which is miR-186 may regulate both inflammatory response and platelet activation. Addition to this, we found that, 4 common miRNAs namely, miR-15b-5p, miR-15a-5p, miR-16-5p, and miR-129-5p may regulate both blood coagulation and platelet activation. Interestingly, 1 common miRNA, miR-19a-3p was found, that is involved in inflammatory response, blood coagulation, and platelet activation (See Figure 5b). Thus, we found that miR-19a-3p can be the crucial miRNA which regulates all of these three processes. MiR-19a is a member of the miR-17-92 cluster, and its importance was shown in breast cancer cells [84]. Besides, it was found that members of miR-17-92 cluster are involved in coronary artery disease and vascular functions, including ischemia responses and angiogenesis [85–87]. Liu et al. investigated the function of the miR17-92 cluster in adult neural progenitor cells after experimental stroke. They found that stroke substantially up-regulated miR17-92 cluster expression in neural progenitor cells of the adult mouse and inhibition of individual members of the miR17-92 cluster, such as miR-19a, suppressed cell proliferation and increased apoptosis [52]. As we discussed, Jinkling et al. demonstrate that, low circulating miR-19a levels in the blood were found in patients with IS compared to controls [67]. This can be the reason of potential cell protective effect of miR-19a in ischemia. It is worth underlining that only two studies out of 25 analyzed miR-19a-3p, and the only study which used miRNA profiling was found the significant relation with IS. There are several reasons that could help to understand such a discrepancy; (i) different inclusion and exclusion criteria implemented in each study, (ii) differences in population origin, (ii) lack of homogeneity of IS population due to inclusion of different IS subtypes. These weak points should be considered, and better-designed studies should confirm the actual biological role and clinical usefulness of miR-19a-3p in IS. Additionally, in our analysis we demonstrated ten the most significant genes targeted by miRNAs associated with blood coagulation, platelet activation, and inflammation process in IS, namely HMGB1, YWHAZ, PIK3R1, STAT3, MAPK1, CBX5, CAPZB, THBS1, TNFRSF10B, RCOR1 (High-mobility group box 1- HMGB1; tyrosine 3 monooxygenase/tryptophan 5 monooxygenase activating protein, zeta polypeptide- YWHAZ; phosphoinositide-3-kinase regulatory subunit 1- PIK3R1; signal transducers and activators of transcription 3- STAT3; mitogen-activated protein kinase 1- MAPK1; chromobox homolog 5- CBX5; capping actin protein of muscle Z-line subunit beta- CAPZB; thrombospondin-1- THBS1; TNF receptor superfamily member 10b- TNFRSF10B; REST corepresor 1- RCOR1) (See Figure 5a).

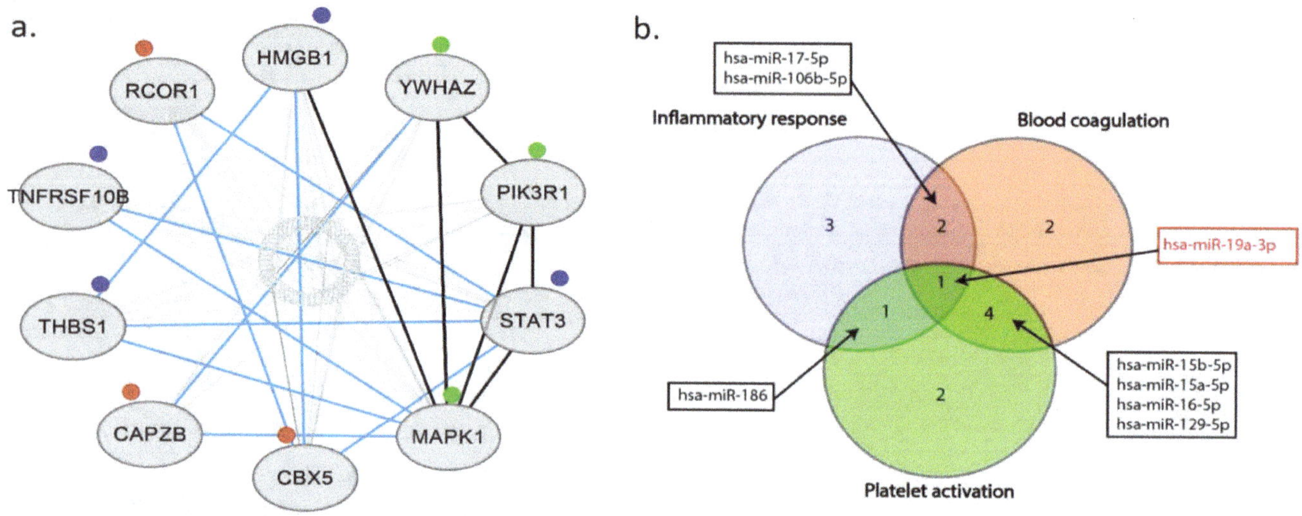

Figure 5. Most important genes targeted by miRNAs and overlapped miRNAs in GO process in IS. (**a**) Top ten genes targeted by miRNAs associated with IS. The circle in the middle indicates the stroke type miRNAs, the ellipses indicate target genes. Red, green and blue marks represent specific GO process-blood coagulation, platelet activation, and inflammation process, respectively. Black edges represent the high confidence connections between genes, blue edges represent low confidence connections. (**b**) Overlapped miRNAs in inflammatory response, blood coagulation, and platelet activation. 2 common miRNAs observed in inflammatory response and blood coagulation namely, miR-17-5p miR-106b-5p. 1 common miRNA observed in inflammatory response and platelet activation namely, miR-186. 4 common miRNAs observed in blood coagulation and platelet activation namely, miR-15b-5p, miR-15a-5p, miR-16-5p, miR-129-5p. 1 common miRNA observed in inflammatory response, blood coagulation and platelet activation namely, miR-19a-3p. miR, microRNA.

HMGB1 is released from necrotic brain tissue and its differential redox forms attract and activate immune cells after ischemic brain injury. Its concentrations correlate with disease severity and outcome after brain injury [88]. Our results support the importance of HMGB in IS. The role of YWHAZ gene was shown in several neurodegenerative diseases such as, Huntington's disease, Alzheimer's disease, and amyotrophic lateral sclerosis [89–91]. As we found in our bioinformatic analysis a similar study showed the importance of YWHAZ gene in CE [92]. Besides, the role of PIK3R1 gene in stroke is yet to be published. On the other hand, one study showed that PIK3R1 is the target of miR-221 in endothelial progenitor cells (EPCs). EPCs assist angiogenesis and have been linked to ischemia-related disorders, including coronary artery disease [93]. Abovementioned, human studies showed that, miR-221 is a potential diagnostic biomarker in IS, thus the role of PIK3R1 and relation with miR-221 in IS should be confirmed by future studies [55,57,64]. As we showed the relation of STAT3 with IS in our analysis, lately Liang et al. reviewed several significant cerebral ischemic and HS treatments that target the STAT3 signaling pathway, including pharmacological and physical therapies [94]. Additionally, Bam et al. investigated miRNA profile in the peripheral blood mononuclear cells of IS patients. Further, they performed bioinformatic analysis and they showed that pro-inflammatory genes like STAT3, IL12A, and IL12B are some of the highly predicted targets for the dysregulated miRNAs [95]. Moreover, studies showed that MAPK can be an important regulator in ischemic and hemorrhagic cerebral vascular disease and raising the possibility that it might be a drug discovery target for stroke. In our bioinformatics analysis we confirmed previous results that showed the importance of MAPK1 gene and its signaling pathway in IS [44,68,96]. CBX5 is a member of the heterochromatin protein 1 (HP1) family, HPs are closely related to the development of major vascular injuries, such as atherosclerosis and hypertension. To the best of our knowledge, only one study showed that association of serum levels of antibodies against CBX5 could potentially represent useful tools for the diagnosis of TIA and predicting the onset of acute-phase of cerebral infarction [97]. In our analysis we showed that CBX5 can be related with IS targeted by miRNAs. THBS1 is a potent regulator of angiogenesis and previous studies showed the importance of THBS1 in stroke [98,99]. Gao et al. showed that plasma THBS1 protein concentrations are elevated and are highly associated with long-term outcome of IS [98]. Based on our bioinformatic analysis we also demonstrated the relation between THBS1 gene and IS targeted by miRNAs. It was shown that decreased cardiac CAPZ protein protects hearts against acute ischemia-reperfusion injury [100]. However, there is no published study describing the role of CAPZB in IS. Similarly, the role of RCOR1 and TNFRSF10B genes in stroke is yet to be published. In our analysis for the first time we found that CAPZB, RCOR1, and TNFRSF10B may have association with IS targeted by miRNAs.

5. Conclusions

The present article provides an up-to-date and comprehensive overview of all circulating miRNAs studied so far as diagnostic or prognostic markers in IS. As we deeply reviewed above, studies showed the impact of more than 1000 different miRNAs on IS, and their association with biochemical and hematological parameters in stroke patients. These studies clarify the value of miRNAs in IS the ischemic pathophysiological process and help us to better understand processes involved in IS pathophysiology. Through a bioinformatic analysis, we associated key biological function and signaling pathways related to the miRNAs with most prominent differential expression pattern in IS compared to controls. Using this approach, we found miR-19a-3p as the single miRNA involved in all three main biological processes selected for the bioinformatic analysis: inflammation, blood coagulation and platelet activation. Furthermore, results of bioinformatics analysis shows the following most significant genes targeted by miRNAs in IS: HMGB1, YWHAZ, PIK3R1, STAT3, MAPK1, CBX5, CAPZB, THBS1, TNFRSF10B, and RCOR1. Moreover, it is worth mentioning that CRP and IL-6 are the most common targets within the inflammatory ontology. This latter finding is particularly interesting, as both CRP and IL-6 are well-known prognostic markers in atherosclerotic vascular disease [101]. In addition, CRP might represent a key link between vascular inflammation and arterial thrombosis or platelet activation [102–104]. Finally, the finding that multiple miRNAs are correlated to inflammation supports the use of a panel of multiple miRNAs.

In conclusion, since many known biomarkers related with IS are not disease specific and have been also associated with other types of brain injury, novel and specific diagnostic/prognostic biomarkers in IS are needed. In this context, circulating miRNAs are very promising biomarkers in stroke. Since specific miRNAs showed significant modulation well before the development of the acute and irreversible cerebral damage, they may potentially allow early stroke diagnosis and/or the identification of subjects at risk before they develop an acute stroke, however future studies are needed.

Author Contributions: C.E.: Analysis of literature search, collection, analysis, and interpretation of data, preparing manuscript, graphical preparation, acceptance of the final version of the manuscript, preparation of the revision. Z.W.: Bioinformatic analysis, graphical preparation, acceptance of the final version of the manuscript, preparation of the revision. S.D.R.: Preparing manuscript, acceptance of the final version of the manuscript, preparation of the revision. D.M.-G.: Analysis of literature search, acceptance of the final version of the manuscript. A.S.: Analysis of literature search, preparation of the revision. C.I.: Analysis of literature search, acceptance of the final version of the manuscript. I.J.-K.: Analysis of literature search, acceptance of the final version of the manuscript. A.C.: Acceptance of the final version of the manuscript, critical revision of the article. M.P.: Analysis of literature search, Collection and interpretation of data, preparing manuscript, acceptance of the final version of the manuscript, preparation of the revision. All authors agreed to be accountable for all aspects of the work in ensuring that questions related to the accuracy or integrity of any part of the work are appropriately investigated and resolved.

References

1. Benjamin, E.J.; Blaha, M.J.; Chiuve, S.E.; Cushman, M.; Das, S.R.; Deo, R.; de Ferranti, S.D.; Floyd, J.; Fornage, M.; Gillespie, C.; et al. American Heart Association Statistics Committee and Stroke Statistics Subcommittee. Heart Disease and Stroke Statistics-2017 Update: A Report from the American Heart Association. *Circulation* **2017**, *135*, 146–603. [CrossRef] [PubMed]
2. De Rosa, S.; Sievert, H.; Sabatino, J.; Polimeni, A.; Sorrentino, S.; Indolfi, C. Percutaneous Closure Versus Medical Treatment in Stroke Patients with Patent Foramen Ovale: A Systematic Review and Meta-analysis. *Ann. Intern. Med.* **2018**, *168*, 343–350. [CrossRef] [PubMed]
3. *European Cardiovascular Disease Statistics*, 2017th ed.; European Heart Network: Brussels, Belgium, 2017.
4. Luengo-Fernandez, R.; Gray, A.M.; Bull, L.; Welch, S.; Cuthbertson, F.; Rothwell, P.M. Quality of life after TIA and stroke: Ten-year results of the Oxford Vascular Study. *Neurology* **2013**, *81*, 1588–1595. [CrossRef] [PubMed]
5. Beal, C.C. Gender and stroke symptoms: A review of the current literature. *J. Neurosci. Nurs.* **2010**, *42*, 80–87. [CrossRef] [PubMed]
6. Chen, Y.; Xiao, Y.; Lin, Z.; Xiao, X.; He, C.; Bihl, J.C.; Zhao, B.; Ma, X.; Chen, Y. The Role of Circulating Platelets Microparticles and Platelet Parameters in Acute Ischemic Stroke Patients. *J. Stroke Cerebrovasc. Dis.* **2015**, *24*, 2313–32320. [CrossRef] [PubMed]
7. Pordzik, J.; Pisarz, K.; De Rosa, S.; Jones, A.D.; Eyileten, C.; Indolfi, C.; Malek, L.; Postula, M. The Potential Role of Platelet-Related microRNAs in the Development of Cardiovascular Events in High-Risk Populations, Including Diabetic Patients: A Review. *Front. Endocrinol.* **2018**, *9*, 74. [CrossRef] [PubMed]
8. Janssen, H.L.; Reesink, H.W.; Lawitz, E.J.; Zeuzem, S.; Rodriguez-Torres, M.; Patel, K.; van der Meer, A.J.; Patick, A.K.; Chen, A.; Zhou, Y.; et al. Treatment of HCV infection by targeting microRNA. *N. Engl. J. Med.* **2013**, *368*, 1685–1694. [CrossRef] [PubMed]
9. Bader, A.G. miR-34—A microRNA replacement therapy is headed to the clinic. *Front. Genet.* **2012**, *3*, 120. [CrossRef] [PubMed]
10. Bushati, N.; Cohen, S.M. MicroRNAs in neurodegeneration. *Curr. Opin. Neurobiol.* **2008**, *18*, 292–296. [CrossRef] [PubMed]
11. Nelson, P.T.; Wang, W.X.; Rajeev, B.W. MicroRNAs (miRNAs) in neurodegenerative diseases. *Brain Pathol.* **2008**, *18*, 130–138. [CrossRef] [PubMed]
12. Lee, H.-J. Exceptional stories of microRNAs. *Exp. Biol. Med.* **2013**, *238*, 339–343. [CrossRef] [PubMed]
13. De Rosa, S.; Indolfi, C. Circulating microRNAs as Biomarkers in Cardiovascular Diseases. In *Experientia Supplementum*; Springer: Basel, Switzerland, 2015; Volume 106, pp. 139–149.
14. Di Ieva, A.; Butz, H.; Niamah, M.; Rotondo, F.; De Rosa, S.; Sav, A.; Yousef, G.M.; Kovacs, K.; Cusimano, M.D. MicroRNAs as biomarkers in pituitary tumors. *Neurosurgery* **2014**, *75*, 181–189, discussion 188–189. [CrossRef] [PubMed]
15. De Rosa, S.; Curcio, A.; Indolfi, C. Emerging role of microRNAs in cardiovascular diseases. *Circ. J.* **2014**, *78*, 567–575. [CrossRef] [PubMed]
16. Fichtlscherer, S.; De Rosa, S.; Fox, H.; Schwietz, T.; Fischer, A.; Liebetrau, C.; Weber, M.; Hamm, C.W.; Röxe, T.; Müller-Ardogan, M.; et al. Circulating microRNAs in patients with coronary artery disease. *Circ. Res.* **2010**, *107*, 677–684. [CrossRef] [PubMed]
17. De Rosa, R.; De Rosa, S.; Leistner, D.; Boeckel, J.N.; Keller, T.; Fichtlscherer, S.; Dimmeler, S.; Zeiher, A.M. Transcoronary Concentration Gradient of microRNA-133a and Outcome in Patients with Coronary Artery Disease. *Am. J. Cardiol.* **2017**, *120*, 15–24. [CrossRef] [PubMed]
18. Sluijter, J.P.; Doevendans, P.A. Circulating microRNA profiles for detection of peripheral arterial disease: Small new biomarkers for cardiovascular disease. *Circ. Cardiovasc. Genet.* **2013**, *6*, 441–443. [CrossRef]
19. Sorrentino, S.; Iaconetti, C.; De Rosa, S.; Polimeni, A.; Sabatino, J.; Gareri, C.; Passafaro, F.; Mancuso, T.; Tammè, L.; Mignogna, C.; et al. Hindlimb Ischemia Impairs Endothelial Recovery and Increases Neointimal Proliferation in the Carotid Artery. *Sci. Rep.* **2018**, *8*, 761. [CrossRef]
20. De Rosa, S.; Fichtlscherer, S.; Lehmann, R.; Assmus, B.; Dimmeler, S.; Zeiher, A.M. Transcoronary concentration gradients of circulating microRNAs. *Circulation* **2011**, *124*, 1936–1944. [CrossRef]
21. Corsten, M.F.; Dennert, R.; Jochems, S.; Kuznetsova, T.; Devaux, Y.; Hofstra, L.; Wagner, D.R.; Staessen, J.A.;

Heymans, S.; Schroen, B. Circulating MicroRNA-208b and MicroRNA-499 reflect myocardial damage in cardiovascular disease. *Circ. Cardiovasc. Genet.* **2010**, *3*, 499–506. [CrossRef]
22. De Rosa, S.; Eposito, F.; Carella, C.; Strangio, A.; Ammirati, G.; Sabatino, J.; Abbate, F.G.; Iaconetti, C.; Liguori, V.; Pergola, V.; et al. Transcoronary concentration gradients of circulating microRNAs in heart failure. *Eur. J. Heart Fail.* **2018**, *20*, 1000–1010. [CrossRef]
23. Tijsen, A.J.; Creemers, E.E.; Moerland, P.D.; de Windt, L.J.; van der Wal, A.C.; Kok, W.E.; Pinto, Y.M. MiR423-5p as a circulating biomarker for heart failure. *Circ. Res.* **2010**, *106*, 1035–1039. [CrossRef] [PubMed]
24. Jaguszewski, M.; Osipova, J.; Ghadri, J.R.; Napp, L.C.; Widera, C.; Franke, J.; Fijalkowski, M.; Nowak, R.; Fijalkowska, M.; Volkmann, I.; et al. A signature of circulating microRNAs differentiates takotsubo cardiomyopathy from acute myocardial infarction. *Eur. Heart J.* **2014**, *35*, 999–1006. [CrossRef] [PubMed]
25. Gareri, C.; De Rosa, S.; Indolfi, C. MicroRNAs for Restenosis and Thrombosis After Vascular Injury. *Circ. Res.* **2016**, *118*, 1170–1184. [CrossRef] [PubMed]
26. Polimeni, A.; De Rosa, S.; Indolfi, C. Vascular miRNAs after balloon angioplasty. *Trends Cardiovasc. Med.* **2013**, *23*, 9–14. [CrossRef] [PubMed]
27. De Rosa, S.; Arcidiacono, B.; Chiefari, E.; Brunetti, A.; Indolfi, C.; Foti, D.P. Type 2 Diabetes Mellitus and Cardiovascular Disease: Genetic and Epigenetic Links. *Front. Endocrinol.* **2018**, *9*, 2. [CrossRef] [PubMed]
28. Zampetaki, A.; Kiechl, S.; Drozdov, I.; Willeit, P.; Mayr, U.; Prokopi, M.; Mayr, A.; Weger, S.; Oberhollenzer, F.; Bonora, E.; et al. Plasma microRNA profiling reveals loss of endothelial miR-126 and other microRNAs in type 2 diabetes. *Circ. Res.* **2010**, *107*, 810–817. [CrossRef] [PubMed]
29. Carino, A.; De Rosa, S.; Sorrentino, S.; Polimeni, A.; Sabatino, J.; Caiazzo, G.; Torella, D.; Spaccarotella, C.; Mongiardo, A.; Strangio, A.; et al. Modulation of Circulating MicroRNAs Levels during the Switch from Clopidogrel to Ticagrelor. *Biomed. Res. Int.* **2016**, *2016*, 3968206. [CrossRef] [PubMed]
30. Streit, S.; Michalski, C.W.; Erkan, M.; Kleeff, J.; Friess, H. Northern blot analysis for detection and quantification of RNA in pancreatic cancer cells and tissues. *Nat. Protoc.* **2009**, *4*, 37–43. [CrossRef] [PubMed]
31. Várallyay, E.; Burgyán, J.; Havelda, Z. MicroRNA detection by northern blotting using locked nucleic acid probes. *Nat. Protoc.* **2008**, *3*, 190–196. [CrossRef] [PubMed]
32. Ferraro, D.; Champ, J.; Teste, B.; Serra, M.; Malaquin, L.; Viovy, J.L.; de Cremoux, P.; Descroix, S. Microfluidic platform combining droplets and magnetic tweezers: Application to HER2 expression in cancer diagnosis. *Sci. Rep.* **2016**, *6*, 25540. [CrossRef] [PubMed]
33. Eminaga, S.; Christodoulou, D.C.; Vigneault, F.; Church, G.M.; Seidman, J.G. Quantification of microRNA expression with next-generation sequencing. *Curr. Protoc. Mol. Biol.* **2013**, *103*, 4–17. [CrossRef] [PubMed]
34. Chugh, P.; Dittmer, D.P. Potential Pitfalls in microRNA Profiling. *Wiley Interdiscip. Rev. RNA* **2012**, *3*, 601–616. [CrossRef] [PubMed]
35. Yang, H.; Hui, A.; Pampalakis, G.; Soleymani, L.; Liu, F.F.; Sargent, E.H.; Kelley, S.O. Direct, electronic microRNA detection for the rapid determination of differential expression profiles. *Angew. Chem. Int. Ed. Engl.* **2009**, *48*, 8461–8464. [CrossRef] [PubMed]
36. Roy, S.; Soh, J.H.; Gao, Z. A microfluidic-assisted microarray for ultrasensitive detection of miRNA under an optical microscope. *Lab Chip* **2011**, *11*, 1886–1894. [CrossRef] [PubMed]
37. Dong, H.; Jin, S.; Ju, H.; Hao, K.; Xu, L.P.; Lu, H.; Zhang, X. Trace and label-free microRNA detection using oligonucleotide encapsulated silver nanoclusters as probes. *Anal. Chem.* **2012**, *84*, 8670–8674. [CrossRef] [PubMed]
38. Tran, H.V.; Piro, B.; Reisberg, S.; Tran, L.D.; Duc, H.T.; Pham, M.C. Label-free and reagentless electrochemical detection of microRNAs using a conducting polymer nanostructured by carbon nanotubes: Application to prostate cancer biomarker miR-141. *Biosens. Bioelectron.* **2013**, *49*, 164–169. [CrossRef] [PubMed]
39. Chen, C.; Ridzon, D.A.; Broomer, A.J.; Zhou, Z.; Lee, D.H.; Nguyen, J.T.; Barbisin, M.; Xu, N.L.; Mahuvakar, V.R.; Andersen, M.R.; et al. Real-time quantification of microRNAs by stem-loop RT-PCR. *Nucleic Acids Res.* **2005**, *33*, e179. [CrossRef] [PubMed]
40. Kang, K.; Peng, X.; Luo, J.; Gou, D. Identification of circulating miRNA biomarkers based on global quantitative real-time PCR profiling. *J. Anim. Sci. Biotechnol.* **2012**, *3*, 4. [CrossRef]
41. Long, G.; Wang, F.; Li, H.; Yin, Z.; Sandip, C.; Lou, Y.; Wang, Y.; Chen, C.; Wang, D.W. Circulating miR-30a, miR-126 and let-7b as biomarker for ischemic stroke in humans. *BMC Neurol.* **2013**, *13*, 178. [CrossRef]

42. Li, Y.; Maegdefessel, L. My heart will go on-beneficial effects of anti-MiR-30 after myocardial infarction. *Ann. Transl. Med.* **2016**, *4*, 144. [CrossRef]
43. Peng, G.; Yuan, Y.; Wu, S.; He, F.; Hu, Y.; Luo, B. MicroRNA let-7e is a potential circulating biomarker of acute stage ischemic stroke. *Transl. Stroke Res.* **2015**, *6*, 437–445. [CrossRef] [PubMed]
44. Huang, S.; Lv, Z.; Guo, Y.; Li, L.; Zhang, Y.; Zhou, L.; Yang, B.; Wu, S.; Zhang, Y.; Xie, C.; et al. Identification of blood Let-7e-5p as a biomarker for ischemic stroke. *PLoS ONE* **2016**, *11*, e0163951. [CrossRef] [PubMed]
45. Gong, Z.; Zhao, S.; Zhang, J.; Xu, X.; Guan, W.; Jing, L.; Liu, P.; Lu, J.; Teng, J.; Peng, T.; et al. Initial research on the relationship between let-7 family members in the serum and massive cerebral infarction. *J. Neurol. Sci.* **2016**, *361*, 150–157. [CrossRef] [PubMed]
46. Panda, A.C.; Grammatikakis, I.; Kim, K.M.; De, S.; Martindale, J.L.; Munk, R.; Yang, X.; Abdelmohsen, K.; Gorospe, M. Identification of senescence-associated circular RNAs (SAC-RNAs) reveals senescence suppressor CircPVT1. *Nucleic Acids Res.* **2017**, *45*, 4021–4035. [CrossRef] [PubMed]
47. Leung, L.Y.; Chan, C.P.; Leung, Y.K.; Jiang, H.L.; Abrigo, J.M.; Wang, D.F.; Chung, J.S.; Rainer, T.H.; Graham, C.A. Comparison of miR-124-3p and miR-16 for early diagnosis of hemorrhagic and ischemic stroke. *Clin. Chim. Acta* **2014**, *433*, 139–144. [CrossRef] [PubMed]
48. Wu, J.; Du, K.; Lu, X. Elevated expressions of serum miR-15a, miR-16, and miR-17-5p are associated with acute ischemic stroke. *Int. J. Clin. Exp. Med.* **2015**, *8*, 21071–21079.
49. Spinetti, G.; Fortunato, O.; Caporali, A.; Shantikumar, S.; Marchetti, M.; Meloni, M.; Descamps, B.; Floris, I.; Sangalli, E.; Vono, R.; et al. MicroRNA-15a and microRNA-16 impair human circulating proangiogenic cell functions and are increased in the proangiogenic cells and serum of patients with critical limb ischemia. *Circ. Res.* **2013**, *112*, 335–346. [CrossRef]
50. Liu, L.F.; Liang, Z.; Lv, Z.R.; Liu, X.H.; Bai, J.; Chen, J.; Chen, C.; Wang, Y. MicroRNA-15a/b are upregulated in response to myocardial ischemia/reperfusion injury. *J. Geriatr. Cardiol.* **2012**, *9*, 28–32.
51. Yin, K.J.; Fan, Y.; Hamblin, M.; Zhang, J.; Zhu, T.; Li, S.; Hawse, J.R.; Subramaniam, M.; Song, C.Z.; Urrutia, R.; et al. KLF11 mediates PPARgamma cerebrovascular protection in ischaemic stroke. *Brain* **2013**, *136*, 1274–1287. [CrossRef]
52. Liu, X.S.; Chopp, M.; Wang, X.L.; Zhang, L.; Hozeska-Solgot, A.; Tang, T.; Kassis, H.; Zhang, R.L.; Chen, C.; Xu, J.; et al. MicroRNA-17-92 cluster mediates the proliferation and survival of neural progenitor cells after stroke. *J. Biol. Chem.* **2013**, *288*, 12478–12488. [CrossRef]
53. Jin, F.; Xing, J. Circulating pro-angiogenic and anti-angiogenic microRNA expressions in patients with acute ischemic stroke and their association with disease severity. *Neurol. Sci.* **2017**, *38*, 2015–2023. [CrossRef] [PubMed]
54. Gan, C.S.; Wang, C.W.; Tan, K.S. Circulatory microRNA-145 expression is increased in cerebral ischemia. *Genet. Mol. Res.* **2012**, *11*, 147–152. [CrossRef] [PubMed]
55. Jia, L.; Hao, F.; Wang, W.; Qu, Y. Circulating miR-145 is associated with plasma high-sensitivity C-reactive protein in acute ischemic stroke patients. *Cell. Biochem. Funct.* **2015**, *33*, 314–319. [CrossRef] [PubMed]
56. He, W.; Chen, S.; Chen, X.; Li, S.; Chen, W. Bioinformatic analysis of potential microRNAs in ischemic stroke. *J. Stroke Cerebrovasc. Dis.* **2016**, *25*, 1753–1759. [CrossRef] [PubMed]
57. Tsai, P.C.; Liao, Y.C.; Wang, Y.S.; Lin, H.F.; Lin, R.T.; Juo, S.H. Serum microRNA-21 and microRNA-221 as potential biomarkers for cerebrovascular disease. *J. Vasc. Res.* **2013**, *50*, 346–354. [CrossRef] [PubMed]
58. Zhou, J.; Zhang, J. Identification of miRNA-21 and miRNA-24 in plasma as potential early stage markers of acute cerebral infarction. *Mol. Med. Rep.* **2014**, *10*, 971–976. [CrossRef] [PubMed]
59. Wang, Y.; Zhang, Y.; Huang, J.; Chen, X.; Gu, X.; Wang, Y.; Zeng, L.; Yang, G.Y. Increase of circulating miR-223 and insulin-like growth factor-1 is associated with the pathogenesis of acute ischemic stroke in patients. *BMC Neurol.* **2014**, *14*, 77. [CrossRef] [PubMed]
60. Chevillet, J.R.; Kang, Q.; Ruf, I.K.; Briggs, H.A.; Vojtech, L.N.; Hughes, S.M.; Cheng, H.H.; Arroyo, J.D.; Meredith, E.K.; Gallichotte, E.N.; et al. Quantitative and stoichiometric analysis of the microRNA content of exosomes. *Proc. Natl. Acad. Sci. USA* **2014**, *111*, 14888–14893. [CrossRef] [PubMed]
61. Chen, Y.; Song, Y.; Huang, J.; Qu, M.; Zhang, Y.; Geng, J.; Zhang, Z.; Liu, J.; Yang, G.Y. Increased circulating exosomal miRNA-223 is associated with acute ischemic stroke. *Front. Neurol.* **2017**, *8*, 57. [CrossRef]
62. Ji, Q.; Ji, Y.; Peng, J.; Zhou, X.; Chen, X.; Zhao, H.; Xu, T.; Chen, L.; Xu, Y. Increased brain-specific miR-9 and miR-124 in the serum exosomes of acute ischemic stroke patients. *PLoS ONE* **2016**, *11*, e0163645. [CrossRef]
63. Liu, Y.; Zhang, J.; Han, R.; Liu, H.; Sun, D.; Liu, X. Downregulation of serum brain specific microRNA is

associated with inflammation and infarct volume in acute ischemic stroke. *J. Clin. Neurosci.* **2015**, *22*, 291–295. [CrossRef] [PubMed]

64. Chen, Z.; Wang, K.; Huang, J.; Zheng, G.; Lv, Y.; Luo, N.; Liang, M.; Huang, L. Upregulated serum miR-146b serves as a biomarker for acute ischemic stroke. *Cell. Physiol. Biochem.* **2018**, *45*, 397–405. [CrossRef] [PubMed]
65. Wang, Y.; Barbacioru, C.; Hyland, F.; Xiao, W.; Hunkapiller, K.L.; Blakem, J.; Chan, F.; Gonzalez, C.; Zhang, L.; Samaha, R.R. Large scale real-time PCR validation on gene expression measurements from two commercial long-oligonucleotide microarrays. *BMC Genom.* **2006**, *7*, 59. [CrossRef]
66. Moldovan, L.; Batte, K.E.; Trgovcich, J.; Wisler, J.; Marsh, C.B.; Piper, M. Methodological challenges in utilizing miRNAs as circulating biomarkers. *J. Cell. Mol. Med.* **2014**, *18*, 371–390. [CrossRef]
67. Jickling, G.C.; Ander, B.P.; Zhan, X.; Noblett, D.; Stamova, B.; Liu, D. MicroRNA expression in peripheral blood cells following acute ischemic stroke and their predicted gene targets. *PLoS ONE* **2014**, *9*, e99283. [CrossRef]
68. Li, P.; Teng, F.; Gao, F.; Zhang, M.; Wu, J.; Zhang, C. Identification of circulating microRNAs as potential biomarkers for detecting acute ischemic stroke. *Cell. Mol. Neurobiol.* **2015**, *35*, 433–447. [CrossRef]
69. Zeng, L.; He, X.; Wang, Y.; Tang, Y.; Zheng, C.; Cai, H.; Liu, J.; Wang, Y.; Fu, Y.; Yang, G.Y. MicroRNA-210 overexpression induces angiogenesis and neurogenesis in the normal adult mouse brain. *Gene Ther.* **2014**, *21*, 37–43. [CrossRef]
70. Ouyang, Y.B.; Xu, L.; Yue, S.; Liu, S.; Giffard, R.G. Neuroprotection by astrocytes in brain ischemia: Importance of microRNAs. *Neurosci. Lett.* **2014**, *565*, 53–58. [CrossRef]
71. Ouyang, Y.B.; Giffard, R.G. MicroRNAs affect BCL-2 family proteins in the setting of cerebral ischemia. *Neurochem. Int.* **2014**, *77C*, 2–8. [CrossRef]
72. Tan, K.S.; Armugam, A.; Sepramaniam, S.; Lim, K.Y.; Setyowati, K.D.; Wang, C.W.; Jeyaseelan, K. Expression profile of MicroRNAs in young stroke patients. *PLoS ONE* **2009**, *4*, e7689. [CrossRef]
73. Sepramaniam, S.; Tan, J.R.; Tan, K.S.; DeSilva, D.A.; Tavintharan, S.; Woon, F.P.; Wang, C.W.; Yong, F.L.; Karolina, D.S.; Kaur, P.; et al. Circulating microRNAs as biomarkers of acute stroke. *Int. J. Mol. Sci.* **2014**, *15*, 1418–1432. [CrossRef] [PubMed]
74. Wang, W.H.; Guan, S.; Zhang, L.Y.; Lei, S.; Zeng, Y.J. Circulating microRNAs as novel potential biomarkers for early diagnosis of acute stroke in humans. *J. Stroke Cerebrovasc. Dis.* **2014**, *23*, 2607–2613. [CrossRef]
75. Tian, C.; Li, Z.; Yang, Z.; Huang, Q.; Liu, J.; Hong, B. Plasma microRNA-16 is a biomarker for diagnosis, stratification, and prognosis of hyperacute cerebral infarction. *PLoS ONE* **2016**, *11*, e0166688. [CrossRef]
76. Tiedt, S.; Prestel, M.; Malik, R.; Schieferdecker, N.; Duering, M.; Kautzky, V.; Stoycheva, I.; Böck, J.; Northoff, B.H.; Klein, M.; et al. RNA-Seq identifies circulating miR-125a-5p, miR-125b-5p, and miR-143-3p as potential biomarkers for acute ischemic stroke. *Circ. Res.* **2017**, *121*, 970–980. [CrossRef]
77. Mick, E.; Shah, R.; Tanriverdi, K.; Murthy, V.; Gerstein, M.; Rozowsky, J.; Kitchen, R.; Larson, M.G.; Levy, D.; Freedman, J.E. Stroke and circulating extracellular RNAs. *Stroke* **2017**, *48*, 828–834. [CrossRef] [PubMed]
78. Wu, J.; Fan, C.L.; Ma, L.J.; Liu, T.; Wang, C.; Song, J.X.; Lv, Q.S.; Pan, H.; Zhang, C.N.; Wang, J.J. Distinctive expression signatures of serum microRNAs in ischaemic stroke and transient ischaemic attack patients. *Thromb. Haemost.* **2017**, *117*, 992–1001. [CrossRef]
79. Agarwal, V.; Bell, G.W.; Nam, J.; Bartel, D.P. Predicting effective microRNA target sites in mammalian mRNAs. *eLife* **2015**, *4*, e05005. [CrossRef] [PubMed]
80. Chou, C.H.; Shrestha, S.; Yang, C.D.; Chang, N.W.; Lin, Y.L.; Liao, K.W.; Huang, W.C.; Sun, T.H.; Tu, S.J.; Lee, W.H.; et al. miRTarBase update 2018: A resource for experimentally validated microRNA-target interactions. *Nucleic Acids Res.* **2018**, *46*, D296–D302. [CrossRef]
81. Durinck, S.; Spellman, P.; Birney, E.; Huber, W. Mapping identifiers for the integration of genomic datasets with the R/Bioconductor package biomaRt. *Nat. Protoc.* **2009**, *4*, 1184–1191. [CrossRef]
82. Shannon, P.; Markiel, A.; Ozier, O.; Baliga, N.S.; Wang, J.T.; Ramage, D.; Amin, N.; Schwikowski, B.; Ideker, T. Cytoscape: A software environment for integrated models of biomolecular interaction networks. *Genome Res.* **2003**, *13*, 2498–2504. [CrossRef]
83. Szklarczyk, D.; Morris, J.H.; Cook, H.; Kuhn, M.; Wyder, S.; Simonovic, M.; Santos, A.; Doncheva, N.T.; Roth, A.; Bork, P.; et al. The STRING database in 2017: Quality-controlled protein-protein association networks, made broadly accessible. *Nucleic Acids Res.* **2017** *45*, D362–D368. [CrossRef]
84. Zhang, X.; Yu, H.; Lou, J.R.; Zheng, J.; Zhu, H.; Popescu, N.I.; Lupu, F.; Lind, S.E.; Ding, W.Q. MicroRNA-19

(miR-19) regulates tissue factor expression in breast cancer cells. *J. Biol. Chem.* **2011**, *286*, 1429–1435. [CrossRef]
85. Weber, M.; Baker, M.B.; Patel, R.S.; Quyyumi, A.A.; Bao, G.; Searles, C.D. MicroRNA expression profile in CAD patients and the impact of ACEI/ARB. *Cardiol. Res. Pract.* **2011**, *2011*, 532915. [CrossRef]
86. Doebele, C.; Bonauer, A.; Fischer, A.; Scholz, A.; Reiss, Y.; Urbich, C.; Hofmann, W.K.; Zeiher, A.M.; Dimmeler, S. Members of the microRNA-17-92 cluster exhibit a cell-intrinsic antiangiogenic function in endothelial cells. *Blood* **2010**, *115*, 4944–4950. [CrossRef]
87. Bonauer, A.; Carmona, G.; Iwasaki, M.; Mione, M.; Koyanagi, M.; Fischer, A.; Burchfield, J.; Fox, H.; Doebele, C.; Ohtani, K.; et al. MicroRNA-92a controls angiogenesis and functional recovery of ischemic tissues in mice. *Science* **2009**, *324*, 1710–1713. [CrossRef]
88. Singh, V.; Roth, S.; Veltkamp, R.; Liesz, A. HMGB1 as a key mediator of immune mechanisms in ischemic stroke. *Antioxid. Redox Signal.* **2016**, *24*, 635–651. [CrossRef]
89. Diamanti, D.; Lahiri, N.; Tarditi, A.; Magnoni, L.; Fondelli, C.; Morena, E.; Malusa, F.; Pollio, G.; Diodato, E.; Tripepi, G.; et al. Reference genes selection for transcriptional profiling in blood of HD patients and R6/2 mice. *J. Huntingtons Dis.* **2013**, *2*, 185–200. [CrossRef]
90. Dervishi, I.; Gozutok, O.; Murnan, K.; Gautam, M.; Heller, D.; Bigio, E.; Ozdinler, P.H. Protein-protein interactions reveal key canonical pathways, upstream regulators, interactome domains, and novel targets in ALS. *Sci. Rep.* **2018**, *8*, 14732. [CrossRef]
91. Miller, J.A.; Oldham, M.C.; Geschwind, D.H. A systems level analysis of transcriptional changes in Alzheimer's disease and normal aging. *J. Neurosci.* **2008**, *28*, 1410–1420. [CrossRef]
92. Wu, C.C.; Chen, B.S. Key immune events of the pathomechanisms of early cardioembolic stroke: Multi-database mining and systems biology approach. *Int. J. Mol. Sci.* **2016**, *17*, 305. [CrossRef]
93. Chang, T.Y.; Huang, T.S.; Wang, H.W.; Chang, S.J.; Lo, H.H.; Chiu, Y.L.; Wang, Y.L.; Hsiao, C.D.; Tsai, C.H.; Chan, C.H.; et al. miRNome traits analysis on endothelial lineage cells discloses biomarker potential circulating microRNAs which affect progenitor activities. *BMC Genom.* **2014**, *15*, 802. [CrossRef]
94. Liang, Z.; Wu, G.; Fan, C.; Xu, J.; Jiang, S.; Yan, X.; Di, S.; Ma, Z.; Hu, W.; Yang, Y. The emerging role of signal transducer and activator of transcription 3 in cerebral ischemic and hemorrhagic stroke. *Prog. Neurobiol.* **2016**, *137*, 1–16. [CrossRef]
95. Bam, M.; Yang, X.; Sen, S.; Zumbrun, E.E.; Dennis, L.; Zhang, J.; Nagarkatti, P.S.; Nagarkatti, M. Characterization of dysregulated miRNA in peripheral blood mononuclear cells from ischemic stroke patients. *Mol. Neurobiol.* **2018**, *55*, 1419–1429. [CrossRef]
96. Sun, J.; Nan, G. The mitogen-activated protein kinase (MAPK) signaling pathway as a discovery target in stroke. *J. Mol. Neurosci.* **2016**, *59*, 90–98; Erratum in: *J. Mol. Neurosci.* **2016**, *59*, 430. [CrossRef]
97. Wang, H.; Zhang, X.M.; Tomiyoshi, G.; Nakamura, R.; Shinmen, N.; Kuroda, H.; Kimura, R.; Mine, S.; Kamitsukasa, I.; Wada, T.; et al. Association of serum levels of antibodies against MMP1, CBX1, and CBX5 with transient ischemic attack and cerebral infarction. *Oncotarget* **2017**, *9*, 5600–5613. [CrossRef]
98. Gao, J.B.; Tang, W.D.; Wang, H.X.; Xu, Y. Predictive value of thrombospondin-1 for outcomes in patients with acute ischemic stroke. *Clin. Chim. Acta* **2015**, *450*, 176–180. [CrossRef]
99. Cevik, O.; Baykal, A.T.; Sener, A. Platelets proteomic profiles of acute ischemic stroke patients. *PLoS ONE* **2016**, *11*, e0158287. [CrossRef]
100. Yang, F.H.; Pyle, W.G. Reduced cardiac CapZ protein protects hearts against acute ischemia-reperfusion injury and enhances preconditioning. *J. Mol. Cell. Cardiol.* **2012**, *52*, 761–772. [CrossRef]
101. Ridker, P.M.; Everett, B.M.; Thuren, T.; MacFadyen, J.G.; Chang, W.H.; Ballantyne, C.; Fonseca, F.; Nicolau, J.; Koenig, W.; Anker, S.D.; et al. CANTOS Trial Group. Antiinflammatory Therapy with Canakinumab for Atherosclerotic Disease. *N. Engl. J. Med.* **2017**, *377*, 1119–1131. [CrossRef]
102. Cirillo, P.; Golino, P.; Calabrò, P.; Calì, G.; Ragni, M.; De Rosa, S.; Cimmino, G.; Pacileo, M.; De Palma, R.; Forte, L.; et al. C-reactive protein induces tissue factor expression and promotes smooth muscle and endothelial cell proliferation. *Cardiovasc. Res.* **2005**, *68*, 47–55. [CrossRef]
103. De Rosa, S.; Cirillo, P.; Pacileo, M.; Di Palma, V.; Paglia, A.; Chiariello, M. Leptin stimulated C-reactive protein production by human coronary artery endothelial cells. *J. Vasc. Res.* **2009**, *46*, 609–617. [CrossRef] [PubMed]
104. Cirillo, P.; Golino, P.; Calabrò, P.; Ragni, M.; Forte, L.; Piro, O.; De Rosa, S.; Pacileo, M.; Chiariello, M. Activated platelets stimulate tissue factor expression in smooth muscle cells. *Thromb. Res.* **2003**, *112*, 51–57. [CrossRef] [PubMed]

3

Perspectives on microRNAs and Phased Small Interfering RNAs in Maize (*Zea mays* L.): Functions and Big Impact on Agronomic Traits Enhancement

Zhanhui Zhang [1,*], Sachin Teotia [1,2,3], Jihua Tang [1] and Guiliang Tang [1,2,*]

1 State Key Laboratory of Wheat and Maize Crop Science, Henan Agricultural University, Zhengzhou 450002, China; steotia@mtu.edu (S.T.); tangjihua1@163.com (J.T.)
2 Department of Biological Sciences, Michigan Technological University, Houghton, MI 49931, USA
3 Department of Biotechnology, Sharda University, Greater Noida 201306, India
* Correspondence: zhanhuiz15@icloud.com (Z.Z.); gtang1@mtu.edu (G.T.);

Abstract: Small RNA (sRNA) population in plants comprises of primarily micro RNAs (miRNAs) and small interfering RNAs (siRNAs). MiRNAs play important roles in plant growth and development. The miRNA-derived secondary siRNAs are usually known as phased siRNAs, including phasiRNAs and tasiRNAs. The miRNA and phased siRNA biogenesis mechanisms are highly conserved in plants. However, their functional conservation and diversification may differ in maize. In the past two decades, lots of miRNAs and phased siRNAs have been functionally identified for curbing important maize agronomic traits, such as those related to developmental timing, plant architecture, sex determination, reproductive development, leaf morphogenesis, root development and nutrition, kernel development and tolerance to abiotic stresses. In contrast to *Arabidopsis* and rice, studies on maize miRNA and phased siRNA biogenesis and functions are limited, which restricts the small RNA-based fundamental and applied studies in maize. This review updates the current status of maize miRNA and phased siRNA mechanisms and provides a survey of our knowledge on miRNA and phased siRNA functions in controlling agronomic traits. Furthermore, improvement of those traits through manipulating the expression of sRNAs or their targets is discussed.

Keywords: maize (*Zea mays* L.); miRNA; phasiRNA; tasiRNA; agronomic traits; crop improvement

1. Introduction

Plant and animal small RNAs (sRNAs) are short noncoding regulatory RNAs in the size range of ~20 to 30 nucleotides (nt) [1,2]. These sRNAs play crucial roles in various biological regulatory processes through mediating gene silencing at both transcriptional and posttranscriptional levels [1,3]. According to the origin and biogenesis, plant sRNAs can be categorized into several major classes, micro RNAs (miRNAs), heterochromatic small interfering RNAs (hc-siRNAs), phased small interfering RNAs (phased siRNAs), and natural antisense transcript small interfering RNAs (NAT-siRNAs) [4].

Plants miRNAs are processed from long *MIRNA* transcripts by a microprocessor and dicing complexes [5–7]. Compared to animal miRNAs, plant miRNAs tend to have fewer targets that mainly encode transcription factors and F-box proteins [8]. This indicates that miRNA is at the central position of gene expression regulatory networks of plant growth and development. The accumulating studies proved miRNAs to be key regulators of various biological regulatory processes in plants, including developmental timing, plant architecture, organ polarity, inflorescence development and responses to biotic and abiotic stresses [9,10]. Additionally, miRNAs also drive secondary siRNA generation that are defined as phased siRNAs. Such secondary siRNAs, including canonical phased siRNAs (phasiRNAs) and phased trans-acting siRNAs (tasiRNAs), also play key roles in plant development [11–13]. Moreover, manipulation of mRNA transcript abundance via miRNA control provides a unique strategy for the improvement of the complex agronomic traits of crops [14,15]. Thus, understanding the functions of miRNAs and related secondary siRNAs in various plant species, especially in crops like maize, is essential for crop improvement.

Maize is not only a model plant genetic system, but is also an important crop species for food, fuel and feed [16]. Like *Drosophila* and the worm *Caenorhabditis elegans*, maize has been a significant contributor to a number of important discoveries, including the so-called "jumping genes" (transposons), activator/dissociation (Ac/Ds) and Mutator, as well as the epigenetic phenomenon termed paramutation [17,18]. However, there have been limited studies on the roles of miRNAs and miRNA-derived secondary siRNAs in maize metabolism, development and stress responses [12,19–34], making it far from utilized in agronomic traits improvement through genetic engineering. Compared to *Arabidopsis* and rice, maize sRNA and RNAi mechanisms remain only partially resolved.

This review examines the current status of our understanding of the biogenesis and functions of miRNA and phased siRNA in maize, with a focus on their key components and the missing links of the pathways. Such study can help evaluate the potential roles of maize sRNAs in the enhancement of agronomic traits. First, we survey the recent findings regarding miRNA and phased siRNA working mechanisms in maize. We further compare the mechanistic differences for those mechanisms between maize and the model plants *Arabidopsis* and rice highlighting the missing links in maize. Furthermore, we review the identified miRNA and phased siRNA functions in regulating important agronomic traits in maize. Finally, we discuss the potential applications of these small regulatory RNAs or of their target genes in agronomic traits enhancement.

2. MiRNA and phasiRNA Biogenesis in Maize

2.1. Core Components of sRNA Biogenesis in Plants

In plants, the sRNAs biogenesis and gene silencing mainly depends on the activities of three kinds of proteins, dicer or dicer-like proteins (DCLs), argonautes (AGOs) and RNA-dependent RNA polymerases (RDRs) [1,32,35,36]. The sRNA-mediated gene silencing is initiated by double stranded RNA (dsRNA) generation by RDRs or the folding of *MIRNA* gene transcripts [1,36]. The dsRNA is processed into sRNAs, 20–30 nt in length, by the cleavages of microprocessors, DCLs [1,5,36]. Different classes of sRNAs are recognized by specific AGOs to assemble RNA-induced silencing complex (RISC) [1,36]. In *Arabidopsis*, 4 DCLs, 10 AGOs and six RDRs are encoded, while in rice, eight DCLs, 19 AGOs and five RDRs are encoded [35]. In maize, these major components have also been identified mainly based on their orthologs in *Arabidopsis* and rice genomes, which include five ZmDCL, 17 ZmAGO and five ZmRDR genes (Table 1, Figure 1A) [32,37]. Among them, only a few ones have been experimentally verified so far, including *fuzzy tassel* (*fzt, ZmDCL1*) [22], *ragged seedling2* (*rgd2, ZmAGO7*) [38], *ZmRDR1* [37], *mediator of paramutation 1* (*mop1, ZmRDR2*) [39].

2.2. MiRNA-Mediated Gene Silencing in Maize

In plants, 21-nt miRNA biogenesis includes four steps (Figure 1): (1) *MIRNA*s transcription; (2) precursor miRNA (pre-miRNA) generation by dicer-like RNase III protein I (DCL1) cleavage; (3) miRNA duplexes release; (4) miRNA duplexes methylation and export to cytoplasm and miRNA-RISC assembly [6,7,40–42]. In the cytoplasm, the miRNA-RISCs mediate their target mRNA degradation [43,44], or translational inhibition in plants [45]. In contrast to the 21 nt miRNAs, a class of 24 nt miRNAs was discovered in plants (Figure 1). These 24 nt miRNAs are processed by DCL3 during their biogenesis [46,47]. In RNA-directed DNA methylation (RdDM), these 24 nt miRNAs are sorted into AGO4 to direct DNA methylation at the loci of their origin, thus regulating their target genes in trans [46,47]. After their biogenesis, miRNAs are also subjected to catabolism [48], in which demethylated or uridylated miRNAs are degraded by small RNA degrading nucleases (SDNs) [40,49].

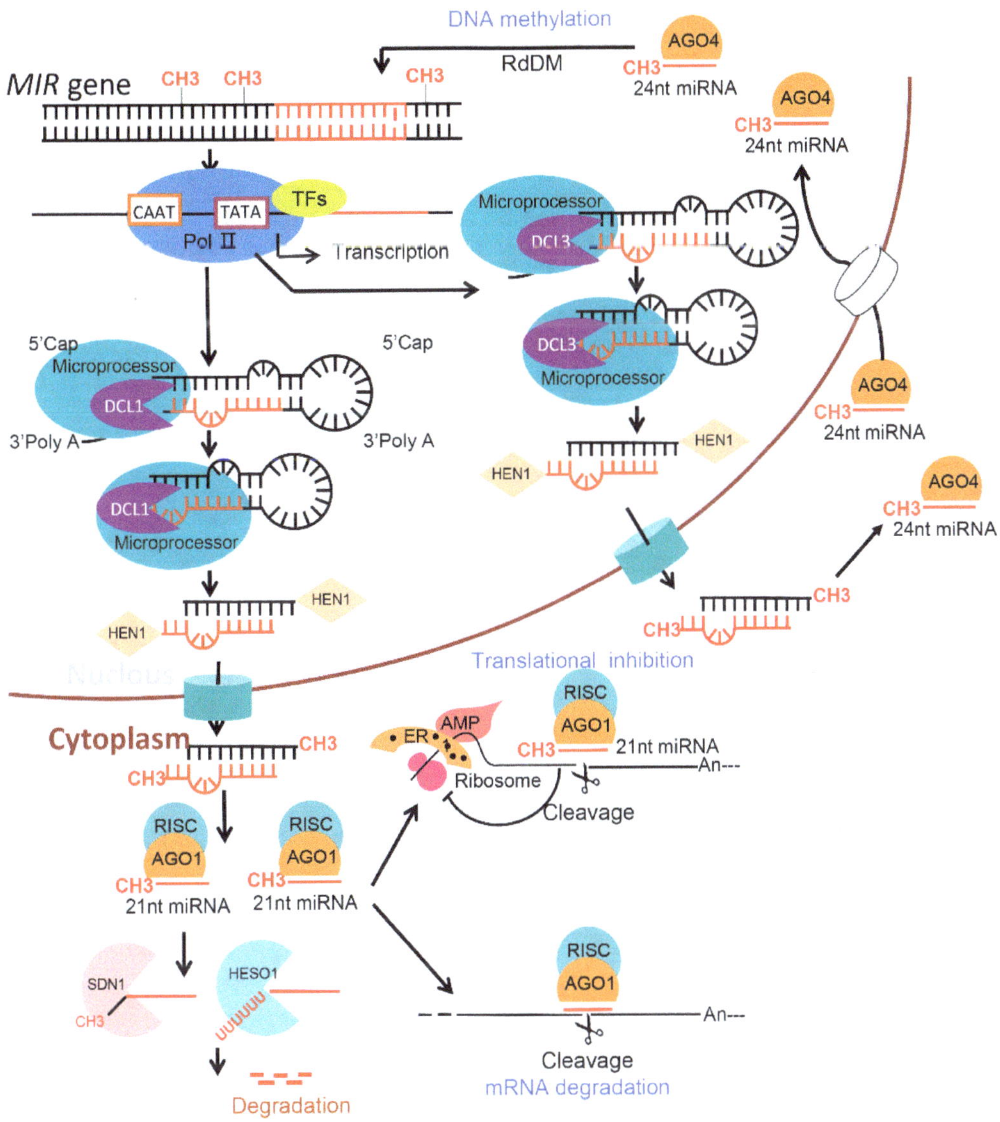

Figure 1. An overview showing micro RNA (miRNA) biogenesis and functioning in plants. *MIRNA* genes are transcribed to form primary miRNAs, from which 21 and 24 nt miRNAs are processed by DCL1 and DCL3, respectively. Their 3′ ends are methylated by HEN1. While the 21 nt species are involved in cleavage or translational inhibition of the target mRNAs, the 24 nt miRNAs are involved in DNA methylation.

2.3. Origin and Biogenesis of phasiRNAs in Maize

Generally, phasiRNA biogenesis is initiated by cleavage of single-stranded *PHAS* loci transcripts by 22 nt miRNAs. Then, those cleaved single-stranded RNAs are used to generate dsRNAs by RDRs. DCLs further phase dsRNAs to produce 21 or 24 nt phasiRNAs. PhasiRNAs are subsequently loaded to AGOs to regulate gene expression network [11,50] (Figure 2A,B). In grasses, including maize, phasiRNA precursors, *PHAS* loci transcripts, are transcribed by RNA polymerase II. These long noncoding precursor transcripts are internally cleaved, guided by 22 nt miR2118 to generate the 21 nt phasiRNAs or by miR2275 for the 24 nt phasiRNA. Such special class of small RNAs are specifically expressed in reproductive organs, conferring male fertility [11,12,51]. In maize, the biogenesis of 21 and 24 nt phasiRNAs are regulated by DCL4 and DCL5, respectively (Figure 2A,B). Next, 21 and 24 nt phasiRNAs are recruited by AGO5c and AGO18b, respectively, to assemble RISC and regulate gene expression [12].

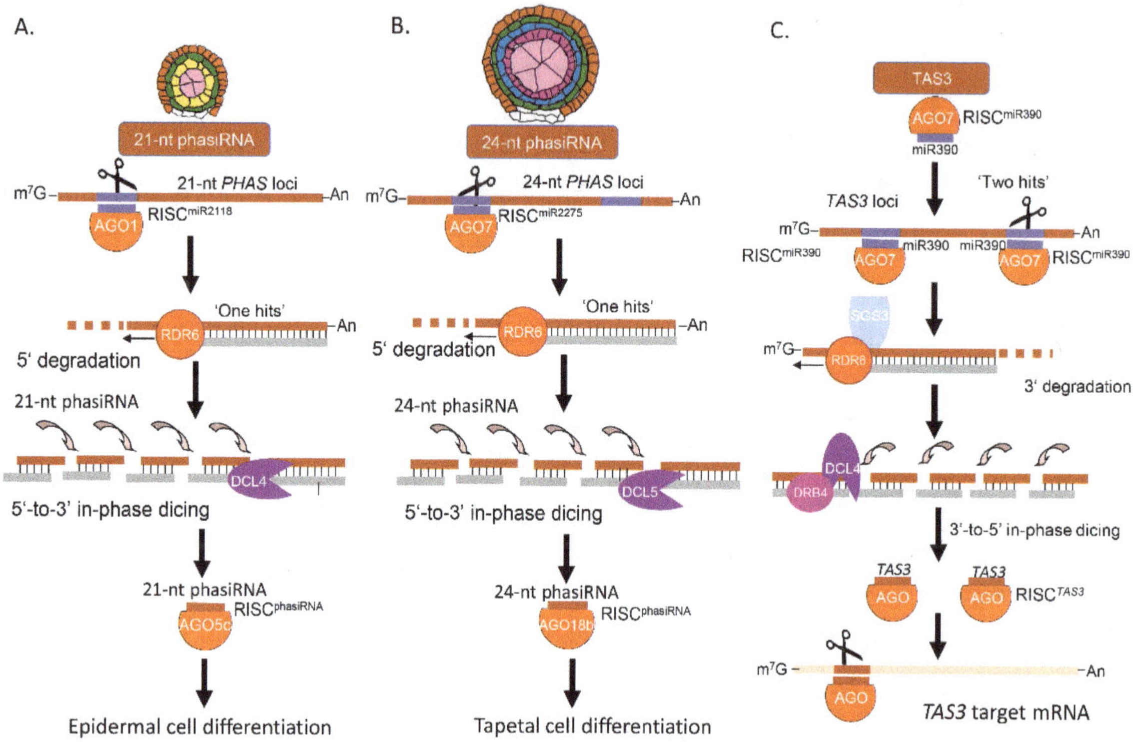

Figure 2. Phased small interfering RNA (phasiRNA) and trans-acting simple interfering RNA (tasiRNA) biogenesis pathways in maize. (**A**) The 21 nt phasiRNA biogenesis pathway. (**B**) The 24 nt phasiRNA biogenesis pathway. The regulatory mechanism of 24 nt has not been fully uncovered. (**C**) *TAS3*-tasiRNA biogenesis pathway.

The production of the 21-nt tasiRNAs is initiated by a miRNA through RDR6 and DCL4 [52]. In *Arabidopsis*, miR173, miR390 and miR828 trigger the production of *TAS1a-c*/*TAS2*, *TAS3*, and *TAS4* siRNAs, respectively [53,54]. The maize *TAS3* pathway has been identified through the mutations, *leafbladeless1* (*ldl1*) and *ragged seedling2* (*rgd2*), which encode the orthologs of SGS3 and AGO7 of *Arabidopsis* (Figure 2C) [55]. After *TAS3* siRNAs is generated, they are recruited by AGO7 to assemble RISC and induce ARF3 gene silencing by targeting mRNA transcripts.

2.4. Functional Redundancy and Divergence of the Key Components in Maize sRNA Biogenesis Pathways

2.4.1. DCLs

Based on the phylogenetic analysis, different DCLs from *Arabidopsis*, rice and maize were classified into four subgroups (Table 1, Figure 3A,B). ZmDCL1 showed high similarity with *Arabidopsis* AtDCL1 and rice OsDCL1a–1c; ZmDCL3a and ZmDCL5/3b are similar to *Arabidopsis* DCL3 and rice OsDCL3a–3b; and ZmDCL2 and ZmDCL4 are most similar to AtDCL2 and AtDCL4, respectively [23,32]. In *Arabidopsis*, AtDCL1 produces mature miRNAs [56]; AtDCL2 is involved in virus defense-related siRNA generation and has functional redundancy with AtDCL4 [57]; while AtDCL3 catalyzes the production of 24-nt siRNAs [58]; and AtDCL4 is mainly for the production of tasiRNAs [59]. Although the DCL family proteins are largely functionally conserved among the three plant species, DCL3a and DCL3b are considered specific to monocots and predate the divergence of rice and maize [60].

Table 1. Known and putative components of the micro RNA (miRNA) and simple interfering (siRNA) pathways in maize.

Gene	Accession Number	Chromosomal Location (5′-3′)	Type
1. *ZmDCLs*			
ZmDCL1	GRMZM2G040762_P01	Chr. 1: 4,600,841–4,608,248	DCL1
ZmDCL2	GRMZM2G301405_P01	Chr. 5: 19,916,753–19,927,967	DCL2
ZmDCL3a	GRMZM5G814985_P01	Chr. 3: 164,415,209–164,418,189	DCL3
ZmDCL5/3b	GRMZM2G413853_P01	Chr. 1: 229,801,762–229,819,069	DCL3
ZmDCL4	GRMZM2G160473_P01	Chr. 10: 129,990,456–129,992,917	DCL4
2. *ZmAGOs*			
ZmAGO1a	GRMZM2G441583_P01	Chr. 6: 43,253,105–43,261,555	AGO1
ZmAGO1b	AC209206.3_FGP011	Chr. 10: 137,506,877–137,513,415	AGO1
ZmAGO1c	GRMZM2G039455_P01	Chr. 2: 17,563,301–17,573,156	AGO1
ZmAGO1d	GRMZM2G361518_P01	Chr. 5: 64,791,077–64,796,881	AGO1
ZmAGO2a	GRMZM2G007791_P01	Chr. 2: 9,973,816–9,981,340	ZIPPY
ZmAGO2b	GRMZM2G354867_P01	Chr. 1:142,397,812–142,403,450	ZIPPY
ZmAGO4	GRMZM2G589579_P01	Chr. 8: 2,511,663–2,519,008	AGO4
ZmAGO5a	GRMZM2G461936_P02	Chr. 5: 13,611,800–13,618,698	MEL1
ZmAGO5b	GRMZM2G059033_P01	Chr. 2: 233,385,077–233,392,000	MEL1
ZmAGO5c	GRMZM2G347402_P01	Chr. 7: 72,044,775–72,053,779	MEL1
ZmAGO5d	GRMZM2G123063_P01	Chr. 5:4,000,995–4,009,425	MEL1
ZmAGO7	GRMZM2G354867_P01	Chr. 10: 141,823,070–141,828,449	ZIPPY
ZmAGO9	GRMZM2G141818_P03	Chr. 6: 168,642,369–168,650,358	AGO4
ZmAGO10a	AC189879.3_FG003	Chr. 9: 87,408,375–87,414,276	AGO1
ZmAGO10b	GRMZM2G079080_P02	Chr. 6: 103,286,236–103,293,200	AGO1
ZmAGO18a	GRMZM2G105250_P01	Chr. 2: 199,510,528–199,516,085	OsAGO18
ZmAGO18b	GRMZM2G457370_P01	Chr. 1: 250,132,189–250,137,737	OsAGO18
ZmAGO18c	GRMZM2G457370_P02	Chr. 1: 250,132,189–250,137,737	OsAGO18
3. *ZmRDRs*			
ZmRDR1	GRMZM2G481730_P01	Chr. 5: 205,385,818–205,389,710	RDR1
ZmMOP1	GRMZM2G042443_P01	Chr. 2: 41,131,324–41,136,928	RDR2
ZmRDR6a	GRMZM2G357825_P01	Chr. 9: 109,055,576–109,093,885	RDR6
ZmRDR6b	GRMZM2G145201_P01	Chr. 3: 102,532,883–102,536,036	RDR6
ZmRDR6c	GRMZM2G347931_P01	Chr. 9: 106,302,354–106,306,175	RDR6

Note: this information for maize dicer-like (DCL), argonaute (AGO), and RNA-dependent RNA polymerase (RDR), including accession number, chromosomal location and ORFs, was retrieved from the B73 maize sequence database (http://www.maizesequence.org/index.html).

2.4.2. AGOs

In *Arabidopsis*, AtAGO1 is associated with miRNA-mediated gene silencing [61]; AtAGO7 is preferentially associated with a single miRNA, miR390, to trigger production of *TAS3* [52]; and AtAGO5 is a putative germline-specific Argonaute complex associated with miRNAs in mature *Arabidopsis* pollen [62]. In addition, AtAGO2 was identified to have a stand-in role for AtAGO1 in antivirus defense when AGO1-targeted silencing is overcome by viral suppressors [63], AtAGO4 is associated with endogenous siRNAs that direct DNA methylation [64].

In maize, 17 genes encoding 18 AGO family proteins were identified, almost double the number reported in *Arabidopsis* (Table 1, Figure 3A,C) [24,32]. These ZmAGOs were divided phylogenetically into five subgroups: AGO1 (ZmAGO1a-1d and ZmAGO10a, b), MEL1/AGO5 (ZmAGO5a-5d), AGO7 (ZmAGO2 and ZmAGO7), AGO4 (ZmAGO4), and finally the ZmAGO18 (ZmAGO18a-c) [32]. The ZmAGO18 subgroup, ZmAGO18a, ZmAGO18b and ZmAGO18c, are encoded by two genes (*GRMZM2G105250* encodes ZmAGO18a, and ZmAGO18b and ZmAGO18c are encoded by two transcripts of *GRMZM2G457370*) [32]. They displayed high structural similarity to OsAGO18, whose expression is strongly induced by viral infection in rice and confers broad-spectrum virus resistance by sequestering the OsmiR168 from targeting OsAGO1 [65]. Nonetheless, ZmAGO18a is highly expressed in ears, while ZmAGO18b is mostly enriched in tassels, suggesting that ZmAGO18 family may have functional diversities from the OsAGO18 [66]. In fact, ZmAGO18b was proposed to bind the 24 nt phasiRNAs that are suggested to be the products of ZmDCL5/3b in the phasiRNA pathway, based on their concurrent spatial and temporal expression in developing maize ear/tassel development [12]. The mutant *ragged seedling2* (*rgd2*) has been identified to encode an AGO7-like protein required to produce *TAS3* [38], and its functions are highly conserved among *Arabidopsis*, rice and maize [67–69].

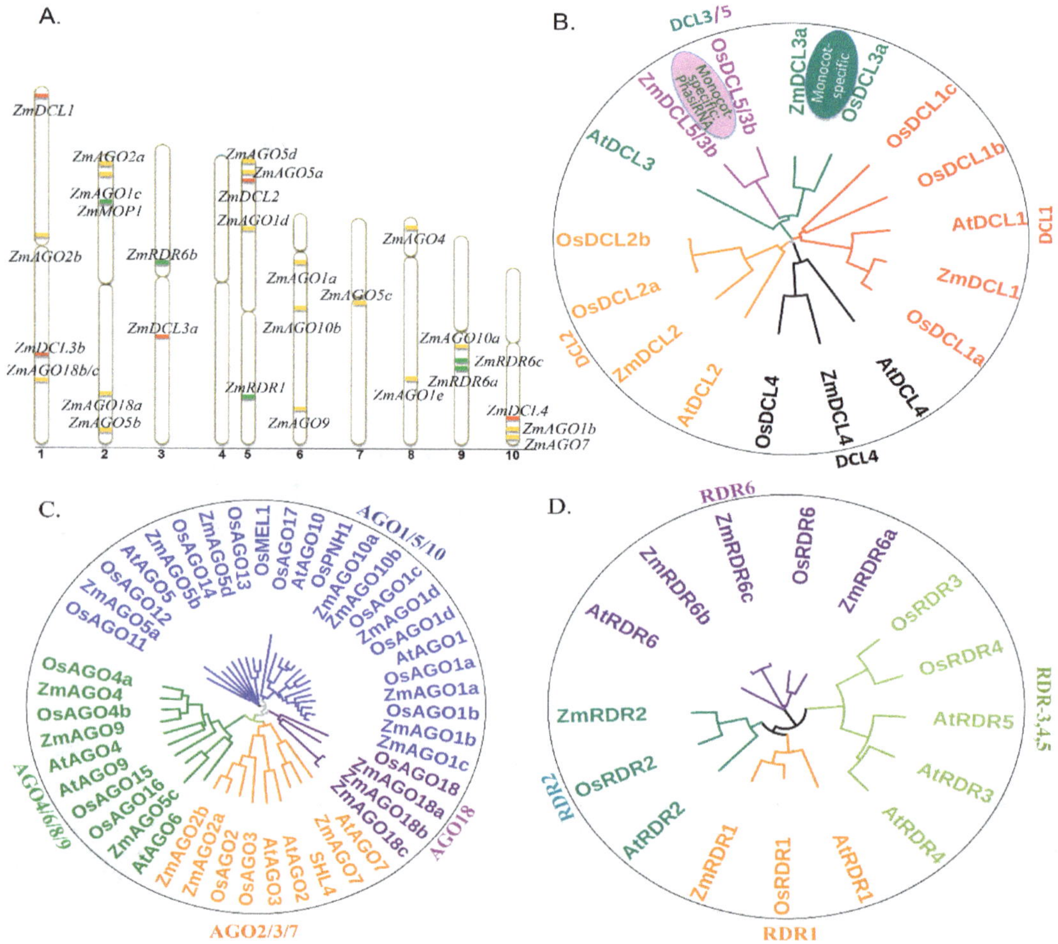

Figure 3. Chromosomal locations and phylogenetic analysis of known and putative components of the RNAi and miRNA pathways in maize. The protein sequences of dicer-like (DCL), argonautes (AGOs) and RNA-dependent RNA polymerase (RDR) in *Arabidopsis*, rice and maize AGOs were obtained from protein database (http://www.ncbi.nlm.nih.gov/protein). The neighbor-joining tree was constructed using Clustal omega [70] and iTol online software [71]. (**A**) The abbreviation of *At* represents *Arabidopsis thaliana*, *Os* for *Oryza sativa*, and *Zm* for *Zea mays*. Red bars indicate the chromosomal locations of *ZmDCLs*, yellow bars for *ZmAGOs*, and green bars for *ZmRDRs*. (**B**) Plants have four types of DCL proteins. There are 4 DCLs encoded in *Arabidopsis* genome, 4 DCL family members in rice, and 4 DCLs in maize. Of these DCL proteins, DCL3a and DCL3b are considered specific to monocots and predate the divergence of rice and maize. (**C**) 10, 19 and 18 AGOs are encoded by *Arabidopsis*, rice and maize, respectively, that can be divided phylogenetically into five subgroups in maize: AGO1, MEL1/AGO5, AGO7, AGO4, and AGO18. AGO18 subgroup has three members in maize. ZmAGO18a-c are considered specific to monocots along with OsAGO18. (**D**) Plants have six types of functionally distinct RDRs. While *Arabidopsis* has all the six types, rice lacks RDR5 and maize lacks RDR3, 4, and 5. In contrast to *Arabidopsis* and rice which have only single member RDR6 family, maize has a multiple member RDR6 family, which is composed of ZmRDR6a, ZmRDR6b, and ZmRDR6c.

2.4.3. RDRs

Six, five and five RDRs have been identified in *Arabidopsis*, rice and maize, respectively (Table 1, Figure 1D) [32,35]. These RDRs were divided phylogenetically into four subgroups: RDR1, RDR2, RDR3/4/5, and RDR6. AtRDR1 and its homolog in maize, ZmRDR1, have been reported to be involved in antiviral defense [37]. AtRDR2 plays a crucial role in RNA-directed DNA methylation and repressive chromatin modifications of certain transgenes, endogenous genes and centromeric repeats that correlate with the production of 24 nt interfering sRNAs [72]. In maize, MOP1 (a homolog of AtRDR2) has proven to be essential for a siRNA-directed gene-silencing pathway, and is also involved in the maintenance of transposon silencing and paramutation [39]. The remaining three RDR homologs of *Arabidopsis* RDR6 in maize, ZmRDR6a-c, are involved in tasiRNA biogenesis [67,73]. We tentatively renamed these three RDRs, previously known as ZmRDR3 and ZmRDR4 [32], to be ZmRDR6a, ZmRDR6b, and ZmRDR6c. ZmRDR6, such a multiple membered family, can be better revealed by identifying their double/triple mutants. Similar to how RDR6 was identified to be important in production of tasiRNAs, an unidentified RDR is expected to play a key role in production of phasiRNAs in maize [12].

3. Functions of miRNAs and phasiRNAs in Maize

3.1. *The Interaction of miR156 and miR172 Fine Tunes Plant Developmental Timing*

In maize, the transition from juvenile to adult leaves is marked by changes in cell shape, the production of epidermal wax deposits and of specialized cell types like leaf hairs, and a change in the identity of organs that grow from their axillary meristems. In maize and *Arabidopsis*, the roles of miR156 and miR172 interaction in developmental transitions have been widely explored [25,27,74–76]. MiR156 expression levels decrease with leaf age, while that of miR172 increase (Figure 5A). Their targets, encoding squamosa promoter binding protein-like (SBP-Like) and Apetala 2 (AP2) transcription factors, respectively, are expressed in complementary patterns. The mutant *Corngrass1* (*Cg1*) with increased levels of miR156 and reduced miR172 activity, displays restrained developmental transitions, prolonged juvenile features and delayed flowering (Figure 4) [25,77]. In turn, releasing *SPLs* from miR156 regulation leads to premature acquisition of adult leaf features and early flowering, resembling phenotypes of *glossy15* (*gl15*) plants, with reduced activity of miR172 targets (Figure 4) [27,29].

3.2. *Plant Architecture Modulated by miR156 and miR319*

In maize, plant architecture is mainly determined by tillers, plant height, leaf number, leaf angle and tassel branches. Compared with its ancestor, teosinte (*Zea mays* ssp. parviglumis), maize exhibits a profound increase in apical dominance with a single tiller [78]. Previous researches have proved *teosinte branched1* (*tb1*) gene, encoding a TCP transcription factor that is targeted by miR319, as a major contributor to this domestication change in maize (Figures 4 and 5B) [79,80]. By increasing JA levels, the *tb1* mutant of maize causes a complete loss of apical dominance, allowing the unrestrained outgrowth of axillary buds and inflorescent architectural alterations [79,81]. MiR156 has been proved to be the important regulator in maize and rice plant architecture formation [25,82]. The dominant *Corngrass1* (*Cg1*) mutant of maize has phenotypic changes that are present in the grass-like ancestors of maize, exhibiting numerous tillers, inflorescent architectural alterations and erect leaves (Figure 5B) [25]. The research by Lu et al. [83] in rice revealed that the *ideal plant architecture1* (*IPA1*, *OsSPL14*) could directly bind to the promoter of rice *teosinte branched1* (*Ostb1*), to suppress rice tillering. Likewise, the maize tillering related *ZmSPL* (miR156 target) gene is possible at the upstream of *tb1* in related regulatory pathway. The roles of miR156 in leaf angle and inflorescent architectural modulation have been identified in the corresponding *ZmSPL* mutants, such as *LIGULELESS1* (*LG1*), *tasselsheath4* (*tsh4*, *ZmSBP2*), *UNBRANCHED 2* (*UB2*) and *UB3* [30,33,84,85](Figure 4; Figure 5B).

3.3. *Roles of miR172, miR156 and miR159 in Sex Determination*

In maize, inflorescence development and sex determination are key factors for grain yield. MiR172 has been identified to play important roles in inflorescence development and sex determination (Figure 4) [86]. Especially, the interplay of miR156 and miR172 contributes largely in maize sex determination and meristem cell fate. In *Cg1* mutant, increased levels of miR156 cause similar phenotypic alterations as seen in *ts4* mutants [25]. Moreover, *STTMmiR172* and *ts4* mutants have reduced expression of miR172 and increased expression of at least two of its targets, *ids1* (*indeterminate spikelet1*) and *sid1* (*sister of indeterminate spikelet1*). These mutants displayed irregular branching within the inflorescence and feminization of the tassel caused by a lack of pistil abortion [86–88]. Decreased levels of miR156 have been detected in feminized tassels of maize *mop1* and *ts1* (*tasselseed1*), implying the missing link of miR156-SPLs with sex-determination genes *ts1*, *ts2*, *ts4*, *Ts6*, and *mop1* [34,89]. Additionally, the mutants of *fuzzy tassel* (encoding dicer-like1 protein) exhibit indeterminate meristems, fasciation, and alterations in sex determination [22]. Such reproductive development alterations are possibly associated with miR159-*GAMYB* pathway, with miR159 and its targets playing the important roles in another development [21,90].

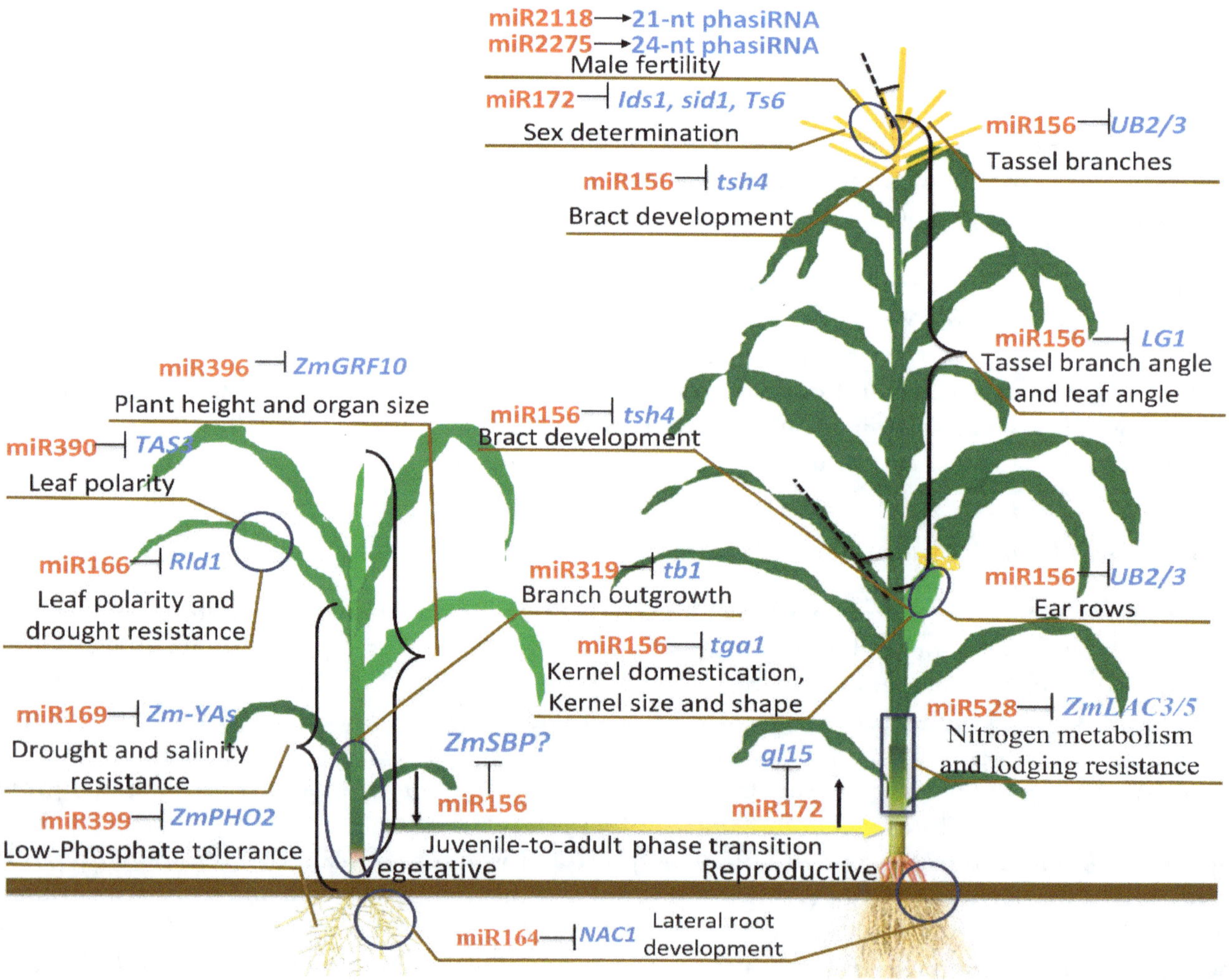

Figure 4. Summary of functionally validated miRNAs and their targets in maize. Nine miRNAs (in red font) and three phasiRNAs (in red font) that regulate specific agronomic traits (in black font) by inducing their targets (in blue font) gene silencing. Question marks indicate that specific *ZmSBP* taking part in juvenile-to-adult phase transition is not known.

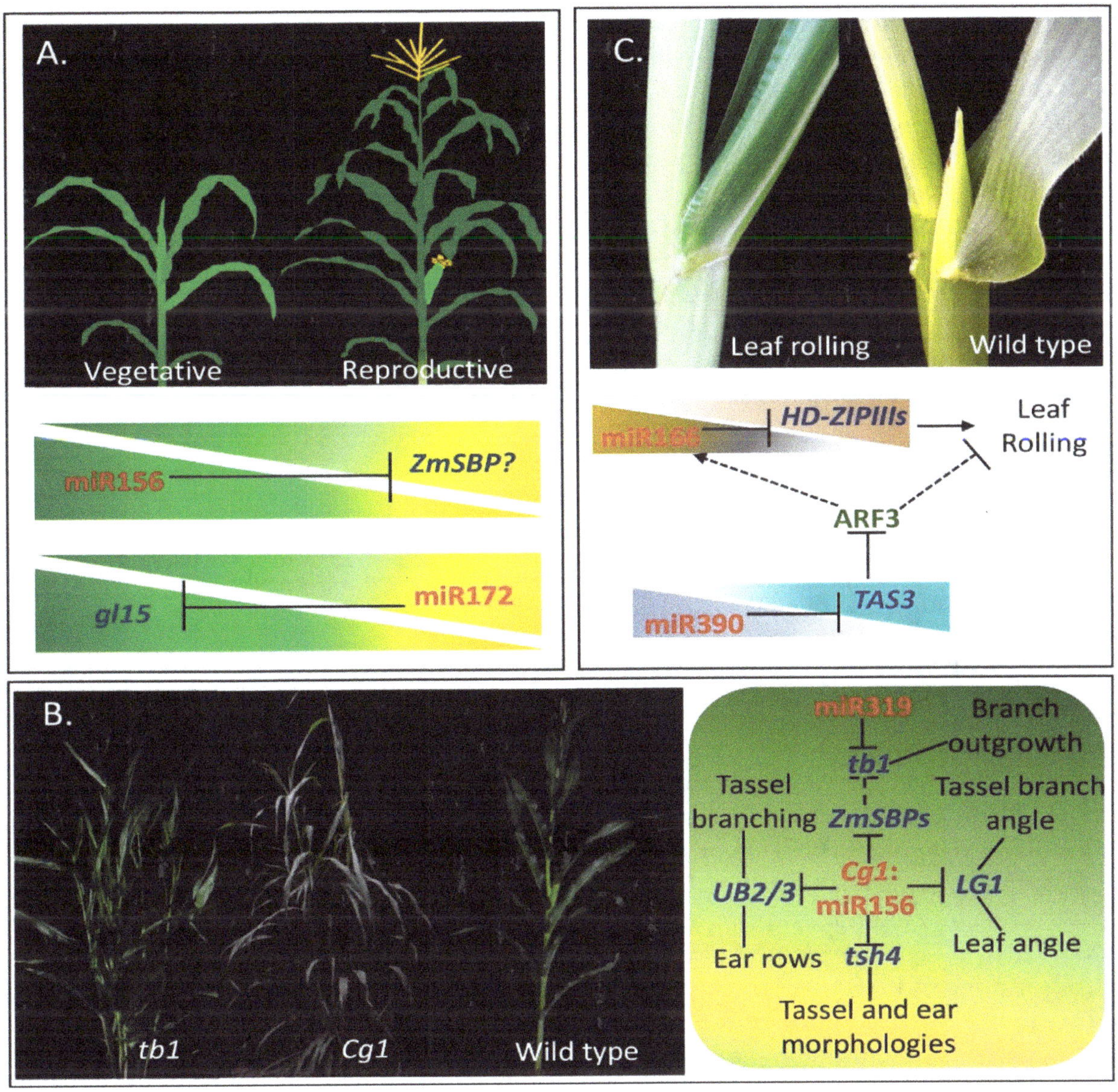

Figure 5. miRNAs or miRNA-phasiRNA interactions in agronomic traits. (**A**) Maize plant developmental timing is fine-tuned by the interaction between miR156 and miR172. However, the misslink between the two miRNAs still need to be addressed; (**B**) plant architectural modulation by the interaction of miR156 and miR319. Representative plants, *tb1* (leaf), *Cg1* (**middle**), and wild type (**right**) (all in the background of Chinese inbred line Zheng58) (Unpublished data), are shown. The potential *ZmSBP* gene probably connects the phenotype of apical dominance loss between maize mutants *tb1* and *Cg1*. In this context, the connection between *ZmSBPs* and *tb1* still need to be experimentally identified; (**C**) leaf shapes are being regulated by miR166 and miR390-TAS3 regulatory networks. *STTMmiR166* mutants have rolling leaf phenotype (**left**), the wild type is ZZC01 (**right**). In this context, the connection between miR166 and ARF3 is still unclear.

3.4. Leaf Patterns Are Shaped by miR166 and miR390-TAS3

Leaves are the most important photosynthetic organs in land plants, which are nearly flat organs designed to efficiently capture light and perform photosynthesis. In maize the specification of abaxial/adaxial polarity was found to be intimately associated with sRNAs, such as miR166, miR390 and *TAS3* (Figure 4; Figure 5C) [26,38,91,92]. The miR166 targets belong to class III homeodomain/leucine zipper (*HD-ZIPIII*) genes. The maize miR166 knockdown and miR166 target over-expression mutants, *STTMmiR166* and *rolled leaf1* (*rld1*), displays an upward curling of the leaf blade that causes adaxialization or partial reversal of leaf polarity [26,88]. The roles of miR166 and *HD-ZIPIII* in leaf polarity are conserved between *Arabidopsis* and maize [93,94]. In plants, miR390 triggers *TAS3*-tasiRNA biogenesis, which interplay with ARF3 to take part in plant development regulation [54,67]. In maize, the mutants of tasiRNA biogenesis pathway components exhibit leaf polarity alterations, *ragged seedling2* (*rgd2*) or *leaf bladeless1* (*lbl1*) [13]. Moreover, several researches proposed that miR390-*TAS3* define the adaxial side of the leaf by restricting the expression domain of miR166, which in turn demarcates the abaxial side of leaves by restricting the expression of adaxial determinants [38,92,95].

3.5. PhasiRNAs and Maize Male Fertility

In hybrid maize, male sterility has been widely studied due to both its biological significance and commercial use in hybrid seed production [96]. Maize male fertility is determined by dozens of genes and sRNAs, especially phasiRNAs (Figures 2A,B and 4) [11,12,96]. Indeed, a study reported that two classes of phasiRNAs, 21 and 24 nt in length, were detected to be highly expressed in maize anthers and confer male fertility [12]. The mutant lacking 21 nt phasiRNA, *ocl4*, showed male sterility due to defects in epidermal signaling. Meanwhile, the mutant lacking 24 nt phasiRNA lacking mutants also showed male sterility for due to defective anther subepidermis. This indicated that two types of phasiRNAs regulate anther development independently, with 21 nt premeiotic phasiRNAs regulating epidermal and 24 nt meiotic phasiRNAs regulating tapetal cell differentiation [12].

3.6. Other miRNA Functions in Maize

Several miRNAs have been identified to regulate important maize agronomic traits, such as kernel development, plant growth, abiotic stress tolerance, root development, and nutrition metabolism (Figure 4). The miR156 target, *tga1*, not only confers the domestication of maize naked grains, but also determines the maize kernel shape and size [97,98]. A report on *ZmGRF10*, a miR396 target, indicated that this miRNA is a potential regulator for maize leaf size and plant height [99]. The overexpression *ZmGRF10* mutant displayed reduction in leaf size and plant height by decreasing cell proliferation. Other studies have shown that drought and salinity stresses induce aberrant expression of many miRNAs in maize, for example miR166 and miR169. In maize, miR169 plays a critical role during plant drought, salt and ABA stress response by targeting *NUCLEAR FACTOR-Y subunit A* (*NF-YA*) genes [28]. In *Arabidopsis* and rice, miR166-*HD-ZIP IIIs* have been proven to be associated with drought and ABA stress resistance through maintaining ABA homeostasis [100,101]. Based on our unpublished data, the maize miR166 probably affects tolerance to drought and salinity stresses like in rice and *Arabidopsis*. In maize, miR164 was experimentally identified to be an important regulator in lateral root development by targeting *ZmNAC1* [31,102].

A recent research identified miR528, a monocot-specific miRNA, to be an important regulator for maize nitrogen metabolism in maize. In the miR528 knock-down mutant of maize, under nitrogen-luxury conditions, targets of miR528 are upregulated and mediate increase in lignin content along with superior lodging resistance [20]. The miR399 was identified to regulate the low-phosphate responses in maize [103]. The transgenic plant with miR399 over-expression showed significant phosphorus-toxicity phenotypes, indicating that miR399 is functionally conserved in monocots and dicots.

4. Exploiting the Roles of Maize Small RNAs in Important Agronomic Traits Improvement

Most of agronomic traits are quantitative traits, which are controlled by multiple loci and complex regulatory networks. MiRNAs and phased siRNAs are important participants in these complex regulatory networks. Manipulating the expression levels of miRNAs, miRNA targets, and phased siRNAs is a possible way for agronomic traits improvement. With grain yield increasing, the agronomic traits of maize have been improved through genetic selection [104], which is probably consistent with the elite allele selection of miRNAs and their targets in breeding. Compared with old maize varieties, modern varieties usually have reduced stature, more upright leaves, decreased tassel size, rolling leaf, superior staygreen, less tillers, shorter anthesis-silking interval, less ears per plant and superior stress resistance [104]. Based upon the knowledge about miRNA and phased siRNA functions, manipulating the expression of these small regulatory RNAs and their targets is a possible approach for agronomic traits improvement.

Flowering time represents the developmental transition from vegetative to reproductive phase. Maize spread from its origin to worldwide places with the gradually adaption of flowering time to the local climate [105]. Flowering time determines the length of vegetative phase, biomass and grain yield in maize. The interplay between miR156 and miR172 fine tunes the maize developmental timing and tillering [25,86,88]. Increasing the expression levels of miR156 can elongate the vegetative phase and tillering in maize, which is important to achieve high biomass for silage feed. MiR156 silencing or miR172 over-expression is able to impel maize flowering and precocity, which is in favor of maize mechanized harvest in special regions.

Ideal plant architecture is highly associated with maize planting density and lodging resistance, thereby achieving higher yield. Maize miR156 also regulates plant architectural traits through binding its target genes, such as *tsh4*, *LG1*, *UB2* and *UB3* [30,33,84,85,106,107]. Manipulating the expression of these *ZmSBPs* at optimal levels is needed for idea plant architectural traits. For instance, decreased expression of *LG1* can promote the leaf angle and reduce tassel branches. Furthermore, manipulating the expression of *UB2* and *UB3* in tassel branches and ear rows by using tissue-specific promotor is helpful to get ideal tassel and ear architecture. Additionally, repressing the expression of miR166 or increasing the expression of its targets will increase the leaf rolling, which can be helpful for improving the leaf shapes [88].

In global maize production, lodging and drought are two main abiotic stresses that accounts for large yield loss annually. In a recent research, miR528 has been proved to affect lodging resistance through regulating lignin biosynthesis [20]. Gene silencing of miR528 or overexpression of its targets is helpful for enhancing maize lodging resistance. Knock-down of miR164 promotes maize lateral root development, which can help toward drought and lodging resistance [31]. In the response toward abiotic stress, such as drought, ABA and salinity, miR169 and its targets (NF-YAs) contribute the major regulatory roles through ABA signaling [108]. Lowering the expression of miR169, or increasing that of NF-YAs, can facilitate maize resistance to drought. MiR166 silencing confers resistance against drought in rice and *Arabidopsis*, which is likely conserved in maize too [88,100,101].

5. Future Perspectives

As discussed above, enhanced knowledge on miRNA and phased siRNA functions will be helpful for improving some agronomic traits, including developmental timing, plant architecture, and abiotic stress resistance. Genetic engineering for elite maize germplasms and hybrids still face several hurdles. First, only a small proportion of miRNAs and phased siRNAs have been studied in maize, their complex regulatory networks remain largely unknown. The functional identification of sRNAs is largely dependent upon creating mutants. In maize, the abundant genetic variations or mutations in germplasm pools can provide useful raw materials for the study of these regulatory sRNAs or their targets [109]. Creating new mutants for specific sRNA using artificial miRNA, Short tandem target mimic (STTM) or target mimic (TM) techniques, are efficient strategies for uncovering the functions of these regulatory sRNAs in maize [110–112]. Second, plant miRNA and phased siRNA usually express in spatial and temporal manner. Thus, manipulating the expression of miRNA and phased siRNA in specific tissues and developmental stages can precisely target the traits for improvement. This can be achieved by expressing the transgene expression using tissue- or development-specific promoters, or inducible promoters. Fine genome editing of miRNAs, phased siRNAs and target genes by the CRISPR/Cas9 system can facilitate more subtle manipulations for the target agronomic traits, which is an alternative strategy. Third, for maize hybrids worldwide planted, screening elite hybrid is the most important mask in maize breeding. Usually, the ideal phenotypes in parental inbred lines do not always transfer to the corresponding hybrid. Screening of an elite hybrid is bit of an art and magic, which requires all the yield related traits to reach a balance, and with high heterosis and stress resistance. The current theory of heterosis model facilitate the breeders to make hybrid crosses with high heterosis. Screening the inbred lines with elite genotype/haplotype of miRNAs, phased siRNAs and their targets is fundamental in breeding. Introgressing the elite genotype or haplotype into inbred lines based on heterosis model/heterotic groups will enable the parental elite phenotypes get transferred to their hybrids.

Author Contributions: G.T. and J.T. provided the guideline, Z.Z. collected related references and wrote the manuscript, Z.Z. drew the Figures, J.T., G.T. and S.T. modified the Figures and revised the manuscript.

Acknowledgments: We are thankful to anonymous reviewers for their valuable suggestions to improve this article.

References

1. Chen, X. Small RNAs and their roles in plant development. *Annu. Rev. Cell Dev. Biol.* **2009**, *25*, 21–44. [CrossRef] [PubMed]
2. Kim, V.N.; Han, J.; Siomi, M.C. Biogenesis of small RNAs in animals. *Nat. Rev. Mol. Cell Biol.* **2009**, *10*, 126–139. [CrossRef] [PubMed]
3. Fei, Q.; Xia, R.; Meyers, B.C. Phased, secondary, small interfering RNAs in posttranscriptional regulatory networks. *Plant Cell* **2013**, *25*, 2400–2415. [CrossRef] [PubMed]
4. Axtell, M.J. Classification and comparison of small RNAs from plants. *Annu. Rev. Plant Biol.* **2013**, *64*, 137–159. [CrossRef] [PubMed]
5. Wang, Z.; Ma, Z.; Castillo-González, C.; Sun, D.; Li, Y.; Yu, B.; Zhao, B.; Li, P.; Zhang, X. SWI2/SNF2 ATPase CHR2 remodels pri-miRNAs via Serrate to impede miRNA production. *Nature* **2018**, *557*, 516–521. [CrossRef]
6. Rogers, K.; Chen, X. Biogenesis, turnover, and mode of action of plant microRNAs. *Plant Cell* **2013**, *25*, 2383–2399. [CrossRef] [PubMed]
7. Voinnet, O. Origin, biogenesis, and activity of plant microRNAs. *Cell* **2009**, *136*, 669–687. [CrossRef]
8. Rhoades, M.W.; Reinhart, B.J.; Lim, L.P.; Burge, C.B.; Bartel, B.; Bartel, D.P. Prediction of plant microRNA targets. *Cell* **2002**, *110*, 513–520. [CrossRef]
9. Cuperus, J.T.; Fahlgren, N.; Carrington, J.C. Evolution and functional diversification of *MIRNA* Genes. *Plant Cell* **2011**, *23*, 431–442. [CrossRef]
10. Jones-Rhoades, M.W.; Bartel, D.P.; Bartel, B. MicroRNAs and their regulatory roles in plants. *Annu. Rev. Plant Biol.* **2006**, *57*, 19–53. [CrossRef]
11. Yu, Y.; Zhou, Y.; Zhang, Y.; Chen, Y. Grass phasiRNAs and male fertility. *Sci. China Life Sci.* **2018**, *61*, 148–154. [CrossRef] [PubMed]
12. Zhai, J.; Zhang, H.; Arikit, S.; Huang, K.; Nan, G.L.; Walbot, V.; Meyers, B.C. Spatiotemporally dynamic, cell-type-dependent premeiotic and meiotic phasiRNAs in maize anthers. *Proc. Natl. Acad. Sci. USA* **2015**, *112*, 3146–3151. [CrossRef]
13. Juarez, M.T.; Twigg, R.W.; Timmermans, M.C. Specification of adaxial cell fate during maize leaf development. *Development* **2004**, *131*, 4533–4544. [CrossRef] [PubMed]
14. Wang, H.; Wang, H. The miR156/SPL module, a regulatory hub and versatile toolbox, gears up crops for enhanced agronomic traits. *Mol. Plant* **2015**. [CrossRef] [PubMed]
15. Tang, J.; Chu, C. MicroRNAs in crop improvement: Fine-tuners for complex traits. *Nat. Plants* **2017**, *3*, 17077. [CrossRef] [PubMed]
16. Gore, M.A.; Chia, J.M.; Elshire, R.J.; Sun, Q.; Ersoz, E.S.; Hurwitz, B.L.; Peiffer, J.A.; McMullen, M.D.; Grills, G.S.; Ross-Ibarra, J.; et al. A first-generation haplotype map of maize. *Science* **2009**, *326*, 1115–1117. [CrossRef] [PubMed]
17. Brutnell, T.P. Transposon tagging in maize. *Funct. Integr. Genom.* **2002**, *2*, 4–12. [CrossRef]
18. Arteaga-Vazquez, M.A.; Chandler, V.L. Paramutation in maize: RNA mediated trans-generational gene silencing. *Curr. Opin. Genet. Dev.* **2010**, *20*, 156–163. [CrossRef] [PubMed]
19. Sun, W.; Xiang, X.; Zhai, L.; Zhang, D.; Cao, Z.; Liu, L.; Zhang, Z. AGO18b negatively regulates determinacy of spikelet meristems on the tassel central spike in maize. *J. Integr. Plant Biol.* **2018**, *60*, 65–78. [CrossRef] [PubMed]

20. Sun, Q.; Liu, X.; Yang, J.; Liu, W.; Du, Q.; Wang, H.; Fu, C.; Li, W.X. MicroRNA528 affects lodging resistance of maize by regulating lignin biosynthesis under nitrogen-luxury conditions. *Mol. Plant* **2018**, *11*, 806–814. [CrossRef]
21. Field, S.; Thompson, B. Analysis of the Maize dicer-like1 Mutant, fuzzy tassel, Implicates MicroRNAs in Anther Maturation and Dehiscence. *PLoS ONE* **2016**, *11*, e0146534. [CrossRef] [PubMed]
22. Thompson, B.E.; Basham, C.; Hammond, R.; Ding, Q.; Kakrana, A.; Lee, T.F.; Simon, S.A.; Meeley, R.; Meyers, B.C.; Hake, S. The dicer-like1 homolog fuzzy tassel is required for the regulation of meristem determinacy in the inflorescence and vegetative growth in maize. *Plant Cell* **2014**, *26*, 4702–4717. [CrossRef] [PubMed]
23. Petsch, K.; Manzotti, P.S.; Tam, O.H.; Meeley, R.; Hammell, M.; Consonni, G.; Timmermans, M.C. Novel DICER-LIKE1 siRNAs bypass the requirement for DICER-LIKE4 in maize development. *Plant Cell* **2015**, *27*, 2163–2177. [CrossRef] [PubMed]
24. Xu, D.; Yang, H.; Zou, C.; Li, W.X.; Xu, Y.; Xie, C. Identification and functional characterization of the *AGO1* ortholog in maize. *J. Integr. Plant Biol.* **2016**, *58*, 749–758. [CrossRef] [PubMed]
25. Chuck, G.; Cigan, A.M.; Saeteurn, K.; Hake, S. The heterochronic maize mutant *Corngrass1* results from overexpression of a tandem microRNA. *Nat. Genet.* **2007**, *39*, 544–549. [CrossRef] [PubMed]
26. Juarez, M.T.; Kui, J.S.; Thomas, J.; Heller, B.A.; Timmermans, M.C. MicroRNA-mediated repression of *rolled leaf1* specifies maize leaf polarity. *Nature* **2004**, *428*, 84–88. [CrossRef] [PubMed]
27. Lauter, N.; Kampani, A.; Carlson, S.; Goebel, M.; Moose, S.P. MicroRNA172 down-regulates *glossy15* to promote vegetative phase change in maize. *Proc. Natl. Acad. Sci. USA* **2005**, *102*, 9412–9417. [CrossRef] [PubMed]
28. Luan, M.; Xu, M.; Lu, Y.; Zhang, L.; Fan, Y.; Wang, L. Expression of zma-miR169 miRNAs and their target *ZmNF-YA* genes in response to abiotic stress in maize leaves. *Gene* **2015**, *555*, 178–185. [CrossRef] [PubMed]
29. Xu, D.; Wang, X.; Huang, C.; Xu, G.; Liang, Y.; Chen, Q.; Wang, C.; Li, D.; Tian, J.; Wu, L.; et al. *Glossy15* plays an important role in the divergence of the vegetative transition between maize and its progenitor, teosinte. *Mol. Plant* **2017**, *10*, 1579–1583. [CrossRef] [PubMed]
30. Chuck, G.S.; Brown, P.J.; Meeley, R.; Hake, S. Maize SBP-box transcription factors *unbranched2* and *unbranched3* affect yield traits by regulating the rate of lateral primordia initiation. *Proc. Natl. Acad. Sci. USA* **2014**, *111*, 18775–18780. [CrossRef] [PubMed]
31. Li, J.; Guo, G.; Guo, W.; Guo, G.; Tong, D.; Ni, Z.; Sun, Q.; Yao, Y. miRNA164-directed cleavage of *ZmNAC1* confers lateral root development in maize (*Zea mays* L.). *BMC Plant Biol.* **2012**, *12*, 220. [CrossRef] [PubMed]
32. Qian, Y.; Cheng, Y.; Cheng, X.; Jiang, H.; Zhu, S.; Cheng, B. Identification and characterization of Dicer-like, Argonaute and RNA-dependent RNA polymerase gene families in maize. *Plant Cell Rep.* **2011**, *30*, 1347–1363. [CrossRef] [PubMed]
33. Chuck, G.; Whipple, C.; Jackson, D.; Hake, S. The maize SBP-box transcription factor encoded by *tasselsheath4* regulates bract development and the establishment of meristem boundaries. *Development* **2010**, *137*, 1243–1250. [CrossRef] [PubMed]
34. Hultquist, J.F.; Dorweiler, J.E. Feminized tassels of maize *mop1* and *ts1* mutants exhibit altered levels of miR156 and specific SBP-box genes. *Planta* **2008**, *229*, 99–113. [CrossRef] [PubMed]
35. Kapoor, M.; Arora, R.; Lama, T.; Nijhawan, A.; Khurana, J.P.; Tyagi, A.K.; Kapoor, S. Genome-wide identification, organization and phylogenetic analysis of Dicer-like, Argonaute and RNA-dependent RNA Polymerase gene families and their expression analysis during reproductive development and stress in rice. *BMC Genom.* **2008**, *9*, 451. [CrossRef] [PubMed]
36. Matzke, M.A.; Birchler, J.A. RNAi-mediated pathways in the nucleus. *Nat. Rev. Genet.* **2005**, *6*, 24–35. [CrossRef]
37. He, J.; Dong, Z.; Jia, Z.; Wang, J.; Wang, G. Isolation, expression and functional analysis of a putative RNA-dependent RNA polymerase gene from maize (*Zea mays* L.). *Mol. Biol. Rep.* **2010**, *37*, 865–874. [CrossRef]
38. Douglas, R.N.; Wiley, D.; Sarkar, A.; Springer, N.; Timmermans, M.C.; Scanlon, M.J. *Ragged seedling2* Encodes an ARGONAUTE7-like protein required for mediolateral expansion, but not dorsiventrality, of maize leaves. *Plant Cell* **2010**, *22*, 1441–1451. [CrossRef]
39. Dorweiler, J.E.; Carey, C.C.; Kubo, K.M.; Hollick, J.B.; Kermicle, J.L.; Chandler, V.L. *Mediator of paramutation1* is required for establishment and maintenance of paramutation at multiple maize loci. *Plant Cell* **2000**, *12*, 2101–2118. [CrossRef]

40. Ramachandran, V.; Chen, X. Small RNA metabolism in *Arabidopsis*. *Trends Plant Sci.* **2008**, *13*, 368–374. [CrossRef]
41. Yu, B.; Yang, Z.; Li, J.; Minakhina, S.; Yang, M.; Padgett, R.W.; Steward, R.; Chen, X. Methylation as a crucial step in plant microRNA biogenesis. *Science* **2005**, *307*, 932–935. [CrossRef] [PubMed]
42. Reinhart, B.J.; Weinstein, E.G.; Rhoades, M.W.; Bartel, B.; Bartel, D.P. MicroRNAs in plants. *Genes Dev.* **2002**, *16*, 1616–1626. [CrossRef] [PubMed]
43. Llave, C.; Xie, Z.; Kasschau, K.D.; Carrington, J.C. Cleavage of Scarecrow-like mRNA targets directed by a class of *Arabidopsis* miRNA. *Science* **2002**, *297*, 2053–2056. [CrossRef]
44. Zhang, Z.; Hu, F.; Sung, M.W.; Shu, C.; Castillo-Gonzalez, C.; Koiwa, H.; Tang, G.; Dickman, M.; Li, P.; Zhang, X. RISC-interacting clearing 3′-5′exoribonucleases (RICEs) degrade uridylated cleavage fragments to maintain functional RISC in *Arabidopsis thaliana*. *eLife* **2017**, *6*, e24466. [CrossRef] [PubMed]
45. Huntzinger, E.; Izaurralde, E. Gene silencing by microRNAs: Contributions of translational repression and mRNA decay. *Nat. Rev. Genet.* **2011**, *12*, 99–110. [CrossRef]
46. Pontes, O.; Costa-Nunes, P.; Vithayathil, P.; Pikaard, C.S. RNA polymerase V functions in *Arabidopsis* interphase heterochromatin organization independently of the 24-nt siRNA-directed DNA methylation pathway. *Mol. Plant* **2009**, *2*, 700–710. [CrossRef] [PubMed]
47. Teotia, S.; Singh, D.; Tang, G. DNA Methylation in plants by microRNAs. In *Plant Epigenetics*; Rajewsky, N., Jurga, S., Barciszewski, J., Eds.; Springer International Publishing: Cham, Switzerland, 2017; pp. 247–262.
48. Axtell, M.J. Lost in translation? microRNAs at the rough ER. *Trends Plant Sci.* **2017**, *22*, 273–274. [CrossRef] [PubMed]
49. Yu, Y.; Ji, L.; Le, B.H.; Zhai, J.; Chen, J.; Luscher, E.; Gao, L.; Liu, C.; Cao, X.; Mo, B.; et al. ARGONAUTE10 promotes the degradation of miR165/6 through the SDN1 and SDN2 exonucleases in Arabidopsis. *PLoS Biol.* **2017**, *15*, e2001272. [CrossRef] [PubMed]
50. Borges, F.; Martienssen, R.A. The expanding world of small RNAs in plants. *Nat. Rev. Mol. Cell Biol.* **2015**, *16*, 727–741. [CrossRef]
51. Arikit, S.; Zhai, J.; Meyers, B.C. Biogenesis and function of rice small RNAs from non-coding RNA precursors. *Curr. Opin. Plant Biol.* **2013**, *16*, 170–179. [CrossRef] [PubMed]
52. Axtell, M.J.; Jan, C.; Rajagopalan, R.; Bartel, D.P. A two-hit trigger for siRNA biogenesis in plants. *Cell* **2006**, *127*, 565–577. [CrossRef] [PubMed]
53. Allen, E.; Howell, M.D. miRNAs in the biogenesis of trans-acting siRNAs in higher plants. *Semin. Cell Dev. Biol.* **2010**, *21*, 798–804. [CrossRef] [PubMed]
54. De Felippes, F.F.; Marchais, A.; Sarazin, A.; Oberlin, S.; Voinnet, O. A single miR390 targeting event is sufficient for triggering TAS3-tasiRNA biogenesis in Arabidopsis. *Nucleic Acids Res.* **2017**, *45*, 5539–5554. [CrossRef]
55. Dotto, M.C.; Petsch, K.A.; Aukerman, M.J.; Beatty, M.; Hammell, M.; Timmermans, M.C. Genome-wide analysis of *leafbladeless1*-regulated and phased small RNAs underscores the importance of the TAS3 ta-siRNA pathway to maize development. *PLoS Genet.* **2014**, *10*, e1004826. [CrossRef] [PubMed]
56. Park, W.; Li, J.; Song, R.; Messing, J.; Chen, X. CARPEL FACTORY, a Dicer homolog, and HEN1, a novel protein, act in microRNA metabolism in *Arabidopsis thaliana*. *Curr. Biol.* **2002**, *12*, 1484–1495. [CrossRef]
57. Parent, J.S.; Bouteiller, N.; Elmayan, T.; Vaucheret, H. Respective contributions of *Arabidopsis* DCL2 and DCL4 to RNA silencing. *Plant J. Cell Mol. Biol.* **2015**, *81*, 223–232. [CrossRef]
58. Xie, Z.; Johansen, L.K.; Gustafson, A.M.; Kasschau, K.D.; Lellis, A.D.; Zilberman, D.; Jacobsen, S.E.; Carrington, J.C. Genetic and functional diversification of small RNA pathways in plants. *PLoS Biol.* **2004**, *2*, e104. [CrossRef]
59. Xie, Z.; Allen, E.; Wilken, A.; Carrington, J.C. DICER-LIKE 4 functions in trans-acting small interfering RNA biogenesis and vegetative phase change in *Arabidopsis thaliana*. *Proc. Natl. Acad. Sci. USA* **2005**, *102*, 12984–12989. [CrossRef]
60. Margis, R.; Fusaro, A.F.; Smith, N.A.; Curtin, S.J.; Watson, J.M.; Finnegan, E.J.; Waterhouse, P.M. The evolution and diversification of Dicers in plants. *FEBS Lett.* **2006**, *580*, 2442–2450. [CrossRef]
61. Vaucheret, H.; Vazquez, F.; Crete, P.; Bartel, D.P. The action of *ARGONAUTE1* in the miRNA pathway and its regulation by the miRNA pathway are crucial for plant development. *Genes Dev.* **2004**, *18*, 1187–1197. [CrossRef]
62. Borges, F.; Pereira, P.A.; Slotkin, R.K.; Martienssen, R.A.; Becker, J.D. MicroRNA activity in the *Arabidopsis* male germline. *J. Exp. Bot.* **2011** *62*, 1611–1620. [CrossRef] [PubMed]

63. Harvey, J.J.; Lewsey, M.G.; Patel, K.; Westwood, J.; Heimstadt, S.; Carr, J.P.; Baulcombe, D.C. An antiviral defense role of AGO2 in plants. *PLoS ONE* **2011**, *6*, e14639. [CrossRef]
64. Zilberman, D.; Cao, X.; Johansen, L.K.; Xie, Z.; Carrington, J.C.; Jacobsen, S.E. Role of *Arabidopsis ARGONAUTE4* in RNA-directed DNA methylation triggered by inverted repeats. *Curr. Biol.* **2004**, *14*, 1214–1220. [CrossRef] [PubMed]
65. Wu, J.; Yang, Z.; Wang, Y.; Zheng, L.; Ye, R.; Ji, Y.; Zhao, S.; Ji, S.; Liu, R.; Xu, L.; et al. Viral-inducible Argonaute18 confers broad-spectrum virus resistance in rice by sequestering a host microRNA. *eLife* **2015**, *4*. [CrossRef] [PubMed]
66. Zhai, L.; Sun, W.; Zhang, K.; Jia, H.; Liu, L.; Liu, Z.; Teng, F.; Zhang, Z. Identification and characterization of Argonaute gene family and meiosis-enriched Argonaute during sporogenesis in maize. *J. Integr. Plant Biol.* **2014**, *56*, 1042–1052. [CrossRef] [PubMed]
67. Marin, E.; Jouannet, V.; Herz, A.; Lokerse, A.S.; Weijers, D.; Vaucheret, H.; Nussaume, L.; Crespi, M.D.; Maizel, A. miR390, *Arabidopsis TAS3* tasiRNAs, and their AUXIN RESPONSE FACTOR targets define an autoregulatory network quantitatively regulating lateral root growth. *Plant Cell* **2010**, *22*, 1104–1117. [CrossRef] [PubMed]
68. Montgomery, T.A.; Howell, M.D.; Cuperus, J.T.; Li, D.; Hansen, J.E.; Alexander, A.L.; Chapman, E.J.; Fahlgren, N.; Allen, E.; Carrington, J.C. Specificity of ARGONAUTE7-miR390 interaction and dual functionality in *TAS3* trans-acting siRNA formation. *Cell* **2008**, *133*, 128–141. [CrossRef] [PubMed]
69. Xia, R.; Meyers, B.C.; Liu, Z.; Beers, E.P.; Ye, S.; Liu, Z. MicroRNA superfamilies descended from miR390 and their roles in secondary small interfering RNA Biogenesis in Eudicots. *Plant Cell* **2013**, *25*, 1555–1572. [CrossRef] [PubMed]
70. Sievers, F.; Higgins, D.G. Clustal Omega for making accurate alignments of many protein sequences. *Protein Sci.* **2018**, *27*, 135–145. [CrossRef]
71. Letunic, I.; Bork, P. Interactive Tree of Life v2: Online annotation and display of phylogenetic trees made easy. *Nucleic Acids Res.* **2011**, *39*, W475–W478. [CrossRef]
72. Zaratiegui, M.; Irvine, D.V.; Martienssen, R.A. Noncoding RNAs and gene silencing. *Cell* **2007**, *128*, 763–776. [CrossRef] [PubMed]
73. Cho, S.H.; Coruh, C.; Axtell, M.J. miR156 and miR390 regulate tasiRNA accumulation and developmental timing in *Physcomitrella patens*. *Plant Cell* **2012**, *24*, 4837–4849. [CrossRef]
74. Aukerman, M.J.; Sakai, H. Regulation of flowering time and floral organ identity by a microRNA and its *APETALA2*-like target genes. *Plant Cell* **2003**, *15*, 2730–2741. [CrossRef] [PubMed]
75. Jung, J.H.; Seo, Y.H.; Seo, P.J.; Reyes, J.L.; Yun, J.; Chua, N.H.; Park, C.M. The *GIGANTEA*-regulated microRNA172 mediates photoperiodic flowering independent of *CONSTANS* in *Arabidopsis*. *Plant Cell* **2007**, *19*, 2736–2748. [CrossRef] [PubMed]
76. Wu, G.; Poethig, R.S. Temporal regulation of shoot development in *Arabidopsis thaliana* by miR156 and its target *SPL3*. *Development* **2006**, *133*, 3539–3547. [CrossRef] [PubMed]
77. Chuck, G.S.; Tobias, C.; Sun, L.; Kraemer, F.; Li, C.; Dibble, D.; Arora, R.; Bragg, J.N.; Vogel, J.P.; Singh, S.; et al. Overexpression of the maize *Corngrass1* microRNA prevents flowering, improves digestibility, and increases starch content of switchgrass. *Proc. Natl. Acad. Sci. USA* **2011**, *108*, 17550–17555. [CrossRef] [PubMed]
78. Doebley, J.; Stec, A.; Hubbard, L. The evolution of apical dominance in maize. *Nature* **1997**, *386*, 485–488. [CrossRef] [PubMed]
79. Kidner, C.A.; Martienssen, R.A. Spatially restricted microRNA directs leaf polarity through ARGONAUTE1. *Nature* **2004**, *428*, 81–84. [CrossRef] [PubMed]
80. Palatnik, J.F.; Allen, E.; Wu, X.; Schommer, C.; Schwab, R.; Carrington, J.C.; Weigel, D. Control of leaf morphogenesis by microRNAs. *Nature* **2003**, *425*, 257–263. [CrossRef] [PubMed]
81. Schommer, C.; Palatnik, J.F.; Aggarwal, P.; Chetelat, A.; Cubas, P.; Farmer, E.E.; Nath, U.; Weigel, D. Control of jasmonate biosynthesis and senescence by miR319 targets. *PLoS Biol.* **2008**, *6*, e230. [CrossRef]
82. Jiao, Y.; Wang, Y.; Xue, D.; Wang, J.; Yan, M.; Liu, G.; Dong, G.; Zeng, D.; Lu, Z.; Zhu, X.; et al. Regulation of *OsSPL14* by OsmiR156 defines ideal plant architecture in rice. *Nat. Genet.* **2010**, *42*, 541–544. [CrossRef] [PubMed]
83. Lu, Z.; Yu, H.; Xiong, G.; Wang, J.; Jiao, Y.; Liu, G.; Jing, Y.; Meng, X.; Hu, X.; Qian, Q.; et al. Genome-wide binding analysis of the transcription activator ideal plant architecture1 reveals a complex network regulating rice plant architecture. *Plant Cell* **2013**, *25*, 3743–3759. [CrossRef] [PubMed]

84. Moreno, M.A.; Harper, L.C.; Krueger, R.W.; Dellaporta, S.L.; Freeling, M. *liguleless1* encodes a nuclear-localized protein required for induction of ligules and auricles during maize leaf organogenesis. *Genes Dev.* **1997**, *11*, 616–628. [CrossRef] [PubMed]
85. Liu, L.; Du, Y.; Shen, X.; Li, M.; Sun, W.; Huang, J.; Liu, Z.; Tao, Y.; Zheng, Y.; Yan, J.; et al. *KRN4* controls quantitative variation in maize kernel row number. *PLoS Genet.* **2015**, *11*, e1005670. [CrossRef] [PubMed]
86. Chuck, G.; Meeley, R.; Irish, E.; Sakai, H.; Hake, S. The maize *tasselseed4* microRNA controls sex determination and meristem cell fate by targeting *Tasselseed6/indeterminate spikelet1*. *Nat. Genet.* **2007**, *39*, 1517–1521. [CrossRef] [PubMed]
87. Banks, J.A. MicroRNA, sex determination and floral meristem determinacy in maize. *Genome Biol.* **2008**, *9*, 204. [CrossRef] [PubMed]
88. Peng, T.; Qiao, M.; Liu, H.; Teotia, S.; Zhang, Z.; Zhao, Y.; Wang, B.; Zhao, D.; Shi, L.; Zhang, C.; et al. A resource for inactivation of microRNAs using Short Tandem Target Mimic technology in model and crop plants. *Mol. Plant* **2018**, *11*, 1400–1417. [CrossRef] [PubMed]
89. Donaire, L.; Barajas, D.; Martinez-Garcia, B.; Martinez-Priego, L.; Pagan, I.; Llave, C. Structural and genetic requirements for the biogenesis of tobacco rattle virus-derived small interfering RNAs. *J. Virol.* **2008**, *82*, 5167–5177. [CrossRef] [PubMed]
90. Tsuji, H.; Aya, K.; Ueguchi-Tanaka, M.; Shimada, Y.; Nakazono, M.; Watanabe, R.; Nishizawa, N.K.; Gomi, K.; Shimada, A.; Kitano, H.; et al. GAMYB controls different sets of genes and is differentially regulated by microRNA in aleurone cells and anthers. *Plant J. Cell Mol. Biol.* **2006**, *47*, 427–444. [CrossRef] [PubMed]
91. Archak, S.; Nagaraju, J. Computational prediction of rice (*Oryza sativa*) miRNA targets. *Genom. Proteom. Bioinform.* **2007**, *5*, 196–206. [CrossRef]
92. Nogueira, F.T.; Madi, S.; Chitwood, D.H.; Juarez, M.T.; Timmermans, M.C. Two small regulatory RNAs establish opposing fates of a developmental axis. *Genes Dev.* **2007**, *21*, 750–755. [CrossRef] [PubMed]
93. McConnell, J.R.; Emery, J.; Eshed, Y.; Bao, N.; Bowman, J.; Barton, M.K. Role of *PHABULOSA* and *PHAVOLUTA* in determining radial patterning in shoots. *Nature* **2001**, *411*, 709–713. [CrossRef] [PubMed]
94. Emery, J.F.; Floyd, S.K.; Alvarez, J.; Eshed, Y.; Hawker, N.P.; Izhaki, A.; Baum, S.F.; Bowman, J.L. Radial patterning of *Arabidopsis* shoots by class III HD-ZIP and KANADI genes. *Curr. Biol.* **2003**, *13*, 1768–1774. [CrossRef] [PubMed]
95. Nogueira, F.T.; Chitwood, D.H.; Madi, S.; Ohtsu, K.; Schnable, P.S.; Scanlon, M.J.; Timmermans, M.C. Regulation of small RNA accumulation in the maize shoot apex. *PLoS Genet.* **2009**, *5*, e1000320. [CrossRef] [PubMed]
96. Shkibbe, D.S.; Schnable, P.S. Male sterility in maize. *Maydica* **2005**, *50*, 367–376.
97. Wang, H.; Nussbaum-Wagler, T.; Li, B.; Zhao, Q.; Vigouroux, Y.; Faller, M.; Bomblies, K.; Lukens, L.; Doebley, J.F. The origin of the naked grains of maize. *Nature* **2005**, *436*, 714–719. [CrossRef]
98. Wang, H.; Studer, A.J.; Zhao, Q.; Meeley, R.; Doebley, J.F. Evidence that the origin of naked kernels during maize domestication was caused by a single amino acid substitution in *tga1*. *Genetics* **2015**, *200*, 965–974. [CrossRef]
99. Wu, L.; Zhang, D.; Xue, M.; Qian, J.; He, Y.; Wang, S. Overexpression of the maize *GRF10*, an endogenous truncated growth-regulating factor protein, leads to reduction in leaf size and plant height. *J. Integr. Plant Biol.* **2014**, *56*, 1053–1063. [CrossRef]
100. Yan, J.; Zhao, C.; Zhou, J.; Yang, Y.; Wang, P.; Zhu, X.; Tang, G.; Bressan, R.A.; Zhu, J.K. The miR165/166 mediated regulatory module plays critical roles in ABA homeostasis and response in *Arabidopsis thaliana*. *PLoS Genet.* **2016**, *12*, e1006416. [CrossRef]
101. Zhang, J.; Zhang, H.; Srivastava, A.K.; Pan, Y.; Bai, J.; Fang, J.; Shi, H.; Zhu, J.K. Knockdown of rice microRNA166 confers drought resistance by causing leaf rolling and altering stem xylem development. *Plant Physiol.* **2018**, *176*, 2082–2094. [CrossRef]
102. Hochholdinger, F.; Yu, P.; Marcon, C. Genetic control of root system development in maize. *Trends Plant Sci.* **2018**, *23*, 79–88. [CrossRef] [PubMed]
103. Du, Q.; Wang, K.; Zou, C.; Xu, C.; Li, W.X. The *PILNCR1*-miR399 regulatory module is important for low phosphate tolerance in maize. *Plant Physiol.* **2018**, *177*, 1743–1753. [CrossRef]
104. Duvick, D.N. The contribution of breeding to yield advances in maize (*Zea mays* L.). *Adv. Agron.* **2005**, *86*, 83–145. [CrossRef]

105. Buckler, E.S.; Holland, J.B.; Bradbury, P.J.; Acharya, C.B.; Brown, P.J.; Browne, C.; Ersoz, E.; Flint-Garcia, S.; Garcia, A.; Glaubitz, J.C.; et al. The genetic architecture of maize flowering time. *Science* **2009**, *325*, 714–718. [CrossRef]
106. Lackey, E.; Ng, D.W.; Chen, Z.J. RNAi-mediated down-regulation of *DCL1* and *AGO1* induces developmental changes in resynthesized *Arabidopsis* allotetraploids. *New Phytol.* **2010**, *186*, 207–215. [CrossRef] [PubMed]
107. Moon, J.; Candela, H.; Hake, S. The Liguleless narrow mutation affects proximal-distal signaling and leaf growth. *Development* **2013**, *140*, 405–412. [CrossRef] [PubMed]
108. Luan, M.; Xu, M.; Lu, Y.; Zhang, Q.; Zhang, L.; Zhang, C.; Fan, Y.; Lang, Z.; Wang, L. Family-wide survey of miR169s and *NF-YAs* and their expression profiles response to abiotic stress in maize roots. *PLoS ONE* **2014**, *9*, e91369. [CrossRef]
109. Portwood, J.L., 2nd; Woodhouse, M.R.; Cannon, E.K.; Gardiner, J.M.; Harper, L.C.; Schaeffer, M.L.; Walsh, J.R.; Sen, T.Z.; Cho, K.T.; Schott, D.A.; et al. MaizeGDB 2018: The maize multi-genome genetics and genomics database. *Nucleic Acids Res.* **2019**, *47*, D1146–D1154. [CrossRef] [PubMed]
110. Yan, J.; Gu, Y.; Jia, X.; Kang, W.; Pan, S.; Tang, X.; Chen, X.; Tang, G. Effective small RNA destruction by the expression of a short tandem target mimic in *Arabidopsis*. *Plant Cell* **2012**, *24*, 415–427. [CrossRef]
111. Franco-Zorrilla, J.M.; Valli, A.; Todesco, M.; Mateos, I.; Puga, M.I.; Rubio-Somoza, I.; Leyva, A.; Weigel, D.; Garcia, J.A.; Paz-Ares, J. Target mimicry provides a new mechanism for regulation of microRNA activity. *Nat. Genet.* **2007**, *39*, 1033–1037. [CrossRef]
112. Jeong, D.H.; Park, S.; Zhai, J.; Gurazada, S.G.; De Paoli, E.; Meyers, B.C.; Green, P.J. Massive analysis of rice small RNAs: Mechanistic implications of regulated microRNAs and variants for differential target RNA cleavage. *Plant Cell* **2011**, *23*, 4185–4207. [CrossRef] [PubMed]

4

MicroRNAs in Cardiac Autophagy: Small Molecules and Big Role

Teng Sun [1], Meng-Yang Li [2], Pei-Feng Li [2] and Ji-Min Cao [1,*]

[1] Key Laboratory of Cellular Physiology, Ministry of Education, Department of Physiology, Shanxi Medical University, Taiyuan 030001, China; tengsun@qdu.edu.cn

[2] Institute for Translational Medicine, Qingdao University, Qingdao 266021, China; limengyang@qdu.edu.cn (M.-Y.L.); peifeng@ioz.ac.cn (P.-F.L.)

* Correspondence: caojimin@126.com;

Abstract: Autophagy, which is an evolutionarily conserved process according to the lysosomal degradation of cellular components, plays a critical role in maintaining cell homeostasis. Autophagy and mitochondria autophagy (mitophagy) contribute to the preservation of cardiac homeostasis in physiological settings. However, impaired or excessive autophagy is related to a variety of diseases. Recently, a close link between autophagy and cardiac disorders, including myocardial infarction, cardiac hypertrophy, cardiomyopathy, cardiac fibrosis, and heart failure, has been demonstrated. MicroRNAs (miRNAs) are a class of small non-coding RNAs with a length of approximately 21–22 nucleotides (nt), which are distributed widely in viruses, plants, protists, and animals. They function in mediating the post-transcriptional gene silencing. A growing number of studies have demonstrated that miRNAs regulate cardiac autophagy by suppressing the expression of autophagy-related genes in a targeted manner, which are involved in the pathogenesis of heart diseases. This review summarizes the role of microRNAs in cardiac autophagy and related cardiac disorders. Furthermore, we mainly focused on the autophagy regulation pathways, which consisted of miRNAs and their targeted genes.

Keywords: microRNAs; autophagy; mitophagy; cardiac diseases; biomarker

1. Overview of Autophagy and MicroRNAs

1.1. Autophagy

Autophagy is an evolutionarily conserved process of the lysosome-dependent degradation of cytoplasm components and damaged organelles, such as endoplasmic reticulum, peroxisomes, and mitochondria [1]. Factors that induce cellular senility and stress, including infections, toxics, hypoxia, and nutrient starvation, can induce autophagy, which plays a role in protecting cells and maintaining homeostasis. When it is stimulated by pathological factors, autophagy is either disrupted or contributes to autophagic cell death. Abnormal autophagy is related to a variety of pathological disorders. Autophagy involves the following steps: autography induction, vesicle nucleation, vesicle elongation and autophagosomes formation, and retrieval and fusion between autophagosomes and lysosomes [2]. Mitochondria produce energy in the form of ATP and play important roles in cellular homeostasis, signaling, apoptosis, autophagy, and metabolism. Damaged and dysfunctional mitochondria are deleterious to cells and can even lead to various types of diseases [3–8]. Therefore, quality control of mitochondria is crucial. Autophagy is the main form of mitochondria degradation, which is a process that is called mitophagy [3,4]. Mitophagy is an autophagic response that specifically targets damaged and hence potentially cytotoxic mitochondria. Putative kinase 1 (PINK1)-Parkin axis is a well-known regulation pathway in mitophagy [9].

Furthermore, Bcl-2 interacting protein 3 (BNIP3) [10], Bcl2-like protein 13 (Bcl2-L-13) [11], and FUN14 domain containing protein1 (FUNDC1) [12] are reported to be involved in mitophagy regulation [3].

Autophagy is widely implicated in cardiac homeostasis in health and diseases, such as myocardial infarction, myocardial hypertrophy, cardiac fibrosis, cardiomyopathy, and heart failure [13–17]. Autophagy positively or negatively regulates myocardial infarction and myocardial hypertrophy. The modulation of autophagy processes affects cardiac function [13,18–20]. Cardiac fibrosis and heart failure are associated with excessive autophagy. The suppression of autophagy could attenuate fibrosis and improve heart function [16,17]. Additionally, autophagy is implicated in cardiomyopathy. It is reported that autophagy is suppressed in high-fat diet-induced obesity cardiomyopathy [15,21], while autophagy is activated in doxorubicin-induced cardiomyopathy and cardiotoxicity [22].

1.2. MicroRNAs

MicroRNAs (miRNAs) are small non-coding RNAs with a length of approximately 21–22 nucleotides (nt), which are distributed widely in viruses, plants, protists, and animals. They function to mediate post-transcriptional gene silencing. MiRNAs usually exhibit a high conservation level and a very low rate of evolution [23,24]. Mature miRNAs are generated via a two-step processing by Drosha and Dicer. The primary transcripts with a length of hundreds of nucleotides, which are named Pri-miRNAs, are initially generated in the nucleus. The Drasha complex cleaves pri-miRNA into pre-miRNA. Pre-miRNA is transported to the cytoplasm by Exportin-5. In the cytoplasm, pre-miRNA is processed by RNase III Dicer. Eventually, one arm is generated as the mature miRNA with a length of about 22 nt, while the other one is degraded [25–28]. The mature miRNAs are assembled into the ribonucleoprotein (RNP) complex, which are called miRNPs or miRISCs. By interacting with the 3′-untranslated region (UTR), 5′-untranslated region (UTR), or coding sequence (CDS) region of the target genes, miRNAs mediate gene silencing at the post-transcriptional or translational levels. Argonaute proteins (AGOs), which are the key components of miRNPs, compete with translation initiation factors (e.g., elf4E) by binding to their m7G cap to inhibit translation initiation. AGOs also recruit the 60S ribosomal subunit binding protein elF6 to prevent the binding of the 60S subunit, which triggers translation initiation. At the post-initiation stage, either miRNAs or AGOs become associated with polysomes to play a supposed role in elongating inhibition or the dropping off of the ribosome. Moreover, the miRNPs could recruit the CCR4-NOT complex to mediate the deadenylation of mRNAs during the translation process [29,30]. Based on their function in mediating gene silencing, miRNAs are deeply linked to numerous biological processes and related diseases [18,31,32].

2. MicroRNAs Regulate the Core Autophagy Signaling Cascades

Accumulated research has indicated that miRNAs play critical roles in the autophagy processes. Autophagy proceeds in several successive stages, including induction, vesicle nucleation, vesicle elongation, and maturation [31,33]. MiRNAs are deeply implicated in the regulation of the main autophagy signaling cascades [31,34], which usually occurs through modulating the expression of autophagy-related genes (Figure 1).

Autophagy induction is mainly regulated by the unc-51-like kinase (ULK) complex, which is composed of ULK1/2, autophagy-related gene 13 (ATG13), ATG101, and focal adhesion kinase family interacting protein with a mass of 200 kDa (FIP200). The mammalian target of rapamycin (mTOR) is one of the most important upstream regulators for the induction of autophagy. Under normal conditions, the mammalian target of rapamycin complex 1 (mTORC1) interacts with and inactivates the ULK1/2. Once exposed to autophagy-induced factors, such as myocardial ischemia, anoxia, or other myocardial damage stimuli, mTORC1 is dissociated from the ULK complex, which dephosphorylates the ULK1/2. The activated ULK1/2 subsequently phosphorylates ATG13 and TIP200, which leads to the initiation of autophagy [33,34]. It was reported that miR-372, miR-17-5p, and miR-106a inhibit autophagy by down-regulating ULK1 [35–37]. MiR-885-3p was found to directly target ULK2 mRNA to control the cell viability and autophagy [38]. ULK2 is suppressed by miR-26b in a targeted manner, which interrupts the autophagy initiation [39,40]. Additionally, miR-20a and miR-20b negatively regulate autophagy via targeting FIP200 [41].

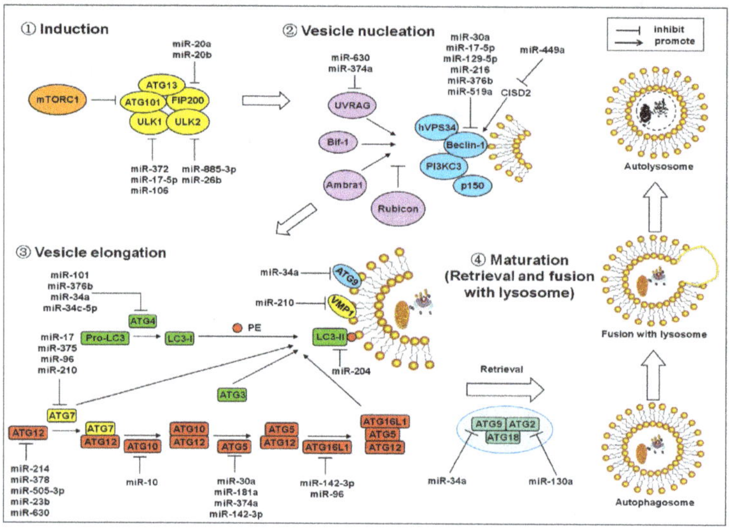

Figure 1. MicroRNAs regulate the core autophagy signaling cascades. Autophagy proceeds in four successive stages including induction, vesicle nucleation, vesicle elongation, and maturation. MicroRNAs are implicated in the processes. See text for detailed explanations. Arrows represent the promotion effect. T bars represent the inhibition effect.

Vesicle nucleation is the step that forms the double membrane-bond vesicles, which are called autophagosomes. The activation of the PI3K complex, which is composed of PI3KC3, hVPS34, Beclin-1, and p150, triggers the vesicle nucleation. In addition, several binding partners including Ambra1, Bif-1, UV irradiation resistance-associated gene (UVRAG), and Rubicon positively or negatively regulate this complex. In response to ample nutrients, the Bcl-2 family anti-apoptotic proteins interact with Beclin-1 and inhibit autophagy [31,42]. It was reported that miR-630 and miR-374a could repress the expression of UVRAG [43]. Regarding the PI3K complex, miR-30a, miR-17-5p, miR-129-5p, miR-216, miR-376b, and miR-519a prevent the autophagy process through directly targeting Beclin-1 [44–48]. MiR-449a indirectly promotes Beclin-1-mediated autophagy via suppressing the CDGSH iron sulfur domain 2 (CISD2) [49]. MiR-449 or miR-146a targeting Bcl-2 is also involved in autophagy regulation [50,51]. ATG12 system and LC3 system, the two ubiquitin-like protein conjugation pathways, are responsible for the vesicle elongation. In the ATG12 pathway, ATG12 interacts with ATG5 by binding ATG7 and ATG10 successively. The ATG12-ATG5 complex finally conjugates with ATG16L1 to form a large multimeric complex. In the LC3 pathway, ATG4 cleaves pro-LC3 to LC3-I. LC3-I is subsequently conjugated to phosphatidylethanolamine (PE) under the catalysis of ATG7 and E2-like ligase ATG3. Mediated by the ATG16L1-ATG5-ATG12 complex, which functions as an E3 ligase, LC3-I is processed to the phagophore membrane-bound LC3-II [31,33]. These two conjugation pathways could be regulated by microRNAs. The ATG12-mediated autophagy is targetedly inhibited by miR-214, miR-378, miR-505-3p, miR-23b, and miR-630 [43,52–55]. MiR-181a/miR-30a/miR-374a targeting ATG5 [43,56,57], miR-142-3p targeting ATG5/ATG16L [58], miR-96 targeting ATG7/ATG16L1 [59], miR-210/miR-17/miR-375 targeting ATG7 [60–62], and miR-20 targeting ATG10 [63] have been demonstrated to regulate autophagy. Additionally, the down-regulation of ATG4B is mediated by miR-34a/34c-5p, which suppresses the rapamycin-induced autophagy [64]. ATG4C and ATG4D are silenced by miR-376b and miR-101, respectively [48,65]. MiR-204 prevents autophagy by targeting LC3-II [66]. VMP1, a critical transmembrane protein for phagophore formation, is a direct target of miR-210 [67].

The maturation of autophagy includes retrieval and fusion with lysosomes. The process is poorly studied in mammals and humans. The retrieval step is mainly mediated by the ATG9-ATG2-AGT18 complex, which recruits lipids and proteins to the growing phagophore. Lastly, autophagosomes fuse to lysosomes, forming autolysosomes where the inner membrane of the former autophagosome and the engulfed cargo are degraded by acid hydrolases [31,68,69]. ATG9A is a direct target of miR-34a in autophagy maturation step [70]. MiR-130a interferes with the ATG9-ATG2-AGT18 complex formation via down-regulating ATG2B [71].

3. The Role of MicroRNAs in Cardiac Autophagy

The importance of autophagy for the preservation of cardiac homeostasis in physiological settings has been demonstrated. When they are subjected to the cardiomyocyte-specific deletion of autophagy-related gene 5 (Atg5), which is required for optimal autophagic responses, mice will develop cardiac hypertrophy, left ventricular dilatation, contractile dysfunction, and premature death. This is accompanied by a disorganized sarcomere structure, mitochondrial misalignment, and aggregation [72–74]. MiR-19a-3p/19b-3p inhibits autophagy in cardiomyocytes by targeting the TGF-β II-Atg 5 pathway [16]. MiR-33 reduces lipid droplet catabolism by negatively regulating Atg5 and miR-214-3p, which reduces the oxidized low-density lipoprotein-initiated autophagy by directly targeting Atg5 mRNA. This ultimately contributes to atherosclerosis [75,76]. In human aortic smooth muscle cells, the forced expression of miR-221/222 silences phosphatase and tensin homolog deleted on chromosome ten (PTEN) and subsequently activates the Akt signaling. This eventually inhibits autophagy by down-regulating the expression of LC3II and ATG5, as well as by elevating the expression of SQSTM1/p62 [77]. MiR-30b disrupts autophagy in vascular smooth muscle cells by decreasing autophagy-related genes, such as Atg5 and LC3II [78]. MiR-212 and miR-132 inhibit autophagy by negatively regulating the pro-autophagic transcription factor forkhead box O3 (FOXO3). The mice engineered to overexpress miR-212 and miR-132 in cardiomyocytes prematurely succumb because of pathological cardiac hypertrophy and heart failure [20,79]. Another critical microRNA that regulates cardiac autophagy is miR-199a. MiR-199a represses autophagy by indirectly activating the mechanistic target of rapamycin complex 1 (mTORC1), which leads to cardiac hypertrophy. In this model, tissue degeneration could be partially reversed by a potent autophagy inducer, rapamycin [14]. In addition, the silencing of miR-143 promotes the autophagy of the c-kit$^+$ cardiac progenitor cells in response to oxidative stress. Autophagy-related gene 7 (Atg7) is identified as the target gene of miR-143 in the autophagy pathway [80].

Mitophagy, which is responsible for the quality control of mitochondria to match the metabolic or developmental demands, is critical for cardiac homeostasis. Mitophagy is usually executed by a series of mitochondria functional molecules, such as Bcl2 interacting protein 3 (BNIP3L, best known as NIX), FUNDC1, Parkin, PTEN-induced putative kinase 1 (PINK 1), and mitofusin 2 (MFN2) [81–84]. MicroRNAs have been found to be involved in the mitophagy flux. MiR-137 markedly inhibits mitophagy without affecting global autophagy in response to hypoxia in neurocytes. A further exploration of the mechanism shows that miR-137 targets two mitophagy receptors, FUNDC1 and NIX, and downregulates their expression. The suppression of mitophagy mediated by miR-137 could be reversed by the forced expression of FUNDC1 and NIX [85]. MiR-27a and miR-27b prevent mitophagic influx by suppressing PINK1 expression at the translational level, which subsequently decreases ubiquitin phosphorylation, Parkin translocation, and LC3 II accumulation in damaged mitochondria. Furthermore, this inhibits the lysosomal degradation of the damaged mitochondria in neurocytes [86]. MiR-181a blocks the colocalization of mitochondria and autophagosomes/lysosomes by targeting Parkin. The overexpression of miR-181a inhibits the mitochondrial uncoupling agents, which induces mitophagy without affecting global autophagy. In contrast, the knockdown of miR-181 accelerates the mitophagy in neuroblastoma cells [87]. The role of microRNAs in cardiac mitophagy has been rarely reported. In the cardiac system, miR-410 inhibits the excessive mitophagy from cardiac ischemia/reperfusion injuries by modulating the heat shock protein B1 activity via a direct interaction with the 3′-untranslated region of high-mobility group box 1 protein [88].

The degradation of damaged cells, organelles, or intracellular components by autophagy has been demonstrated to prevent the aging of cardiomyocytes. It was found that the senescent myocardium is associated with impaired autophagy. The induction of autophagy could reverse age-dependent cardiac hypertrophy and diastolic dysfunction in old mice [89,90]. During the processes, AMP-activated protein kinase (AMPK) strongly promotes an autophagy flux, which consists of unc-51 like autophagy activating kinase (ULK) 1, Beclin 1, and phosphatidylinositol 3-kinase catalytic subunit type 3 (PIK3C3) [72,91–93]. MiR-20a and miR-106b reduce autophagy via the suppression of the ULK1 expression in myoblasts. The cells with the knockdown of endogenous miR-20a and miR-106b exhibit a normal autophagy activity [94]. MiR-19a-3p/19b-3p targeting the Beclin 1 pathway has been demonstrated to inhibit autophagy in cardiomyocytes [16]. MiR-221/222 inhibits autophagy by indirectly regulating Beclin 1 [77]. In vascular smooth muscle cells, Beclin 1 is a target of miR-30b to mediate autophagy [78]. MiR-155 plays a beneficial role in the aging myocardium by negatively regulating the autophagy-inhibitor, mTOR [95].

In summary, autophagy and mitophagy are deeply implicated in the cardiac homeostasis processes, including heart development, compensation, and aging, which could be widely regulated by microRNAs.

4. MicroRNAs in Autophagy-Related Heart Diseases

Autophagy is widely involved in cardiac physiology, and impaired autophagy or abnormal autophagy usually contribute to cardiac pathological disorders [96]. Recent research has suggested a close link between autophagy and heart diseases, including myocardial infarction [18], ischemia-reperfusion injury [66], myocardial hypertrophy [97], cardiac fibrosis [16], cardiomyopathy [98], and heart failure [17]. During the above processes, a variety of microRNAs have been demonstrated to play a critical role (Figure 2). MicroRNAs participate in regulating heart diseases through modulating the cardiac autophagic cell death. Many critical autophagy-related genes, such as ATG7, ATG9, LC3, p62, Beclin-1, AMPK, mTOR, BAG3, TNFα, and PARP-1, have been demonstrated to be direct or indirect targets of microRNAs in the autophagy-signaling pathways, which are involved in the pathogenesis of heart diseases [14,18,19,99–105].

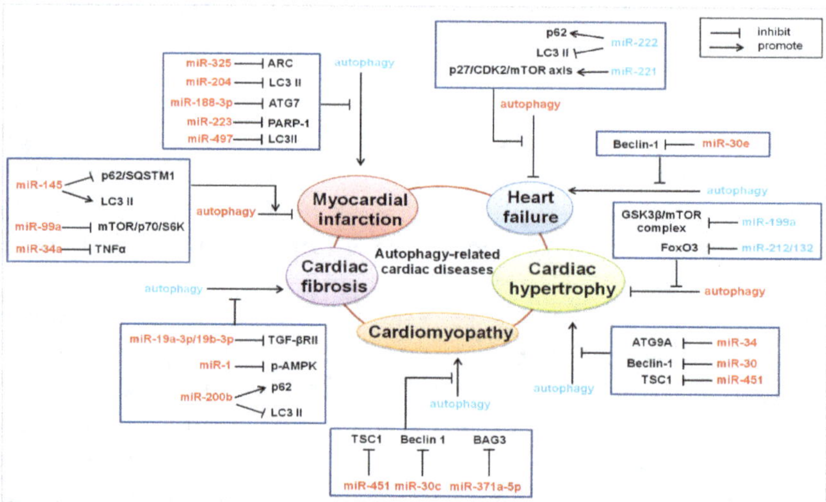

Figure 2. Summary of miRNAs regulating cardiac autophagy and related heart diseases. The molecular mechanisms of the miRNAs regulating the autophagy and autophagy-related heart diseases are shown in the figure. See text for detailed explanations. Arrows represent the promotion effect. T bars represent the inhibition effect. Red represents the negative regulation in diseases. Blue represents the positive regulation in diseases.

4.1. Myocardial Infarction

Myocardial infarction (MI) is pathologically defined as myocardial cell death due to prolonged ischemia or anoxia, which is the most severe manifestation of coronary heart disease. Cardiac cell death plays a decisive role in the pathogenesis of MI, because of the terminal differentiation and loss of regenerative ability of cardiomyocytes [106]. Autophagic cell death has been demonstrated to greatly contribute to the pathogenesis of MI. As a class of important autophagy regulators, microRNAs promote or inhibit MI by mediating the autophagy pathway.

A famous microRNA that is involved in autophagic cell death and MI is miR-325. MiR-325 is upregulated both in the cardiomyocytes treated with anoxia/reoxygenation, and in the hearts subjected to ischemia/reperfusion injury. The cardiomyocyte-specific overexpression of miR-325 potentiates excessive autophagic responses and myocardial infarct sizes, whereas the knockdown of miR-325 inhibited the autophagic cell death and MI. The apoptosis repressor with caspase recruit domain (ARC) is identified as the downstream mediator and the transcription factor, E2F1, as the upstream regulator of miR-325 in the autophagy program. A novel autophagic regulating model composed of E2F1, miR-325, and ARC in MI has been clarified in the literature [107]. The outstanding role of miR-30e in autophagic cell death and related MI has been demonstrated. MiR-30e protects hearts from MI. The cardioprotective mechanism of miR-30e has been explored. The level of miR-30e is dramatically decreased in the animal models of MI. The silencing of miR-30e significantly inhibits cellular apoptosis by modulating the apoptosis-related gene, Bax, and caspase-3, and meanwhile activates the autophagic flux and Notch1/Hes1/Akt signaling pathway. The effect of the knockdown of miR-30e on apoptosis and oxidative stress damage could be reversed significantly by autophagy inhibitor 3-methyladenine [108]. MiR-34a could inhibit autophagic cell death in the hearts that are subjected to ischemia/reperfusion via regulating tumor necrosis factor α (TNFα), thereby reducing myocardial injury [104]. MiR-145 plays a cardioprotective effect in myocardial infarction by promoting cardiac autophagy. Rabbits administrated with miR-145 mimics exhibit a significantly smaller infarct size and improved cardiac function than that of the control group upon ischemia/reperfusion injury. Further study shows that miR-145 promotes autophagic cell death in cardiomyocytes by directly targeting fibroblast growth factor receptor substrate 2 (FRS2) mRNA, and it subsequently accelerates the transition of LC3B I to II and down-regulates p62/SQSTM1, which inhibits MI [19]. MiR-223 is up-regulated in rat hearts that have undergone coronary ligation. The overexpression of miR-223 inhibits excessive cardiac autophagy. Further mechanistic study reveals that miR-223 protects cardiomyocytes from excessive autophagy via the Akt/mTOR pathway by targeting poly (ADP-ribose) polymerase 1 (PARP-1) [105]. MiR-188-3p is down-regulated in the myocardial infarction model. The forced expression of miR-188-3p suppresses autophagic cell death by targeting ATG7, which subsequently reduces the infarct sizes and improves cardiac function [18]. MiR-204 could inhibit autophagy by targeting LC3 II upon ischemia-reperfusion injury [66]. MiR-99a plays a cardioprotective role in the post-MI left ventricle remodeling by preventing cell apoptosis and increasing autophagy via an mTOR/p70/S6K pathway, which improves both the cardiac function and survival rate in a murine model of MI [109]. Myocardial miR-497 is dramatically down-regulated in the murine hearts subjected to MI, and in the cardiomyocytes subjected to hypoxia/reoxygenation. The overexpression of miR-497 induces apoptosis by targeting the anti-apoptosis gene, Bcl-2, and inhibits autophagy by targeting the autophagy gene, LC3BII, while the silencing of miR-497 exhibits the opposite effect in response to MI [110].

4.2. Cardiac Hypertrophy

A growing number of studies have revealed a critical pathogenic role of the altered activity of autophagy in cardiac hypertrophy. It was reported that cardiac autophagy was increased in a maladaptive manner in the hearts that were subjected to a pressure overload. However, some other studies have suggested that myocardial autophagy is insufficient, which results from a chronic pressure overload and contributes to maladaptive cardiac remodeling and heart failure [111]. MicroRNAs have been demonstrated to participate in regulating autophagic cell death in cardiac hypertrophy. In angiotensin-induced cardiac hypertrophy, miR-30 and miR-34 play an anti-hypertrophy role by preventing autophagy [97,99]. MiR-451 has been demonstrated to be involved in suppressing the autophagosome formation by targeting tuberous sclerosis complex1 (TSC1), which inhibits abnormal autophagy upon hypertrophic stimuli. By controlling the autophagy process, the ectopic overexpression of miR-451 attenuates cardiac hypertrophy, while the knockdown of miR-451 accelerates hypertrophy [112]. The MiR-212/132 family promotes pathological cardiac hypertrophy and heart failure by directly targeting the anti-hypertrophic and pro-autophagic transcription factor, FoxO3. The MiR-212/132 null mice do not frequently develop hypertrophy and heart failure, whereas the overexpression of miR-212/132 leads to the hyperactivation of pro-hypertrophic calcineurin/NFAT signaling and an impaired autophagic response in response to starvation [20]. Cardiac specific miR-199a transgenic mice suffer cardiac hypertrophy, accompanied with decreased autophagy levels. The enhancement of autophagy by the forced expression of Atg5 attenuates the hypertrophic effects of overexpression of miR-199a on cardiomyocytes. In exploring the molecular mechanism, miR-199a has been demonstrated to target the glycogen synthase kinase 3β (GSK3β)/mTOR complex signaling pathway in modulating autophagy and cardiac hypertrophy [14].

4.3. Cardiac Fibrosis

A link between autophagy and cardiac fibroblasts has been demonstrated recently. However, the exploration on the molecular mechanism, including the role of microRNAs in autophagy-related cardiac fibrosis, is insufficient. It was reported that miR-19a-3p/19b-3p could inhibit epithelial mesenchymal transition, extracellular matrix production, and the invasion of human cardiac fibroblasts by targeting TGF-β RII. Moreover, the enhancement of autophagy rescues the inhibition effect of miR-19a-3p/19b-3p on cardiac fibroblasts. These results suggest that miR-19a-3p/19b-3p exhibits an anti-fibroblast effect through regulating cardiac autophagy [16]. MiR-1 promotes high glucose-induced cardiac fibrosis by down-regulating pro-autophagic p-AMPK [102]. MiR-200b controls cardiac fibroblast autophagy during cardiac fibrosis. The expression level of miR-200b is decreased in the cardiac fibrosis model. The MiR-200b mimic inhibits the LC3BII/I ratio, increases p62, and alleviates cardiac fibroblast autophagy, whereas the knockdown of miR-200b exhibits an opposite effect [113].

4.4. Cardiomyopathy

MicroRNA-mediated autophagy is involved in the pathogenesis of several cardiomyopathy. Until now, miR-30c, miR-371a-5p, and miR-451 have been precisely demonstrated to regulate cardiac autophagy in cardiomyopathy [98,103,112]. The depletion of miR-30c enhances autophagic cell death in diabetic hearts. The overexpression of miR-30c inhibits autophagy in diabetic hearts and subsequently improves cardiac function and structure in the diabetic mice model. In exploring the molecular mechanisms, it was found that miR-30c suppressed autophagy in diabetic cardiomyopathy via targeting BECN1, through direct binding to BECN1 3′ UTR [98]. Bcl-2-associated athanogene 3 (BAG3), an autophagy pathway mediator, is a direct target of miR-371a-5p in healthy donors. Mutations and polymorphisms, including a frequent nucleotide change g2252c in the BAG3 3′-untranslated region (3′-UTR) of Takotsubo patients, leads to the loss of binding to miR-371a-5p, which probably contributes to the Takotsubo cardiomyopathy (TTC) pathogenesis [103]. MiR-451 is down-regulated in the heart tissues from hypertrophic cardiomyopathy (HCM) patients. The ectopic overexpression of miR-451 inhibits the autophagosome formation through targetedly suppressing tuberous sclerosis complex 1 (TSC1), and subsequently decreasing the size of the cardiomyocytes [112].

4.5. Heart Failure

Emerging evidence confirms that restoring autophagy, which improves bulk protein degradation, is proved to be beneficial in heart failure [5]. The cardioprotective effect of autophagy on heart failure could be mediated by microRNAs. MiR-221 has emerged as a representative regulator of autophagy-mediated heart failure. Research suggests that the cardiac-specific overexpression of miR-221 in mice leads to cardiac dysfunction and heart failure, and reduces the autophagic flux by inhibiting the autophagic vesicle formation in the meanwhile. Further study showed that miR-221 inhibits autophagy and promotes heart failure by modulating the p27/CDK2/mTOR axis [17]. MiR-222, which shares the same gene cluster with miR-221, exhibits a potential function in autophagic cell death and heart failure. Transgenic mice with a cardiac-specific expression of miR-222 significantly develop heart failure and are accompanied with autophagy inhibition. MiR-221 downregulates LC3 II, upregulates p62, and activates the mTOR pathway, all of which are critical autophagy regulators [100]. Another microRNA regulating cardiac autophagy in heart failure is miR-30e. MiR-30e inhibits the autophagy of cardiomyocytes by down-regulating the Beclin-1 expression, and subsequently mediates the cardioprotection of the angiotensin-converting enzyme 2 (ACE2) in the rats with Dox-induced heart failure [101].

In summary, microRNAs targeting the autophagy signaling pathways in cardiac disorders have been demonstrated by numerous evidence. The autophagic cell death mediated by microRNAs plays a critical role in the pathogenesis of heart diseases.

5. Conclusions and Perspectives

As outlined in this review, miRNAs function as pro- or anti-regulators in cardiac autophagy, by targeting extensive signaling pathways. This knowledge has improved our understanding of the role and mechanism of autophagy in cardiac diseases. Mitophagy is one critical part of the autophagy processes, which plays a decisive role in cardiac physiology and pathology. However, studies on the role of microRNAs in cardiac mitophagy are greatly insufficient, and thus more explorations need to be carried out in the future. Considering that, usually, one miRNA targets several genes or several miRNAs target one gene in autophagy regulation, the crosslink between the different pathways needs to be further explored. Autophagy is an evolutionarily conserved self-protective process and plays a critical role in maintaining cell homeostasis. However, excessive autophagy usually contributes to the pathogenesis of a variety of diseases, including cardiac disorders. Up until now, it has been a mystery as to how beneficial autophagy turns into a harmful autophagic cell death. With respect to MI as well as cardiac hypertrophy and heart failure, the miRNAs-regulated autophagy probably functions as the promoter or inhibitor in the pathogenesis. For instance, miR-145 plays a cardioprotective effect in MI by promoting cardiac autophagy [19], while miR-188-3p inhibits MI via inactivating the autophagy pathway [18]. MiR-451 functions as a cardiac hypertrophy inhibitor by activating autophagy [112], while the knockdown of miR-199a suppresses cardiac hypertrophy by promoting autophagic cell death [14]. MiR-221 inhibits autophagy and promotes heart failure [100], while miR-30e plays a cardioprotective role in heart failure by modulating autophagy [101]. Whether miRNA-regulated autophagy plays the role of an angel or devil in cardiac pathology, remains controversial. Cardiac cell death is the cytological basis of cardiac disorders. Apart from autophagy, several other types of cell death, such as apoptosis, are involved in the cardiac pathology. In the exploration of miRNA regulating cardiac autophagy, the apoptosis pathway is in crosslink with the autophagy pathway. For example, miR-223, miR-99a, and miR-30e regulate both the cardiac autophagy process and the apoptosis process [101,105,109]. Therefore, research focusing on the relationship of different types of cardiac cell death is necessary in the future. The important role of microRNAs in regulating cardiac autophagy has been demonstrated by accumulated evidence from studies. However, there are still many fuzzy and controversial issues that need to be explored in the future.

Author Contributions: Conceptualization, T.S. and J.-M.C.; Methodology, T.S. and M.-Y.L.; Software, T.S. and M.-Y.L.; Validation, T.S. and J.-M.C.; Formal Analysis, T.S., P.-F.L. and J.-M.C.; Investigation, T.S., M.-Y.L. and P.-F.L.; Resources, T.S.; Data Curation, T.S. and J.-M.C.; Writing-Original Draft Preparation, T.S., M.-Y.L. and J.-M.C.; Writing-Review & Editing, T.S. and J.-M.C.; Visualization, T.S. and J.-M.C.; Supervision, T.S. and J.-M.C.; Project Administration, T.S. and J.-M.C.; Funding Acquisition, T.S. and J.-M.C.

Acknowledgments: This work was supported by the China Postdoctoral Science Foundation (2016M592134), the Qingdao Postdoctoral Application Research Project (2016074), the National Natural Science Foundation of China (81670313), and the fund for Shanxi '1331 Project' Key Subjects Construction (1331KSC).

Abbreviations

ACE2	angiotensin-converting enzyme 2
AGOs	argonaute proteins
AMPK	AMP- activated protein kinase
ARC	apoptosis repressor with caspase recruit domain
ATG	autophagy-related gene
BAG3	Bcl-2-associated athanogene 3
Bcl2-L-13	Bcl2-like protein 13
BNIP3	Bcl-2 interacting protein 3
BNIP3L	Bcl2 interacting protein 3 like
CDS	coding sequence
CISD2	CDGSH iron sulfur domain 2
FIP200	focal adhesion kinase family interacting protein of 200 kDa
FOXO3	forkhead box O3
FUNDC1	FUN14 domain containing protein 13
GSK3β	glycogen synthase kinase 3β
HCM	hypertrophic cardiomyopathy
MFN2	mitofusion2
miRNAs	microRNAs
mTOR	mammalian target of rapamycin
mTORC1	mTOR complex 1
PARP-1	poly (ADP-ribose) polymerase 1
PE	phosphatidylethanolamine
PIK3C3	phosphatidylinositol 3-kinase catalytic subunit type 3
PINK1	putative kinase 1
PTEN	phosphatase and tensin homolog deleted on chromosome ten
TSC1	tuberous sclerosis complex 1
TTC	Takatsubo cardiomyopathy
ULK	Unc-51-like kinase
UTR	untranslated region
UVRAG	UV irradiation resistance-associated gene

References

1. He, C.; Klionsky, D.J. Regulation mechanisms and signaling pathways of autophagy. *Annu. Rev. Genet.* **2009**, *43*, 67–93. [CrossRef] [PubMed]
2. Legakis, J.E.; Yen, W.L.; Klionsky, D.J. A cycling protein complex required for selective autophagy. *Autophagy* **2007**, *3*, 422–432. [CrossRef] [PubMed]
3. Kiriyama, Y.; Nochi, H. Intra- and intercellular quality control mechanisms of mitochondria. *Cells* **2017**, *7*, 1. [CrossRef] [PubMed]
4. Harper, J.W.; Ordureau, A.; Heo, J.M. Building and decoding ubiquitin chains for mitophagy. *Nat. Rev. Mol. Cell Biol.* **2018**, *19*, 93–108. [CrossRef] [PubMed]
5. Ghosh, R.; Pattison, J.S. Macroautophagy and chaperone-mediated autophagy in heart failure: The known and the unknown. *Oxid. Med. Cell. Longev.* **2018**, *2018*. [CrossRef] [PubMed]
6. Liu, H.; Dai, C.; Fan, Y.; Guo, B.; Ren, K.; Sun, T.; Wang, W. From autophagy to mitophagy: The roles of p62 in neurodegenerative diseases. *J. Bioenerg. Biomembr.* **2017**, *49*, 413–422. [CrossRef] [PubMed]
7. Yan, C.; Li, T.S. Dual role of mitophagy in cancer drug resistance. *Anticancer Res.* **2018**, *38*, 617–621. [PubMed]
8. Herst, P.M.; Rowe, M.R.; Carson, G.M.; Berridge, M.V. Functional mitochondria in health and disease. *Front. Endocrinol.* **2017**, *8*. [CrossRef] [PubMed]

9. Cummins, N.; Gotz, J. Shedding light on mitophagy in neurons: What is the evidence for pink1/parkin mitophagy in vivo? *Cell. Mol. Life Sci.* **2018**, *75*, 1151–1162. [CrossRef] [PubMed]
10. Zhang, J.; Liu, L.; Xue, Y.; Ma, Y.; Liu, X.; Li, Z.; Li, Z.; Liu, Y. Endothelial monocyte-activating polypeptide-ii induces bnip3-mediated mitophagy to enhance temozolomide cytotoxicity of glioma stem cells via down-regulating mir-24-3p. *Front. Mol. Neurosci.* **2018**, *11*, 92. [CrossRef] [PubMed]
11. Murakawa, T.; Yamaguchi, O.; Hashimoto, A.; Hikoso, S.; Takeda, T.; Oka, T.; Yasui, H.; Ueda, H.; Akazawa, Y.; Nakayama, H.; et al. Bcl-2-like protein 13 is a mammalian atg32 homologue that mediates mitophagy and mitochondrial fragmentation. *Nat. Commun.* **2015**, *6*, 7527. [CrossRef] [PubMed]
12. Zhou, H.; Zhu, P.; Wang, J.; Zhu, H.; Ren, J.; Chen, Y. Pathogenesis of cardiac ischemia reperfusion injury is associated with ck2alpha-disturbed mitochondrial homeostasis via suppression of fundc1-related mitophagy. *Cell Death Differ.* **2018**, *25*, 1080. [CrossRef] [PubMed]
13. Liu, C.Y.; Zhang, Y.H.; Li, R.B.; Zhou, L.Y.; An, T.; Zhang, R.C.; Zhai, M.; Huang, Y.; Yan, K.W.; Dong, Y.H.; et al. Lncrna caif inhibits autophagy and attenuates myocardial infarction by blocking *p*53-mediated myocardin transcription. *Nat. Commun.* **2018**, *9*, 29. [CrossRef] [PubMed]
14. Li, Z.; Song, Y.; Liu, L.; Hou, N.; An, X.; Zhan, D.; Li, Y. Mir-199a impairs autophagy and induces cardiac hypertrophy through mtor activation. *Cell Death Differ.* **2017**, *24*, 1205–1213. [CrossRef] [PubMed]
15. Wang, S.; Wang, C.; Turdi, S.; Richmond, K.L.; Zhang, Y.; Ren, J. ALDH2 protects against high fat diet-induced obesity cardiomyopathy and defective autophagy: Role of CaM kinase II, histone H3K9 methyltransferase SUV39H, Sirt1, and PGC-1α deacetylation. *Int. J. Obes.* **2018**, *42*, 1073–1087. [CrossRef] [PubMed]
16. Zou, M.; Wang, F.; Gao, R.; Wu, J.; Ou, Y.; Chen, X.; Wang, T.; Zhou, X.; Zhu, W.; Li, P.; et al. Autophagy inhibition of hsa-miR-19a-3p/19b-3p by targeting TGF-β R II during TGF-β1-induced fibrogenesis in human cardiac fibroblasts. *Sci. Rep.* **2016**, *6*, 24747. [CrossRef] [PubMed]
17. Su, M.; Wang, J.; Wang, C.; Wang, X.; Dong, W.; Qiu, W.; Wang, Y.; Zhao, X.; Zou, Y.; Song, L.; et al. MicroRNA-221 inhibits autophagy and promotes heart failure by modulating the p27/CDK2/mTOR axis. *Cell Death Differ.* **2015**, *22*, 986–999. [CrossRef] [PubMed]
18. Wang, K.; Liu, C.Y.; Zhou, L.Y.; Wang, J.X.; Wang, M.; Zhao, B.; Zhao, W.K.; Xu, S.J.; Fan, L.H.; Zhang, X.J.; et al. APF lncRNA regulates autophagy and myocardial infarction by targeting miR-188-3p. *Nat. Commun.* **2015**, *6*, 6779. [CrossRef] [PubMed]
19. Higashi, K.; Yamada, Y.; Minatoguchi, S.; Baba, S.; Iwasa, M.; Kanamori, H.; Kawasaki, M.; Nishigaki, K.; Takemura, G.; Kumazaki, M.; et al. MicroRNA-145 repairs infarcted myocardium by accelerating cardiomyocyte autophagy. *Am. J. Physiology-Heart Circ. Physiol.* **2015**, *309*, H1813–H1826. [CrossRef] [PubMed]
20. Ucar, A.; Gupta, S.K.; Fiedler, J.; Erikci, E.; Kardasinski, M.; Batkai, S.; Dangwal, S.; Kumarswamy, R.; Bang, C.; Holzmann, A.; et al. The miRNA-212/132 family regulates both cardiac hypertrophy and cardiomyocyte autophagy. *Nat. Commun.* **2012**, *3*, 1078. [CrossRef] [PubMed]
21. Chen, Z.; Li, Y.; Wang, Y.; Qian, J.; Ma, H.; Wang, X.; Jiang, G.; Liu, M.; An, Y.; Ma, L.; et al. Cardiomyocyte-restricted low density lipoprotein receptor-related protein 6 (LRP6) deletion leads to lethal dilated cardiomyopathy partly through Drp1 signaling. *Theranostics* **2018**, *8*, 627–643. [CrossRef] [PubMed]
22. Carresi, C.; Musolino, V.; Gliozzi, M.; Maiuolo, J.; Mollace, R.; Nucera, S.; Maretta, A.; Sergi, D.; Muscoli, S.; Gratteri, S.; et al. Anti-oxidant effect of bergamot polyphenolic fraction counteracts doxorubicin-induced cardiomyopathy: Role of autophagy and c-kit^{pos}CD45^{neg}CD31^{neg} cardiac stem cell activation. *J. Mol. Cell. Cardiol.* **2018**, *119*, 10–18. [CrossRef] [PubMed]
23. Ambros, V. The functions of animal microRNAs. *Nature* **2004**, *431*, 350–355. [CrossRef] [PubMed]
24. Bartel, D.P. Micrornas: Genomics, biogenesis, mechanism, and function. *Cell* **2004**, *116*, 281–297. [CrossRef]
25. Lee, Y.; Ahn, C.; Han, J.; Choi, H.; Kim, J.; Yim, J.; Lee, J.; Provost, P.; Radmark, O.; Kim, S.; et al. The nuclear RNase III drosha initiates microRNA processing. *Nature* **2003**, *425*, 415–419. [CrossRef] [PubMed]
26. Han, J.; Lee, Y.; Yeom, K.H.; Nam, J.W.; Heo, I.; Rhee, J.K.; Sohn, S.Y.; Cho, Y.; Zhang, B.T.; Kim, V.N. Molecular basis for the recognition of primary micrornas by the Drosha-DGCR8 complex. *Cell* **2006**, *125*, 887–901. [CrossRef] [PubMed]
27. Lund, E.; Guttinger, S.; Calado, A.; Dahlberg, J.E.; Kutay, U. Nuclear export of microRNA precursors. *Science* **2004**, *303*, 95–98. [CrossRef] [PubMed]
28. Hutvagner, G.; McLachlan, J.; Pasquinelli, A.E.; Balint, E.; Tuschl, T.; Zamore, P.D. A cellular function for

the RNA-interference enzyme Dicer in the maturation of the let-7 small temporal RNA. *Science* **2001** *293* 834–838. [CrossRef] [PubMed]

29. Filipowicz, W.; Bhattacharyya, S.N.; Sonenberg, N. Mechanisms of post-transcriptional regulation by microRNAs: Are the answers in sight? *Nat. Rev. Genet.* **2008**, *9*, 102–114. [CrossRef] [PubMed]
30. Eulalio, A.; Huntzinger, E.; Izaurralde, E. Getting to the root of miRNA-mediated gene silencing. *Cell* **2008**, *132*, 9–14. [CrossRef] [PubMed]
31. Su, Z.; Yang, Z.; Xu, Y.; Chen, Y.; Yu, Q. Micrornas in apoptosis, autophagy and necroptosis. *Oncotarget* **2015**, *6*, 8474–8490. [CrossRef] [PubMed]
32. Llorens, F.; Thune, K.; Marti, E. Regional and subtype-dependent miRNA signatures in sporadic Creutzfeldt-Jakob disease are accompanied by alterations in miRNA silencing machinery and biogenesis. *PloS Pathog.* **2018**, *14*, e1006802. [CrossRef] [PubMed]
33. Zhou, L.; Ma, B.; Han, X. The role of autophagy in angiotensin II induced pathological cardiac hypertrophy. *J. Mol. Endocrinol.* **2016**, *57*, R143–R152. [CrossRef] [PubMed]
34. Frankel, L.B.; Lund, A.H. Microrna regulation of autophagy. *Carcinogenesis* **2012**, *33*, 2018–2025. [CrossRef] [PubMed]
35. Chen, H.; Zhang, Z.; Lu, Y.; Song, K.; Liu, X.; Xia, F.; Sun, W. Downregulation of UKL1 by microRNA-372 inhibits the survival of human pancreatic adenocarcinoma cells. *Cancer Sci.* **2017**, *108*, 1811–1819. [CrossRef] [PubMed]
36. Duan, X.; Zhang, T.; Ding, S.; Wei, J.; Su, C.; Liu, H.; Xu, G. MicroRNA-17-5p modulates bacille calmette-guerin growth in RAW264.7 cells by targeting ULK1. *PloS ONE* **2015**, *10*, e0138011. [CrossRef] [PubMed]
37. Rothschild, S.I.; Gautschi, O.; Batliner, J.; Gugger, M.; Fey, M.F.; Tschan, M.P. MicroRNA-106a targets autophagy and enhances sensitivity of lung cancer cells to Src inhibitors. *Lung Cancer* **2017**, *107*, 73–83. [CrossRef] [PubMed]
38. Huang, Y.; Chuang, A.Y.; Ratovitski, E.A. Phospho-ΔNp63α/miR-885–3p axis in tumor cell life and cell death upon cisplatin exposure. *Cell Cycle* **2011**, *10*, 3938–3947. [CrossRef] [PubMed]
39. Li, Z.; Li, J.; Tang, N. Long noncoding RNA Malat1 is a potent autophagy inducer protecting brain microvascular endothelial cells against oxygen-glucose deprivation/reoxygenation-induced injury by sponging miR-26b and upregulating ULK2 expression. *Neuroscience* **2017**, *354*, 1–10. [CrossRef] [PubMed]
40. John Clotaire, D.Z.; Zhang, B.; Wei, N.; Gao, R.; Zhao, F.; Wang, Y.; Lei, M.; Huang, W. Mir-26b inhibits autophagy by targeting ULK2 in prostate cancer cells. *Biochem. Biophys. Res. Commun.* **2016**, *472*, 194–200. [CrossRef] [PubMed]
41. Li, S.; Qiang, Q.; Shan, H.; Shi, M.; Gan, G.; Ma, F.; Chen, B. MiR-20a and miR-20b negatively regulate autophagy by targeting RB1CC1/FIP200 in breast cancer cells. *Life Sci.* **2016**, *147*, 143–152. [CrossRef] [PubMed]
42. Cheng, Y.; Ren, X.; Hait, W.N.; Yang, J.M. Therapeutic targeting of autophagy in disease: Biology and pharmacology. *Pharmacol. Rev.* **2013**, *65*, 1162–1197. [CrossRef] [PubMed]
43. Huang, Y.; Guerrero-Preston, R.; Ratovitski, E.A. Phospho-ΔNp63α-dependent regulation of autophagic signaling through transcription and micro-RNA modulation. *Cell Cycle* **2012**, *11*, 1247–1259. [CrossRef] [PubMed]
44. Chen, J.; Yu, Y.; Li, S.; Liu, Y.; Zhou, S.; Cao, S.; Yin, J.; Li, G. Micro RNA-30a ameliorates hepatic fibrosis by inhibiting Beclin1-mediated autophagy. *J. Cell. Mol. Med.* **2017**, *21*, 3679–3692. [CrossRef] [PubMed]
45. Hou, W.; Song, L.; Zhao, Y.; Liu, Q.; Zhang, S. Inhibition of Beclin-1-mediated autophagy by microRNA-17-5p enhanced the radiosensitivity of glioma cells. *Oncol. Res. Featur. Preclin. Clin. Cancer Ther.* **2017**, *25*, 43–53. [CrossRef] [PubMed]
46. Zhao, K.; Zhang, Y.; Kang, L.; Song, Y.; Wang, K.; Li, S.; Wu, X.; Hua, W.; Shao, Z.; Yang, S.; et al. Methylation of microRNA-129-5p modulates nucleus pulposus cell autophagy by targeting Beclin-1 in intervertebral disc degeneration. *Oncotarget* **2017**, *8*, 86264–86276. [CrossRef] [PubMed]
47. Zhang, X.; Shi, H.; Lin, S.; Ba, M.; Cui, S. MicroRNA-216a enhances the radiosensitivity of pancreatic cancer cells by inhibiting beclin-1-mediated autophagy. *Oncol. Rep.* **2015**, *34*, 1557–1564. [CrossRef] [PubMed]
48. Korkmaz, G.; le Sage, C.; Tekirdag, K.A.; Agami, R.; Gozuacik, D. miR-376b controls starvation and mTOR inhibition-related autophagy by targeting ATG4C and BECN1. *Autophagy* **2012**, *8*, 165–176. [CrossRef] [PubMed]

49. Sun, A.G.; Meng, F.G.; Wang, M.G. Cisd2 promotes the proliferation of glioma cells via suppressing beclin1mediated autophagy and is targeted by microRNA449a. *Mol. Med. Rep.* **2017**, *16*, 7939–7948. [CrossRef] [PubMed]
50. Han, R.; Ji, X.; Rong, R.; Li, Y.; Yao, W.; Yuan, J.; Wu, Q.; Yang, J.; Yan, W.; Han, L.; et al. MiR-449a regulates autophagy to inhibit silica-induced pulmonary fibrosis through targeting Bcl2. *J. Mol. Med.* **2016**, *94*, 1267–1279. [CrossRef] [PubMed]
51. Chen, G.; Gao, X.; Wang, J.; Yang, C.; Wang, Y.; Liu, Y.; Zou, W.; Liu, T. Hypoxia-induced microRNA-146a represses Bcl-2 through Traf6/IRAK1 but not Smad4 to promote chondrocyte autophagy. *Biol. Chem.* **2017**, *398*, 499–507. [CrossRef] [PubMed]
52. Hu, J.L.; He, G.Y.; Lan, X.L.; Zeng, Z.C.; Guan, J.; Ding, Y.; Qian, X.L.; Liao, W.T.; Ding, Y.Q.; Liang, L. Inhibition of ATG12-mediated autophagy by miR-214 enhances radiosensitivity in colorectal cancer. *Oncogenesis* **2018**, *7*, 16. [CrossRef] [PubMed]
53. Tan, D.; Zhou, C.; Han, S.; Hou, X.; Kang, S.; Zhang, Y. MicroRNA-378 enhances migration and invasion in cervical cancer by directly targeting autophagy-related protein 12. *Mol. Med. Rep.* **2018**, *17*, 6319–6326. [CrossRef] [PubMed]
54. Yang, K.; Yu, B.; Cheng, C.; Cheng, T.; Yuan, B.; Li, K.; Xiao, J.; Qiu, Z. *Mir505-3p* regulates axonal development via inhibiting the autophagy pathway by targeting *Atg12*. *Autophagy* **2017**, *13*, 1679–1696. [CrossRef] [PubMed]
55. Sun, L.; Liu, A.; Zhang, J.; Ji, W.; Li, Y.; Yang, X.; Wu, Z.; Guo, J. miR-23b improves cognitive impairments in traumatic brain injury by targeting ATG12-mediated neuronal autophagy. *Behav. Brain Res.* **2018**, *340*, 126–136. [CrossRef] [PubMed]
56. Yang, J.; He, Y.; Zhai, N.; Ding, S.; Li, J.; Peng, Z. MicroRNA-181a inhibits autophagy by targeting Atg5 in hepatocellular carcinoma. *Front. Biosci. (Landmark edition)* **2018**, *23*, 388–396.
57. Fu, X.T.; Shi, Y.H.; Zhou, J.; Peng, Y.F.; Liu, W.R.; Shi, G.M.; Gao, Q.; Wang, X.Y.; Song, K.; Fan, J.; et al. MicroRNA-30a suppresses autophagy-mediated anoikis resistance and metastasis in hepatocellular carcinoma. *Cancer Lett.* **2018**, *412*, 108–117. [CrossRef] [PubMed]
58. Zhang, K.; Chen, J.; Zhou, H.; Chen, Y.; Zhi, Y.; Zhang, B.; Chen, L.; Chu, X.; Wang, R.; Zhang, C. Pu.1/microRNA-142-3p targets ATG5/ATG16L1 to inactivate autophagy and sensitize hepatocellular carcinoma cells to sorafenib. *Cell Death Dis.* **2018**, *9*, 312. [CrossRef] [PubMed]
59. Gan, J.; Cai, Q.; Qu, Y.; Zhao, F.; Wan, C.; Luo, R.; Mu, D. MiR-96 attenuates status epilepticus-induced brain injury by directly targeting Atg7 and Atg16l1. *Sci. Rep.* **2017**, *7*, 10270. [CrossRef] [PubMed]
60. Wang, C.; Zhang, Z.Z.; Yang, W.; Ouyang, Z.H.; Xue, J.B.; Li, X.L.; Zhang, J.; Chen, W.K.; Yan, Y.G.; Wang, W.J. MiR-210 facilitates ECM degradation by suppressing autophagy via silencing of ATG7 in human degenerated NP cells. *Biomed. Pharmacother.* **2017**, *93*, 470–479. [CrossRef] [PubMed]
61. Comincini, S.; Allavena, G.; Palumbo, S.; Morini, M.; Durando, F.; Angeletti, F.; Pirtoli, L.; Miracco, C. MicroRNA-17 regulates the expression of ATG7 and modulates the autophagy process, improving the sensitivity to temozolomide and low-dose ionizing radiation treatments in human glioblastoma cells. *Cancer Biol. Ther.* **2013**, *14*, 574–586. [CrossRef] [PubMed]
62. Liu, L.; Shen, W.; Zhu, Z.; Lin, J.; Fang, Q.; Ruan, Y.; Zhao, H. Combined inhibition of EGFR and c-ABL suppresses the growth of fulvestrant-resistant breast cancer cells through miR-375-autophagy axis. *Biochem. Bioph. Res. Commun.* **2018**, *498*, 559–565. [CrossRef] [PubMed]
63. He, W.; Cheng, Y. Inhibition of miR-20 promotes proliferation and autophagy in articular chondrocytes by PI3K/AKT/mTOR signaling pathway. *Biomed. Pharmacother.* **2018**, *97*, 607–615. [CrossRef] [PubMed]
64. Wu, Y.; Dai, X.; Ni, Z.; Yan, X.; He, F.; Lian, J. The downregulation of ATG4B mediated by microRNA-34a/34c-5p suppresses rapamycin-induced autophagy. *Irani. J. Basic Med. Sci.* **2017**, *20*, 1125–1130.
65. Frankel, L.B.; Wen, J.; Lees, M.; Hoyer-Hansen, M.; Farkas, T.; Krogh, A.; Jaattela, M.; Lund, A.H. MicroRNA-101 is a potent inhibitor of autophagy. *EMBO J.* **2011**, *30*, 4628–4641. [CrossRef] [PubMed]
66. Xiao, J.; Zhu, X.; He, B.; Zhang, Y.; Kang, B.; Wang, Z.; Ni, X. MiR-204 regulates cardiomyocyte autophagy induced by ischemia-reperfusion through LC3-II. *J. Biomed. Sci.* **2011**, *18*, 35. [CrossRef] [PubMed]
67. Ying, Q.; Liang, L.; Guo, W.; Zha, R.; Tian, Q.; Huang, S.; Yao, J.; Ding, J.; Bao, M.; Ge, C.; et al. Hypoxia-inducible microRNA-210 augments the metastatic potential of tumor cells by targeting vacuole membrane protein 1 in hepatocellular carcinoma. *Hepatology* **2011**, *54*, 2064–2075. [CrossRef] [PubMed]

68. Rotter, D.; Rothermel, B.A. Targets, trafficking, and timing of cardiac autophagy. *Pharmacol. Res.* **2012**, *66*, 494–504. [CrossRef] [PubMed]
69. Wang, Z.V.; Rothermel, B.A.; Hill, J.A. Autophagy in hypertensive heart disease. *J. Biol. Chem.* **2010**, *285*, 8509–8514. [CrossRef] [PubMed]
70. Yang, J.; Chen, D.; He, Y.; Melendez, A.; Feng, Z.; Hong, Q.; Bai, X.; Li, Q.; Cai, G.; Wang, J.; et al. MiR-34 modulates caenorhabditis elegans lifespan via repressing the autophagy gene atg9. *Age* **2013**, *35*, 11–22. [CrossRef] [PubMed]
71. Kovaleva, V.; Mora, R.; Park, Y.J.; Plass, C.; Chiramel, A.I.; Bartenschlager, R.; Dohner, H.; Stilgenbauer, S.; Pscherer, A.; Lichter, P.; et al. MircoRNA-130a targets ATG2b and DICER1 to inhibit autophagy and trigger killing of chronic lymphocytic leukemia cells. *Cancer Res.* **2012**, *72*, 1763–1772. [CrossRef] [PubMed]
72. Bravo-San Pedro, J.M.; Kroemer, G.; Galluzzi, L. Autophagy and mitophagy in cardiovascular disease. *Circ. Res.* **2017**, *120*, 1812–1824. [CrossRef] [PubMed]
73. Nakai, A.; Yamaguchi, O.; Takeda, T.; Higuchi, Y.; Hikoso, S.; Taniike, M.; Omiya, S.; Mizote, I.; Matsumura, Y.; Asahi, M.; et al. The role of autophagy in cardiomyocytes in the basal state and in response to hemodynamic stress. *Nat. Med.* **2007**, *13*, 619–624. [CrossRef] [PubMed]
74. Taneike, M.; Yamaguchi, O.; Nakai, A.; Hikoso, S.; Takeda, T.; Mizote, I.; Oka, T.; Tamai, T.; Oyabu, J.; Murakawa, T.; et al. Inhibition of autophagy in the heart induces age-related cardiomyopathy. *Autophagy* **2010**, *6*, 600–606. [CrossRef] [PubMed]
75. Ouimet, M.; Ediriweera, H.; Afonso, M.S.; Ramkhelawon, B.; Singaravelu, R.; Liao, X.; Bandler, R.C.; Rahman, K.; Fisher, E.A.; Rayner, K.J.; et al. MicroRNA-33 regulates macrophage autophagy in atherosclerosis. *Arterioscleros. Thromb. Vasc. Biol.* **2017**, *37*, 1058–1067. [CrossRef] [PubMed]
76. Wang, J.; Wang, W.N.; Xu, S.B.; Wu, H.; Dai, B.; Jian, D.D.; Yang, M.; Wu, Y.T.; Feng, Q.; Zhu, J.H.; et al. MicroRNA-214-3p: A link between autophagy and endothelial cell dysfunction in atherosclerosis. *Acta Physiol.* **2018**, *222*, e12973. [CrossRef] [PubMed]
77. Li, L.; Wang, Z.; Hu, X.; Wan, T.; Wu, H.; Jiang, W.; Hu, R. Human aortic smooth muscle cell-derived exosomal miR-221/222 inhibits autophagy via a PTEN/Akt signaling pathway in human umbilical vein endothelial cells. *Biochem. Biophys. Res. Commun.* **2016**, *479*, 343–350. [CrossRef] [PubMed]
78. Wang, J.; Sun, Y.T.; Xu, T.H.; Sun, W.; Tian, B.Y.; Sheng, Z.T.; Sun, L.; Liu, L.L.; Ma, J.F.; Wang, L.N.; et al. MicroRNA-30b regulates high phosphorus level-induced autophagy in vascular smooth muscle cells by targeting BECN1. *Cell. Physiol. Biochem.* **2017**, *42*, 530–536. [CrossRef] [PubMed]
79. Pietrocola, F.; Izzo, V.; Niso-Santano, M.; Vacchelli, E.; Galluzzi, L.; Maiuri, M.C.; Kroemer, G. Regulation of autophagy by stress-responsive transcription factors. *Semin. Cancer Biol.* **2013**, *23*, 310–322. [CrossRef] [PubMed]
80. Ma, W.; Ding, F.; Wang, X.; Huang, Q.; Zhang, L.; Bi, C.; Hua, B.; Yuan, Y.; Han, Z.; Jin, M.; et al. By targeting *Atg7* microRNA-143 mediates oxidative stress-induced autophagy of c-kit$^+$ mouse cardiac progenitor cells. *EBioMedicine* **2018**, *32*, 182–191. [CrossRef] [PubMed]
81. Dorn, G.W., 2nd. Mitochondrial pruning by Nix and BNip3: An essential function for cardiac-expressed death factors. *J. Cardiovasc. Transl. Res.* **2010**, *3*, 374–383. [CrossRef] [PubMed]
82. Gong, G.; Song, M.; Csordas, G.; Kelly, D.P.; Matkovich, S.J.; Dorn, G.W., 2nd. Parkin-mediated mitophagy directs perinatal cardiac metabolic maturation in mice. *Science* **2015**, *350*, aad2459. [CrossRef] [PubMed]
83. Andres, A.M.; Hernandez, G.; Lee, P.; Huang, C.; Ratliff, E.P.; Sin, J.; Thornton, C.A.; Damasco, M.V.; Gottlieb, R.A. Mitophagy is required for acute cardioprotection by simvastatin. *Antioxid. Redox Signal.* **2014**, *21*, 1960–1973. [CrossRef] [PubMed]
84. Billia, F.; Hauck, L.; Konecny, F.; Rao, V.; Shen, J.; Mak, T.W. PTEN-inducible kinase 1 (PINK1)/Park6 is indispensable for normal heart function. *Proc. Natl. Acad. Sci. USA* **2011**, *108*, 9572–9577. [CrossRef] [PubMed]
85. Li, W.; Zhang, X.; Zhuang, H.; Chen, H.G.; Chen, Y.; Tian, W.; Wu, W.; Li, Y.; Wang, S.; Zhang, L.; et al. MicroRNA-137 is a novel hypoxia-responsive microRNA that inhibits mitophagy via regulation of two mitophagy receptors FUNDC1 and NIX. *J. Biol. Chem.* **2014**, *289*, 10691–10701. [CrossRef] [PubMed]
86. Kim, J.; Fiesel, F.C.; Belmonte, K.C.; Hudec, R.; Wang, W.X.; Kim, C.; Nelson, P.T.; Springer, W.; Kim, J. MiR-27a and miR-27b regulate autophagic clearance of damaged mitochondria by targeting PTEN-induced putative kinase 1 (PINK1). *Mol. Neurodegener.* **2016**, *11*, 55. [CrossRef] [PubMed]
87. Cheng, M.; Liu, L.; Lao, Y.; Liao, W.; Liao, M.; Luo, X.; Wu, J.; Xie, W.; Zhang, Y.; Xu, N.

MicroRNA-181a suppresses parkin-mediated mitophagy and sensitizes neuroblastoma cells to mitochondrial uncoupler-induced apoptosis. *Oncotarget* **2016**, *7*, 42274–42287. [CrossRef] [PubMed]

88. Yang, F.; Li, T. MicroRNA-410 is involved in mitophagy after cardiac ischemia/reperfusion injury by targeting high-mobility group box 1 protein. *J. Cell. Biochem.* **2018**, *119*, 2427–2439. [CrossRef] [PubMed]
89. Verjans, R.; van Bilsen, M.; Schroen, B. MiRNA deregulation in cardiac aging and associated disorders. *Int. Rev. Cell Mol. Biol.* **2017**, *334*, 207–263. [PubMed]
90. Yan, L.; Gao, S.; Ho, D.; Park, M.; Ge, H.; Wang, C.; Tian, Y.; Lai, L.; De Lorenzo, M.S.; Vatner, D.E.; et al. Calorie restriction can reverse, as well as prevent, aging cardiomyopathy. *Age* **2013**, *35*, 2177–2182. [CrossRef] [PubMed]
91. Egan, D.F.; Shackelford, D.B.; Mihaylova, M.M.; Gelino, S.; Kohnz, R.A.; Mair, W.; Vasquez, D.S.; Joshi, A.; Gwinn, D.M.; Taylor, R.; et al. Phosphorylation of ULK1 (hATG1) by AMP-activated protein kinase connects energy sensing to mitophagy. *Science* **2011**, *331*, 456–461. [CrossRef] [PubMed]
92. Kim, J.; Kim, Y.C.; Fang, C.; Russell, R.C.; Kim, J.H.; Fan, W.; Liu, R.; Zhong, Q.; Guan, K.L. Differential regulation of distinct Vps34 complexes by AMPK in nutrient stress and autophagy. *Cell* **2013**, *152*, 290–303. [CrossRef] [PubMed]
93. Russell, R.C.; Tian, Y.; Yuan, H.; Park, H.W.; Chang, Y.Y.; Kim, J.; Kim, H.; Neufeld, T.P.; Dillin, A.; Guan, K.L. ULK1 induces autophagy by phosphorylating Beclin-1 and activating VPS34 lipid kinase. *Nat. Cell Biol.* **2013**, *15*, 741–750. [CrossRef] [PubMed]
94. Wu, H.; Wang, F.; Hu, S.; Yin, C.; Li, X.; Zhao, S.; Wang, J.; Yan, X. MiR-20a and miR-106b negatively regulate autophagy induced by leucine deprivation via suppression of ULK1 expression in C2C12 myoblasts. *Cell. signal.* **2012**, *24*, 2179–2186. [CrossRef] [PubMed]
95. Wan, G.; Xie, W.; Liu, Z.; Xu, W.; Lao, Y.; Huang, N.; Cui, K.; Liao, M.; He, J.; Jiang, Y.; et al. Hypoxia-induced miR155 is a potent autophagy inducer by targeting multiple players in the mtor pathway. *Autophagy* **2014**, *10*, 70–79. [CrossRef] [PubMed]
96. Martins-Marques, T.; Ribeiro-Rodrigues, T.; Pereira, P.; Codogno, P.; Girao, H. Autophagy and ubiquitination in cardiovascular diseases. *DNA Cell Biol.* **2015**, *34*, 243–251. [CrossRef] [PubMed]
97. Pan, W.; Zhong, Y.; Cheng, C.; Liu, B.; Wang, L.; Li, A.; Xiong, L.; Liu, S. MiR-30-regulated autophagy mediates angiotensin II-induced myocardial hypertrophy. *PloS ONE* **2013**, *8*, e53950. [CrossRef] [PubMed]
98. Chen, C.; Yang, S.; Li, H.; Yin, Z.; Fan, J.; Zhao, Y.; Gong, W.; Yan, M.; Wang, D.W. MiR30c is involved in diabetic cardiomyopathy through regulation of cardiac autophagy via BECN1. *Mol. Therapy-Nucleic Acids* **2017**, *7*, 127–139. [CrossRef] [PubMed]
99. Huang, J.; Sun, W.; Huang, H.; Ye, J.; Pan, W.; Zhong, Y.; Cheng, C.; You, X.; Liu, B.; Xiong, L.; et al. MiR-34a modulates angiotensin ii-induced myocardial hypertrophy by direct inhibition of atg9a expression and autophagic activity. *PloS ONE* **2014**, *9*, e94382. [CrossRef] [PubMed]
100. Su, M.; Chen, Z.; Wang, C.; Song, L.; Zou, Y.; Zhang, L.; Hui, R.; Wang, J. Cardiac-specific overexpression of miR-222 induces heart failure and inhibits autophagy in mice. *Cell. Physiol. Biochem.* **2016**, *39*, 1503–1511. [CrossRef] [PubMed]
101. Lai, L.; Chen, J.; Wang, N.; Zhu, G.; Duan, X.; Ling, F. MiRNA-30e mediated cardioprotection of ACE2 in rats with doxorubicin-induced heart failure through inhibiting cardiomyocytes autophagy. *Life Sci.* **2017**, *169*, 69–75. [CrossRef] [PubMed]
102. Qiu, J.; Wang, A.; Xu, Y.; Qiao, S.; An, J.; Li, H.; Wang, C. Role of microRNA-1-mediated AMP-activated protein kinase pathway in cardiac fibroblasts induced by high glucose in rats. *Zhonghua wei zhong bing ji jiu yi xue* **2018**, *30*, 145–150. [PubMed]
103. D'Avenia, M.; Citro, R.; De Marco, M.; Veronese, A.; Rosati, A.; Visone, R.; Leptidis, S.; Philippen, L.; Vitale, G.; Cavallo, A.; et al. A novel miR-371a-5p-mediated pathway, leading to BAG3 upregulation in cardiomyocytes in response to epinephrine, is lost in takotsubo cardiomyopathy. *Cell Death Dis.* **2015**, *6*, e1948. [CrossRef] [PubMed]
104. Shao, H.; Yang, L.; Wang, L.; Tang, B.; Wang, J.; Li, Q. MicroRNA-34a protects myocardial cells against ischemia-reperfusion injury through inhibiting autophagy via regulating TNAα expression. *Biochem. Cell Biol.* **2018**, *96*, 349–354. [CrossRef] [PubMed]
105. Liu, X.; Deng, Y.; Xu, Y.; Jin, W.; Li, H. MicroRNA-223 protects neonatal rat cardiomyocytes and h9c2 cells from hypoxia-induced apoptosis and excessive autophagy via the akt/mtor pathway by targeting parp-1. *J. Mol. Cell. Cardiol.* **2018** *118*, 133–146. [CrossRef] [PubMed]

106. Sun, T.; Dong, Y.H.; Du, W.; Shi, C.Y.; Wang, K.; Tariq, M.A.; Wang, J.X.; Li, P.F. The role of microRNAs in myocardial infarction: From molecular mechanism to clinical application. *Int. J. Mol. Sci.* **2017**, *18*, 745. [CrossRef] [PubMed]
107. Bo, L.; Su-Ling, D.; Fang, L.; Lu-Yu, Z.; Tao, A.; Stefan, D.; Kun, W.; Pei-Feng, L. Autophagic program is regulated by miR-325. *Cell Death Differ.* **2014**, *21*, 967–977. [CrossRef] [PubMed]
108. Zheng, J.; Li, J.; Kou, B.; Yi, Q.; Shi, T. MicroRNA-30e protects the heart against ischemia and reperfusion injury through autophagy and the notch1/Hes1/Akt signaling pathway. *Int. J. Mol. Med.* **2018**, *41*, 3221–3230. [CrossRef] [PubMed]
109. Li, Q.; Xie, J.; Li, R.; Shi, J.; Sun, J.; Gu, R.; Ding, L.; Wang, L.; Xu, B. Overexpression of microRNA-99a attenuates heart remodelling and improves cardiac performance after myocardial infarction. *J. Cell. Mol. Med.* **2014**, *18*, 919–928. [CrossRef] [PubMed]
110. Li, X.; Zeng, Z.; Li, Q.; Xu, Q.; Xie, J.; Hao, H.; Luo, G.; Liao, W.; Bin, J.; Huang, X.; et al. Inhibition of microRNA-497 ameliorates anoxia/reoxygenation injury in cardiomyocytes by suppressing cell apoptosis and enhancing autophagy. *Oncotarget* **2015**, *6*, 18829–18844. [CrossRef] [PubMed]
111. Wang, X.; Cui, T. Autophagy modulation: A potential therapeutic approach in cardiac hypertrophy. *Am. J. Physiol. Heart Circ. Physiol.* **2017**, *313*, H304–H319. [CrossRef] [PubMed]
112. Song, L.; Su, M.; Wang, S.; Zou, Y.; Wang, X.; Wang, Y.; Cui, H.; Zhao, P.; Hui, R.; Wang, J. MiR-451 is decreased in hypertrophic cardiomyopathy and regulates autophagy by targeting tsc1. *J. Cell. Mol. Med.* **2014**, *18*, 2266–2274. [CrossRef] [PubMed]
113. Zhao, X.D.; Qin, R.H.; Yang, J.J.; Xu, S.S.; Tao, H.; Ding, X.S.; Shi, K.H. DNMT3A controls miR-200b in cardiac fibroblast autophagy and cardiac fibrosis. *Inflamm. Res.* **2018**, *67*, 681–690. [CrossRef] [PubMed]

5

Unleashing the Full Potential of Oncolytic Adenoviruses against Cancer by Applying RNA Interference: The Force Awakens

Tereza Brachtlova and Victor W. van Beusechem *

Amsterdam UMC, Vrije Universiteit Amsterdam, Medical Oncology, Cancer Center Amsterdam, De Boelelaan 1117, 1007 MB Amsterdam, The Netherlands; t.brachtlova@vumc.nl

* Correspondence: VW.vanBeusechem@vumc.nl;

Abstract: Oncolytic virus therapy of cancer is an actively pursued field of research. Viruses that were once considered as pathogens threatening the wellbeing of humans and animals alike are with every passing decade more prominently regarded as vehicles for genetic and oncolytic therapies. Oncolytic viruses kill cancer cells, sparing healthy tissues, and provoke an anticancer immune response. Among these viruses, recombinant adenoviruses are particularly attractive agents for oncolytic immunotherapy of cancer. Different approaches are currently examined to maximize their therapeutic effect. Here, knowledge of virus–host interactions may lead the way. In this regard, viral and host microRNAs are of particular interest. In addition, cellular factors inhibiting viral replication or dampening immune responses are being discovered. Therefore, applying RNA interference is an attractive approach to strengthen the anticancer efficacy of oncolytic viruses gaining attention in recent years. RNA interference can be used to fortify the virus' cancer cell-killing and immune-stimulating properties and to suppress cellular pathways to cripple the tumor. In this review, we discuss different ways of how RNA interference may be utilized to increase the efficacy of oncolytic adenoviruses, to reveal their full potential.

Keywords: RNA interference; small interfering RNA; microRNA; oncolytic virotherapy; conditionally replicating adenovirus (CRAd)

---

1. Oncolytic Virotherapy

Despite many different types of treatments used to combat cancer, cancer is still one of the leading causes of death worldwide. While on one hand the specific traits cancer cells acquire, such as infinite replicative potential and evading apoptosis [1], provide angles for anticancer treatment, these traits on the other hand complicate treatment and, in many cases, contribute to treatment failure. Constant replication and genetic instability allow for adaptation to medication and emerged resistance. There is, therefore, a great need for new treatments that are also effective against therapy-resistant cancers. One promising approach is oncolytic virus therapy (OVT), using viruses that selectively replicate in and kill cancer cells [2,3]. There are several reasons why OVT is considered an elegant option for cancer treatment. Firstly, oncolytic viruses exhibit specific replication in cancer cells, often dependent on the genetic changes that discriminate cancer cells from non-malignant cells [4,5]. Hence, genetic changes underlying malignant cell growth are utilized for selective therapy. Secondly, oncolytic viruses kill their host cell as an integral part of their life cycle. Replication in a host cell inevitably leads to death of the host cell. Oncolytic viruses, thus, have profound direct anticancer cell activity. Thirdly, selective replication in cancer cells produces large numbers of progeny viruses that upon their release are capable of infecting new cancer cells causing subsequent cycles of lytic replication. Thus, the anticancer effect is amplified in situ. On top of these direct anticancer effects, oncolytic viruses engage immune system responses of the host [6,7]. Oncolytic virus replication in a tumor can help override the immune suppressive conditions that exist in the tumor microenvironment [7]. Upon virus infection, an antitumor immune response is induced by the production of immune-stimulatory cytokines by virus-infected cancer cells and the concerted release of endogenous danger-associated molecular patterns, virus-derived pathogen-associated molecular patterns, and tumor-specific or tumor-associated antigens from lysed cancer cells in a pro-inflammatory context. This may induce tumor-specific immunity, also against cancer cells at distant sites not subjected to virus infection [7].

Several different viruses are considered for use in OVT. Some of them advanced to clinical trials. Of these, an oncolytic herpes simplex virus (HSV) was recently approved for treatment of melanoma by the competent authorities in the United States and Europe [8,9] and an oncolytic adenovirus was already approved in 2005 in China for treatment of head and neck cancer in combination with chemotherapy [10]. Generally, oncolytic viruses can be divided into three main groups. The first group contains viruses with a natural propensity to preferentially replicate in cancer cells, while being non-pathogenic in humans. Many of these viruses are of non-human origin. This group comprises autonomous parvovirus, myxoma virus, Newcastle disease virus, and reovirus [11–13]. Many of the second group consists of viruses that were attenuated by propagation in vitro. These attenuated strains of, e.g., measles virus or vaccinia viruses were used safely in humans as vaccine vectors and were, therefore, considered safe starting points to build oncolytic viruses. The third group consists of viruses that were genetically engineered by introducing mutations in their genome to ensure selective replication in cancer cells. The latter group includes oncolytic viruses derived from adenovirus, HSV, and vesicular stomatitis virus (VSV), as well as an attenuated and genetically modified strain of poliovirus [11–13]. Oncolytic adenoviruses, often also referred to as conditionally replicating adenoviruses (CRAds), represent one of the most extensively researched oncolytic viruses. This review focuses on oncolytic adenoviruses, but the general concepts discussed apply to other oncolytic viruses as well.

Oncolytic Adenoviruses

Adenoviruses are small, non-enveloped double-stranded DNA viruses (34–36 kb), continuously present in the population and responsible for self-limiting infections in immune-competent individuals, including mild respiratory, conjunctival, and gastrointestinal diseases, or cystitis. Infections in immunocompromised patients are much more severe, including infections in liver (hepatitis), lung (pneumonia), or heart (myocarditis) that can result in death (reviewed by Lion [14]). Adenoviral replication is strongly dependent on expression of the immediate early E1A region. To establish a successful replication in a host cell, the adenovirus seizes control over retinoblastoma protein (pRb) and p53 protein (using E1A and E1B viral proteins), which suspends the host cell cycle in the synthesis (S) phase [15].

Non-replicating adenoviruses have been used as gene delivery vectors in cancer gene therapy strategies for many years. Replication-deficient adenovirus vectors lack genes from the E1 region and may also have E3 genes deleted to accommodate newly introduced genes [16]. Although these adenoviruses showed low toxicity and promising results compared to conventional therapy, their efficacy in clinical trials remained low, which minimizes their use.

Current research focuses mainly on CRAds that are considered much more promising agents for the treatment of cancer. Replication of CRAds is controlled either by regulating mostly E1A gene expression with a tumor-specific promoter or by deletions in the viral genome that require cellular factors for compensation [17]. Examples of the first strategy include human telomerase (hTERT) promoter-driven oncolytic adenoviruses [18–20]. The first CRAd to be evaluated for cancer treatment, ONYX-015, was made following the second strategy. This adenovirus variant was designed to selectively replicate in p53-deficient cancer cells by deletion of the *E1B55K* gene [21]. Although ONYX-015 showed cancer cell-selective replication, its efficacy was disappointing [22]. Since then, newer generations CRAds with improved selectivity and potency were developed, including Ad5-Δ24 and ICOVIR-5 [23,24]. Nevertheless, despite very encouraging results from in vitro and animal studies, the anticancer efficacy of CRAds, as well as of other oncolytic viruses, as a single agent in humans is generally modest [25]. Thus, there is a clear need to increase the efficacy of OVT. This could be achieved using more effective delivery methods or by enhancing the potency of CRAds to kill cancer cells or to induce an antitumor immune response. In addition, while most efforts are on improving anticancer treatment efficacy, studies are also undertaken to more stringently control CRAd replication in healthy cells.

2. Strategies to Increase the Efficacy of Oncolytic Virus Therapy with CRAds

2.1. Achieving More Effective Delivery of Oncolytic Adenovirus to Tumors

Effective OVT with CRAds requires that viruses are delivered to tumors in the human body and that they enter cancer cells to initiate oncolysis. Notably, cancer cells are sometimes resistant to CRAd infection due to low expression of the primary receptor molecule coxsackie-adenovirus receptor (CAR) [26]. Typical neoplasms in which downregulation of CAR expression was observed include prostate, colon, and kidney cancers [27]. Retargeting strategies allow overcoming this obstacle, specifically by diversion of the virus to other cell surface receptors. Strategies that were successfully followed to accomplish this were, e.g., incorporation of a cyclic RGD4C peptide motif in the adenovirus fiber knob to allow entry via $\alpha_V\beta3$ and $\alpha_V\beta5$ integrins [28], pseudotyping the viral capsid with proteins from other serotype adenoviruses or with chimeric capsid proteins [29,30], or expressing bispecific adapter molecules from the CRAd genome targeting virus entry via an alternative cell surface receptor [31]. Generally, these modifications resulted in more effective CRAds with broader applicability in OVT.

The administration route to deliver the virus to tumor cells in the human body poses another challenge. Systemic administration of CRAds was proven quite ineffective since most injected virions are eliminated before they reach their target. Much research is put into the development of methods to chemically modify viral capsids to shield them from sequestration in the liver and inactivation by the immune system [32]. Another interesting approach is to use carrier cells as temporary virus hosts delivering oncolytic viruses, including CRAds, to tumor sites. This Trojan horse concept is very attractive, because it not only hides the virus from the immune system, but also exploits the capacity of cells to extravasate from the circulation and home to tissues [33,34]. However, several major challenges remain, including premature expression of viral proteins in the carrier cell, complicated timing of the delivery, acquired adaptive immunity to carrier cells, or the inability to pass through capillaries, which results in the accumulation in, e.g., lungs, and subsequent release of the virus before delivering it to the tumor [33,35,36]. Moreover, there is a contradiction in delivering a virus with cancer-selective replication properties using a non-malignant carrier cell. At least a single virus lifecycle should be completed in this cell to allow release of infectious progeny virus at the tumor site. This means that either the virus should not be entirely cancer-selective, or the carrier cell should have cancer cell-like properties, such as a deregulation in growth control. Both options may raise safety concerns that need to be addressed.

2.2. Improving Oncolytic Adenovirus Specificity by Employing microRNA-Dependent Replication

A novel strategy to make CRAds safer by further limiting their replication in non-malignant cells exploits microRNA (miRNA) technology. Here, a miRNA target sequence is incorporated in the 3′ untranslated region (UTR) of the transcript encoding a viral protein essential for replication, usually E1A. Infection of cells expressing the relevant miRNA results in silencing of the essential protein and, consequently, suppression of viral replication. In recognition of the fact that human adenoviruses have strong tropism for liver cells and are efficiently delivered to liver cells when administered systemically, which could lead to liver toxicity, particular efforts were made to inhibit CRAd replication in liver cells. Ylösmäki et al. [37] constructed Δ24-type CRAds carrying three target sites for the liver-specific miRNA miR-122 in the Δ24-E1A 3′ UTR. While this reduced Δ24-E1A expression in miR-122-expressing hepatoma cells by more than 90%, this did not inhibit virus replication. By introducing a second mutation in the E1A messenger RNA (mRNA) that reduces overall expression levels, this problem was successfully addressed. The resulting virus was severely attenuated, failing to induce cytopathic effects in hepatoma cells and producing more than 10,000-fold reduced titers in hepatoma cells compared to non-small lung cancer cells. The same approach was taken independently by others; CRAds with miR-122 binding sites exhibited reduced E1A expression and virus replication in liver cells and almost absent liver toxicity upon injection in mice [38,39].

Knowledge of differential expression of miRNAs in cancer cells versus non-malignant cells was used similarly to construct CRAds with reduced replication in non-cancer cells. In particular, miR-199, which is downregulated in cancer cells, was considered for this purpose [19,40]. Significantly reduced virus replication and toxicity was observed in non-malignant cell cultures and in livers of newborn mice if miR-199 target sequences were incorporated into the E1A 3′ UTR. By combining target sequences for miR-199a and liver-specific miR-122a in the 3′ UTR of an hTERT promoter-driven E1A gene, a most selective CRAd was obtained [19]. Oncolytic potency in cancer cell lines was fully retained. Others translated the general concept of miRNA-dependent suppression of CRAd replication to more specific applications, where viral toxicity might arise upon local tissue administration. MiRNA-regulated CRAds were made in a wild-type adenovirus background for selective treatment of pancreatic cancer and glioma [41,42]. The viruses were effective in animal models and exhibited strongly reduced toxicity profiles. For this, insertion of four miRNA target sites appeared sufficient, but this might depend on host miRNA expression levels and the strength of the viral gene promoter. Thus, preclinical data suggest that host cell miRNA-regulation of adenovirus replication alone can already provide sufficient CRAd selectivity. However, further development of this type of CRAd will likely be done in the background of CRAds carrying mutations with proven clinical safety.

Another strategy to restrict CRAd replication in healthy cells was taken by Gürlevik et al. [43]. They expressed multiple artificial miRNAs targeting essential adenovirus genes from a single polycistronic transcript, driven by a tumor suppressor p53-dependent promoter, inserted in the adenovirus genome. In cells with functional p53 expression, but not in p53 mutant cells, the artificial miRNAs were expressed and silenced their target genes. Although this attenuated viral replication in p53 wild-type cells, virus progeny production was not prevented. The p53-dependent attenuated phenotype was also observed in mice injected with the virus, where p53 wild-type mice carried lower virus copy numbers in their liver than did p53 knockout mice. The inhibition of virus replication that could be achieved in this study appeared inferior to that achieved when relying on host miRNAs. This can possibly be explained by the fact that the latter are already highly expressed in the host cell at the time of infection, whereas CRAd-encoded miRNAs need to be produced after cell entry and their level of expression might be replication-dependent, as copy numbers increase tremendously during replication. Hence, functional knockdown of viral genes by virus-encoded miRNAs is probably not reached before viral replication is initiated.

2.3. *Strategies for Improving the Potency of Oncolytic Adenoviruses*

The strong selectivity of CRAd replication in cancer cells that is obtained with currently available technology, providing excellent safety, allows exploration of methods to enhance the cancer cell-killing potency of CRAds. Apart from the tropism modifications mentioned above, investigated approaches include the introduction of potency-enhancing genome mutations and the expression of potency-enhancing transgenes. The former is usually based on identification of such mutations in virus variants that are either naturally occurring or were selected using molecular evolution methods. A good example of this approach is the genetic bioselection of a virus with a truncating mutation in the endoplasmatic retention domain of the E3/19K protein that promotes virus progeny release from infected cells [44]. Incorporation of this T1 mutation in an Ad5-Δ24 backbone with RGD4C fiber modification produced a CRAd with strong antitumor activity in vitro and in vivo, without compromising its selectivity toward cancer cells [45]. Another example is ColoAd1, a recombination product between two naturally occurring serotypes selected for rapid replication on human cancer cell lines, which exhibits increased potency in particular on colorectal carcinoma cells [46]. Transgenes that can be carried by the virus to increase its potency include genes encoding fusogenic proteins or tissue-remodeling peptides, enzymes for use in suicide gene therapy approaches, pro-apoptotic or tumor suppressor proteins, and immune-modulating proteins. Expression of a fusogenic protein induces cell–cell fusions, forming syncytia consisting of CRAd-infected cells and adjacent non-infected cells, leading to death also of the fused non-infected cells and, thus, more effective tumor eradication [47]. Another way to promote spread of CRAds through tumor tissues is to express molecules that remodel the extracellular matrix, allowing better virus dispersion in a solid tumor mass [48–50]. The suicide gene therapy concept, where an enzyme is expressed that converts a non-toxic pro-drug into a toxic compound, was also applied in the context of CRAds [51,52]. This was met with various levels of success, as premature death of the host cell inhibits virus propagation. Careful timing of virus and pro-drug administration is, thus, essential, which makes the design of clinical treatment schedules challenging. The same can be said about expression of apoptosis-inducing proteins, where premature cell death inhibits, and properly timed cell death promotes virus spread [53]. In contrast, expression of tumor suppressor protein p53 was not met with this limitation, presumably because adenovirus-encoded E1B55K protein interacts with p53 and regulates its activity during adenovirus replication [54,55]. In the late phase of the virus lifecycle, p53 promotes cell death, thereby accelerating virus progeny release [54,56]. In addition, p53 increases late adenoviral gene expression, presumably by enhancing transcription from the virus major late promoter (MLP) [57] (Figure 1). Exogenous p53 expression augmented CRAd replication in the majority of cancer cell lines, primary tumor cell cultures, and xenograft tumor models tested [54,55,58–63]. Notably, transgene expression from the adenovirus genome during virus replication in host cells can be delayed by coupling its transcription to the MLP. Apart from avoiding premature transgene expression, this strategy offers the additional advantage of confining transgene expression to cancer cells in which the CRAd replicates selectively [64]. In addition to the aforementioned ways of enhancing the primary tumor-eradicating potency of CRAds, there is currently much interest in promoting the secondary, immune-stimulating effect of these viruses, by incorporating genes encoding immune-stimulatory molecules, such as cytokines, chemokines, and checkpoint inhibitors (reviewed by de Gruijl et al. [7]). Altogether, these developments led to substantially improved CRAd variants with more potent cancer cell-killing properties than earlier generation viruses. Nevertheless, one area of potential CRAd improvement was long overlooked. While cancers may lack certain molecules needed for effective OVT that can be supplemented by transgene expression from the CRAd genome, they could also express molecules inhibiting OVT that should, thus, be depleted. Currently available gene suppression techniques allow investigating this and constructing even more powerful CRAds capable of overcoming obstacles for OVT in cancer cells. The next section focuses on this new development.

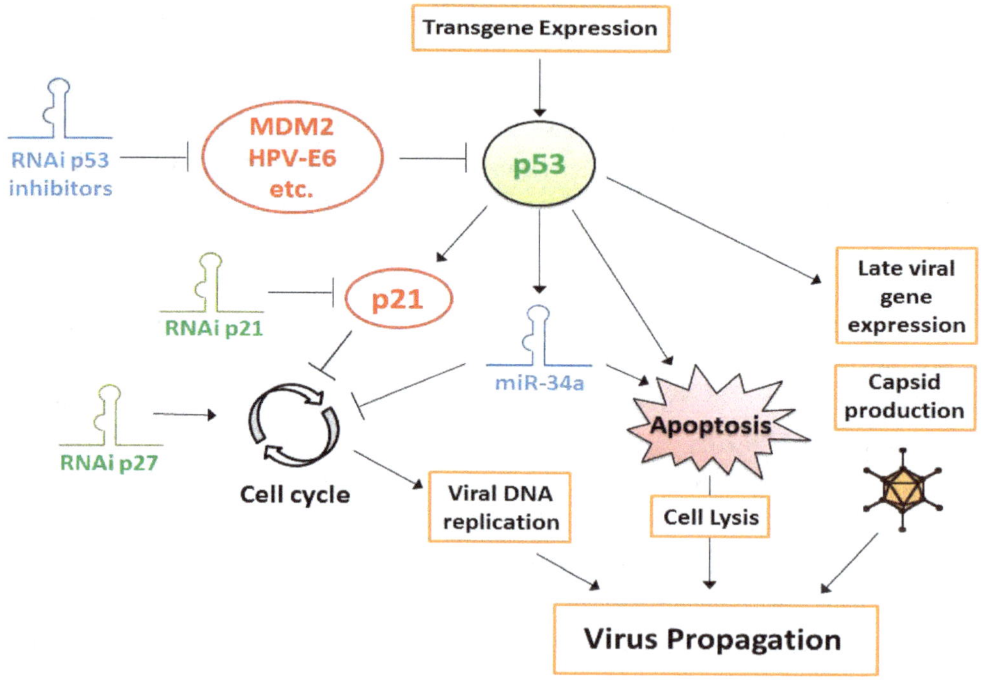

Figure 1. Tumor suppressor p53-dependent strategies to increase the potency of conditionally replicating adenoviruses (CRAds). Expression of functional p53 in cancer cells promotes CRAd propagation by stimulating late viral gene expression and accelerating cell lysis to release virus progeny. In contrast, direct p53 target p21 inhibits the cell cycle and, thereby, viral DNA replication. Silencing cyclin-dependent kinase inhibitors p21 or p27 through RNA interference (RNAi) alleviates cell cycle arrest and, thus, promotes viral DNA replication. Potential other means of promoting CRAd propagation include silencing of p53 inhibitors, which are known to inhibit CRAd replication, and silencing of p53 transcriptional target microRNA (miR)-34a. Molecules in red depict confirmed inhibitors of CRAd efficacy; molecules in green depict confirmed stimulators of CRAd efficacy; molecules in blue depict proposed, but not yet experimentally confirmed modulators of CRAd efficacy; orange delineated viral processes can be influenced to increase potency; cellular processes that are affected by viruses during infection and that can be modulated by transgenes and RNAi molecules expressed by armed CRAds are delineated in black.

3. Gene Suppression to Make Oncolytic Viruses More Effective

There are two requirements for improving OVT by suppressing inhibitors of CRAds in cancer cells, i.e., (i) identification of the inhibitory gene products in cancer cells, and (ii) design of an effective way of suppressing genes, using molecules that can be carried by CRAds. For the former, one may build on available knowledge of virus–host cell interactions or use high-throughput functional screening approaches. The latter, we know now, can be successfully achieved by using RNA interference (RNAi).

Currently, RNAi is the most widely used method to specifically downregulate functional gene expression. RNAi is a conserved cellular surveillance system in eukaryotes that activates a sequence-specific degradation of RNA species homologous to short non-coding double-stranded RNA (dsRNA) molecules [65]. These dsRNA molecules, including miRNAs and short interfering RNAs (siRNAs), are products of the RNase III enzymes Drosha and Dicer [66]. One of the strands of the miRNAs and siRNAs, the active or leading strand, is incorporated into a multiprotein complex called RNA-induced silencing complex (RISC) and directs the target RNA recognition leading to its degradation. In addition, miRNAs direct translational repression and mRNA deadenylation, further contributing to gene silencing [67]. The mechanism of this strongly conserved regulation of cellular mRNAs by miRNAs was thoroughly reviewed in several publications [65,68,69].

A means of producing exogenous siRNAs or miRNAs in a cell is to express small RNA molecules as single transcripts that form a stem-loop structure, generally referred to as short hairpin RNAs (shRNAs) and short hairpin miRNAs (shmiRNAs). Their structure is very similar to that of pre-miRNA, i.e., the Drosha-cleaved product in endogenous miRNA biogenesis. The dsRNA in the stem of these molecules is processed by Dicer to form mature siRNA and miRNA duplexes, respectively, that will then incorporate their active strand into RISC and direct RNAi. Successful RNAi induced by adenovirus-encoded exogenous shRNA was achieved initially using replication-defective vectors [70]. This opened the possibility of silencing genes in cancer cells with a therapeutic intent.

Because the efficiency of gene delivery with replication-defective vectors is generally considered inadequate for effective cancer treatment, interest rose in using CRAds as shRNA delivery vectors. However, achieving effective RNAi in cells infected with a replication-competent adenovirus was not considered trivial. RNAi is recognized as a cellular defense mechanism against virus infection and this led many viruses to evolve molecules that inhibit RNAi [71,72]. Although the importance of RNAi as an antiviral response in mammalian cells is heavily debated (e.g., References [73,74]), there was reason to assume that the process of RNAi would be hampered in cells in which a virus is replicating. In fact, several independent reports showed that adenovirus-encoded virus-associated RNAs (VA RNAs) inhibit the RNAi machinery (Figure 2). VA RNAs (I and II) were found to inhibit the host cell RNAi pathway at the level of nuclear export, by competing with miRNA precursor molecules for binding to Exportin 5; at the level of processing RNAi precursor molecules, by inhibiting the activity of Dicer; and at the effector level, by competing with siRNA and miRNA for incorporation into RISC [75,76]. In addition, it was later reported that the titration of Exportin 5 by VA RNA also reduced Dicer mRNA nuclear export and, thereby, Dicer protein levels [77]. Apart from this direct inhibition of the RNAi machinery by virus-encoded RNA, it was recently reported that the type I interferon response evoked by adenovirus vectors in cells inhibits the processing of shRNAs by Dicer, resulting in a less effective knockdown of target genes [78]. Altogether, there were reasons to expect that effective CRAd replication and RNAi could be incompatible.

Nevertheless, in a proof-of-concept study using firefly luciferase as target, Carette et al. showed that CRAd-induced shRNA can induce proper gene silencing in human cancer cells [79]. That suggested that, although interferon-responses and VA RNAs inhibit the RNAi machinery, they do not prohibit RNAi brought about by adenovirus-encoded shRNA. Still, Machitani et al. showed that RNAi mediated by adenovirus-encoded shRNA can be improved by deleting VA RNA sequences [80]. Notably, this study was done with replication-defective adenovirus vectors. Since VA RNA expression is highly replication-dependent, competitive incorporation into the RNAi machinery is expectedly even higher in a replication competent context. Thus, to achieve highly effective RNAi using CRAds it could be considered to (partly) delete VA RNA sequences. However, adenovirus lacking major VA RNA I cannot efficiently translate viral mRNA in the late phase of the infection, resulting in poor replication [81]. In addition, it was recently reported that VA RNA II-encoded viral miRNA promotes adenovirus replication, presumably via suppression of cullin 4-mediated inhibition of Jun N-terminal kinase (JNK) signaling [82]. VA RNA-depleted CRAds are, therefore, considered less attractive agents for OVT. However, observations reported by Kamel et al. suggested that VA RNA I expression is required primarily to counteract cellular anti-viral interferon responses, not so much the RNAi machinery [83]. This notion is supported by the observation that processing of VA RNAs by Dicer into viral miRNAs inhibits adenovirus replication, apparently by depleting VA RNAs capable of inhibiting double-stranded RNA-activated kinase (PKR) [84]. Thus, the RNAi machinery appears to contribute to the cellular defense against adenovirus infection indirectly, by inhibiting the capacity of the virus to suppress the major anti-viral interferon–PKR response. This implies that if the PKR and RNAi inhibitory functions of VA RNA I can be dissected by introducing specific mutations, or, alternatively, VA RNA I is deleted and the interferon response is inhibited through other means, such as an shRNA targeting PKR, more effectively replicating and gene-silencing CRAds could perhaps be made.

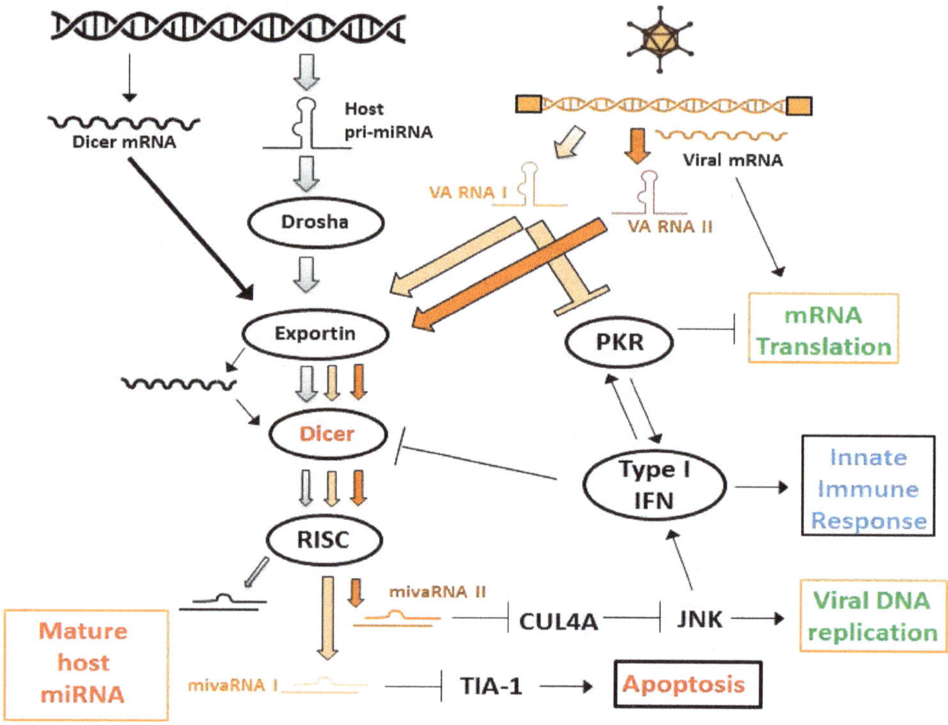

Figure 2. Interaction between adenovirus-encoded small RNA molecules and the host cell RNAi machinery and interferon–double-stranded RNA-activated kinase (PKR) response in the late phase of infection. Virus-associated (VA) RNAs I and II inhibit endogenous miRNA processing via competition at several levels in the RNAi machinery, as well as by reducing Dicer messenger RNA (mRNA) nuclear export. Upon processing of VA RNAs into viral miRNAs (mivaRNAs), they silence cellular mRNAs to inhibit apoptosis and promote viral DNA replication. VA RNA I directly inhibits PKR to promote (viral) mRNA translation and, thus, protein expression for capsid production. Regulation of type I interferons evoking innate immune responses is complex, with on one hand mivaRNA II-mediated stimulation and on the other hand VA RNA I-mediated inhibition of PKR-dependent stimulation. The type I interferon response inhibits Dicer function, contributing to inhibition of miRNA processing. Colors illustrate the different effects in cells in the late phase of infection: red, inhibited; green, stimulated; blue, complex regulation; orange, virus-encoded and virus-related; black, cellular.

Although the efficacy of CRAd-mediated RNAi can probably be improved, the findings by Carette et al. [79] fueled the development of CRAds capable of silencing a variety of target genes in cancer cells to achieve more effective anticancer treatment. In many cases, the target genes were not necessarily expected to impact the efficacy of OVT, but were chosen merely for their known or anticipated anticancer effects. Examples include, among others, silencing an oncogene to inhibit tumorigenicity, silencing a pro-angiogenic growth factor to inhibit angiogenesis, or silencing an apoptosis inhibitor to promote cell death [85–87]. These RNAi-inducing CRAds could, thus, be considered therapeutic agents providing a combination of OVT and targeted therapy. The two independent activities provided by CRAd RNAi were often additive, resulting in a more effective treatment in preclinical models than could be achieved with unarmed control CRAds or replication-defective RNAi vectors. Below, we focus on approaches to achieve more effective OVT.

3.1. Combining OVT with Suppression of CRAd-Inhibitory Target Genes in Cancer Cells

As a logical extension of their above-described observation that Dicer inhibits adenovirus replication, Machitani et al. developed a CRAd expressing an shRNA against Dicer [88]. While replicating in cancer cells, the virus silenced Dicer expression, allowing an increase in VA RNA copy numbers available for inhibiting PKR activity. Consequently, the virus produced higher genome copy numbers and functional progeny titers than did a control virus silencing luciferase. This resulted in a more effective lysis of cancer cells in vitro and a more effective inhibition of tumor growth in vivo. Apparently, in this respect, the gain of more effective VA RNA I-mediated PKR inhibition was more important than the loss of VA RNA II-mediated JNK activation. Although expression of the shRNA was driven by the H1 mammalian polymerase (Pol) III promoter, which is not expected to provide cancer cell selectivity, Dicer expression was not significantly reduced in non-malignant cells. This can best be explained by the fact that CRAds do not replicate in non-cancer cells and, consequently, genome copy numbers in these cells remain very low. Thus, while it can be argued that inhibition of the RNAi machinery in cells could interfere with endogenous miRNA processing causing unpredictable and possibly detrimental effects, the efficiency of shRNA expression in the absence of viral replication was apparently low enough to avoid such unwanted effects.

Another approach that was successfully taken to increase the potency of CRAds is to suppress cell-cycle inhibition in cancer cells. Since CRAds, by design, lost their capacity to take control over the cell cycle, to induce the S phase for efficient genome replication, their replication efficiency depends on the spontaneous progression of cancer cells through the cell cycle. Although proper regulation of the cell cycle is lost in cancer cells, partial regulation of cell-cycle progression may still exist. It could, therefore, be conceived that cell-cycle inhibition, in particular at the Gap 1 (G1) checkpoint, inhibits adenovirus replication. This notion was supported by the finding that overexpression of the major cyclin-dependent kinase (CDK) inhibitor $p21^{WAF1/CIP1}$ in cancer cells inhibited CRAd replication [89] and by the observation that CRAds replicated more efficiently, inducing stronger cytopathic effects, in $p21^{WAF1/CIP1}$ knockout cells [90]. In experiments where CRAds were combined with siRNAs targeting the CDK inhibitors $p21^{WAF1/CIP1}$ or $p27^{KIP}$, virus progeny production and cell-killing potency were increased, the latter depending on cellular p53 status [90] (Figure 1). These observations provided incentive to construct a prostate-specific CRAd expressing an shRNA targeting $p21^{WAF1/CIP1}$ [91]. As expected, the virus exhibited increased replication in and oncolysis of prostate cancer cells in vitro and provided enhanced tumor suppression and survival in a tumor xenograft animal model. Interestingly, apart from the sought effect on adenovirus replication via cell-cycle control, a less anticipated positive effect was observed as well. Silencing $p21^{WAF1/CIP1}$ increased expression of the androgen receptor in prostate cancer cells, thereby stimulating expression of the viral E1A gene driven by the used androgen receptor-dependent prostate-specific promoter. While the relative contributions of both effects are difficult to dissect, the total effect of expressing a $p21^{WAF1/CIP1}$-silencing shRNA from the CRAd genome on the efficacy of OVT was evident.

A foreseeable further improvement of CRAd potency could be to arm the CRAd with both a p53 gene and a $p21^{WAF1/CIP1}$-silencing shRNA. As discussed above, the utility of expressing p53 from the CRAd genome to increase its potency was shown in a variety of cancer models. Since $p21^{WAF1/CIP1}$ is a direct transcriptional target of p53, an unwanted side effect of expressing p53 could be induction of $p21^{WAF1/CIP1}$ expression, limiting CRAd replication. Combined expression of p53 and an shRNA targeting $p21^{WAF1/CIP1}$ could then overcome this limitation. A replication-defective adenovirus vector with cocistronic expression of p53 and a $p21^{WAF1/CIP1}$-targeting artificial miRNA was already tested for its anticancer properties [92]. This viral vector caused more effective apoptosis induction in cancer cells and decreased xenograft tumor growth compared to control vectors expressing only p53 or $p21^{WAF1/CIP1}$-targeting miRNA. A similar favorable combination effect of expressing p53 and silencing $p21^{WAF1/CIP1}$ is to be expected in the context of OVT with a CRAd and, therefore, deserves to be explored.

Along the same line of reasoning, CRAds expressing p53 could probably be made more effective by silencing p53 inhibitors (Figure 1). Many p53 inhibitors are known and confirmed active in cancer cells. Paradigm examples are human papillomavirus (HPV)-E6 protein in cervical cancer cells and amplified mouse double minute 2 homolog (MDM2) in sarcomas. We already showed that CRAds expressing p53 variants incapable of binding to HPV-E6 or MDM2 are more effective in these cancers [61,62]. A limitation of this approach could be that the introduction of mutations in the p53 protein to abrogate binding to its inhibitor might compromise its transcriptional activity. Therefore, combining the expression of wild-type p53 with an shRNA targeting the inhibitor could perhaps be more effective. In addition, cancer cells often express multiple p53 inhibitors. This can be deduced from functional screens for p53 modulators done on cancer cell lines [93–95]. Preventing binding to a single inhibitor might, therefore, not be sufficient in many cases. It will be difficult, if not impossible, to prevent binding of p53 to multiple inhibitors by mutating its amino-acid sequence. In contrast, since their small size allows for multiple shRNAs to be inserted into a single CRAd genome, it should not be too difficult to simultaneously silence multiple p53 inhibitors, maximizing the activity of p53-expressing CRAds.

3.2. *Combining OVT with Targeting Immune Suppression*

As mentioned above, stimulation of the immune system to evoke an anti-tumor response is currently considered a crucial property of oncolytic viruses. This inherent property can be further enhanced by arming the virus with, e.g., immunostimulatory cytokines [7]. While this resulted in much improved OVT systems, active immune suppression in the tumor microenvironment was not fully addressed. Many immune-suppressive molecules, expressed by cancer cells or by suppressive cells of the immune system, were already identified in the tumor microenvironment. Surprisingly, with the technology to combine CRAds with RNAi at hand, arming CRAds with shRNAs targeting such molecules is still quite an unexplored approach. The exception is a CRAd encoding a combination of MART1 melanoma antigen, granulocyte-macrophage colony-stimulating factor (GM-CSF) cytokine, and an shRNA silencing transforming growth factor β2 (TGF-β2) [96]. The rationale for the combination was that, while GM-CSF has many immune-stimulating properties, such as inducing maturation of dendritic cells (DCs), expansion and differentiation of lymphocytes, and recruitment of natural killer (NK) cells, it is also known to enhance the expansion of myeloid-derived suppressor cells (MDSC). TGF-β2 antagonizes GM-CSF-induced DC maturation, inhibits tumor antigen-specific T-cell activation, and stimulates MDSC. Concomitant expression of GM-CSF and suppression of TGF-β2 was, thus, expected to provide a more potent induction of antitumor immunity than expressing GM-CSF alone. This CRAd was tested in a prime-boost therapeutic vaccination strategy, using immune-competent mice and mouse melanoma cells engineered to allow human adenovirus replication. A plasmid expressing MART1 was injected intramuscularly to generate MART1-specific memory T cells and the CRAd was injected intratumorally to boost the immune response. The strategy significantly increased tumor infiltration with cytotoxic T cells, NK cells, NKT cells, (mature) DCs, and macrophages, and decreased regulatory T cells. Immune cell activity was evident from high levels of interferon-γ (IFN-γ) expressed. Together with the oncolytic activity of the virus, this resulted in delayed tumor growth. The contribution of silencing TGF-β2 to the total effect was, however, not entirely clear from the reported results.

Another CRAd that could reactivate antitumor immunity through suppressing an immune suppressive target gene in cancer cells was made, but not tested as such. This CRAd expressing antisense RNA against signal transducer and activator of transcription 3 (STAT3) was shown capable of depleting STAT3 protein in infected cancer cells, as well as its downstream targets [97]. The virus exhibited increased replication and more potent antitumor activity, presumably through counteracting the effects of STAT3 on cell survival, cell migration, and angiogenesis. STAT3 is, however, also involved in immune evasion. It would, therefore, be very interesting to investigate the effects of the STAT3-suppressing virus on the antitumor immune response. The host specificity of human adenoviruses is, however, a major challenge in the design of such experiments. It is very difficult to develop an immune-competent animal model with available tools for immunological studies, that allows replication of human adenoviruses. Although Syrian hamsters are semi-permissive for human adenovirus replication and, thus, provide a useful immune-competent animal model for CRAd efficacy studies [98], and CRAds armed with immune-stimulating transgenes were successfully tested in these animals (e.g., References [99,100]), only few antibodies are available to study hamster immune cell subsets, precluding in-depth analysis of immune responses. For this, ex vivo cultures of human whole tumor-derived single-cell suspensions or fresh tissue slides, or novel mouse models carrying patient-derived xenograft tumors and reconstituted with immune cells from the same donor could prove highly valuable.

3.3. Exploiting Virus–Host Interactions via MicroRNAs

Although, as mentioned above, a physiological role for RNAi in response to virus infection in mammals is being challenged, miRNA-mediated virus–host interactions were described for a number of mammalian viruses (reviewed in References [101–103]). This includes anti-viral or pro-viral effects of host miRNAs, as well as virus-encoded miRNAs that modify host cell biology to establish an environment conducive to completion of the virus lifecycle or that exploit the host RNAi machinery to regulate viral gene expression. Paradigm examples of human miRNAs that promote virus replication include miR-122 and miR-132 that promote hepatitis C virus replication and human immunodeficiency virus 1 (HIV-1) replication, respectively [104,105]. Inhibitory effects of human miRNAs on viruses are, however, more common. They could be part of the host innate antiviral response to recover from an infection, or alternatively, be part of a strategy of the virus to establish a persistent infection [106].

In the case of adenovirus, the VA RNAs constitute the virus-encoded miRNAs (Figure 2). They are processed by the RNAi machinery and are functional in directing silencing of target sequences [76,107,108]. As mentioned above, VA RNA II-encoded viral miRNA was shown to promote adenovirus replication [82]. The functional importance of VA RNA I acting as miRNAs is, however, questioned. While specific inhibition of miRNA target site binding using chemically modified oligonucleotides (i.e., antagomirs) reduced adenovirus replication [76], introduction of mutations in the miRNA seed sequence did not appear to influence adenovirus propagation [83]. Nevertheless, specific cellular target genes for VA RNA I-encoded miRNA, including T-cell-restricted intracellular antigen 1 (TIA-1), were identified and their silencing was confirmed in virus-infected cells [109]. The relevance of silencing these genes for the virus lifecycle is so far mostly unresolved. However, it seems unlikely that adenoviruses evolved specific silencing of target genes in host cells without any functional importance.

Adenoviruses also modulate the expression of host miRNAs. We found that, on top of a general suppression of miRNA levels in adenovirus-infected cancer cells, which is in agreement with the described interferon- and VA RNA-dependent effects on the RNAi machinery, adenovirus serotype 5 significantly reduced expression of a subset of human miRNAs, in particular miR-27, miR-100, miR-155, miR-181, and miR-222 in the late phase of replication [110]. Interestingly, others reported that these miRNAs were also downregulated in cells infected with other viruses, suggesting that these miRNAs could be involved in general antiviral responses, and that viruses evolved specific countermeasures against them. In contrast, only very few miRNAs were induced upon adenovirus infection [110]. Interestingly, among these was miR-132, which was also reported overexpressed in human monocytes infected with several other viruses, suppressing innate antiviral immunity and promoting viral replication [111]. Hence, different viruses exhibit striking similarities in their communication with host cells via miRNAs. As miRNAs usually have many targets, modulation of a single host miRNA by a virus can have complex biological effects, silencing multiple genes that together impact virus replication. In addition, not all target genes predicted on the basis of sequence complementarity are validated as genuine miRNA targets. They may not or only under certain circumstances be silenced. This makes it very difficult to dissect virus–host interactions via miRNAs and identify the crucial host genes contributing to virus biology.

Since miRNA expression is deregulated in cancer cells, CRAds that are replicating in cancer cells might encounter miRNA expression patterns that differ from the ones that human adenoviruses evolved to cope with in their natural differentiated host cells. Consequently, CRAds might not exploit host miRNA expression to their full benefit, and host antiviral miRNA responses to viral infection might not always be effectively counteracted. It can be envisioned, therefore, that overexpression or inhibition of certain miRNAs in cancer cells increases CRAd replication and, thus, oncolytic potency. The identification of miRNAs or miRNA inhibitors for this purpose could perhaps be based on known virus–host interactions and deregulated miRNA expression patterns. For example, tumor suppressor miR-34a expression is often absent or low in cancer cells as a consequence of chromosomal deletion or p53 deficiency. As a direct transcriptional target of p53 and silencer of genes required for cell proliferation and survival, miR-34 mimics part of the p53 activities [112]. In view of the established effects of expressing p53 on CRAd efficacy, it could, thus, be envisioned that expressing miR-34a from the CRAd genome similarly promotes CRAd propagation and cancer cell killing (see Figure 1). MiR-34a-expressing CRAds were presented in two studies. In the first study, hTERT promoter-driven E1A-Δ24-type CRAds were armed with miR-34a and/or interleukin-24 (IL-24) and tested in hepatocellular carcinoma models [20]. The cytokine IL-24 was chosen for its pro-apoptotic and anti-angiogenic properties. Knockdown of miR-34a target genes was confirmed in infected cells. The effects of miR-34a on cytotoxicity in vitro and tumor growth inhibition in vivo were minimal. In contrast, the effects of expressing IL-24 were significant and most prominent anti-tumor effects, including complete tumor regression, were achieved with the virus carrying both miR-34a and IL-24. This suggests that the augmented efficacy of the CRAds depended primarily on the anti-tumor activities of IL-24. In the second study, carcinoembryonic antigen promoter-driven E1A-type CRAds were armed with miR-34a and/or miR-126 and tested in pancreatic cancer models [113]. These two tumor suppressor miRNAs were chosen because of their low expression in pancreatic cancer and confirmed anti-tumor effects. Also in this study the effect of expressing miR-34a was very small. The significance of miR-34a expression for tumor growth control in comparison to a control CRAd was not demonstrated. Interestingly, the miR-34a-expressing CRAd appeared to produce somewhat lower virus titers in pancreatic cancer cell lines than the control virus. In this respect, the virus, thus, did not reproduce previous observations made with p53-expressing CRAds.

Apart from rationally choosing miRNAs to empower CRAds, one may also take an unbiased approach to identify useful miRNAs through functional screening. The availability of genome-wide miRNA mimic and inhibitor libraries allows global analysis of miRNA effects on the lifecycle of viruses [110,114]. Screening for miRNA effects on the replication of a number of human and mouse herpes viruses, including human and mouse cytomegalovirus (CMV), as well as validation on an evolutionary unrelated RNA virus, identified several miRNAs with broad proviral or antiviral properties [114]. In particular, miR-30 broadly stimulated virus replication, and miR-199a and miR-214 inhibited virus replication. Since these observations were made on a diversity of viruses, the miRNAs are not likely to act on viral elements. Instead, most probably, they regulate host cell pathways that support or protect against virus replication. Interestingly, miR-199a and miR-214 are expressed from the same intronic cluster [115] and were found downregulated in CMV-infected cells. Thus, these viruses appear to counteract two antiviral miRNAs by suppressing a single transcript. While miR-30 was not reported differentially expressed in this study, several other viruses were shown to induce the expression of miR-30 variants [116,117]. Intriguingly, on one hand, this upregulation appears IFN-dependent [116], whereas, on the other hand, miR-30 inhibited type-I IFN signaling, allowing increased virus replication [117]. In any event, these findings suggest that certain virus–host interactions via miRNAs are broadly conserved. Hence, observations made in miRNA screens with one virus might be applicable to other viruses as well.

We performed screens for adenovirus replication and cytotoxicity on prostate cancer cells using the same miRNA libraries as were used in the aforementioned study [110]. We identified several miRNAs that stimulate adenovirus-induced cell death. Some of these inhibited adenovirus replication, possibly as a consequence of premature cell death occurring before adenovirus replication was completed. In general, strong induction of cell death was associated with reduced virus production. The exception was miR-181a. Expression of this miRNA strongly induced cell death and slightly increased virus production. The most striking results were obtained with miR-26b. Transfection of miR-26b mimics increased adenovirus-induced cell death, as well as infectious virus production and release. Consequently, adenovirus propagated more rapidly in human prostate cancer cell lines, producing larger plaque sizes. This makes miR-26 a promising candidate for incorporation in the genome of CRAds, to augment OVT of prostate cancer and perhaps other cancers. Construction and characterization of such a novel CRAd is, therefore, warranted. An important question that needs to be answered then is if the timing and levels of miRNA expression obtained during CRAd replication in cancer cells will reproduce the functional effects obtained by combining miRNA mimic transfection and virus infection.

3.4. RNAi Screening for Inhibitors of Oncolytic Virus Efficacy in Cancer Cells

Clearly, virus–host interactions are complex and our current knowledge of the cellular pathways involved is incomplete. There are, therefore, probably many more opportunities to improve the potency of oncolytic viruses than we can rationally hypothesize. The advent of high-throughput screening technology with whole-genome RNAi libraries offered the opportunity to functionally interrogate every single gene, one by one, for its biological role. Many investigations used this technology to identify cellular factors involved in susceptibility to virus infection (reviewed in Reference [118]). These screens helped identify genes responsible for viral entry, uncoating of the virus, replication of the genome, particle assembly, and spread of progeny viruses. Not surprisingly, these studies focused primarily on viruses that constitute a threat to public health, where the knowledge obtained can be used to develop novel antiviral treatments. However, genome-wide RNAi screening can also be used to identify determinants of oncolytic virus anticancer efficacy. With the ultimate goal of increasing the efficacy of OVT by arming the virus with an shRNA in mind, the most important factors of the virus–host interaction to be studied are those involved in virus genome replication and infectious progeny production, host cell killing, and modulation of the immune response. Early steps in the virus lifecycle, such as entry, uncoating, and early gene expression are of less relevance, since these processes cannot be influenced by virus-encoded shRNAs.

Functional genomic screens usually have a single readout of biological effect. Although high-content (image) analyses can measure more complex phenotypes, such as multiple descriptors of cellular morphology, most screens measure a single parameter, such as cell viability, reporter gene expression, or binding/uptake/secretion of a (fluorescent) molecule. Hence, the most appropriate readout for oncolytic virus efficacy needs to be carefully chosen. In addition, it is crucial that the screening effort is followed by thorough examination of the effects of hits on other aspects of the virus lifecycle that are not measured in the discovery screen. For example, in our miRNA library screens described above [110], we identified a number of miRNAs that strongly promoted adenovirus-induced cell death, but this was at the expense of reducing infectious progeny virus production. Thus, interference with host cell gene expression can have opposing effects on different steps in the virus lifecycle, negating the overall effect of gene silencing on OVT.

Until now, there are a few examples reported to show that genome-wide RNAi screening can be used effectively to increase our understanding of the interactions between oncolytic viruses and their host cancer cells and to identify determinants of their anticancer efficacy [119–121].

In the first example, an arrayed siRNA screen was performed, measuring effects on the killing of three cancer cell lines by Maraba virus [119]. Although Maraba is not a widely used oncolytic virus, it is related to the more commonly used VSV, and the major findings in this study could be validated for both viruses. Among the hits identified in the screens, genes within endoplasmic reticulum (ER) stress response pathways were enriched. Knocking down a key gene in the ER stress response pathway activated caspase-dependent apoptosis in virus-infected cells. This increased oncolytic virus potency in many cancer cell lines, reaching up to four orders of magnitude more effective killing. Importantly, augmented killing was much less pronounced on non-malignant cells, providing a high therapeutic index. Notably, in the screen design used, siRNAs were transfected three days before virus infection. Thus, ER stress responses were already inhibited at the time of infection. In follow-up time scheduling experiments with chemical compounds targeting key proteins in the ER stress response pathway, it was found that virus-induced cell killing was augmented when ER stress responses were inhibited prior to infection, but not when they were inhibited during infection. This pointed at a preconditioning process, where inhibition of ER stress response rewired cellular signaling, leading to increased cell death upon subsequent virus infection. Hence, while the findings of this study can be used to develop combination treatments consisting of preconditioning with a chemical compound followed by virus administration, they do not provide footholds for arming viruses with transgenes or shRNAs. For this purpose, it is probably better to delay gene silencing relative to virus infection in the screen design.

In the second example, a pooled shRNA cell depletion screen was done on breast cancer cells that were infected with an oncolytic HSV [120]. This identified a component of the RNA splicing machinery, arginine-rich splicing factor 2 (SRSF2). Although the effects of silencing SRSF2 were not very strong, they could be reproduced on different cell lines and with different reagents. Silencing SRSF2 changed the abundance of pro- and anti-apoptotic mRNAs, and increased apoptosis. In contrast, it did not stimulate virus replication. In fact, a trend toward inhibition of infectious virus production was observed. SRSF2 depletion appeared to dampen HSV-induced activation of mammalian target of rapamycin (mTOR) signaling, thereby probably inhibiting multiple processes, including cell cycle, virus replication, and cellular antiviral responses. Thus, the effects of targeting SRSF2 could promote, as well as inhibit, the HSV oncolytic potency. Apparently, under the applied experimental conditions the overall effect was more potent cytotoxicity. Importantly, also in this study, gene knockdown or knockout was already established before cells were subjected to the virus. It is difficult to predict if delayed silencing of SRSF2 after virus entry will also tip the scales toward increased oncolysis. Interestingly, treatment of cancer cells with an mTOR inhibitor increased the oncolytic potency of CRAds [122,123]. This suggests a similarity in the virus–host interactions of different oncolytic viruses that could allow translation of the findings to broader applications.

The third example is the systematic analysis of human cell factors affecting replication of Myxoma virus [121]. Genome-wide arrayed siRNA screens were done on a breast cancer cell line that was infected with virus three days after siRNA transfection. A recombinant virus was used that expresses a marker gene in the late phase of replication. Marker gene expression served as a surrogate marker for virus replication. The screens yielded a large database of genes that either enhance or inhibit Myxoma virus replication. Follow-up experiments zooming in on certain cellular pathways provided support for their role in virus replication, but interpretation was difficult and some observations seemed contradictory. This is illustrated, e.g., by the observations made when silencing the Raf/mitogen-activated protein kinase kinase (MEK)/extracellular signal-regulated kinase (ERK) pathway. Whereas silencing of many genes upstream of MEK/ERK enhanced virus replication, silencing more downstream pathway genes, including ERK1, inhibited virus replication. Effects of targeting single genes were usually modest. This is in agreement with the concept that viruses modulate cellular pathways by converting multiple switches in a concerted manner, and that host cells respond to virus infection by activating many systems that collectively create antiviral defenses. Finally, based on the known involvement of RNA helicases in permissiveness of cells for replication of a variety of viruses, Rahman et al. [124] performed a focused siRNA library screen on cervical cancer cells using marker gene-expressing Myxoma viruses, followed by confirmation experiments in other cell lines. This effort identified several RNA helicases that consistently increased or decreased Myxoma virus production. An intriguing finding was that knockdown of retinoic acid-induced gene 1 (RIG-1) reduced Myxoma replication. This was highly unexpected, because of the known role of RIG-1 in antiviral innate immune responses. Knockdown of five RNA helicases stimulated Myxoma virus production, without an apparent effect on viral gene expression or replication. Although their mechanism of action remains to be resolved, these genes could be appropriate targets to construct more potent oncolytic Myxoma viruses.

Similar screens, as far as we are aware, are yet to be reported for CRAds. In addition, despite the recognized crucial role of the immune system in the anticancer effects of oncolytic viruses, RNAi screens with oncolytic viruses have not studied this important aspect. While the design of relevant high-throughput assays for modulation of the immune system will be a challenge, such efforts are warranted. They are expected to provide highly valuable information to design next-generation CRAds for more effective OVT.

4. Conclusions

For the past decades, molecular cancer and virology research advanced forward with incredible speed and efficiency. This brought OVT from the bench to the bedside. Novel molecular targets were identified, oncolytic virus constructs designed, and combined therapies proposed or tested in clinical trials. So far, only a few OVTs obtained market authorization, but others are likely to follow. Oncolytic virus therapy moved from a mere concept to a functioning approach. RNA interference became one of the successfully applied methods to make OVT more effective. Here, knowledge of miRNA expression in cancer and healthy cells, as well as of miRNA-mediated virus–host interactions, is implemented. Clearly, a great amount of work needs to be done before OVT will reach its full potential. The contribution of RNAi technology to reaching this goal is difficult to predict. However, although the battle is not over, new research results are constantly made that bring us closer to victory.

Acknowledgments: The authors would like to thank Tanja D. de Gruijl for critical reading of the manuscript and helpful suggestions.

References

1. Hanahan, D.; Weinberg, R.A. The hallmarks of cancer. *Cell* **2000**, *100*, 57–70. [CrossRef]
2. Parato, K.A.; Senger, D.; Forsyth, P.A.J.; Bell, J.C. Recent progress in the battle between oncolytic viruses and tumours. *Nat. Rev. Cancer* **2005**, *5*, 965–976. [CrossRef] [PubMed]
3. Vähä-Koskela, M.J.V.; Heikkilä, J.E.; Hinkkanen, A.E. Oncolytic viruses in cancer therapy. *Cancer Lett.* **2007**, *254*, 178–216. [CrossRef] [PubMed]
4. Sinkovics, J.G.; Horvath, J.C. Natural and genetically engineered viral agents for oncolysis and gene therapy of human cancers. *Arch. Immunol. Ther. Exp. (Warsz.)* **2008**, *56*, 1–59. [CrossRef] [PubMed]
5. Nguyen, T.L.A.; Tumilasci, V.F.; Singhroy, D.; Arguello, M.; Hiscott, J. The emergence of combinatorial strategies in the development of RNA oncolytic virus therapies. *Cell. Microbiol.* **2009**, *11*, 889–897. [CrossRef] [PubMed]
6. Diaconu, I.; Cerullo, V.; Hirvinen, M.L.M.; Escutenaire, S.; Ugolini, M.; Pesonen, S.K.; Bramante, S.; Parviainen, S.; Kanerva, A.; Loskog, A.S.I.; et al. Immune response is an important aspect of the antitumor effect produced by a CD40L-encoding oncolytic adenovirus. *Cancer Res.* **2012**, *72*, 2327–2338. [CrossRef] [PubMed]
7. de Gruijl, T.D.; Janssen, A.B.; van Beusechem, V.W. Arming oncolytic viruses to leverage antitumor immunity. *Expert Opin. Biol. Ther.* **2015**, *15*, 959–971. [CrossRef] [PubMed]
8. Pol, J.; Buqué, A.; Aranda, F.; Bloy, N.; Cremer, I.; Eggermont, A.; Erbs, P.; Fucikova, J.; Galon, J.; Limacher, J.-M.; et al. Trial Watch—Oncolytic viruses and cancer therapy. *Oncoimmunology* **2016**, *5*, e1117740. [CrossRef] [PubMed]
9. Andtbacka, R.H.I.; Kaufman, H.L.; Collichio, F.; Amatruda, T.; Senzer, N.; Chesney, J.; Delman, K.A.; Spitler, L.E.; Puzanov, I.; Agarwala, S.S.; et al. Talimogene Laherparepvec Improves Durable Response Rate in Patients With Advanced Melanoma. *J. Clin. Oncol.* **2015**, *33*, 2780–2788. [CrossRef] [PubMed]
10. Yu, W.; Fang, H. Clinical Trials with Oncolytic Adenovirus in China. *Curr. Cancer Drug Targets* **2007**, *7*, 141–148. [CrossRef] [PubMed]
11. Chiocca, E.A.; Rabkin, S.D. Oncolytic viruses and their application to cancer immunotherapy. *Cancer Immunol. Res.* **2014**, *2*, 295–300. [CrossRef] [PubMed]
12. Russell, S.J.; Peng, K.-W.; Bell, J.C. Oncolytic virotherapy. *Nat. Biotechnol.* **2012**, *30*, 658–670. [CrossRef] [PubMed]
13. Cattaneo, R.; Miest, T.; Shashkova, E.V.; Barry, M.A. Reprogrammed viruses as cancer therapeutics: Targeted, armed and shielded. *Nat. Rev. Microbiol.* **2008** *6*, 529–540. [CrossRef] [PubMed]

14. Lion, T. Adenovirus Infections in Immunocompetent and Immunocompromised Patients. *Clin. Microbiol. Rev.* **2014**, *27*, 441–462. [CrossRef] [PubMed]
15. Ben-Israel, H.; Kleinberger, T. Adenovirus and cell cycle control. *Front. Biosci.* **2002**, *7*, 1369–1395. [CrossRef]
16. Robert-Guroff, M. Replicating and non-replicating viral vectors for vaccine development. *Curr. Opin. Biotechnol.* **2007**, *18*, 546–556. [CrossRef] [PubMed]
17. Bauerschmitz, G.J.; Barker, S.D.; Hemminki, A. Adenoviral gene therapy for cancer: From vectors to targeted and replication competent agents (review). *Int. J. Oncol.* **2002**, *21*, 1161–1174. [CrossRef] [PubMed]
18. Doloff, J.C.; Waxman, D.J. Dual E1A oncolytic adenovirus: Targeting tumor heterogeneity with two independent cancer-specific promoter elements, DF3/MUC1 and hTERT. *Cancer Gene Ther.* **2011**, *18*, 153–166. [CrossRef] [PubMed]
19. Sugio, K.; Sakurai, F.; Katayama, K.; Tashiro, K.; Matsui, H.; Kawabata, K.; Kawase, A.; Iwaki, M.; Hayakawa, T.; Fujiwara, T.; et al. Enhanced Safety Profiles of the Telomerase-Specific Replication-Competent Adenovirus by Incorporation of Normal Cell-Specific microRNA-Targeted Sequences. *Clin. Cancer Res.* **2011**, *17*, 2807–2818. [CrossRef] [PubMed]
20. Lou, W.; Chen, Q.; Ma, L.; Liu, J.; Yang, Z.; Shen, J.; Cui, Y.; Bian, X.; Qian, C. Oncolytic adenovirus co-expressing miRNA-34a and IL-24 induces superior antitumor activity in experimental tumor model. *J. Mol. Med.* **2013**, *91*, 715–725. [CrossRef] [PubMed]
21. Bischoff, J.R.; Kirn, D.H.; Williams, A.; Heise, C.; Horn, S.; Muna, M.; Ng, L.; Nye, J.A.; Sampson-Johannes, A.; Fattaey, A.; et al. An adenovirus mutant that replicates selectively in p53-deficient human tumor cells. *Science* **1996**, *274*, 373–376. [CrossRef] [PubMed]
22. Nemunaitis, J.; Khuri, F.; Ganly, I.; Arseneau, J.; Posner, M.; Vokes, E.; Kuhn, J.; McCarty, T.; Landers, S.; Blackburn, A.; et al. Phase II trial of intratumoral administration of ONYX-015, a replication-selective adenovirus, in patients with refractory head and neck cancer. *J. Clin. Oncol.* **2001**, *19*, 289–298. [CrossRef] [PubMed]
23. Cascallo, M.; Alonso, M.M.; Rojas, J.J.; Perez-Gimenez, A.; Fueyo, J.; Alemany, R. Systemic Toxicity–Efficacy Profile of ICOVIR-5, a Potent and Selective Oncolytic Adenovirus Based on the pRB Pathway. *Mol. Ther.* **2007**, *15*, 1607–1615. [CrossRef] [PubMed]
24. Fueyo, J.; Gomez-Manzano, C.; Alemany, R.; Lee, P.S.Y.; McDonnell, T.J.; Mitlianga, P.; Shi, Y.-X.; Levin, V.A.; Yung, W.K.A.; Kyritsis, A.P. A mutant oncolytic adenovirus targeting the Rb pathway produces anti-glioma effect in vivo. *Oncogene* **2000**, *19*, 2–12. [CrossRef] [PubMed]
25. Eager, R.M.; Nemunaitis, J. Clinical development directions in oncolytic viral therapy. *Cancer Gene Ther.* **2011**, *18*, 305–317. [CrossRef] [PubMed]
26. Li, Y.; Pong, R.C.; Bergelson, J.M.; Hall, M.C.; Sagalowsky, A.I.; Tseng, C.P.; Wang, Z.; Hsieh, J.T. Loss of adenoviral receptor expression in human bladder cancer cells: A potential impact on the efficacy of gene therapy. *Cancer Res.* **1999**, *59*, 325–330. [PubMed]
27. Reeh, M.; Bockhorn, M.; Görgens, D.; Vieth, M.; Hoffmann, T.; Simon, R.; Izbicki, J.R.; Sauter, G.; Schumacher, U.; Anders, M. Presence of the Coxsackievirus and Adenovirus Receptor (CAR) in human neoplasms: A multitumour array analysis. *Br. J. Cancer* **2013**, *109*, 1848–1858. [CrossRef] [PubMed]
28. Suzuki, K.; Fueyo, J.; Krasnykh, V.; Reynolds, P.N.; Curiel, D.T.; Alemany, R. A conditionally replicative adenovirus with enhanced infectivity shows improved oncolytic potency. *Clin. Cancer Res.* **2001**, *7*, 120–126. [CrossRef] [PubMed]
29. Guse, K.; Ranki, T.; Ala-Opas, M.; Bono, P.; Särkioja, M.; Rajecki, M.; Kanerva, A.; Hakkarainen, T.; Hemminki, A. Treatment of metastatic renal cancer with capsid-modified oncolytic adenoviruses. *Mol. Cancer Ther.* **2007**, *6*, 2728–2736. [CrossRef] [PubMed]
30. Kanerva, A.; Mikheeva, G.V.; Krasnykh, V.; Coolidge, C.J.; Lam, J.T.; Mahasreshti, P.J.; Barker, S.D.; Straughn, M.; Barnes, M.N.; Alvarez, R.D.; et al. Targeting adenovirus to the serotype 3 receptor increases gene transfer efficiency to ovarian cancer cells. *Clin. Cancer Res.* **2002**, *8*, 275–280. [PubMed]
31. van Beusechem, V.W.; Mastenbroek, D.C.J.; van den Doel, P.B.; Lamfers, M.L.M.; Grill, J.; Würdinger, T.; Haisma, H.J.; Pinedo, H.M.; Gerritsen, W.R. Conditionally replicative adenovirus expressing a targeting adapter molecule exhibits enhanced oncolytic potency on CAR-deficient tumors. *Gene Ther.* **2003**, *10*, 1982–1991. [CrossRef] [PubMed]
32. Choi, J.-W.; Lee, Y.S.; Yun, C.-O.; Kim, S.W. Polymeric oncolytic adenovirus for cancer gene therapy. *J. Control. Release* **2015**, *219*, 181–191. [CrossRef] [PubMed]

33. Power, A.T.; Bell, J.C. Taming the Trojan horse: Optimizing dynamic carrier cell/oncolytic virus systems for cancer biotherapy. *Gene Ther.* **2008**, *15*, 772–779. [CrossRef] [PubMed]
34. Yong, R.L.; Shinojima, N.; Fueyo, J.; Gumin, J.; Vecil, G.G.; Marini, F.C.; Bogler, O.; Andreeff, M.; Lang, F.F. Human Bone Marrow-Derived Mesenchymal Stem Cells for Intravascular Delivery of Oncolytic Adenovirus 24-RGD to Human Gliomas. *Cancer Res.* **2009**, *69*, 8932–8940. [CrossRef] [PubMed]
35. Ong, H.T.; Hasegawa, K.; Dietz, A.B.; Russell, S.J.; Peng, K.-W. Evaluation of T cells as carriers for systemic measles virotherapy in the presence of antiviral antibodies. *Gene Ther.* **2007**, *14*, 324–333. [CrossRef] [PubMed]
36. Power, A.T.; Wang, J.; Falls, T.J.; Paterson, J.M.; Parato, K.A.; Lichty, B.D.; Stojdl, D.F.; Forsyth, P.A.J.; Atkins, H.; Bell, J.C. Carrier cell-based delivery of an oncolytic virus circumvents antiviral immunity. *Mol. Ther.* **2007**, *15*, 123–130. [CrossRef] [PubMed]
37. Ylosmaki, E.; Hakkarainen, T.; Hemminki, A.; Visakorpi, T.; Andino, R.; Saksela, K. Generation of a Conditionally Replicating Adenovirus Based on Targeted Destruction of E1A mRNA by a Cell Type-Specific MicroRNA. *J. Virol.* **2008**, *82*, 11009–11015. [CrossRef] [PubMed]
38. Cawood, R.; Chen, H.H.; Carroll, F.; Bazan-Peregrino, M.; van Rooijen, N.; Seymour, L.W. Use of Tissue-Specific MicroRNA to Control Pathology of Wild-Type Adenovirus without Attenuation of Its Ability to Kill Cancer Cells. *PLoS Pathog.* **2009**, *5*, e1000440. [CrossRef] [PubMed]
39. Leja, J.; Nilsson, B.; Yu, D.; Gustafson, E.; Åkerström, G.; Öberg, K.; Giandomenico, V.; Essand, M. Double-Detargeted Oncolytic Adenovirus Shows Replication Arrest in Liver Cells and Retains Neuroendocrine Cell Killing Ability. *PLoS ONE* **2010**, *5*, e8916. [CrossRef] [PubMed]
40. Callegari, E.; Elamin, B.K.; D'Abundo, L.; Falzoni, S.; Donvito, G.; Moshiri, F.; Milazzo, M.; Altavilla, G.; Giacomelli, L.; Fornari, F.; et al. Anti-Tumor Activity of a miR-199-dependent Oncolytic Adenovirus. *PLoS ONE* **2013**, *8*, e73964. [CrossRef] [PubMed]
41. Bofill-De Ros, X.; Gironella, M.; Fillat, C. MiR-148a- and miR-216a-regulated Oncolytic Adenoviruses Targeting Pancreatic Tumors Attenuate Tissue Damage Without Perturbation of miRNA Activity. *Mol. Ther.* **2014**, *22*, 1665–1677. [CrossRef] [PubMed]
42. Yao, W.; Guo, G.; Zhang, Q.; Fan, L.; Wu, N.; Bo, Y. The application of multiple miRNA response elements enables oncolytic adenoviruses to possess specificity to glioma cells. *Virology* **2014**, *458*, 69–82. [CrossRef] [PubMed]
43. Gürlevik, E.; Woller, N.; Schache, P.; Malek, N.P.; Wirth, T.C.; Zender, L.; Manns, M.P.; Kubicka, S.; Kühnel, F. p53-dependent antiviral RNA-interference facilitates tumor-selective viral replication. *Nucleic. Acids Res.* **2009**, *37*, e84. [CrossRef] [PubMed]
44. Gros, A.; Martínez-Quintanilla, J.; Puig, C.; Guedan, S.; Molleví, D.G.; Alemany, R.; Cascallo, M. Bioselection of a gain of function mutation that enhances adenovirus 5 release and improves its antitumoral potency. *Cancer Res.* **2008**, *68*, 8928–8937. [CrossRef] [PubMed]
45. Dong, W.; van Ginkel, J.-W.H.; Au, K.Y.; Alemany, R.; Meulenberg, J.J.M.; van Beusechem, V.W. ORCA-010, a Novel Potency-Enhanced Oncolytic Adenovirus, Exerts Strong Antitumor Activity in Preclinical Models. *Hum. Gene Ther.* **2014**, *25*, 897–904. [CrossRef] [PubMed]
46. Kuhn, I.; Harden, P.; Bauzon, M.; Chartier, C.; Nye, J.; Thorne, S.; Reid, T.; Ni, S.; Lieber, A.; Fisher, K.; et al. Directed Evolution Generates a Novel Oncolytic Virus for the Treatment of Colon Cancer. *PLoS ONE* **2008**, *3*, e2409. [CrossRef] [PubMed]
47. Guedan, S.; Grases, D.; Rojas, J.J.; Gros, A.; Vilardell, F.; Vile, R.; Mercade, E.; Cascallo, M.; Alemany, R. GALV expression enhances the therapeutic efficacy of an oncolytic adenovirus by inducing cell fusion and enhancing virus distribution. *Gene Ther.* **2012**, *19*, 1048–1057. [CrossRef] [PubMed]
48. Guedan, S.; Rojas, J.J.; Gros, A.; Mercade, E.; Cascallo, M.; Alemany, R. Hyaluronidase expression by an oncolytic adenovirus enhances its intratumoral spread and suppresses tumor growth. *Mol. Ther.* **2010**, *18*, 1275–1283. [CrossRef] [PubMed]
49. Kim, J.H.; Lee, Y.S.; Kim, H.; Huang, J.H.; Yoon, A.R.; Yun, C.O. Relaxin expression from tumor-targeting adenoviruses and its intratumoral spread, apoptosis induction, and efficacy. *J. Natl. Cancer Inst.* **2006**, *98*, 1482–1493. [CrossRef] [PubMed]
50. Choi, I.-K.; Lee, Y.-S.; Yoo, J.Y.; Yoon, A.-R.; Kim, H.; Kim, D.-S.; Seidler, D.G.; Kim, J.-H.; Yun, C.-O. Effect of decorin on overcoming the extracellular matrix barrier for oncolytic virotherapy. *Gene Ther.* **2010**, *17*, 190–201. [CrossRef] [PubMed]

51. Morris, J.C.; Wildner, O. Therapy of Head and Neck Squamous Cell Carcinoma with an Oncolytic Adenovirus Expressing HSV-tk. *Mol. Ther.* **2000**, *1*, 56–62. [CrossRef] [PubMed]
52. Oosterhoff, D.; Pinedo, H.M.; Witlox, M.A.; Carette, J.E.; Gerritsen, W.R.; van Beusechem, V.W. Gene-directed enzyme prodrug therapy with carboxylesterase enhances the anticancer efficacy of the conditionally replicating adenovirus AdΔ24. *Gene Ther.* **2005**, *12*, 1011–1018. [CrossRef] [PubMed]
53. Mi, J.; Li, Z.-Y.; Ni, S.; Steinwaerder, D.; Lieber, A. Induced Apoptosis Supports Spread of Adenovirus Vectors in Tumors. *Hum. Gene Ther.* **2001**, *12*, 1343–1352. [CrossRef] [PubMed]
54. Van Beusechem, V.W.; Van den Doel, P.B.; Grill, J.; Pinedo, H.M.; Gerritsen, W.R. Conditionally replicative adenovirus expressing p53 exhibits enhanced oncolytic potency. *Cancer Res.* **2002**, *62*, 6165–6171. [PubMed]
55. Sauthoff, H.; Pipiya, T.; Heitner, S.; Chen, S.; Norman, R.G.; Rom, W.N.; Hay, J.G. Late Expression of p53 from a Replicating Adenovirus Improves Tumor Cell Killing and Is More Tumor Cell Specific than Expression of the Adenoviral Death Protein. *Hum. Gene Ther.* **2002**, *13*, 1859–1871. [CrossRef] [PubMed]
56. Hall, A.R.; Dix, B.R.; O'Carroll, S.J.; Braithwaite, A.W. p53-dependent cell death/apoptosis is required for a productive adenovirus infection. *Nat. Med.* **1998**, *4*, 1068–1072. [CrossRef] [PubMed]
57. Royds, J.A.; Hibma, M.; Dix, B.R.; Hananeia, L.; Russell, I.A.; Wiles, A.; Wynford-Thomas, D.; Braithwaite, A.W. p53 promotes adenoviral replication and increases late viral gene expression. *Oncogene* **2006**, *25*, 1509–1520. [CrossRef] [PubMed]
58. Wang, X.; Su, C.; Cao, H.; Li, K.; Chen, J.; Jiang, L.; Zhang, Q.; Wu, X.; Jia, X.; Liu, Y.; et al. A novel triple-regulated oncolytic adenovirus carrying p53 gene exerts potent antitumor efficacy on common human solid cancers. *Mol. Cancer Ther.* **2008**, *7*, 1598–1603. [CrossRef] [PubMed]
59. Geoerger, B.; Vassal, G.; Opolon, P.; Dirven, C.M.F.; Morizet, J.; Laudani, L.; Grill, J.; Giaccone, G.; Vandertop, W.P.; Gerritsen, W.R.; et al. Oncolytic Activity of p53-Expressing Conditionally Replicative Adenovirus AdΔ24-p53 against Human Malignant Glioma. *Cancer Res.* **2004**, *64*, 5753–5759. [CrossRef] [PubMed]
60. Geoerger, B.; van Beusechem, V.W.; Opolon, P.; Morizet, J.; Laudani, L.; Lecluse, Y.; Barrois, M.; Idema, S.; Grill, J.; Gerritsen, W.R.; et al. Expression of p53, or targeting towards EGFR, enhances the oncolytic potency of conditionally replicative adenovirus against neuroblastoma. *J. Gene Med.* **2005**, *7*, 584–594. [CrossRef] [PubMed]
61. van Beusechem, V.W.; van den Doel, P.B.; Gerritsen, W.R. Conditionally replicative adenovirus expressing degradation-resistant p53 for enhanced oncolysis of human cancer cells overexpressing murine double minute 2. *Mol. Cancer Ther.* **2005**, *4*, 1013–1018. [CrossRef] [PubMed]
62. Heideman, D.A.M.; Steenbergen, R.D.M.; van der Torre, J.; Scheffner, M.; Alemany, R.; Gerritsen, W.R.; Meijer, C.J.L.M.; Snijders, P.J.F.; van Beusechem, V.W. Oncolytic Adenovirus Expressing a p53 Variant Resistant to Degradation by HPV E6 Protein Exhibits Potent and Selective Replication in Cervical Cancer. *Mol. Ther.* **2005**, *12*, 1083–1090. [CrossRef] [PubMed]
63. Graat, H.C.A.; Carette, J.E.; Schagen, F.H.E.; Vassilev, L.T.; Gerritsen, W.R.; Kaspers, G.J.L.; Wuisman, P.I.J.M.; van Beusechem, V.W. Enhanced tumor cell kill by combined treatment with a small-molecule antagonist of mouse double minute 2 and adenoviruses encoding p53. *Mol. Cancer Ther.* **2007**, *6*, 1552–1561. [CrossRef] [PubMed]
64. Carette, J.E.; Graat, H.C.A.; Schagen, F.H.E.; Abou El Hassan, M.A.I.; Gerritsen, W.R.; van Beusechem, V.W. Replication-dependent transgene expression from a conditionally replicating adenovirus via alternative splicing to a heterologous splice-acceptor site. *J. Gene Med.* **2005**, *7*, 1053–1062. [CrossRef] [PubMed]
65. Hannon, G.J. RNA interference. *Nature* **2002**, *418*, 244–251. [CrossRef] [PubMed]
66. Carmell, M.A.; Hannon, G.J. RNase III enzymes and the initiation of gene silencing. *Nat. Struct. Mol. Biol.* **2004**, *11*, 214–218. [CrossRef] [PubMed]
67. Huntzinger, E.; Izaurralde, E. Gene silencing by microRNAs: Contributions of translational repression and mRNA decay. *Nat. Rev. Genet.* **2011**, *12*, 99–110. [CrossRef] [PubMed]
68. Doench, J.G.; Sharp, P.A. Specificity of microRNA target selection in translational repression. *Genes Dev.* **2004**, *18*, 504–511. [CrossRef] [PubMed]
69. Bushati, N.; Cohen, S.M. microRNA Functions. *Annu. Rev. Cell Dev. Biol.* **2007**, *23*, 175–205. [CrossRef] [PubMed]
70. Xia, H.; Mao, Q.; Paulson, H.L.; Davidson, B.L. siRNA-mediated gene silencing in vitro and in vivo. *Nat. Biotechnol.* **2002** *20*, 1006–1010. [CrossRef] [PubMed]

71. Cullen, B.R. RNA interference: Antiviral defense and genetic tool. *Nat. Immunol.* **2002**, *3*, 597–599. [CrossRef] [PubMed]
72. Roth, B.M.; Pruss, G.J.; Vance, V.B. Plant viral suppressors of RNA silencing. *Virus Res.* **2004**, *102*, 97–108. [CrossRef] [PubMed]
73. Backes, S.; Langlois, R.A.; Schmid, S.; Varble, A.; Shim, J.V.; Sachs, D.; TenOever, B.R. The Mammalian response to virus infection is independent of small RNA silencing. *Cell Rep.* **2014**, *8*, 114–125. [CrossRef] [PubMed]
74. tenOever, B.R. Questioning antiviral RNAi in mammals. *Nat. Microbiol.* **2017**, *2*, 17052. [CrossRef] [PubMed]
75. Lu, S.; Cullen, B.R. Adenovirus VA1 noncoding RNA can inhibit small interfering RNA and MicroRNA biogenesis. *J. Virol.* **2004**, *78*, 12868–12876. [CrossRef] [PubMed]
76. Aparicio, O.; Razquin, N.; Zaratiegui, M.; Narvaiza, I.; Fortes, P. Adenovirus Virus-Associated RNA Is Processed to Functional Interfering RNAs Involved in Virus Production. *J. Virol.* **2006**, *80*, 1376–1384. [CrossRef] [PubMed]
77. Bennasser, Y.; Chable-Bessia, C.; Triboulet, R.; Gibbings, D.; Gwizdek, C.; Dargemont, C.; Kremer, E.J.; Voinnet, O.; Benkirane, M. Competition for XPO5 binding between Dicer mRNA, pre-miRNA and viral RNA regulates human Dicer levels. *Nat. Struct. Mol. Biol.* **2011**, *18*, 323–327. [CrossRef] [PubMed]
78. Machitani, M.; Sakurai, F.; Wakabayashi, K.; Takayama, K.; Tachibana, M.; Mizuguchi, H. Type I Interferons Impede Short Hairpin RNA-Mediated RNAi via Inhibition of Dicer-Mediated Processing to Small Interfering RNA. *Mol. Ther.-Nucleic Acids* **2017**, *6*, 173–182. [CrossRef] [PubMed]
79. Carette, J.E.; Overmeer, R.M.; Schagen, F.H.E.; Alemany, R.; Barski, O.A.; Gerritsen, W.R.; Van Beusechem, V.W. Conditionally Replicating Adenoviruses Expressing Short Hairpin RNAs Silence the Expression of a Target Gene in Cancer Cells. *Cancer Res.* **2004**, *64*, 2663–2667. [CrossRef] [PubMed]
80. Machitani, M.; Sakurai, F.; Katayama, K.; Tachibana, M.; Suzuki, T.; Matsui, H.; Yamaguchi, T.; Mizuguchi, H. Improving adenovirus vector-mediated RNAi efficiency by lacking the expression of virus-associated RNAs. *Virus Res.* **2013**, *178*, 357–363. [CrossRef] [PubMed]
81. Thimmappaya, B.; Weinberger, C.; Schneider, R.J.; Shenk, T. Adenovirus VAI RNA is required for efficient translation of viral mRNAs at late times after infection. *Cell* **1982**, *31*, 543–551. [CrossRef]
82. Wakabayashi, K.; Machitani, M.; Tachibana, M.; Sakurai, F.; Mizuguchi, H. A microRNA derived from adenovirus virus-associated RNAII promotes virus infection *via* post-transcriptional gene silencing. *J. Virol.* **2018**, Epub ahead of print. [CrossRef] [PubMed]
83. Kamel, W.; Segerman, B.; Öberg, D.; Punga, T.; Akusjärvi, G. The adenovirus VA RNA-derived miRNAs are not essential for lytic virus growth in tissue culture cells. *Nucleic Acids Res.* **2013**, *41*, 4802–4812. [CrossRef] [PubMed]
84. Machitani, M.; Sakurai, F.; Wakabayashi, K.; Tomita, K.; Tachibana, M.; Mizuguchi, H. Dicer functions as an antiviral system against human adenoviruses via cleavage of adenovirus-encoded noncoding RNA. *Sci. Rep.* **2016**, *6*, 27598. [CrossRef] [PubMed]
85. Zhang, Y.-A.; Nemunaitis, J.; Samuel, S.K.; Chen, P.; Shen, Y.; Tong, A.W. Antitumor Activity of an Oncolytic Adenovirus-Delivered Oncogene Small Interfering RNA. *Cancer Res.* **2006**, *66*, 9736–9743. [CrossRef] [PubMed]
86. Chu, L.; Gu, J.; Sun, L.; Qian, Q.; Qian, C.; Liu, X. Oncolytic adenovirus-mediated shRNA against Apollon inhibits tumor cell growth and enhances antitumor effect of 5-fluorouracil. *Gene Ther.* **2008**, *15*, 484–494. [CrossRef] [PubMed]
87. Yoo, J.Y.; Kim, J.-H.; Kwon, Y.-G.; Kim, E.-C.; Kim, N.K.; Choi, H.J.; Yun, C.-O. VEGF-specific short hairpin RNA-expressing oncolytic adenovirus elicits potent inhibition of angiogenesis and tumor growth. *Mol. Ther.* **2007**, *15*, 295–302. [CrossRef] [PubMed]
88. Machitani, M.; Sakurai, F.; Wakabayashi, K.; Tachibana, M.; Fujiwara, T.; Mizuguchi, H. Enhanced Oncolytic Activities of the Telomerase-Specific Replication-Competent Adenovirus Expressing Short-Hairpin RNA against Dicer. *Mol. Cancer Ther.* **2017**, *16*, 251–259. [CrossRef] [PubMed]
89. Höti, N.; Chowdhury, W.; Hsieh, J.-T.; Sachs, M.D.; Lupold, S.E.; Rodriguez, R. Valproic Acid, a Histone Deacetylase Inhibitor, Is an Antagonist for Oncolytic Adenoviral Gene Therapy. *Mol. Ther.* **2006**, *14*, 768–778. [CrossRef] [PubMed]

90. Shiina, M.; Lacher, M.D.; Christian, C.; Korn, W.M. RNA interference-mediated knockdown of p21WAF1 enhances anti-tumor cell activity of oncolytic adenoviruses. *Cancer Gene Ther.* **2009**, *16*, 810–819. [CrossRef] [PubMed]
91. Höti, N.; Chowdhury, W.H.; Mustafa, S.; Ribas, J.; Castanares, M.; Johnson, T.; Liu, M.; Lupold, S.E.; Rodriguez, R. Armoring CRAds with p21/Waf-1 shRNAs: The next generation of oncolytic adenoviruses. *Cancer Gene Ther.* **2010**, *17*, 585–597. [CrossRef] [PubMed]
92. Idogawa, M.; Sasaki, Y.; Suzuki, H.; Mita, H.; Imai, K.; Shinomura, Y.; Tokino, T. A Single Recombinant Adenovirus Expressing p53 and p21-targeting Artificial microRNAs Efficiently Induces Apoptosis in Human Cancer Cells. *Clin. Cancer Res.* **2009**, *15*, 3725–3732. [CrossRef] [PubMed]
93. Huang, Q.; Raya, A.; DeJesus, P.; Chao, S.-H.; Quon, K.C.; Caldwell, J.S.; Chanda, S.K.; Izpisua-Belmonte, J.C.; Schultz, P.G. Identification of p53 regulators by genome-wide functional analysis. *Proc. Natl. Acad. Sci. USA* **2004**, *101*, 3456–3461. [CrossRef] [PubMed]
94. Llanos, S.; Efeyan, A.; Monsech, J.; Dominguez, O.; Serrano, M. A High-Throughput Loss-of-Function Screening Identifies Novel p53 Regulators. *Cell Cycle* **2006**, *5*, 1880–1885. [CrossRef] [PubMed]
95. Siebring-van Olst, E.; Blijlevens, M.; de Menezes, R.X.; van der Meulen-Muileman, I.H.; Smit, E.F.; van Beusechem, V.W. A genome-wide siRNA screen for regulators of tumor suppressor p53 activity in human non-small cell lung cancer cells identifies components of the RNA splicing machinery as targets for anticancer treatment. *Mol. Oncol.* **2017**, *11*, 534–551. [CrossRef] [PubMed]
96. Kim, S.Y.; Kang, D.; Choi, H.J.; Joo, Y.; Kim, J.-H.; Song, J.J. Prime-boost immunization by both DNA vaccine and oncolytic adenovirus expressing GM-CSF and shRNA of TGF-β2 induces anti-tumor immune activation. *Oncotarget* **2017**, *8*, 15858–15877. [CrossRef] [PubMed]
97. Han, Z.; Hong, Z.; Chen, C.; Gao, Q.; Luo, D.; Fang, Y.; Cao, Y.; Zhu, T.; Jiang, X.; Ma, Q.; et al. A novel oncolytic adenovirus selectively silences the expression of tumor-associated STAT3 and exhibits potent antitumoral activity. *Carcinogenesis* **2009**, *30*, 2014–2022. [CrossRef] [PubMed]
98. Thomas, M.A.; Spencer, J.F.; La Regina, M.C.; Dhar, D.; Tollefson, A.E.; Toth, K.; Wold, W.S.M. Syrian Hamster as a Permissive Immunocompetent Animal Model for the Study of Oncolytic Adenovirus Vectors. *Cancer Res.* **2006**, *66*, 1270–1276. [CrossRef] [PubMed]
99. Cerullo, V.; Pesonen, S.; Diaconu, I.; Escutenaire, S.; Arstila, P.T.; Ugolini, M.; Nokisalmi, P.; Raki, M.; Laasonen, L.; Särkioja, M.; et al. Oncolytic Adenovirus Coding for Granulocyte Macrophage Colony-Stimulating Factor Induces Antitumoral Immunity in Cancer Patients. *Cancer Res.* **2010**, *70*, 4297–4309. [CrossRef] [PubMed]
100. Wang, P.; Li, X.; Wang, J.; Gao, D.; Li, Y.; Li, H.; Chu, Y.; Zhang, Z.; Liu, H.; Jiang, G.; et al. Re-designing Interleukin-12 to enhance its safety and potential as an anti-tumor immunotherapeutic agent. *Nat. Commun.* **2017**, *8*, 1395. [CrossRef] [PubMed]
101. Schütz, S.; Sarnow, P. Interaction of viruses with the mammalian RNA interference pathway. *Virology* **2006**, *344*, 151–157. [CrossRef] [PubMed]
102. Skalsky, R.L.; Cullen, B.R. Viruses, microRNAs, and Host Interactions. *Annu. Rev. Microbiol.* **2010**, *64*, 123–141. [CrossRef] [PubMed]
103. Grundhoff, A.; Sullivan, C.S. Virus-encoded microRNAs. *Virology* **2011**, *411*, 325–343. [CrossRef] [PubMed]
104. Jopling, C.L. Modulation of Hepatitis C Virus RNA Abundance by a Liver-Specific MicroRNA. *Science* **2005**, *309*, 1577–1581. [CrossRef] [PubMed]
105. Chiang, K.; Liu, H.; Rice, A.P. miR-132 enhances HIV-1 replication. *Virology* **2013**, *438*, 1–4. [CrossRef] [PubMed]
106. Russo, A.; Potenza, N. Antiviral effects of human microRNAs and conservation of their target sites. *FEBS Lett.* **2011**, *585*, 2551–2555. [CrossRef] [PubMed]
107. Andersson, M.G.; Haasnoot, P.C.J.; Xu, N.; Berenjian, S.; Berkhout, B.; Akusjarvi, G. Suppression of RNA Interference by Adenovirus Virus-Associated RNA. *J. Virol.* **2005**, *79*, 9556–9565. [CrossRef] [PubMed]
108. Sano, M.; Kato, Y.; Taira, K. Sequence-specific interference by small RNAs derived from adenovirus VAI RNA. *FEBS Lett.* **2006**, *580*, 1553–1564. [CrossRef] [PubMed]
109. Aparicio, O.; Carnero, E.; Abad, X.; Razquin, N.; Guruceaga, E.; Segura, V.; Fortes, P. Adenovirus VA RNA-derived miRNAs target cellular genes involved in cell growth, gene expression and DNA repair. *Nucleic Acids Res.* **2010**, *38*, 750–763. [CrossRef] [PubMed]
110. Hodzic, J.; Sie, D.; Vermeulen, A.; van Beusechem, V.W. Functional screening identifies human miRNAs

that modulate adenovirus propagation in prostate cancer cells. *Hum. Gene Ther.* **2017**, *2016*, 143. [CrossRef] [PubMed]

111. Lagos, D.; Pollara, G.; Henderson, S.; Gratrix, F.; Fabani, M.; Milne, R.S.B.; Gotch, F.; Boshoff, C. miR-132 regulates antiviral innate immunity through suppression of the p300 transcriptional co-activator. *Nat. Cell Biol.* **2010**, *12*, 513–519. [CrossRef] [PubMed]
112. He, L.; He, X.; Lowe, S.W.; Hannon, G.J. microRNAs join the p53 network—Another piece in the tumour-suppression puzzle. *Nat. Rev. Cancer* **2007**, *7*, 819–822. [CrossRef] [PubMed]
113. Feng, S.-D.; Mao, Z.; Liu, C.; Nie, Y.-S.; Sun, B.; Guo, M.; Su, C. Simultaneous overexpression of miR-126 and miR-34a induces a superior antitumor efficacy in pancreatic adenocarcinoma. *OncoTargets Ther.* **2017**, *10*, 5591–5604. [CrossRef] [PubMed]
114. Santhakumar, D.; Forster, T.; Laqtom, N.N.; Fragkoudis, R.; Dickinson, P.; Abreu-Goodger, C.; Manakov, S.A.; Choudhury, N.R.; Griffiths, S.J.; Vermeulen, A.; et al. Combined agonist-antagonist genome-wide functional screening identifies broadly active antiviral microRNAs. *Proc. Natl. Acad. Sci. USA* **2010**, *107*, 13830–13835. [CrossRef] [PubMed]
115. Lee, Y.-B.; Bantounas, I.; Lee, D.-Y.; Phylactou, L.; Caldwell, M.A.; Uney, J.B. Twist-1 regulates the miR-199a/214 cluster during development. *Nucleic Acids Res.* **2009**, *37*, 123–128. [CrossRef] [PubMed]
116. Thornburg, N.J.; Hayward, S.L.; Crowe, J.E. Respiratory syncytial virus regulates human microRNAs by using mechanisms involving beta interferon and NF-κB. *mBio* **2012**, *3*, e00220-12. [CrossRef] [PubMed]
117. Zhang, Q.; Huang, C.; Yang, Q.; Gao, L.; Liu, H.-C.; Tang, J.; Feng, W. MicroRNA-30c Modulates Type I IFN Responses To Facilitate Porcine Reproductive and Respiratory Syndrome Virus Infection by Targeting JAK1. *J. Immunol.* **2016**, *196*, 2272–2282. [CrossRef] [PubMed]
118. Panda, D.; Cherry, S. Cell-based genomic screening: Elucidating virus–host interactions. *Curr. Opin. Virol.* **2012**, *2*, 784–792. [CrossRef] [PubMed]
119. Mahoney, D.J.; Lefebvre, C.; Allan, K.; Brun, J.; Sanaei, C.A.; Baird, S.; Pearce, N.; Grönberg, S.; Wilson, B.; Prakesh, M.; et al. Virus-Tumor Interactome Screen Reveals ER Stress Response Can Reprogram Resistant Cancers for Oncolytic Virus-Triggered Caspase-2 Cell Death. *Cancer Cell* **2011**, *20*, 443–456. [CrossRef] [PubMed]
120. Workenhe, S.T.; Ketela, T.; Moffat, J.; Cuddington, B.P.; Mossman, K.L. Genome-wide lentiviral shRNA screen identifies serine/arginine-rich splicing factor 2 as a determinant of oncolytic virus activity in breast cancer cells. *Oncogene* **2016**, *35*, 2465–2474. [CrossRef] [PubMed]
121. Teferi, W.M.; Dodd, K.; Maranchuk, R.; Favis, N.; Evans, D.H. A Whole-Genome RNA Interference Screen for Human Cell Factors Affecting Myxoma Virus Replication. *J. Virol.* **2013**, *87*, 4623–4641. [CrossRef] [PubMed]
122. Homicsko, K.; Lukashev, A.; Iggo, R.D. RAD001 (Everolimus) Improves the Efficacy of Replicating Adenoviruses that Target Colon Cancer. *Cancer Res.* **2005**, *65*, 6882–6890. [CrossRef] [PubMed]
123. Alonso, M.M.; Jiang, H.; Yokoyama, T.; Xu, J.; Bekele, N.B.; Lang, F.F.; Kondo, S.; Gomez-Manzano, C.; Fueyo, J. Delta-24-RGD in Combination With RAD001 Induces Enhanced Anti-glioma Effect via Autophagic Cell Death. *Mol. Ther.* **2008**, *16*, 487–493. [CrossRef] [PubMed]
124. Rahman, M.M.; Bagdassarian, E.; Ali, M.A.M.; McFadden, G. Identification of host DEAD-box RNA helicases that regulate cellular tropism of oncolytic Myxoma virus in human cancer cells. *Sci. Rep.* **2017**, *7*, 15710. [CrossRef] [PubMed]

6

Expanding the miRNA Repertoire in Atlantic Salmon: Discovery of IsomiRs and miRNAs Highly Expressed in Different Tissues and Developmental Stages

Nardos Tesfaye Woldemariam [1], Oleg Agafonov [2], Bjørn Høyheim [3], Ross D. Houston [4], John B. Taggart [5] and Rune Andreassen [1,*]

1 Department of Life Sciences and Health, Faculty of Health Sciences, OsloMet–Oslo Metropolitan University, Pilestredet 50, N-0130 Oslo, Norway; nate@oslomet.no
2 Bioinformatics Core Facility, Department of Core Facilities, Institute of Cancer Research, Radium Hospital, Oslo University Hospital, 0379 Oslo, Norway; olegag@ifi.uio.no
3 Department of Basic Sciences and Aquatic Medicine, Faculty of Veterinary Medicine, Norwegian University of Life Sciences, 0454 Oslo, Norway; bjorn.hoyheim@nmbu.no
4 Division of Genetics and Genomics, The Roslin Institute and Royal (Dick) School of Veterinary Studies, University of Edinburgh, Midlothian EH25 9RG, UK; ross.houston@roslin.ed.ac.uk
5 Institute of Aquaculture, University of Stirling, Stirling, FK9 4LA, UK; j.b.taggart@stir.ac.uk
* Correspondence: rune.andreassen@oslomet.no;

Abstract: MicroRNAs (miRNAs) are important post-transcriptional gene expression regulators. Here, 448 different miRNA genes, including 17 novel miRNAs, encoding for 589 mature Atlantic salmon miRNAs were identified after sequencing 111 samples (fry, pathogen challenged fry, various developmental and adult tissues). This increased the reference miRNAome with almost one hundred genes. Prior to isomiR characterization (mature miRNA variants), the proportion of erroneous sequence variants (ESVs) arising in the analysis pipeline was assessed. The ESVs were biased towards 5′ and 3′ end of reads in unexpectedly high proportions indicating that measurements of ESVs rather than Phred score should be used to avoid misinterpreting ESVs as isomiRs. Forty-three isomiRs were subsequently discovered. The biological effect of the isomiRs measured as increases in target diversity was small (<3%). Five miRNA genes showed allelic variation that had a large impact on target gene diversity if present in the seed. Twenty-one miRNAs were ubiquitously expressed while 31 miRNAs showed predominant expression in one or few tissues, indicating housekeeping or tissue specific functions, respectively. The miR-10 family, known to target *Hox* genes, were highly expressed in the developmental stages. The proportion of miR-430 family members, participating in maternal RNA clearance, was high at the earliest developmental stage.

Keywords: Teleostei; embryogenesis; tissue-enriched miRNAs; post-transcriptional gene regulation

---

1. Introduction

MicroRNAs (miRNAs) are short non-coding RNAs that play an important role in post-transcriptional regulation of gene expression [1]. After transcription, the large primary miRNA transcripts (pri-miRNAs) are cleaved into shorter miRNA precursors (pre-miRNAs) that are exported out of the nucleus by Exportin 5. Here they are processed further by Dicer to produce the mature 5p and 3p miRNAs that are about 20–24 nts long. The mature miRNAs are then loaded onto Argonaute proteins and incorporated in the RNA induced silencing complex (miRISC). As part of the miRISC they form partially complementary bindings with their target mRNAs which subsequently leads to degradation or translational repression of the target transcripts [1,2]. The characteristics of these miRNA precursors and the mature miRNAs produced from Dicer cleavage of the miRNA precursors may be utilized to design bioinformatics tools that identify the miRNA genes and their mature miRNAs. High-throughput sequencing of small RNAs that are analyzed with such dedicated bioinformatics tools against a reference genome allows for massive parallel identification of a large number of miRNA genes [3,4].

Several studies have demonstrated that miRNAs are involved in many biological processes such as development, growth, tissue differentiation and apoptosis [5,6]. There is also evidence that some miRNAs are important regulators of immune function and immune responses [7,8]. As their biological function is defined by the sequence of the mature miRNAs, characterization of the miRNA repertoire is a first step to understand how miRNAs participate in the regulation of a species gene networks.

Genomic research in Atlantic salmon (*Salmo salar*) has been carried out due to the cultural and recreational importance of this species, and more importantly, its economic importance as an aquaculture species (www.fao.org/fishery/affris/species-profiles/atlantic-salmon/atlantic-salmon-home/en/). Atlantic salmon miRNAs were first characterized in 2013 by Andreassen et al. [9] and Bekaert et al. [10] utilizing the first assembly of the Atlantic salmon genome sequence. These miRNA reference sequences (miRNAome) has been used to develop RT-qPCR methods to measure miRNA expression, to investigate the role of miRNAs in host-virus responses, miRNAs that may affect sea louse infestation and miRNAs that may affect testis development [11–14].

The development of sequencing technology and the increased sensitivity of deep sequencing has led to the discovery of isomiRs [15–17]. IsomiRs are mature miRNA sequence variants with different 5′ and/or 3′ ends compared to their corresponding canonical mature miRNAs. IsomiRs are assumed to be products of imprecise cleavage of pre-miRNAs, RNA editing and non-templated nucleotide additions at the 3′ end of miRNAs [17,18]. Knowledge about the function of isomiRs is limited, but it has been suggested that isomiRs cooperate with their corresponding canonical miRNAs to target common biological pathways. Any post-transcriptional modification of the 5′ end of a mature miRNA (5′ isomiRs) is of particular importance as such changes affect the "seed" sequences. Any change of "seed" would affect target specificity, and by this, potentially increase the number of transcripts targeted [18,19].

Small-RNA sequencing datasets consist of a huge number of reads. Some differences in length or sequence among reads are due to errors arising in the sequencing pipeline like RNA degradation, sequence errors introduced in cDNA synthesis or during sequencing [20,21]. Such artefacts may be misinterpreted as isomiRs. Therefore, to characterize isomiRs that are true products of biological post-transcriptional processing one need to distinguish isomiRs from such artefacts arising in the deep sequencing pipeline.

So far, there have been no studies to characterize isomiRs in Atlantic salmon. Although the first miRNA discovery studies provided an important species-specific reference [9,10], the materials sequenced were from relative few samples. No samples from different developmental stages or tissues from fish challenged with pathogens were investigated.

In the present study, about ten times more samples have been included compared to the initial study [9]. One hundred and eleven small RNA libraries from different tissues sampled from adult fish, from different developmental stages, from normal fry, as well as from fry challenged with pathogens were sequenced. Thus, the materials investigated has allowed us to identify miRNAs likely to have tissue and developmental stage specific functions. The proportion of isomiR-like artifacts generated in the small-RNA sequencing pipeline was also revealed. On this background, the first characterization of isomiRs in Atlantic salmon was carried out. Nearly one hundred new miRNA genes, both miRNAs conserved in teleosts, as well as novel miRNAs, were discovered. The new reference miRNAome, where miRNA gene locations are assigned to the present genome sequence [22], forms an important updated resource for miRNA expression studies, as well as for comparative studies of miRNA gene evolution.

2. Materials and Methods

2.1. Materials

A total of 111 samples were sequenced in this study. Samples 1–96 in Table S1 were from fry, while samples 97–111 were tissue samples from particular organs from fully developed adults (intestine, gills, gonads, head-kidney or mid-kidney) and from early developmental stages. The samples from different developmental stages comprised of embryos sampled at 4, 19, 39, and 50 days post fertilization (dpf), an eyed-egg 63dpf, and an alevin one-day post hatching (alevin 1dph). These 111 samples were used in the miRdeep2 analysis (version 0.0.7) [4], the analysis of tissue enriched miRNAs and for isomiR characterization (Table S1). Results from the eleven small-RNA sequenced samples in Andreassen et al. [9] were also included in the datasets to identify miRNAs enriched in particular tissues. An additional 24 tissue samples (Table S2) from brain (n = 4), gills (n = 3), heart (n = 4), intestine (n = 5), liver (n = 5) and white muscle (n = 3) were included to investigate miRNAs enriched in those particular tissues. These samples were used for RT-qPCR analysis as described in Section 2.9.

Fry materials, sampled in Scotland were euthanized using a procedure specifically listed on the appropriate Home Office (UK) license, and all experiments were performed under the approval of Cefas ethical review committee and complied with the Animals Scientific Procedures Act. Some of the pathogens challenged fry were part of a challenge study using infectious pancreatic necrosis virus (IPNV) described in in Robledo et al. [23]. Sacrifice procedure of the other fish in the materials sampled in Norway was approved by the official ethics board FOTS (forsøksdyrutvalgets tilsyns-og søknadssystem). Dissection and sampling of materials were performed in agreement with the provisions enforced by the Norwegian Animal Research Authority.

2.2. Small RNA Extraction

Sampling and extraction of RNA from the fry materials (samples 1–96, Table S1) were carried out, as described in Robledo et al. [23]. Total RNA was isolated using TRI reagent (Sigma–Aldrich®, St. Louis, MO, USA) following the manufacturer's instructions. The RNA quantity and quality were determined using spectrophotometry (NanoDrop ND-1000, Thermo Scientific, Wilmington, DE, USA) and agarose gel electrophoresis, respectively. Total RNA from the remaining materials was extracted by using the mirVana Isolation Kit (Ambion, Life Technologies, Carlsbad, CA, USA) according to the manufacturer's protocol. The RNA concentration and purity were determined using spectrophotometry (NanoDrop ND-1000, Thermo Scientific, Wilmington, DE, USA).

2.3. Library Preparation and Small-RNA Sequencing

Small-RNA library construction and the small-RNA sequencing was performed at the Norwegian Genomics Consortium's genomic core facility (NGC). The Illumina NEBNext small RNA Library Prep Set (New England Biolabs, Inc. Ipswich, MA, USA) was used to prepare the 96 libraries from the 96 fry samples, while Illumina® TruSeq Small RNA sample preparation kit (Illumina, San Diego, CA, USA) was used to construct 15 libraries from the tissue and developmental stage specific samples. 1 µg of total RNA was used per sample as input in the library preparation in accordance with the manufacturer's protocols. The sequencing was carried out on the Illumina Genome Analyzer IIx sequencing platform.

2.4. Pre-Processing and Quality Assesment of Small-RNA Sequence Reads

The raw sequencing reads (fastq-files) from each sample were pre-processed to ensure that the raw data used in downstream analysis were of good quality and desired sizes. An assessment of the raw sequence reads was first carried out using FASTQC (v.0.11.5) [24], to ensure that the small RNA read quality was satisfactory before adaptor sequences were removed using cutadapt (v.1.13) [25]. Additional size filtering was carried out to discard reads that were outside the expected size range of mature miRNAs (18–25 nts). The quality of the trimmed and size filtered reads were analyzed by a second FASTQC analysis. Reads that passed this post-trim QC were used in the downstream analysis of the small RNA sequenced samples.

2.5. Identification of Atlantic Salmon miRNA Precursors, Their Mature miRNAs and miRNA Gene Locations

High quality reads from 108 samples (samples 1–97, 99, 102–111, Table S1) were used for identification of miRNAs applying the miRDeep2 software package. Default settings were used [3,4] to predict miRNA precursors along with their 5p and 3p mature miRNAs. Each sample was independently analyzed with miRDeep2 to allow for detection of miRNAs expressed in particular tissues or developmental stages. The present version of the Atlantic salmon genome assembly, ICSASG_v2, GenBank accession number: GCA_000233375.4 [22], and a genome index consisting of the existing reference Atlantic salmon miRNAs [9,10] was also used in the miRdeep2 analysis. The characterization pipeline is illustrated in Figure 1. The reads were mapped to the Atlantic salmon genome sequence using the miRDeep2 mapper module. Guided by the mapped reads genomic sequences that showed features expected from precursor sequences (e.g., ability to form hairpins) and with aligned reads showing expected characteristics of mature miRNAs processed by Dicer/Drosha were identified. These were assigned a log-odds score (the miRDeep2 score) based on an algorithm that integrates the statistics of the read positions, the frequencies of reads within hairpins, and the posterior probability that the hairpin was derived from a true miRNA gene [3]. To prevent false positive detection of miRNA precursors, a miRDeep2 score of ≥ 2 was used as a cut-off. All predicted precursors with miRDeep2 scores equal to or above the threshold were included in the downstream analysis. They were further analyzed by Basic Local Alignment Search Tool (BLAST) searches, using the putative precursor sequences as input, against miRNAs in the miRNA sequence database (miRBase) (version 22) (http://www.mirbase.org/search.shtml) [26]. A significant hit was defined as matches with an e-value of $\leq 1 \times 10^{-7}$ against any hairpin precursor in the database. Those identified as Atlantic salmon miRNA genes in the first study [9] were annotated with new genome locations in the present Atlantic salmon genome. Other matches were identified as Atlantic salmon orthologs of miRNA genes discovered in other teleost species. These were annotated according to the miRBase nomenclature guidelines (ssa-prefix and same number as in other teleosts) [27,28]. The precursors that did not provide significant matches with any of the miRNAs in miRBase were potentially novel miRNAs. These were further analyzed by blastn (http://blast.ncbi.nlm.nih.gov/Blast) searches against RNA databases in GenBank (http://blast.ncbi.nlm.nih.gov/Blast), the small RNA databases Rfam v.13.0. (http://rfam.xfam.org/search) [29], and the functional RNA database fRNAdb v.3.0. (https://dbarchive.biosciencedbc.jp/blast) [30]. Candidates that matched other kinds of small RNAs in these databases were removed (e-value threshold of $\leq 1 \times 10^{-7}$). The remaining precursors were used as queries in blastn analysis against the Atlantic salmon genome sequence. Any putative precursor that provided a significant hit (e-value threshold of $\leq 1 \times 10^{-7}$) against more than 15 loci in the genome reference sequence were annotated as interspersed repeats and removed. RNA secondary structure of the remaining miRNA precursors and their aligned reads were manually inspected. Novel miRNA genes were identified based on passing the following miRBase criteria [26]; (1) detected in at least two independent samples, (2) at least 10 sequence reads of mature and star miRNAs mapped (with no mismatches) to the hairpin precursor, (3) the mature microRNAs were paired with the precursor hairpin with 0–4 nt overhang at their 3′ ends, (4) the reads mapped supported a consistent pre-processing of 5′ end (5′ homogeneity) and (5) more than 60% of the bases of the mature sequences paired in the hairpin structure. Those that passed all these criteria were annotated as novel Atlantic salmon miRNAs. These were designated "ssa-miR-novel-x" with identifying numbers (x; e.g., ssa-mir-novel-1). All miRNAs, including the novel ones discovered in this study have been submitted to miRBase. When assigned unique miRNA identities by miRBase these will be added as Addendum to this paper.

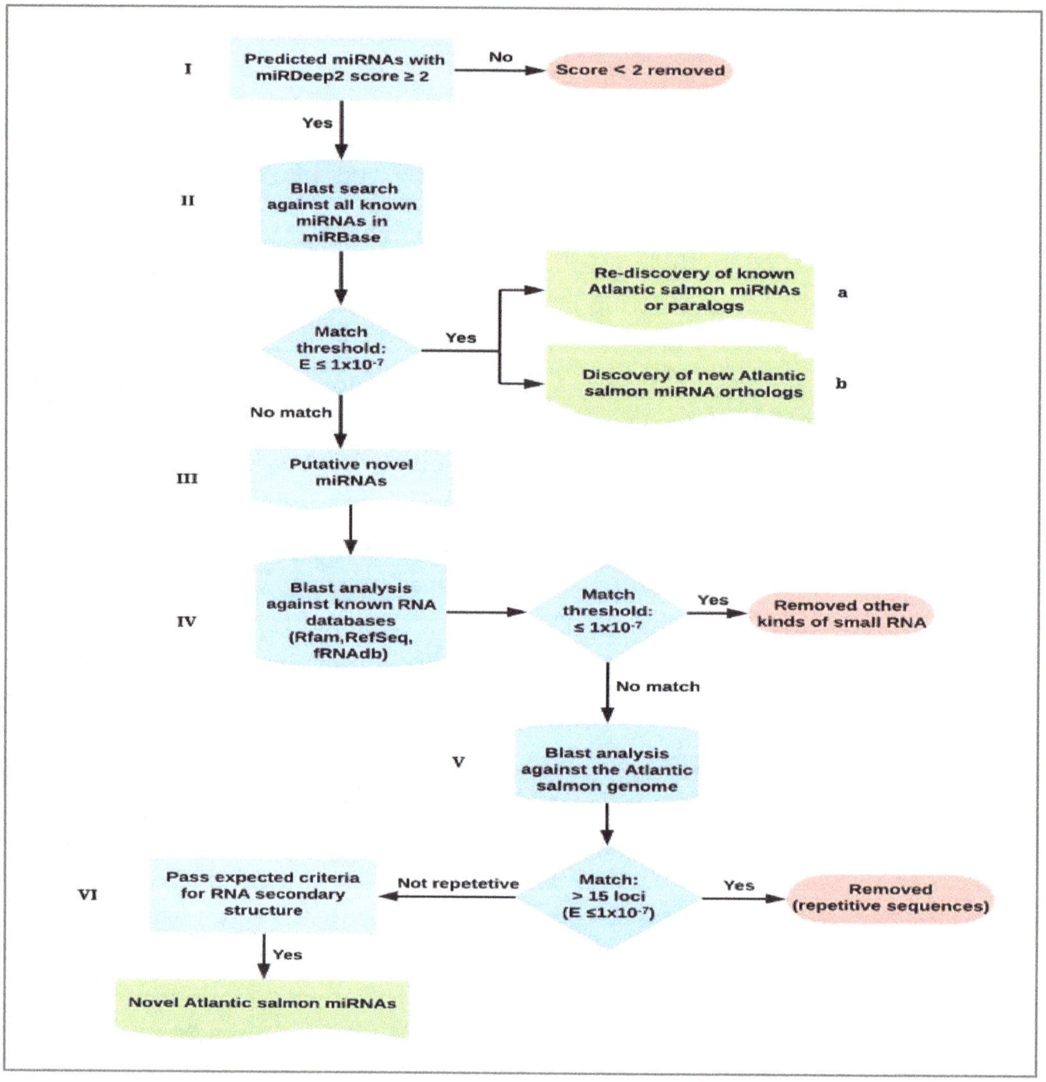

Figure 1. Identification and characterization of Atlantic salmon microRNAs (miRNAs). (I) Candidate miRNAs with miRdeep2 score ≥ 2 were regarded as putative miRNAs. (II) miRNA precursors were used as input for BLAST search against miRBase to identify; (**a**) Atlantic salmon miRNAs already in the database and new paralogs, (**b**) Orthologues of other teleost miRNAs not discovered in prior studies of Atlantic salmon. (III) Candidate miRNAs with no significant matches ($e \leq 1 \times 10^{-7}$) to miRBase were considered as putative novel Atlantic salmon miRNAs. (IV) Putative novel miRNA precursor sequences were BLAST analyzed against other RNA databases (Rfam, NCBI RefSeq and fRNAdb) to remove those that were other kinds of small RNA. (V) Sequences with no matches in these databases were further used for BLAST analysis against the Atlantic salmon reference genome sequence. Significant matches ($e \leq 1 \times 10^{-7}$) to more than 15 loci in the Atlantic salmon genome were annotated as interspersed repeats and removed. (VI) The remaining putative novel miRNAs were compared against expected features of precursors and their mature miRNAs. The candidates that passed all these criteria were regarded as novel Atlantic salmon miRNAs.

2.6. Annotation of Clustered miRNA Genes

The genome location of all Atlantic salmon miRNA genes was identified using the second and improved version of the Atlantic salmon genome assembly. The amount of clustered miRNA genes in the Atlantic salmon genome was examined by comparing their locations in ICSASG_v2. Any miRNA genes located in the same direction in a contig, within a distance of 10 kb or less were annotated as clustered miRNA genes (same definition as used by miRBase). We also compared the miRNA clusters from our study to miRNA clusters in zebrafish (*Danio rerio*) and Atlantic cod (*Gadus morhua*), as given in miRBase to reveal evolutionarily conserved miRNA gene clusters.

2.7. Sequence Errors Arising in the Sequencing Pipeline and IsomiR Detection

Prior to characterizing the abundance of isomiRs the amount and type of reads with erroneous sequences (erroneous sequence variants, ESVs) that arise in the sequencing pipeline (extraction, cDNA synthesis, library prep, small RNA sequencing) must be assessed. High quality reads from 48 fry samples were pooled (441,320,712 reads with Phred quality score above 32). This dataset was collapsed into a total of 7,004,792 unique reads annotated with their read count number. The phiX Control is commonly used as an internal control to measure the proportion of ESVs when larger fragments are sequenced [31,32]. This control can however, not be applied in small RNA sequencing. Instead, we used two highly abundant and ubiquitously expressed Atlantic salmon RNAs; (18S rRNA (Genbank accession: FJ710886.1), and 60S ribosomal protein L37 (GenBank accession: BT058368.1)) for this purpose. The two RNAs, the rRNA (1750 bp) and the short mRNA (540 bp), were used as references as reads that are derived from anywhere out of these larger sequences are not expected to be RNA edited. All reads were aligned against these two references. The aligned reads with bp differences compared to the references would be ESVs (or polymorphism). By aligning the unique sequence reads to those reference sequences in Sequencher software version 5.4.6 (Gene Codes Corporation, Ann Arbor, MI, USA), the ESVs were visualized. Reads that aligned perfectly to the references and reads that represented reads with ESVs were counted. Any bias in error rate within reads (e.g., higher error rate in 3′ end) was also assessed. The ratio between perfectly aligned reads and ESVs were used to set an error threshold. Reads present in lower amount than this threshold and with non-templated sequence variation compared to the canonical reference miRNAs are likely to be ESVs, not isomiR variants that are products of RNA editing.

Detection of isomiRs was carried out by use of isomiR Seed Extension Aligner (isomiR-SEA) software (version 1.60) [33]. The tool detects mature miRNAs and mature miRNA isoforms by a seed-based alignment procedure. The unique reads were aligned to reference miRNAs (mature canonical miRNAs identified in this study) using default settings and commands. The tool classifies miRNA variations into four categories with respect to the reference miRNA sequence. The reads matching perfectly to the reference miRNA (canonical miRNAs), reads with any nucleotide variation in the 5′ end (5′ isomiRs), reads with any nucleotide variation in the 3′ end (3′ isomiRs and 3′ length isomiRs), and reads with mismatches anywhere in their sequences. Count number of the reads identical to canonical miRNA sequences was used to set a lower threshold based on the error ratios revealed in alignments to the reference sequences. If below this threshold, they were removed assuming they were erroneous sequence variants (ESVs) that had arisen in the pipeline, not isomiRs. All reads variants passing the error rate cut-off filtering were aligned to their reference miRNA sequences using Sequencher software. Degraded miRNAs would be identical, but shorter than reference miRNAs. Shorter 5′ and 3′ isomiR variants could therefore not be reliably identified. The total RNA extracted also consists of precursor miRNAs, occasionally present in a larger amount than the mature miRNAs derived from the precursor [34]. Any templated 3′ length isomiRs larger in size than the reference miRNAs could therefore not be reliably detected as such reads could arise from degradation of precursor molecules. These shorter and longer template sequences were excluded from further analysis. The isomiR variants that could be reliably detected in small RNA sequencing data sets were therefore non-templated 5′ isomiRs, non-templated 3′ isomiRs and non-templated 3′ length isomiRs. The identification of miRNAs (Section 2.4) is based on the alignment of reads to the genome sequence. Only reads that aligned perfectly would identify a miRNA gene. As ESVs would be approximately one fifth of all reads aligning, and those not aligning perfectly, they would not interfere with detection of a miRNA gene and it's templated, canonical mature miRNAs (reads aligning perfectly to the template).

2.8. Examinating the Biological Effect on Target Gene Specificity from IsomiR Variation

Target gene predictions of isomiRs and their corresponding canonical mature miRNAs was carried out using the target gene prediction software RNAhybrid version 2.2 [35]. The mature miRNAs (canonical and isomiR variants) were tested against 3′UTRs from all Atlantic salmon mRNA transcripts in the Refseq database of Genbank (https://www.ncbi.nlm.nih.gov/refseq/). The analysis was performed applying the following conditions: Helix constraint 2–8 and no G:U in seed. This allowed only target genes with perfect seed complementarity to be detected. Minimum free energy threshold for RNA hybrids was set to −18 kcal/mol to retrieve results (target site matches) from RNA hybrids with high stability.

2.9. Identification of Tissue Enriched miRNAs

The high-quality reads from 23 tissue samples (samples 6, 7, 97, 98, 100, 102, 104, 106–111 (Table S1) and the 10 samples sequenced in a previous study [9]) were used to provide rough estimations of the expression of individual miRNAs across the different tissues. Each mature miRNA was counted using the reference miRNAome and STAR aligner (version 2.5.2b) [36]. The number of miRNAs in each sample was normalized by reads per million scaling factor (RPM) [37]. In cases where there were biological replicates of same tissue (liver, spleen, kidney, head-kidney and intestine) we used the average RPM values for these tissues. The normalized read counts of individual miRNAs within each tissue were compared to identify miRNAs that were highly expressed. The fifteen most abundant miRNAs in each of the tissues, a total of 43 miRNAs, were identified in this manner. The RPM values from these 43 miRNAs were then compared to identify those miRNAs highly expressed across all tissues (three-fold or less difference) and those highly expressed in one or a few particular tissues (more than ten-fold increase in one or a few tissues). Additional tissue enriched miRNAs that were not among the fifteen most abundant miRNAs from a tissue, were identified in the same manner (RPM comparison). These were also analyzed by RT-qPCR to validate they were enriched in particular tissues (see Section 2.9). Those that showed more than ten times higher expression in a particular tissue (measured by RPM and RT-qPCR or RPM only) was termed highly expressed or tissue enriched.

2.10. RT-qPCR Analysis of Tissue Enriched miRNAs

Nineteen miRNAs (ssa-miR-9a/b-5p, ssa-miR-153a-3p, ssa-miR-122-5p, ssa-miR-499a-5p, ssa-miR-194b-5p and ssa-miR-192a-5p, ssa-miR-96, ssa-miR-129, ssa-miR-132, ssa-miR-135c, ssa-miR-212, ssa-miR-219, ssa-miR-723, ssa-miR-734a, ssa-miR-8163, ssa-miR-736, ssa-miR-459 and ssa-miR-140) were analyzed by RT-qPCR as their RPM values indicated high enrichment in one or few particular tissues. A summary of the samples used for verification of tissue enriched miRNA expression by RT-qPCR, a total of 24 samples, is given in Table S2.

First strand cDNA synthesis was performed using the miScript II RT kit (Qiagen, Hilden, Germany) according to the manufacturer's instructions. miRNAs were then detected using the miScript SYBR® Green PCR kit (Qiagen, Hilden, Germany) as described by the manufacturer. The qPCR assays were carried out by means of custom designed forward primers together with a universal reverse primer provided with the miScript qPCR kit. An overview of the primers specific to each mature miRNA is given in Table S3. The qPCR reactions were performed on Mx3000p qPCR system (Stratagene, Agilent Technologies, LA Jolla, CA, USA) using 96-well plates, with thermocycling conditions of 95 °C for 15 min, followed by 40 cycles of 94 °C for 15 s, 55 °C for 30 s and 70 °C for 30 s. Cq values were obtained using the MX3000p software package (Stratagene, Agilent Technologies, USA). The relative expression of each miRNA was normalized against miR-25-3p and miR-107-3p, based on their good stability across tissues [10]. The mean normalized Cq values for each of the different tissues was calculated. The relative difference in expression between tissues was calculated using the ΔΔCt-method [38]. Statistical significance of the observed relative difference in expression was tested using student's t-test and significance levels corresponding to P-values ≤ 0.05 that were set after Bonferroni-correction based on number of tests.

3. Results

3.1. Small RNA Sequencing and Identification of Atlantic Salmon miRNAs

3.1.1. Generation of RNA Libraries and Results from Small RNA Sequencing

Small RNA libraries were successfully generated for 111 samples from salmon fry (n = 24), pathogen challenged fry (n = 72), tissue samples (intestine, gills, gonads, head-kidney and mid-kidney) from fully developed adults (n = 9), and samples from different developmental stages; embryos sampled at 4, 19, 39, 50 dpf, eyed-egg 63dpf, and an alevin 1dph. After pre-processing the raw reads (see methods), there were a total of 656,748,326 high quality, adapter trimmed and size filtered reads from fry samples and 46,047,362 reads from the different tissues and developmental stage samples. A detailed overview of sample origin, RNA concentration, quality and read numbers are given in Table S1.

3.1.2. Results from Discovery and Characterization of Atlantic Salmon miRNAs

The miRdeep2 algorithm has been shown to be a sensitive and reliable method for identifying miRNAs in different species [3,4]. Here it was successfully used to analyze the processed high-quality reads from each of the 111 samples separately. The results from the miRdeep2 analysis was 941 predicted miRNA precursors with their corresponding 5p and 3p mature reads. These were further processed as described in Figure 1 (see Section 2.4). Out of the 371 different Atlantic salmon miRNA genes already annotated and deposited in the database [9], all but one of each of the miRNA paralogs ssa-mir-210-1 and ssa-mir-29b-4 were re-discovered. In addition, new identical paralogs (i.e., identical precursor miRNA sequences at different genome locations) and new miRNA paralogs with small sequence differences, altogether a total of 533, were discovered. The re-discovery of these miRNAs in this new material provided additional confidence that they are true Atlantic salmon miRNAs. Another 20 miRNAs were identified as orthologues of miRNA genes in other teleosts and annotated as new Atlantic salmon miRNA genes in accordance with miRBase guidelines. Three of these miRNA genes had two copies (paralogs) in Atlantic salmon. We thus identified 556 evolutionary conserved miRNA genes with their corresponding mature 5p and 3p miRNAs (Table S4).

The remaining precursor candidates retrieved from the miRdeep2 analysis showed very low sequence similarity to any of the miRNA genes in miRBase. Seventy-nine of these candidates provided significant matches (e-value $\leq 1 \times 10^{-7}$) to other kinds of small RNA (e.g., rRNA, snoRNAs), and were removed. Hundred and twenty sequences provided multiple hits against the salmon genome sequence indicating they were different kinds of repetitive sequences rather than miRNA genes, and thus removed.

Finally, we applied the miRBase guidelines to identify high confidence novel miRNA genes among the remaining precursors retrieved from the miRdeep2 analysis [26]. Following these guidelines (see Section 2.4), 17 novel miRNA genes were identified. The precursor sequences from all novel miRNAs with their corresponding mature 5p and 3p sequences, annotation of arm dominance and the genome location of each precursor is given in Table 1. One of the novel miRNA genes revealed a perfect match in miRBase, (e-value $\leq 1 \times 10^{-18}$) to a salmon fluke (*Gyrodactylus salaris*) miRNA (gsa-mir-9404). However, this precursor sequence aligned perfectly (100% identity) to the Atlantic salmon genome, while there were no matches to the *Gyrodactylus salaris* genome (GenBank accession: GCA_000715275.1) (e-value > 0.5). This one was also identified as an Atlantic salmon miRNA in Bekaert et al. (miR-new156-5p) [10]. This strongly indicates that this is an Atlantic salmon miRNA rather than originating from *G. salaris*. This miRNA was annotated as ssa-miR-novel-17 in our new reference miRNAome. One novel precursor (ssa-miR-novel-15) had several identical copies clustered at two unique genomic locations, and two novels (ssa-miR-novel-6 and ssa-miR-novel-7) had two and three identical copies respectively, while the other 14 novel miRNA genes were present as single copies.

Table 1. Novel Atlantic salmon miRNAs identified in this study.

miRNA ID [1]	Mature-5p (5′-3′) [2]	Mature-3p (5′-3′) [2]	Precursor Sequence (5′-3′)
ssa-mir-novel-1	UCAGUGAUGUGUACGCCAAAGGU	UCGGCAUACACAUCACUGACA	UCAGUGAUGUGUACGCCAAAGGUGUAAAGCUUCAAGUUCCUCGGCAUACACAUCACUGACA
ssa-mir-novel-2	AGUUUCCCGGACACAGAUUAAGCC	UUUUGUCUGUCUGGGAAACCGG	AGUUUCCCGGACACAGAUUAAGCCUAGUCAUAAUUAUUAUGUUUUGUCUGUCUGGGAAACCGG
ssa-mir-novel-3	UGACGAUACCUUUGGAACAAGA	UUGUACCAAUAGUAAAGUCUGA	UGACGAUACCUUUGGAACAAGAGGUGAAUUACGUCUUAUGCUCUUGUACCAAUAGUAAAGUCUGA
ssa-mir-novel-4	CGGAUCGCUGCGUUCACCAUU	AUGGUGAAUGCAACGAUAAGGC	CGGAUCGCUGCGUUCACCAUUAUAUUUAACUUCAACAGAAUGGUGAAUGCAACGAUAAGGC
ssa-mir-novel-5	UACGGUAUGUACUGUAGGCUAC	UAGGCUACGGUAUGUACUGAAG	UACGGUAUGUACUGUAGGCUACGGUAUGUUAUGUACUGUAGGCUACGGUAUGUACUGAAG
ssa-mir-novel-6	UGAGCCUUGUCCUGGACUAAGA	UCAGUCCAUGACUAGGCUUAAC	UGAGCCUUGUCCUGGACUAAGAAGUACUUCCAAUGGCUAUUUUCAGUCCAUGACUAGGCUUAAC
ssa-mir-novel-7	UUGCUGGUGACACUGUCUGUGA	AAGGCACACUUCACCAGUAUGG	UUGCUGGUGACACUGUCUGUGAUUUAUUUAGAAUUCAAGGCACACUUCACCAGUAUGG
ssa-mir-novel-8	AGACACCUGACACAGCCCCCAUU	UGGGUCUGUGUCUAUUGUCUCU	AGACACCUGACACAGCCCCCAUUCUAUCUCAUAAAAGUGGGUCUGUGUCUAUUGUCUCU
ssa-mir-novel-9	UAGGCGUGUCACUGCGUGUCACA	UGCGCACGGGGCCACGCUCUGC	UAGGCGUGUCACUGCGUGUCACAGUCACUGCUUGCGCACGGGGCCACGCUCUGC
ssa-mir-novel-10	AGGUCUGUUUGUGCUGUCUUCC	GUGACUGCACAAACGGAUCUGG	AGGUCUGUUUGUGCUGUCUUCCAUGGCUUUGGUGACUGCACAAACGGAUCUGG
ssa-mir-novel-11	AUUGUUCAGGGCAUUCAUUUCU	UAAGUGAACCCUUGAGACAAUU	AUUGUUCAGGGCAUUCAUUUCUUGUGAACCAAUCAAUAAGUGAACCCUUGAGACAAUU
ssa-mir-novel-12	UUCGCCCCUGAGGACACACGGU	CCGAAUCCACAGAAGUGAUGC	UUCGCCCCUGAGGACACACGGUGUUUUCUUUUAAUAGCACCGAAUCCACAGAAGUGAUGC
ssa-mir-novel-13	CCUUGACCACGUAACCUGACCA	UUAGGUCAGAUGUGGUCAGGAGA	CCUUGACCACGUAACCUGACCAUAGUUUUCUUGGUUAGGUCAGAUGUGGUCAGGAGA
ssa-mir-novel-14	GGGAAUAUACAUGACUGUGAUU	UCACAGUCGUGUAUAUUCCCUC	GGGAAUAUACAUGACUGUGAUUAUGAUUGAAGAGAAUAAUCACAGUCGUGUAUAUUCCCUC
ssa-mir-novel-15	CAGAGCUCUGCUAUCUGCUGUCU	AAGGAGAAAACAGAGCUCUGCU	CAGAGCUCUGCUAUCUGCUGUCUGUAUCUUGUUAAAGGGGAAGGAGAAAACAGAGCUCUGCU
ssa-mir-novel-16	UUGCUGUUGACACUGUCUGUG	UCAAGGCACACUUAACCAGCAUGG	UUGCUGUUGACACUGUCUGUGAUUUAUUUAAGGCACACUUCAAGGCACACUUAACCAGCAUGG
ssa-mir-novel-17	GCGUCUCAGAGGUCAAACACAGU	UGUGUUAGGCCUCCGAGUCUGA	GCGUCUCAGAGGUCAAACACAGUAAGUCAUAUUAAGCUGUGUUAGGCCUCCGAGUCUGA

[1] Temporary annotation for the novel miRNAs identified. These will be renamed by miRBase when uploaded in the miRNA database. [2] The dominant mature miRNA is given in bold. Genomic location of the novel precursors is given in Table S4.

In summary, there were 448 different miRNA genes, including 17 novel miRNAs discovered. Adding the 102 identical paralogs there were 577 miRNA genes. Their locations in the present Atlantic salmon genome is given in supplemental material (Table S4). The new mature Atlantic salmon miRNAome reference sequences originating from these genes were a total of 589 unique 5p and 3p mature miRNAs. These are given in FASTA format to be used as a reference miRNAome in miRNA expression studies in supplemental file S5.

miRNA genes are often located in clusters [9,39–41] that may be co-transcribed as a single pri-miRNA [2]. Furthermore, miRNA genes located in the same cluster have been shown to work together to control the same gene pathways when regulating various cellular processes [42–44]. With the discovery of more than one hundred new miRNAs and new genome annotation of all miRNA genes, the amount of clustered miRNA genes was re-examined. miRNA gene clusters were defined as suggested by miRBase as two or more miRNA genes located less than 10 kb from each other in the same direction (on the same genomic strand) [45]. Applying this definition, we identified 235 miRNA genes that were grouped into 93 distinct clusters. These clusters account for 40% of all Atlantic salmon miRNA genes. All gene clusters, the miRNA genes in the cluster and their genomic locations are listed in Table S6.

Only six gene clusters were present in single copies (1–6, Table S6). The remaining clusters could be further subdivided into groups where each gene cluster within a group were paralogous copies. There were 23 such groups, including one miRNA gene cluster with a novel miRNA gene (ssa-mir-novel-15, group XXIII, Table S6). In the previous discovery of Atlantic salmon miRNA genes, Andreassen et al. reported a total of 84 miRNA gene clusters [9]. All these, as well as nine additional clusters were revealed in the new genome sequence. A simple comparison of all groups of Atlantic salmon miRNA gene clusters to the orthologue miRNA gene clusters discovered in zebrafish and Atlantic cod (miRBase release 22) showed that 86% of these gene clusters were present in Atlantic cod with at least one copy, while 90% of these were present in zebrafish with at least one copy. Those present in zebrafish and Atlantic cod are also shown in Table S6. The large percentage of orthologous gene clusters in these three species indicates that the majority of gene clusters are evolutionarily conserved in teleosts. In general, most of the evolutionarily conserved miRNA gene clusters showed a higher copy number in Atlantic salmon compared to zebrafish and Atlantic cod. This is in agreement with findings that miRNA genes have been retained in the Atlantic salmon genome in the process of re-diploidization from the evolutionary recent salmonid specific genome duplication [9,46].

3.2. *Characterization of IsomiRs and Polymorphics miRNAs in Atlantic Salmon*

3.2.1. Sequence Errors Arising in the Sequencing Pipeline

The characterization of isomiRs requires that the proportion of reads with ESVs that arise in the sequencing pipeline is measured. A threshold that assures that ESVs are not incorrectly reported as isomiRs can then be set. Alignment of reads to the references (see Section 2.6) allowed us to examine the number of sequence errors arising in all parts of the sequencing pipeline. Figure 2 illustrates the results from the alignment of reads against a small part (bp 150–170) of the reference 18S rRNA.

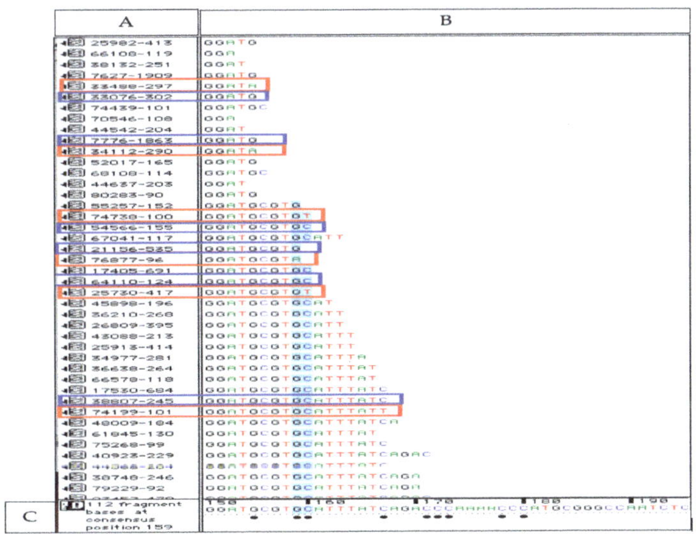

Figure 2. Illustration of aligned reads to the reference (18S rRNA) applying Sequencher software. (**A**) This window shows each collapsed read identified by a unique number followed by the count numbers of this particular read. (**B**) The sequence of each collapsed read is given in this window (**C**) The sequence at the bottom is the reference sequence (18S rRNA). Bullets below bases of the reference indicate that some of the reads differ from the reference at these basepair positions (erroneous sequence variants, ESVs). All collapsed reads shown in red boxes have 3′ ESVs. The collapsed reads shown in blue boxes are reads with correct bp in their 3′ end position. The ratio of reads with 3′ ESVs compared to reads with correct 3′ end bp's in this short part of the sequence (bp 150–bp 170) was 0.10. The other reads below 3′ ESV were all identical to reference at these base positions (e.g., bases in blue color) verifying that the misaligned bases are 3′ ESVs not polymorphic variants.

This revealed the amount of ESVs, as well as the position of the ESVs within reads. The ratio of ESVs distributed across any position within the reads were 0.0004. This was approximately as expected from the Phred quality of reads as a Phred score of 32 means there are less than 0.00063 bp errors due to sequencing alone. However, the alignments revealed that there was a strong bias in ESV position across reads. Almost all ESVs were positioned in the 3′ end (independent of read length). The average ratio of ESVs in these positions (last bases at 3′ end) was 0.21. The most 5′ bases also revealed a higher ratio of ESVs (0.02), but far from the ESV frequency observed at the 3′ end. This showed that other steps in the sequencing pipeline than the sequencing itself, e.g., the cDNA synthesis [19], generated a large amount of ESVs. If not accounted for, these ESVs could be misinterpreted as 3′ or 5′ isomiR variants. Alignment of reads to the two controls also showed that all different sizes of reads (18–25 nts) identical to the references were present in large numbers in the data sets. This showed, as expected, that any characterization of templated size variants (e.g., from imprecise Dicer processing) of canonical miRNAs could not be carried out as they could not be distinguished from those reads with small size variations that were just products of the pipeline itself (e.g., degradation).

3.2.2. IsomiR Characterization

After a series of filtering steps and accounting for the ratio of ESVs (see Section 3.2.1), we identified 41 isomiR variants derived from 37 mature miRNAs, including four isomiR variants of novel miRNAs. Thirty-two of these were non-templated 3′ isomiRs, while eight were non-templated isomiR length variants (3′ length isomiRs). One non-templated 5′ isomiR was also identified (Table S7). The most common variants observed were, thus, the 3′ end variants (98%). Analysis of the ratio of isomiR variants vs. their canonical forms showed, in general, the predominance of the canonical mature miRNAs (Table S7). However, ten isomiR variants that were more abundant than their respective canonical forms were also observed. The most prominent change of non-templated nucleotides was from cytosine (C) to uracil (U). All eight of the 3′ length isomiRs had uracil (U) added in their 3′ ends. Such uridylation is a common RNA modification found for different RNA species, including miRNAs [47,48]. These findings indicated that the editing of mature miRNAs was not random.

The biological significance of these isomiRs was further investigated by predicting the targets of canonical miRNAs and their isomiR forms using RNAhybrid [31]. All the thirty-seven mature miRNAs and their isomiRs were analyzed against the 3′UTRs of Atlantic salmon mRNAs (Refseq, Genbank). The results (Figure 3 and Table S8) showed that the 37 canonical miRNAs together could putatively target 3916 transcripts.

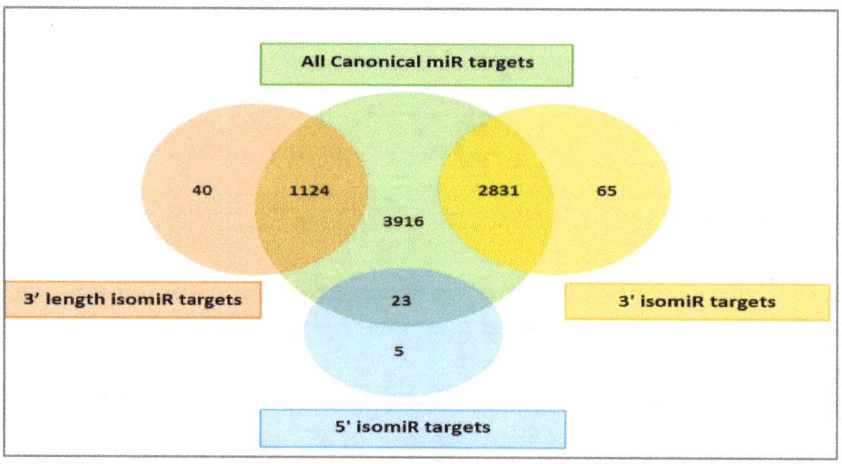

Figure 3. A Venn diagram depicting the overlap and difference between the predicted target genes of canonical miRNAs (green color), 3′ isomiRs (yellow color), 3′ length isomiRs (red color), and 5′ isomiRs (blue color).

The 3′ isomiRs (yellow color, Figure 3) were predicted to share 2831 common targets with their canonical miRNAs. The 3′ length isomiRs (red color, Figure 3) were predicted to share 1124 targets with their canonical miRNAs, while the 5′ isomiRs (blue color, Figure 3) were predicted to share 23 targets with their canonical miRNAs. Furthermore, the 3′ isomiRs were also predicted to target additional 65 transcripts, the 3′ length isomiRs could target 40 additional transcripts, while the 5′ isomiRs could target 5 additional transcripts not predicted as targets of their canonical forms. These comparisons showed that the isomiR variants did not lead to a large increase in putative target transcripts. Only 110 new transcripts (<3%) were added as targets when including the additional transcripts targeted by all isomiR variants. Thus, the biological effect, measured as the increase in the number of targets was small.

3.2.3. Polymorphic Mature miRNAs

Five polymorphic miRNAs (allelic variants) were also identified (Table 2). In all cases, the proportion of reads with the new variant was larger than 40%. In one of the variants (ssa-miR-100a-2-3p), the polymorphism (T to C transition) was in the seed sequence. The biological significance of the variants was, similarly as with isomiRs, investigated by prediction of putative targets of both the reference and the new variants. Figure 4 and Table S9 show the results from this analysis.

Table 2. Polymorphic variants of canonical mature miRNAs.

miRNA	Reference Sequence [1]	Variant Sequence [2]
ssa-miR-16a-1-3p	CCAGTATTGTTCGTGCTGCTGA	CCAGTATTG**C**TCGTGCTGCTGA
ssa-miR-100a-2-3p	ACAAGCTTGTGTCTATAGGTATG	ACAAGCT**C**GTGTCTATAGGTATG
ssa-miR-2188-3p	GCTGTGTGAGGTCAGACCTATC	GCTGTGTGAGGTC**G**GACCTATC
ssa-miR-29b-1-5p	ACTGATTTCTTCTGGTGTTTAGA	ACTGATTTC**C**TCTGGTGTTTAGA
ssa-let-7a-2-3p	CTATACAACTTACTGTCTTTCC	CTATACAAC**A**TACTGTCTTTCC

[1] The reference sequence of mature miRNA. [2] Variant sequence with the polymorphic base given in bold. The most common variant is underlined.

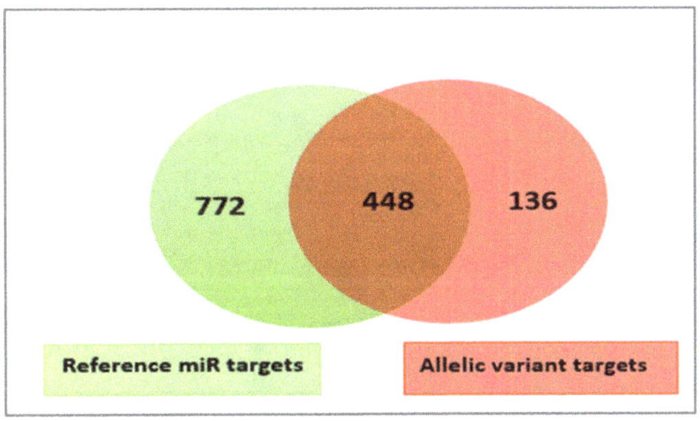

Figure 4. A Venn diagram illustrating the overlap and difference between the predicted target genes of the five reference mature miRNAs (green color) and the new mature miRNA variants identified in this study (light red color).

As shown in Figure 4, the reference mature miRNAs were predicted to target 722 transcripts. The new variants and their reference miRNAs were predicted to share 488 common targets. One hundred-thirty-six transcripts were predicted to be targeted by the new variants (increase of more than 15%) (Table S9a). The effect of allelic variation on target diversity was, thus, larger than the effect observed for isomiRs. However, this was mostly caused by the one variant where the polymorphic site was within the seed. This ssa-miR-100a-2-3p seed variant contributed 81 (60%) of all the additional targets (Table S9b). The reference mature and the new seed variant of the ssa-miR-100a-2-3p were in this case predicted to share only six common target transcripts, but mostly at different target sites on the same transcript. This indicated that a single base change in seed can significantly affect target gene specificity.

3.3. Characterization of miRNA Expression Profiles in Different Tissues and Developmental Stages

3.3.1. Housekeeping miRNAs vs. miRNAs Predominantly Expressed in Particular Tissues

Several studies have demonstrated that miRNAs play an important role in tissue development and/or in the maintenance of tissue specific functions [5,40,49–54]. Such miRNAs are often highly expressed in one or a few tissues. On the other hand, there are miRNAs with ubiquitous high expression across most tissues assumed to maintain housekeeping functions [54]. To identify miRNAs that are likely to have such housekeeping functions the fifteen most abundant miRNAs in each of the tissues brain, liver, heart, head-kidney, muscle, intestine, kidney, spleen, gonads and gills were revealed. Together, there were 43 such miRNAs. Twenty-one of these did not reveal any large expression differences when comparing across tissues (three-fold or less). Further examining the enrichment patterns of these 21 miRNAs showed that seven of these miRNAs were among the top ten most abundant miRNAs in all tissues. These were miR-143-3p, miR-181a-3p, miR-21b-5p, mir-26a-5p, miR-10b-5p, mir-10d-5p and mir-10a-5p (Table S10). Together, these seven miRNAs accounted for more than 30% of all miRNAs expressed in any tissue. Their ubiquitous nature and high expression in many or all tissues suggest that these miRNAs are constitutively expressed and have housekeeping functions.

Other miRNAs showed a predominant expression in particular tissues. Although they were among the 15 most highly expressed miRNAs in one tissue, they showed very little or no expression in the other tissues. All of these showed 10–100 times higher expression in one or few particular tissues. These are shown in the heat map in Figure 5.

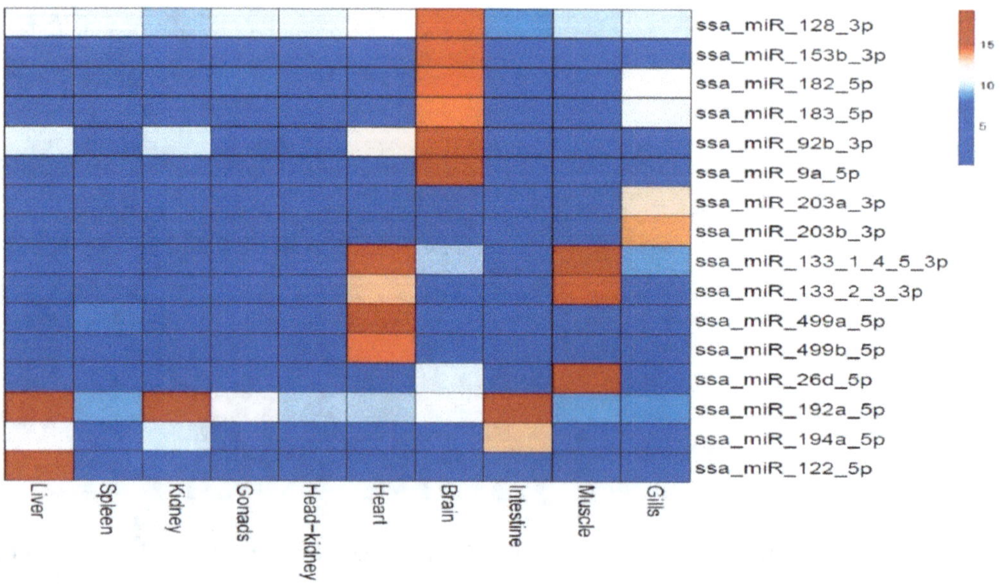

Figure 5. The heatmap illustrates the predominant expression of 16 mature miRNAs in particular tissues. The tissues compared are given at the bottom of the columns, while the miRNAs are given in the horizontal rows to the right. Four of these miRNAs (ssa-miR-9a-5p, ssa-miR-499a-5p, ssa-miR-192a-5p and ssa-miR-122-5p) were also analyzed by RT-qPCR (see Table 2). The expression values (log2) of the miRNAs are depicted in the color scale. Very little or no expression is shown in dark blue, while other colors indicate increased expression (10–100 times) with dark red color as those with the higher expression.

Six of these miRNAs (ssa-miR-128-3p, ssa-miR-153b-3p, ssa-miR-182-5p, ssa-miR-183-5p, ssa-miR-92b-3p and ssa-miR-9a-5p), were predominantly expressed in brain. Two miRNAs from the miR-499-family (ssa-miR-499a/b-5p) showed about 70 times higher expression in cardiac tissue, while all 3p mature miRNAs in the miR-133 family (ssa-miR-133-1-4-5/2-3-3p) showed higher expression in both muscle and cardiac tissue. Another miRNA that was predominantly expressed in muscle tissue was ssa-miR-26d-5p (about 20 times increase). Two miRNAs, ssa-miR-203a/b-3p, were highly expressed in gills. Ssa-miR-192a-5p was predominantly expressed in intestine, but this miRNA also showed an eight to twenty times higher expression in kidney and liver than in all other tissues indicating it was serving a function common to a cell type present in these three tissues. One more miRNA, ssa-miR-122-5p also showed a predominant expression in the liver.

Another 15 miRNAs showed lower abundance (i.e., not among the 15 most abundant miRNAs in one or more tissues), but were still enriched in specific tissues. Additional measurements by RT-qPCR were used to show that there was a significantly higher expression (>10× higher) of these in the particular tissues (p-adjusted ≤ 0.05). The results from RT-qPCR analysis of all these miRNAs (Table 3) agreed with the patterns revealed by the RPM comparisons.

Table 3. Results from RT-qPCR analysis.

miRNA	Tissue [1]	ΔΔCT [2]	Enrichment [3]
ssa-miR-9a-5p [4]	B	−11.13	2241
ssa-miR-9b-3p	B	−10.21	1184
ssa-miR-96-5p	B	−5.18	36
ssa-miR-129-5p	B	−3.56	12
ssa-miR-132-5p	B	−7.46	176
ssa-miR-135c-5p	B	−7.05	133
ssa-miR-153a-3p	B	−10.08	1082
ssa-miR-212ab-3p	B	−7.29	156
ssa-miR-219a-3p	B	−7.66	202
ssa-miR-723-5p	B	−6.83	114
ssa-miR-734a-3p	B	−4.68	26
ssa-miR-122-5p[4]	L	−12.3	5043
ssa-miR-8163-3p	L	−8.6	388
ssa-miR-192a-5p [4]	L	−7.96	249
ssa-miR-499a-5p [4]	H	−9.7	832
ssa-miR-736-3p	H	−14.7	26616
ssa-miR-192a-5p [4]	I	−11.9	3822
ssa-miR-459-5p	I	−12.9	7643
ssa-miR-194b-5p	I	−10.8	1783
ssa-miR-140-3p	G	−3.7	13

[1] Tissue where the particular mature miRNA is highly expressed (B = brain, L = liver, H = heart, I = intestine, G = gills, M = muscle). [2] Logfold difference in highly expressed tissue relative to mean of all other. [3] Times increase in the highly enriched tissue.[4] These four miRNAs are also among those shown in Figure 4. All differences were significant at adjusted p-values ≤ 0.05.

A complete overview of all 31 miRNAs that showed a tissue specific expression pattern is given in Table 4. Brain tissue was the one tissue showing largest number of miRNAs expressed in a tissue specific manner, both among those miRNAs with highly enriched expression (Figure 4), as well as those investigated by RT-qPCR (16 miRNAs, Table 4).

Table 4. miRNAs that show tissue enriched expression patterns.

Gills	Muscle	Intestine	Brain	Heart	Kidney	Liver
ssa-miR-140-3p	ssa-miR-133-1/2-3p	ssa-miR-192a-5p	ssa-miR-128-3p	ssa-miR-133-1/2-3p	ssa-miR-192a-5p	ssa-miR-122-5p
ssa-miR-203a/b-3p	ssa-miR-26d-5p	ssa-miR-194a/b-5p	ssa-miR-129-5p	ssa-miR-499a/b-5p		ssa-miR-192a-5p
		ssa-miR-459-5p	ssa-miR-132-5p	ssa-miR-736-3p		ssa-miR-8163-3p
			ssa-miR-135c-5p			
			ssa-miR-153a/b-3p			
			ssa-miR-182-5p			
			ssa-miR-183-5p			
			ssa-miR-212ab-3p			
			ssa-miR-219a-3p			
			ssa-miR-723-5p			
			ssa-miR-734a-3p			
			ssa-miR-92b-3p			
			ssa-miR-96-5p			
			ssa-miR-9a-5p			
			ssa-miR-9b-3p			

Overview of all mature miRNAs enriched in particular tissues. Different family members that are enriched in the same tissue are indicated by a slash (e.g., ssa-miR-449a/b-5p). Identical family members are shown together as given in the reference miRNAome (Supplementary file S5), (e.g., ssa-miR-212a-3p and ssa-miR-212b-3p are shown as ssa-miR-212ab-3p).

3.3.2. Expression Patterns of miRNAs at Development Specific Stages

It has been reported that miRNAs show developmental stage-specific abundance during the embryonic development of teleosts [32,37]. The miRNA diversity showed, in general, an increase from less than one hundred to close to 600 different miRNAs during development. Embryo 4dpf, corresponding to the earliest developmental stage in our materials, showed the smallest diversity with only 63 different miRNAs expressed (Figure 6). There was a very large increase in the number of miRNAs that were expressed during the next 15 days as the embryo 19dpf sample revealed there were 318 different miRNAs expressed. The number of miRNAs detected increased by another hundred the next 20 days (embryo 39dpf), and at the eyed-egg stage (63dpf) it was at its maximum (585 different miRNAs).

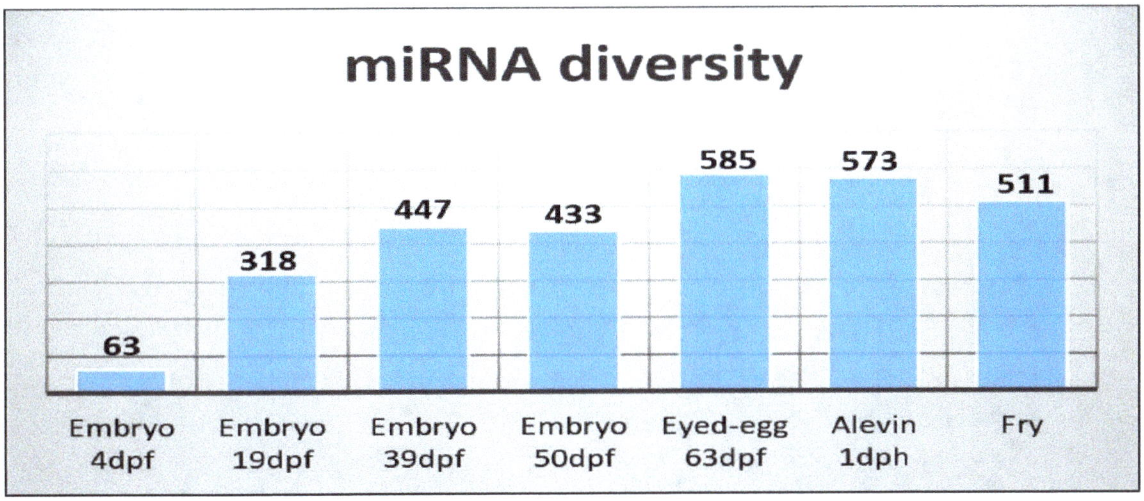

Figure 6. miRNA diversity measured as a number of different miRNAs detected across developmental stages.

Due to the limited number of samples and lack of proper normalization, we could not compare expression differences between the developmental stages. We could however compare the expression differences of miRNAs within each of the developmental stages. A few miRNAs showed high expression at specific stages, while others exhibited a ubiquitous expression pattern and were highly enriched in all the developmental stages. This was the case for the three members of the mir-10 family (ssa-miR-10b-5p, ssa-miR-10d-5p and ssa miR-10a-5p) that constituted a large proportion (>50%) of all miRNAs at most developmental stages (Figure 7A–F), while their abundance was very low in fry (1% of all miRNAs) (Figure 8). The proportion of miRNAs from the ssa-miR-430 cluster was relatively large at the earliest stages of development (embryo 4dpf, Figure 7A), but appeared to decline rapidly as the proportion was 3% at 19dpf and less than 1% at all other stages (Figure 7B–F and Figure 8).

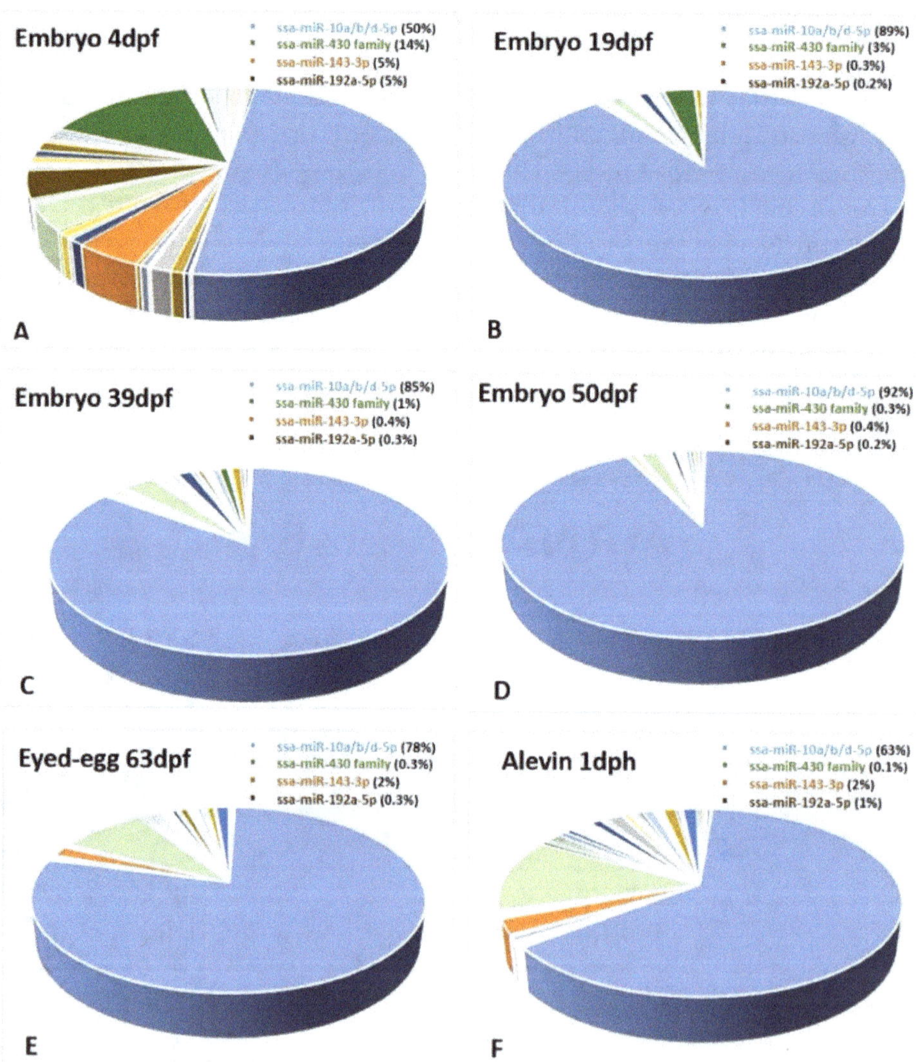

Figure 7. Abundance as a proportion of the total number of miRNAs within each of the developmental stages. The ssa-miR-10a/b/d-5p family, the miR-430 family, ssa-miR-143-3p and ssa-miR-192a-5p are shown in blue, green, orange and brown colors, respectively, in the figures (**A–F**).

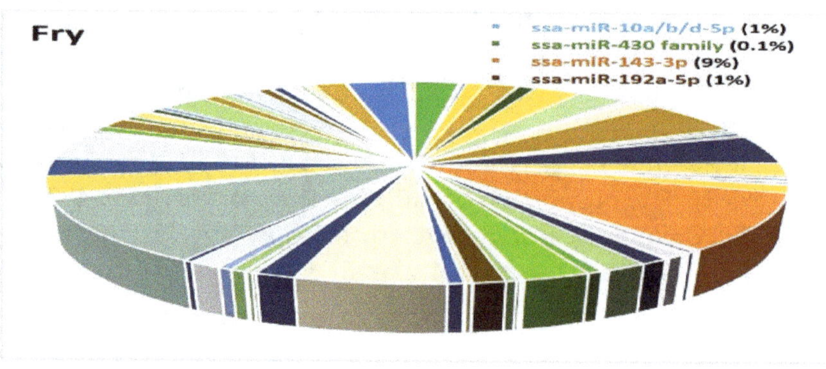

Figure 8. A pie chart demonstrating the miRNA expression diversity (511 miRNAs) in fry. The ssa-miR-10a/b/d-5p family, the miR-430 family, ssa-miR-143-3p and ssa-miR-192a-5p are shown in blue, green, orange and brown colors, respectively, in Figure 8.

The proportion of ssa-miR-143-3p and ssa-miR-192 was about 5% in embryo 4dpf (Figure 7A), but the proportion of these miRNAs declined about a tenfold in the following stages (19dpf to eyed-egg 63dpf, Figure 7B–E). These two miRNAs seemed to increase their expression post hatching (Figure 8).

4. Discussion

4.1. Small RNA Sequencing and Identification of Atlantic Salmon miRNAs

In this study, we performed a comprehensive analysis for miRNA characterization and identified 448 different miRNA genes in Atlantic salmon, including 17 novel miRNAs. Although a high number of Atlantic salmon miRNAs had been identified in 2013, nearly one hundred new miRNA genes, both miRNAs conserved in teleosts, as well as novel miRNAs, were discovered in this study. The increase in number of miRNA genes discovered could largely be attributed to the large number of materials included (111 samples from different tissues, developmental stages, fry and pathogen challenged fry). This miRNAome will be the new, improved reference to apply when investigating differential miRNA expression in Atlantic salmon. Furthermore, the genomic locations of all miRNA genes and their clustering patterns annotated in the updated Atlantic salmon genome sequence will facilitate further studies of the comparative evolution of miRNA genes.

4.2. Characterization of IsomiRs and Polymorphics miRNAs in Atlantic Salmon

With the advancement of high throughput sequencing techniques, many recent studies have reported the presence of a number of mature miRNA sequence variants with different 5′ and/or 3′ ends compared to their corresponding canonical mature miRNAs termed as isomiRs. Small-RNA sequencing projects generate datasets consisting of millions of reads differing in length and quality. Before characterizing isomiRs one need to control size variations and bp errors that arise in the pipeline. The read alignments to the larger sized quality control references indicated that most of the size variations (variation in read length) were artificial and generated in the RNA extraction or the first steps of the sequencing pipeline. There may be some templated isomiR length variants in Atlantic salmon, but due to the platform related high proportion of read length variation, they could not be detected. While there are some studies that report that they have accounted for sequence errors when characterizing isomiRs, many either have used Phred quality score estimates as an error threshold or spike-in controls to measure the sequence errors [15,55]. Despite the fact that Phred score is a good measure of sequence quality, such estimates will only account for the errors caused by the sequencing itself, not other sources of sequence errors that could be generated at other steps of the pipeline (e.g., cDNA synthesis). A high frequency of site-specific bias (especially at the 3′ ends), that would otherwise not be identified by a Phred quality score, was revealed when we estimated the average ratio of ESVs in our data (see Section 3.2.1). As isomiRs are mature miRNA sequence variants with non-templated nucleotide differences in the 5′ or 3′ ends, a large proportion of the erroneous sequence variants revealed could potentially be mistaken for isomiRs. This illustrates the importance of incorporating measurements of error beyond Phred quality to distinguish ESVs from isomiRs. The proportion of ESVs may, however, differ between library preparation methods and sequencing platforms. Nevertheless, controlling the error rate seems crucial.

Most isomiRs identified in our study were the non-templated 3′ isomiRs in the form of nucleotide substitution and/or nucleotide addition. This result is not surprising, as the 3′ isomiR variants are the most common isomiR variants observed in animals and plants [19]. As shown in Figure 3, the isomiRs variants were predicted to cause only very small changes in the number of targeted transcripts. Our findings are consistent with those reported in other studies, as targeting is mainly mediated through complementary binding of the 5′ seed (2–8 nts of mature miRNA sequence) [17,19]. Moreover, the total amount of isomiR variants was small (43 isomiRs). Together, these findings suggest that these types of modifications may have less biological impact than anticipated [56]. Allelic variation in the seed did however have a major effect on target gene diversity. The target gene analysis of the two allelic variants of miR-100a-2-3p (Table 2) showed that the new variant practically acted as a "new miRNA", with a completely new set of target transcripts. The negative selective pressure against such variation, that could allow miRNAs to adapt new regulatory functions, may be less strict when miRNAs are from families where the other members maintain regulation of their original target genes. As the partially tetraploid Atlantic salmon miRNAome has a much larger number of very similar miRNA genes (paralogs) than other diploid fish (e.g., zebrafish, Atlantic cod), this would allow for WGD-derived paralogs to change in seed and develop new functions.

Although the biological significance of 3′ isomiRs and 3′ length isomiRs seems small, these may still cause methodological issues when analyzing miRNAs by quantitative real time PCR (RT-qPCR) [12,16]. This miRNA detection methodology mainly depends on amplification that is initiated with a miRNA-specific primer. A nucleotide difference in the 3′ end (3′ isomiRs and 3′ length variants) could lead to the detection of products with different melting temperatures and thus, affect precise measurements of specific miRNA levels as both the canonical miRNA and the isomiR variant may be cross detected [57]. It is therefore important to be aware of such variants, as they may explain some of the methodological challenges one may come across when measuring miRNA expression by RT-qPCR [16].

4.3. Characterization of miRNA Expression Profiles in Different Tissues and Developmental Stages

Identifying expression patterns of Atlantic salmon miRNAs in different tissues and developmental stages provides important insight into the function of individual miRNAs. The miRNAs discovered in our study showed a wide range of expression profiles in the different tissues and developmental stages. The miRNAs ssa-miR-143-3p, ssa-miR-181a-3p, ssa-miR-21b-5p, ssa-mir-26a-5p, ssa-miR-10b-5p, ssa-mir-10d-5p and ssa-mir-10a-5p were ubiquitously expressed in all tissues tested. The high abundance and ubiquitous expression profile of most of these miRNAs have been reported in several other species [58–61]. The evolutionarily conserved high expression profiles of these miRNAs suggest they are associated with common signaling pathways in vertebrates and have the same housekeeping functions in Atlantic salmon. Other miRNAs showed a tissue specific expression pattern suggesting a specialized role for these miRNAs in tissue differentiation or maintenance of tissue specific functions. The brain-enriched miRNAs, such as the miR-9 family are e.g., known to have important roles in neurogenesis and brain development in other fish and mammals [43,53,54,58]. Also consistent with previous findings in salmon and other teleosts was the tissue specific high enrichment pattern of miR-122-5p in liver [39,51,52,62]. The high expression of miR-192 shown in liver, kidney and intestine tissues have also been reported in other fish species [39,63]. Finally, there was a high expression of the miR-133 family in both muscle and cardiac tissue, whereas the miR-499 family members were only enriched in cardiac tissue. The miR-133 family are amongst others known to regulate cardiomyocyte differentiation and proliferation, cardiac morphogenesis and stress responsive cardiac remodeling process in other species [64,65]. As demonstrated in other species, a majority of the orthologous miRNAs that showed tissue specific expression have specialized functions in different tissues. When revealing same tissue enriched expression in Atlantic salmon it is likely that they also have similar specialized functions in this species.

Several studies have suggested that miRNAs have essential roles in the developmental progression in vertebrates [66,67]. The increase in miRNA diversity along with the different stages of development (Figure 6) indicates that the developmental processes are under miRNA regulation during Atlantic salmon development. The largest change in the proportion of miRNAs within the different developmental stages was observed for the miR-430 family members. This miRNA family were highly abundant during the earliest stages of development, while their abundance decreased rapidly throughout the later stages (Figures 7 and 8). This pattern of expression agreed with those of miR-430 reported in zebrafish. It is assumed that this family of miRNAs are involved in maternal RNA clearance during early embryogenesis [49,68,69]. We also found three members of the miR-10 family that were highly abundant throughout all developmental stages up to one-day post-hatching (Figure 7). Previous studies in other teleosts have identified members of the miR-10 family members as key regulators of *Hox* genes, which are important regulators of embryonic development in vertebrates [50]. The expression of miR-10 family members is, among others, also shown to be associated with mediating cell proliferation and differentiation [70]. The sequences of these Atlantic salmon miRNA family members are highly conserved across species [70]. Together, this implies that they may have similar functions in Atlantic salmon development. The proportion of ssa-miR-143-3p and ssa-miR-192 was higher in embryo 4dpf (Figure 7A) and fry (Figure 8) (Section 3.3) compared to all the other developmental stages. The investigation of tissue specific expression showed that they have very high expression levels in all tissues (ssa-miR-143-3p) or in particular tissues (ssa-miR-192a-5p). In addition, miR-143 accounted for a large proportion (10%) in female gonads, which suggests that these two miRNAs have important functions in adults. Thus, the high abundance of these miRNAs in embryo 4dpf could reflect the maternal contribution rather than a particular function at the earliest stage of development. The rapid disappearance in the later stages could be essential to allow the expression of genes important in development. Further experimental studies are necessary to reveal the particular roles of the miRNAs that showed a specific expression in some tissues and developmental stages.

5. Conclusions

We discovered nearly one hundred new miRNA genes, both miRNAs conserved in teleosts, as well as novel miRNAs, thus, contributing to a major expansion in the number of different miRNAs characterized in Atlantic salmon. The resulting new reference miRNAome provides an important updated resource for miRNA expression studies. Further, a subset of miRNA genes highly abundant in one or more tissues and developmental stages were revealed, suggesting important biological functions of particular miRNAs in the maintenance of tissue specific functions and in the regulation of embryonic development. Together, results from this study provide insight on miRNA regulation that includes those biological processes, and that may be of economic importance to the aquaculture industry.

Author Contributions: Conceived and coordinated the study, R.A. and B.H.; Methodology, R.A.; Software, O.A.; Validation, R.A. and N.T.W.; Formal analysis, R.A., O.A and N.T.W.; Investigation, R.A. and N.T.W.; Resources, R.A., R.D.H. and J.B.T.; Data curation, R.A. and N.T.W.; Writing—original draft preparation, N.T.W.; Writing—review and editing, N.T.W., R.A. and B.H.; Visualization, R.A. and N.T.W.; Supervision, R.A. and B.H.; Project administration, R.A.; Funding acquisition, R.A. All authors revised and approved the final draft.

References

1. Bartel, D.P. MicroRNAs: Genomics, biogenesis, mechanism, and function. *Cell* **2004**, *116*, 281–297. [CrossRef]
2. Ha, M.; Kim, V.N. Regulation of microRNA biogenesis. *Nat. Rev. Mol. Cell Biol.* **2014**, *15*, 509–524. [CrossRef]
3. Friedlander, M.R.; Chen, W.; Adamidi, C.; Maaskola, J.; Einspanier, R.; Knespel, S.; Rajewsky, N. Discovering microRNAs from deep sequencing data using miRDeep. *Nat. Biotechnol.* **2008**, *26*, 407–415. [CrossRef]
4. Friedlander, M.R.; Mackowiak, S.D.; Li, N.; Chen, W.; Rajewsky, N. miRDeep2 accurately identifies known and hundreds of novel microRNA genes in seven animal clades. *Nucleic Acids Res.* **2012**, *40*, 37–52. [CrossRef]
5. Bushati, N.; Cohen, S.M. microRNA functions. *Annu. Rev. Cell Dev. Biol.* **2007**, *23*, 175–205. [CrossRef]
6. Lynam-Lennon, N.; Maher, S.G.; Reynolds, J.V. The roles of microRNA in cancer and apoptosis. *Biol. Rev.* **2009**, *84*, 55–71. [CrossRef]
7. Forster, S.C.; Tate, M.D.; Hertzog, P.J. MicroRNA as Type I Interferon-Regulated Transcripts and Modulators of the Innate Immune Response. *Front. Immunol.* **2015**, *6*, 334. [CrossRef]
8. Sonkoly, E.; Stahle, M.; Pivarcsi, A. MicroRNAs and immunity: Novel players in the regulation of normal immune function and inflammation. *Semin. Cancer Biol.* **2008**, *18*, 131–140. [CrossRef]
9. Andreassen, R.; Worren, M.M.; Hoyheim, B. Discovery and characterization of miRNA genes in Atlantic salmon (*Salmo salar*) by use of a deep sequencing approach. *BMC Genom.* **2013**, *14*, 482. [CrossRef]
10. Bekaert, M.; Lowe, N.R.; Bishop, S.C.; Bron, J.E.; Taggart, J.B.; Houston, R.D. Sequencing and characterisation of an extensive Atlantic salmon (*Salmo salar* L.) microRNA repertoire. *PLoS ONE* **2013**, *8*, e70136. [CrossRef]

11. Andreassen, R.; Woldemariam, N.T.; Egeland, I.O.; Agafonov, O.; Sindre, H.; Hoyheim, B. Identification of differentially expressed Atlantic salmon miRNAs responding to salmonid alphavirus (SAV) infection. *BMC Genom.* **2017**, *18*, 349. [CrossRef]
12. Johansen, I.; Andreassen, R. Validation of miRNA genes suitable as reference genes in qPCR analyses of miRNA gene expression in Atlantic salmon (*Salmo salar*). *BMC Res. Notes* **2014**, *8*, 945. [CrossRef]
13. Valenzuela-Munoz, V.; Novoa, B.; Figueras, A.; Gallardo-Escarate, C. Modulation of Atlantic salmon miRNome response to sea louse infestation. *Dev. Comp. Immunol.* **2017**, *76*, 380–391. [CrossRef]
14. Skaftnesmo, K.O.; Edvardsen, R.B.; Furmanek, T.; Crespo, D.; Andersson, E.; Kleppe, L.; Taranger, G.L.; Bogerd, J.; Schulz, R.W.; Wargelius, A. Integrative testis transcriptome analysis reveals differentially expressed miRNAs and their mRNA targets during early puberty in Atlantic salmon. *BMC Genom.* **2017**, *18*, 801. [CrossRef]
15. Ebhardt, H.A.; Tsang, H.H.; Dai, D.C.; Liu, Y.; Bostan, B.; Fahlman, R.P. Meta-analysis of small RNA-sequencing errors reveals ubiquitous post-transcriptional RNA modifications. *Nucleic Acids Res.* **2009**, *37*, 2461–2470. [CrossRef]
16. Lee, L.W.; Zhang, S.; Etheridge, A.; Ma, L.; Martin, D.; Galas, D.; Wang, K. Complexity of the microRNA repertoire revealed by next-generation sequencing. *RNA* **2010**, *16*, 2170–2180. [CrossRef]
17. Guo, L.; Chen, F. A challenge for miRNA: Multiple isomiRs in miRNAomics. *Gene* **2014**, *544*, 1–7. [CrossRef]
18. Cloonan, N.; Wani, S.; Xu, Q.; Gu, J.; Lea, K.; Heater, S.; Barbacioru, C.; Steptoe, A.L.; Martin, H.C.; Nourbakhsh, E.; et al. MicroRNAs and their isomiRs function cooperatively to target common biological pathways. *Genome Biol.* **2011**, *12*, R126. [CrossRef]
19. Neilsen, C.T.; Goodall, G.J.; Bracken, C.P. IsomiRs-the overlooked repertoire in the dynamic microRNAome. *Trends. Genet.* **2012**, *28*, 544–549. [CrossRef]
20. Lin, W.; Piskol, R.; Tan, M.H.; Li, J.B. Comment on "Widespread RNA and DNA sequence differences in the human transcriptome". *Science* **2012**, *335*, 1302, author reply 1302. [CrossRef]
21. Robasky, K.; Lewis, N.E.; Church, G.M. The role of replicates for error mitigation in next-generation sequencing. *Nat. Rev. Genet.* **2014**, *15*, 56–62. [CrossRef]
22. Lien, S.; Koop, B.F.; Sandve, S.R.; Miller, J.R.; Kent, M.P.; Nome, T.; Hvidsten, T.R.; Leong, J.S.; Minkley, D.R.; Zimin, A.; et al. The Atlantic salmon genome provides insights into rediploidization. *Nature* **2016**, *533*, 200–205. [CrossRef]
23. Robledo, D.; Taggart, J.B.; Ireland, J.H.; McAndrew, B.J.; Starkey, W.G.; Haley, C.S.; Hamilton, A.; Guy, D.R.; Mota-Velasco, J.C.; Gheyas, A.A.; et al. Gene expression comparison of resistant and susceptible Atlantic salmon fry challenged with Infectious Pancreatic Necrosis virus reveals a marked contrast in immune response. *BMC Genom.* **2016**, *17*, 279. [CrossRef]
24. Andrews, S.; Krueger, F.; Seconds-Pichon, A.; Biggins, F.; Wingett, S. *FastQC: A Quality Control Tool for High Throughput Sequence Data*; Babraham Institute: Cambridge, UK, 2012.
25. Martin, M. Cutadapt removes adapter sequences from high-throughput sequencing reads. *EMBnet. J.* **2011**, *17*, 10–12. [CrossRef]
26. Kozomara, A.; Griffiths-Jones, S. miRBase: Annotating high confidence microRNAs using deep sequencing data. *Nucleic Acids Res.* **2014**, *42*, D68–D73. [CrossRef]
27. Ambros, V.; Bartel, B.; Bartel, D.P.; Burge, C.B.; Carrington, J.C.; Chen, X.; Dreyfuss, G.; Eddy, S.R.; Griffiths-Jones, S.; Marshall, M.; et al. A Uniform System for microRNA Annotation. *RNA* **2003**, *9*, 277–279. [CrossRef]
28. Griffiths-Jones, S.; Grocock, R.J.; van Dongen, S.; Bateman, A.; Enright, A.J. miRBase: MicroRNA sequences, targets and gene nomenclature. *Nucleic Acids Res.* **2006**, *34*, D140–D144. [CrossRef]
29. Kalvari, I.; Argasinska, J.; Quinones-Olvera, N.; Nawrocki, E.P.; Rivas, E.; Eddy, S.R.; Bateman, A.; Finn, R.D.; Petrov, A.I. Rfam 13.0: Shifting to a genome-centric resource for non-coding RNA families. *Nucleic Acids Res.* **2018**, *46*, D335–D342. [CrossRef]
30. Mituyama, T.; Yamada, K.; Hattori, E.; Okida, H.; Ono, Y.; Terai, G.; Yoshizawa, A.; Komori, T.; Asai, K. The Functional RNA Database 3.0: Databases to support mining and annotation of functional RNAs. *Nucleic Acids Res.* **2009**, *37*, D89–D92. [CrossRef]
31. Sanger, F.; Air, G.M.; Barrell, B.G.; Brown, N.L.; Coulson, A.R.; Fiddes, J.C.; Hutchison Iii, C.A.; Slocombe, P.M.; Smith, M. Nucleotide sequence of bacteriophage φX174 DNA. *Nature* **1977**, *265*, 687. [CrossRef]

32. Manley, L.J.; Ma, D.; Levine, S.S. Monitoring Error Rates In Illumina Sequencing. *J. Biomol. Tech.* **2016**, *27*, 125–128. [CrossRef]
33. Urgese, G.; Paciello, G.; Acquaviva, A.; Ficarra, E. isomiR-SEA: An RNA-Seq analysis tool for miRNAs/isomiRs expression level profiling and miRNA-mRNA interaction sites evaluation. *BMC Bioinform.* **2016**, *17*, 148. [CrossRef]
34. Schmittgen, T.D.; Lee, E.J.; Jiang, J.; Sarkar, A.; Yang, L.; Elton, T.S.; Chen, C. Real-time PCR quantification of precursor and mature microRNA. *Methods* **2008**, *44*, 31–38. [CrossRef]
35. Rehmsmeier, M.; Steffen, P.; Hochsmann, M.; Giegerich, R. Fast and effective prediction of microRNA/target duplexes. *RNA* **2004**, *10*, 1507–1517. [CrossRef]
36. Dobin, A.; Davis, C.A.; Schlesinger, F.; Drenkow, J.; Zaleski, C.; Jha, S.; Batut, P.; Chaisson, M.; Gingeras, T.R. STAR: Ultrafast universal RNA-seq aligner. *Bioinformatics* **2013**, *29*, 15–21. [CrossRef]
37. Stokowy, T.; Eszlinger, M.; Swierniak, M.; Fujarewicz, K.; Jarzab, B.; Paschke, R.; Krohn, K. Analysis options for high-throughput sequencing in miRNA expression profiling. *BMC Res. Notes* **2014**, *7*, 144. [CrossRef]
38. Schmittgen, T.D.; Livak, K.J. Analyzing real-time PCR data by the comparative C(T) method. *Nat. Protoc* **2008**, *3*, 1101–1108. [CrossRef]
39. Andreassen, R.; Rangnes, F.; Sivertsen, M.; Chiang, M.; Tran, M.; Worren, M.M. Discovery of miRNAs and their corresponding miRNA genes in Atlantic Cod (*Gadus morhua*): Use of stable miRNAs as reference genes reveals subgroups of miRNAs that are highly expressed in particular organs. *PLoS ONE* **2016**, *11*, e0153324. [CrossRef]
40. Chen, P.Y.; Manninga, H.; Slanchev, K.; Chien, M.; Russo, J.J.; Ju, J.; Sheridan, R.; John, B.; Marks, D.S.; Gaidatzis, D.; et al. The developmental miRNA profiles of zebrafish as determined by small RNA cloning. *Genes Dev.* **2005**, *19*, 1288–1293. [CrossRef]
41. Li, S.C.; Chan, W.C.; Ho, M.R.; Tsai, K.W.; Hu, L.Y.; Lai, C.H.; Hsu, C.N.; Hwang, P.P.; Lin, W.C. Discovery and characterization of medaka miRNA genes by next generation sequencing platform. *BMC Genom.* **2010**, *11* (Suppl. S4), S8. [CrossRef]
42. Baskerville, S.; Bartel, D.P. Microarray profiling of microRNAs reveals frequent coexpression with neighboring miRNAs and host genes. *RNA* **2005**, *11*, 241–247. [CrossRef]
43. Giraldez, A.J.; Cinalli, R.M.; Glasner, M.E.; Enright, A.J.; Thomson, J.M.; Baskerville, S.; Hammond, S.M.; Bartel, D.P.; Schier, A.F. MicroRNAs regulate brain morphogenesis in zebrafish. *Science* **2005**, *308*, 833–838. [CrossRef]
44. Sokol, N.S. The role of microRNAs in muscle development. *Curr. Top. Dev. Biol.* **2012**, *99*, 59–78. [CrossRef]
45. Kozomara, A.; Griffiths-Jones, S. miRBase: Integrating microRNA annotation and deep-sequencing data. *Nucleic Acids Res.* **2011**, *39*, D152–D157. [CrossRef]
46. Berthelot, C.; Brunet, F.; Chalopin, D.; Juanchich, A.; Bernard, M.; Noël, B.; Bento, P.; Da Silva, C.; Labadie, K.; Alberti, A.; et al. The rainbow trout genome provides novel insights into evolution after whole-genome duplication in vertebrates. *Nat. Commun.* **2014**, *5*, 3657. [CrossRef]
47. Song, J.B.; Song, J.; Mo, B.X.; Chen, X.M. Uridylation and adenylation of RNAs. *Sci. China Life Sci.* **2015**, *58*, 1057–1066. [CrossRef]
48. Wyman, S.K.; Knouf, E.C.; Parkin, R.K.; Fritz, B.R.; Lin, D.W.; Dennis, L.M.; Krouse, M.A.; Webster, P.J.; Tewari, M. Post-transcriptional generation of miRNA variants by multiple nucleotidyl transferases contributes to miRNA transcriptome complexity. *Genome Res.* **2011**, *21*, 1450–1461. [CrossRef]
49. Giraldez, A.J.; Mishima, Y.; Rihel, J.; Grocock, R.J.; Van Dongen, S.; Inoue, K.; Enright, A.J.; Schier, A.F. Zebrafish MiR-430 promotes deadenylation and clearance of maternal mRNAs. *Science* **2006**, *312*, 75–79. [CrossRef]
50. Giusti, J.; Pinhal, D.; Moxon, S.; Campos, C.L.; Münsterberg, A.; Martins, C. MicroRNA-10 modulates Hox genes expression during Nile tilapia embryonic development. *Mech. Dev.* **2016** *140*, 12–18. [CrossRef]
51. Mennigen, J.A.; Martyniuk, C.J.; Seiliez, I.; Panserat, S.; Skiba-Cassy, S. Metabolic consequences of microRNA-122 inhibition in rainbow trout, Oncorhynchus mykiss. *BMC Genom.* **2014**, *15*, 70. [CrossRef]
52. Mennigen, J.A.; Plagnes-Juan, E.; Figueredo-Silva, C.A.; Seiliez, I.; Panserat, S.; Skiba-Cassy, S. Acute endocrine and nutritional co-regulation of the hepatic omy-miRNA-122b and the lipogenic gene fas in rainbow trout, Oncorhynchus mykiss. *Comp. Biochem. Physiol. B Biochem. Mol. Biol.* **2014**, *169*, 16–24. [CrossRef] [PubMed]

53. Radhakrishnan, B.; Alwin Prem Anand, A. Role of miRNA-9 in Brain Development. *J. Exp. Neurosci.* **2016**, *10*, 101–120. [CrossRef] [PubMed]
54. Ludwig, N.; Leidinger, P.; Becker, K.; Backes, C.; Fehlmann, T.; Pallasch, C.; Rheinheimer, S.; Meder, B.; Stahler, C.; Meese, E.; et al. Distribution of miRNA expression across human tissues. *Nucleic Acids Res.* **2016**, *44*, 3865–3877. [CrossRef] [PubMed]
55. Luo, G.Z.; Hafner, M.; Shi, Z.; Brown, M.; Feng, G.H.; Tuschl, T.; Wang, X.J.; Li, X. Genome-wide annotation and analysis of zebra finch microRNA repertoire reveal sex-biased expression. *BMC Genom.* **2012**, *13*, 727. [CrossRef] [PubMed]
56. Fernandez-Valverde, S.L.; Taft, R.J.; Mattick, J.S. Dynamic isomiR regulation in Drosophila development. *RNA* **2010**, *16*, 1881–1888. [CrossRef]
57. Schamberger, A.; Orban, T.I. 3′ IsomiR species and DNA contamination influence reliable quantification of microRNAs by stem-loop quantitative PCR. *PLoS ONE* **2014**, *9*, e106315. [CrossRef] [PubMed]
58. Bizuayehu, T.T.; Babiak, I. MicroRNA in teleost fish. *Genome Biol. Evol.* **2014**, *6*, 1911–1937. [CrossRef]
59. Lagos-Quintana, M.; Rauhut, R.; Yalcin, A.; Meyer, J.; Lendeckel, W.; Tuschl, T. Identification of tissue-specific microRNAs from mouse. *Curr. Biol.* **2002**, *12*, 735–739. [CrossRef]
60. Herkenhoff, M.E.; Oliveira, A.C.; Nachtigall, P.G.; Costa, J.M.; Campos, V.F.; Hilsdorf, A.W.S.; Pinhal, D. Fishing into the MicroRNA transcriptome. *Front. Genet.* **2018**, *9*, 88. [CrossRef] [PubMed]
61. Juanchich, A.; Bardou, P.; Rue, O.; Gabillard, J.C.; Gaspin, C.; Bobe, J.; Guiguen, Y. Characterization of an extensive rainbow trout miRNA transcriptome by next generation sequencing. *BMC Genom.* **2016**, *17*, 164. [CrossRef]
62. Trattner, S.; Vestergren, A.S. Tissue distribution of selected microRNA in Atlantic salmon. *Eur. J. Lipid Sci. Tech.* **2013**, *115*, 1348–1356. [CrossRef]
63. Xia, J.H.; He, X.P.; Bai, Z.Y.; Yue, G.H. Identification and characterization of 63 MicroRNAs in the Asian seabass Lates calcarifer. *PLoS ONE* **2011**, *6*, e17537. [CrossRef] [PubMed]
64. Mitchelson, K.R.; Qin, W.Y. Roles of the canonical myomiRs miR-1, -133 and -206 in cell development and disease. *World J. Biol. Chem.* **2015**, *6*, 162–208. [CrossRef] [PubMed]
65. Wu, G.; Huang, Z.P.; Wang, D.Z. MicroRNAs in cardiac regeneration and cardiovascular disease. *Sci. China Life Sci.* **2013**, *56*, 907–913. [CrossRef] [PubMed]
66. Alberti, C.; Cochella, L. A framework for understanding the roles of miRNAs in animal development. *Development* **2017**, *144*, 2548–2559. [CrossRef]
67. Rosa, A.; Brivanlou, A.H. MicroRNAs in early vertebrate development. *Cell Cycle* **2009**, *8*, 3513–3520. [CrossRef] [PubMed]
68. Takacs, C.M.; Giraldez, A.J. miR-430 regulates oriented cell division during neural tube development in zebrafish. *Dev. Biol.* **2016**, *409*, 442–450. [CrossRef]
69. Wienholds, E.; Kloosterman, W.P.; Miska, E.; Alvarez-Saavedra, E.; Berezikov, E.; de Bruijn, E.; Horvitz, H.R.; Kauppinen, S.; Plasterk, R.H. MicroRNA expression in zebrafish embryonic development. *Science* **2005**, *309*, 310–311. [CrossRef]
70. Tehler, D.; Hoyland-Kroghsbo, N.M.; Lund, A.H. The miR-10 microRNA precursor family. *RNA Biol.* **2011**, *8*, 728–734. [CrossRef]

7

Tensor Decomposition-Based Unsupervised Feature Extraction can Identify the Universal Nature of Sequence-Nonspecific Off-Target Regulation of mRNA Mediated by MicroRNA Transfection

Y.-H. Taguchi

Department of Physics, Chuo University, Tokyo 112-8551, Japan; tag@granular.com;

Abstract: MicroRNA (miRNA) transfection is known to degrade target mRNAs and to decrease mRNA expression. In contrast to the notion that most of the gene expression alterations caused by miRNA transfection involve downregulation, they often involve both up- and downregulation; this phenomenon is thought to be, at least partially, mediated by sequence-nonspecific off-target effects. In this study, I used tensor decomposition-based unsupervised feature extraction to identify genes whose expression is likely to be altered by miRNA transfection. These gene sets turned out to largely overlap with one another regardless of the type of miRNA or cell lines used in the experiments. These gene sets also overlap with the gene set associated with altered expression induced by a Dicer knockout. This result suggests that the off-target effect is at least as important as the canonical function of miRNAs that suppress translation. The off-target effect is also suggested to consist of competition for the protein machinery between transfected miRNAs and miRNAs in the cell. Because the identified genes are enriched in various biological terms, these genes are likely to play critical roles in diverse biological processes.

Keywords: tensor decomposition; miRNA transfection; sequence-nonspecific off-target regulation

---

1. Introduction

MicroRNA (miRNA) is short noncoding (functional) RNA whose primary function is mRNA degradation and disruption of translation [1]. Thus, it is generally expected that the primary effect of miRNA transfection (or overexpression) on mRNA expression is suppression. Based on this assumption, numerous miRNA transfection and/or overexpression experiments have been conducted to identify genes that are directly targeted by miRNAs [2]; during these analyses, only genes with expression levels inversely related to those of miRNA have been sought. Nevertheless, it was found that many mRNAs whose expression was likely to be altered by miRNA transfection and/or overexpression turned out to positively correlate with miRNA expression. For example, Khan et al. [3] identified multiple genes that are upregulated by miRNA transfection. They reasoned that this effect means competition with endogenous miRNAs because upregulated genes were often targeted by endogenous miRNAs. The protein machinery that binds to endogenous miRNAs was occupied by the transfected miRNAs, and as a result, the genes targeted by endogenous miRNAs were upregulated [4]. In addition, Carroll et al. [5] identified a positive correlation between mRNA expression and transfected miRNA. They theorized that the positive correlations are mediated by interactions with transcription factor E2F1.

Despite these findings, to my knowledge, sequence-nonspecific off-target regulation by miRNA transfection has not been extensively studied to date [2]. Most of the miRNA transfection and/or overexpression experiments have been aimed at identifying canonical targets of miRNAs. Most of these experiments have not been analyzed in the context of sequence-nonspecific off-target regulation by miRNA transfection. Although it is unclear why no one has tried to systematically investigate sequence-nonspecific off-target regulation mediated by miRNA transfection, one possible reason is the lack of a suitable methodology. By definition, miRNA transfection experiments cannot be composed of many samples. Typically, a pair of data points consists of miRNA-transfected cells and mock-transfected cells. Although a few more biological and/or technical replicates are possible, the number of samples available is usually less than 10. This number is often too small to detect significantly altered expression of mRNAs whose total number is up to 10^4. In the case when the aim of a study is identification of canonical interactions between miRNA and mRNA, additional information that can reduce the number of mRNAs under study, e.g., bioinformatically predicted mRNAs targeted by transfected miRNAs, is available. This information can enable researchers to identify significant correlations between transfected miRNAs and mRNAs. Nonetheless, this kind of information is usually not available for the analysis of sequence-nonspecific off-target regulation by miRNA transfection.

In this study, with the aim to resolve this difficulty, I applied tensor decomposition (TD)-based unsupervised feature extraction (FE) to miRNA transfection experiments. TD-based unsupervised FE [6–10] is an extension of principal component analysis (PCA)-based unsupervised FE [11–33], which can identify critical genes even when there is only a small number of samples. TD-based unsupervised FE can also identify critical genes by means of a small number of samples available. Because of the use of this methodology, genes whose mRNA expression was likely to be altered by miRNA transfection were identified here in various combinations of cell lines and transfected miRNAs. Most of sets of genes significantly overlapped with one another regardless of transfected cell types and types of miRNA. These genes also showed altered mRNA expression under the influence of a Dicer knockout (KO). This finding suggests that the primary factor that mediated sequence-nonspecific side effects of miRNA transfection is competition for protein machinery with endogenous miRNAs, as suggested by Khan et al. [3]. In addition, these sets of genes significantly overlapped with various sets of genes whose biological functions and significance have been validated experimentally. Thus, sequence-nonspecific off-target regulation caused by miRNA transfection is expected to also play critical roles in various biological processes.

2. Materials and Methods

2.1. Mathematical Formulation of the Tensor and Tensor Decomposition

Because a tensor or tensor decomposition (TD) is not a popular mathematical concept, I briefly formulate them here. Suppose a three-mode tensor, $x_{ijk} \in \mathbb{R}^{N \times M \times K}$, is the expression level of the ith mRNA when the jth miRNA is transfected into the kth sample. Samples are typically composed of biological replicates and/or treated and untreated (or mock-treated, i.e., control) samples, but situations vary. x_{ijk} can also be formulated as two-mode tensor $x_{i(jk)}$, where (jk) represents a pair of a miRNA and a sample, especially when samples are not paired. x_{ijk} can be decomposed to

$$x_{ijk} = \sum_{\ell_1,\ell_2,\ell_3} G(\ell_1,\ell_2,\ell_3) x_{\ell_1 i} x_{\ell_2 j} x_{\ell_3 k},$$

where $G(\ell_1,\ell_2,\ell_3) \in \mathbb{R}^{N \times M \times K}$ is a core tensor, $x_{\ell_1 i} \in \mathbb{R}^{N \times N}, x_{\ell_2 j} \in \mathbb{R}^{M \times M}$ and $x_{\ell_3 k} \in \mathbb{R}^{K \times K}$ are singular value matrices that are orthogonal. Because this construct is obviously overcomplete, there is no unique TD. In this paper, I employed higher-order singular value decomposition [34] (HOSVD) to perform TD.

2.2. *Using TD-Based Unsupervised FE for Identification of Genes Whose Expression Is Likely to Be Altered by MiRNA Transfection*

To this end, first I need to specify which sample singular value vectors, $x_{\ell_3 k}$, have different values between treated (i.e., miRNA-transfected) samples and control (e.g., mock-transfected) samples. Suppose $x_{\ell'_3 k}$ turned out to represent a dissimilarity between a treated sample and control sample in some ways (see Figure 1B as an example). Next, I need to find miRNA singular value vectors, $x_{\ell_2 j}$, that have constant values for all j (see Figure 1A as an example) because I would like to find genes affected constantly by miRNA transfection independently of the type of transfected miRNA, since it should represent sequence-nonspecific off-target regulation. Let us assume that $x_{\ell'_2 j}$ fulfilled this requirement. Then, I rank core tensors $G(\ell_1, \ell'_2, \ell'_3)$ in the order of absolute values (largest at the top). This approach enables me to select ℓ_1 such that $x_{\ell_1 i}$ is associated with constant sequence-nonspecific off-target regulation. After identifying ℓ'_1 as those associated with larger absolute values of $G(\ell'_1, \ell'_2, \ell'_3)$, is representing larger absolute values of $x_{\ell'_1 i}$ were selected. For this purpose, P-values, P_is (see Figure 1C as an example), were assigned to each i assuming that $x_{\ell'_1 i}$ obeys the χ^2 distribution

$$P_i = P_{\chi^2}\left[> \sum_{\ell'_1} \left(\frac{x_{\ell'_1 i}}{\sigma_{\ell'_1}} \right)^2 \right]$$

where $P_{\chi^2}[> x]$ is cumulative probability that the argument is greater than x, assuming the χ^2 distribution and that $\sigma_{\ell'_1}$ is the standard deviation. The number of degrees of freedom of the χ^2 distribution is equal to the number of ℓ'_1s in the summation. The above equation means that I presume that $x_{\ell'_1 i}$ obeys a multiple Gaussian distribution (null hypothesis). Then, P_is were adjusted via the Benjamini–Hochberg (BH) criterion [35]. Genes associated with adjusted $P_i < 0.01$ were finally selected (see the red bin in Figure 1C as an example).

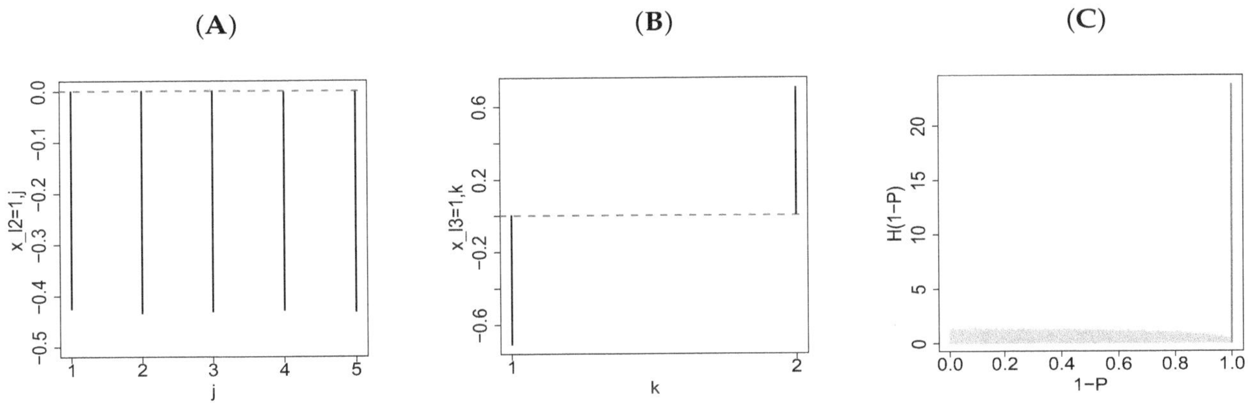

Figure 1. The results on the artificial data: (**A**) $x_{\ell_2=1,j}$ averaged across 100 independent trials. The horizontal red dashed line is $x_{\ell_2=1,j} = 0$ (**B**) $x_{\ell_3=1,k}$ averaged across 100 independent trials. The horizontal red dashed line is $x_{\ell_3=1,k} = 0$ (**C**) A histogram of $1 - P$ computed from $x_{\ell_1=1,i}$. A vertical red segment represents the bin with the smallest P-values.

2.3. *Explanatory Discussion of TD-Based Unsupervised FE*

Readers may wonder why a simple procedure using TD successfully identifies genes whose expression is likely to be altered by sequence-nonspecific off-target regulation mediated by various transfected miRNAs. This result can be explained as follows. Let us say most x_{ijk}s are a random number while a limited number of x_{ijk}s, e.g., $x_{i'jk}$s are coexpressed, i.e.,

$$x_{i'jk} = x_{jk}, i' = 1, \ldots, N'$$

Then, the contribution of $x_{i'jk}$ has the order of magnitude of N', while that of other $N - N'$ (random) $x_{ijk}, i \neq i'$ has the order of magnitude of $\sqrt{N - N'}$. Thus, even if $N' \ll N$, if $N' \approx \sqrt{N - N'}$, then the contribution of $x_{i'jk}$ outperforms that of $x_{ijk}, i \neq i'$. Thus, the contribution of $x_{i'jk}$ should be detected as a singular value vector, $x_{\ell_1 i}$ within which $x_{\ell_1 i'}$ should have a greater contribution than the others ($x_{\ell_1 i}, i \neq i'$). Given that the contribution of $x_{\ell_1 i}, i \neq i'$ is expected to be a Gaussian distribution, $x_{\ell_1 i'}$ can be detected as outliers that do not follow the Gaussian distribution. This is the possible explanation why simple TD-based unsupervised FE successfully identified genes whose expression was likely to be altered by sequence-nonspecific off-target regulation mediated by various transfected miRNAs.

2.4. Artificial Data

To demonstrate the usefulness of TD-based unsupervised FE, I prepared artificial dataset $x_{ijk} \in \mathbb{R}^{N \times M \times 2}$, where N is the number of genes, and M is the number of samples ($k = 1$ for control and $k = 2$ for treated samples). Each treated sample is supposed to be transfected with distinct miRNA, and each control sample is supposed to be either untransfected or transfected with mock miRNA. In these artificial data, at first $x_{ijk} \in \mathcal{N}(0,1)$. After that, they are ordered such that $x_{ijk} > x_{i'jk}$ when $i > i'$ with fixed j and k. This situation introduces complete correlations between distinct pairs of js and ks (e.g., rank correlation coefficients between x_{ijk} and $x_{ij'k'}, j \neq j', k \neq k'$ are always equal to 1.0). Then, $x_{ijk}, i > N_0$ were shuffled at fixed j and k to eliminate the correlation. Next, $x_{ijk} \leftarrow (a-1)x_{ijk} + a\epsilon, \epsilon \in \mathcal{N}(0,1), a > 0$ for $i \leq N_0$ in order to introduce randomness into correlating rows. After that, $x_{ij2} \leftarrow -x_{ij2}$ to introduce a difference between control and treated samples. Finally, $\frac{N_0}{2} < i \leq N_0$ were shuffled at fixed j to generate a sample-specific difference between control and treated samples. This means that $1 \leq i \leq \frac{N_0}{2}$ correspond to a sample-nonspecific (i.e., independent of j) dissimilarity between control and treated samples that represents sequence-nonspecific off-target regulation, and $\frac{N_0}{2} < i \leq N_0$ correspond to a sample-specific (i.e., dependent on j) dissimilarity between control and treated samples that represents sequence-specific regulation. The task at hand is to identify $1 \leq i \leq \frac{N_0}{2}$ as precisely as possible. Specifically, $N = 2000, M = 5, N_0 = 50, a = 0.5$.

2.5. Gene Expression Profiles

In this subsection, I explain 11 analyzed profiles of gene expression (Table 1) in more detail. All of them were retrieved from Gene Expression Omnibus (GEO) [36]. In some cases (Experiments 1, 2, 3, and 5), I used a two-mode tensor, $x_{i(jk)}$, which is simply denoted as x_{ij}, instead of three-mode tensor x_{ijk}, by expanding the second (j) and the third (k) modes into one column (jk) because matched data were not available. In this case, HOSVD is equivalent to simple singular value decomposition. x_{ijk} and $x_{i(jk)}$ were standardized as $\sum_i x_{ijk} = \sum_i x_{i(jk)} = 0$ and $\sum_i x_{ijk}^2 = \sum_i x_{i(jk)}^2 = N$ before TD was applied.

Table 1. Eleven experiments conducted for this analysis. More detailed information is available in the text.

Exp.	GEO ID	Cell Lines (Cancer)	miRNA	Misc
1	GSE26996	BT549 (breast cancer)	miR-200a/b/c	
2	GSE27431	HEY (ovarian cancer)	miR-7/128	mas5
3	GSE27431	HEY (ovarian cancer)	miR-7/128	plier
4	GSE8501	Hela (cervical cancer)	miR-7/9/122a/128a/132/133a/142/148b/181a	
5	GSE41539	CD1 mice	cel-miR-67,hsa-miR-590-3p,hsa-miR-199a-3p	
6	GSE93290	multiple	miR-10a-5p,150-3p/5p,148a-3p/5p,499a-5p,455-3p	
7	GSE66498	multiple	miR-205/29a/144-3p/5p,210,23b,221/222/223	
8	GSE17759	EOC 13.31 microglia cells	miR-146a/b	(KO/OE)
9	GSE37729	HeLa	miR-107/181b	(KO/OE)
10	GSE37729	HEK-293	miR-107/181b	(KO/OE)
11	GSE37729	SH-SY5Y	181b	(KO/OE)

OE: overexpression.

2.5.1. No. 1: GSE26996

File GSE26996_RAW.tar was downloaded and unpacked. Six files, GSM665046_miR200a_1.txt.gz, GSM665047_miR200b_1.txt.gz, GSM665048_miR200c_1.txt.gz, GSM665049_miR200a_2.txt.gz, GSM665050_miR200b_2.txt.gz, and GSM665051_miR200c_2.txt.gz, were loaded into R using the read.csv function. Then, after the exclusion of probes with ControlType=0, gProcessedSignal and rProcessedSignal were extracted as treated and control samples, respectively. Thus, I have $x_{ij}, 1 \leq j \leq 12, 1 \leq i \leq 43376$. Here, $1 \leq j \leq 6$ and $7 \leq j \leq 12$ are treated and control samples, respectively. For this dataset, gene expression profiles are regarded as a two-mode tensor (matrix). Applying PCA to x_{ij}, I found that $x_{\ell_2=2,j}$ are different between treated and control samples (Figure S1) independently of the type of miRNA transfected. Next, genes were selected by means of $x_{\ell_1=2,i}$.

2.5.2. No. 2: GSE27431

File GSE27431_series_matrix.txt.gz was downloaded. It was loaded into R using the read.csv function. Each column corresponds to individual gene expression profiles named as GSEMXXXXXX, which is the GEO ID. Among those, mas5-processed samples are regarded as No. 2. GSM678153 and GSM678154 are miR-7-treated, GSM678156 and GSM678157 are miR-128-treated, whereas GSM678158, GSM678159, and GSM678160 are control samples. Then, $x_{ij}, 1 \leq i \leq 54675, 1 \leq j \leq 7$ are obtained (two-mode tensor). Applying TD to x_{ij}, I found that $x_{\ell_2=2,j}$ reflects the inverse regulation between miR-128 and miR-7 while the control samples are in between (Figure S2). This finding can be interpreted as follows. Target genes of miR-128 (miR-7) are downregulated, but they are upregulated when miR-7 is transfected because of sequence-nonspecific off-target regulation caused by miRNA transfection. Accordingly, I decided to select genes using $x_{\ell_1=2,i}$ again.

2.5.3. No. 3: GSE27431

GSE27431_series_matrix.txt.gz was again downloaded. It was loaded into R using the read.csv function. Each column corresponds to individual gene expression profiles named as GSMXXXXXX, which is a GEO ID as well. Among those, plier-processed samples are regarded as No. 3. GSM678164 and GSM678165 are miR-7-treated, GSM678167 and GSM678168 are miR-128-treated, and GSM678169, GSM678170, GSM678171, GSM678172, GSM678173, and GSM678174 are control samples. Next, x_{ij}, $1 \leq i \leq 54675$, $1 \leq j \leq 10$ are obtained (two-mode tensor). Applying PCA to x_{ij}, I found that $x_{\ell_2=2,j}$ reflects the inverse regulation between miR-128 and miR-7, while the control samples are in between (Figure S3). This result can be interpreted as in No. 2. Thus, I decided to select genes by means of $x_{\ell_1=2,i}$ again.

2.5.4. No. 4: GSE8501

Eighteen raw data files were downloaded from GSMXXXXXX, $210896 \leq XXXXXX \leq 210913$, which are GEO IDs. Columns named as INTENSITY1 and INTENSITY2 are 18 control samples and 18 samples transfected with miR-7/9/122a/128a/132/133a/142/148b/181a (two replicates each), respectively. Then, I generated tensor $x_{ijk}, 1 \leq i \leq 23651, 1 \leq j \leq 18, k = 1, 2$, where js are two replicates of each of nine miRNA-transfected samples and $k = 1, 2$ are control (mock-transfected) and transfected samples, respectively. After applying HOSVD to x_{ijk}, I found that $x_{\ell_3=2,k}$ reflects an inverse relation of expression levels between controls and treated samples, while $x_{\ell_2=1,j}$ reflects constant expression regardless of the type of transfected miRNA (Figure S4). Next, I decided to use $x_{\ell_1 i}$ associated with large absolute values of $G(\ell_1, \ell_2 = 1, \ell_3 = 2)$. Given that $G(\ell_1 = 6, \ell_2 = 1, \ell_3 = 2)$ has the largest absolute value, I decided to use $x_{\ell_1=6,i}$ to select mRNAs.

2.5.5. No. 5: GSE41539

Four files,

GSM1018808_topo_1_empty_trimmed_RNA-Seq.txt.gz,
GSM1018809_topo_2_cel_mir_67_trimmed_RNA-Seq.txt.gz,
GSM1018810_topo_4_mir_590_3p_trimmed_RNA-Seq.txt.gz, and
GSM1018811_topo_3_mir_199a_3p_trimmed_RNA-Seq.txt.gz,

were downloaded from GSE41539. The fourth column ("Unique gene reads") reflected gene expression. Then, I got $x_{ij}, 1 \leq i \leq 36065, 1 \leq j \leq 4$ (two-mode tensor). Applying PCA to x_{ij}, I found that $x_{\ell_2=2,j}$ represents the difference between controls (mock-transfected and cel-miR-67-transfected) and miR-509/199a-3p-transfected samples (Figure S5). After that, I decided to use $x_{\ell_1=2,i}$ for mRNA selection.

2.5.6. No. 6: GSE93290

File GSE93290_RAW.tar was downloaded and unpacked. Sixteen files, from GSM2450420 to GSM2450435, were loaded into R using the read.csv function. In each file, columns named as gProcessedSignal and rProcessedSignal served as controls and treated samples, respectively. Then, I generated three-mode tensor $x_{ijk}, 1 \leq i \leq 62976, 1 \leq j \leq 16, k = 1,2$, where js are 16 samples and $k = 1,2$ are control (mock-transfected) and transfected samples. After applying HOSVD to x_{ijk}, I found that $x_{\ell_3=2,k}$ reflects inversely related expression levels between controls and treated samples, whereas $x_{\ell_2=1,j}$ reflects constant expression regardless of the type of transfected miRNAs (Figure S6). Next, I decided to use $x_{\ell_1 i}$ associated with large absolute values of $G(\ell_1, \ell_2 = 1, \ell_3 = 2)$. Because $G(\ell_1 = 7, \ell_2 = 1, \ell_3 = 2)$ has the largest absolute value, I decided to apply $x_{\ell_1=7,i}$ to select mRNAs.

2.5.7. No. 7: GSE66498

File GSE66498_RAW.tar was downloaded and unpacked. Among these data, 19 files were used, i.e., GSM1623420 to GSM1623422 (miR-205 transfected into cell lines PC3, DU145, and C4-2), GSM1623423 and GSM1623424 (miR-29a transfected into cell lines 786O and A498), GSM1623425 to GSM1623427 (miR-451/144-3p/5p transfected into the T24 cell line), GSM1623434 and GSM1623435 (24 and 48 h after miR-210 transfection into the 786O cell line), GSM1623436 to GSM1623439 (miR-145-5p/3p transfected into BOY and T24 cell lines), GSM1623440 (miR-23b transfected into 786O cells), GSM1623444 to GSM1623446 (miR-221/222/223 transfected into the PC3 cell line), and GSM1623447 (miR-223 transfected into the PC3M cell line). In each file, columns named as gProcessedSignal and rProcessedSignal served as controls and treated samples, respectively. I generated three-mode tensor $x_{ijk}, 1 \leq i \leq 62976, 1 \leq j \leq 19, k = 1,2$, where js are the 19 samples and $k = 1,2$ are control (mock-transfected) and transfected samples. After applying HOSVD to x_{ijk}, I found that $x_{\ell_3=2,k}$ reflects an inverse relation of expression levels between controls and treated samples, whereas $x_{\ell_2=1,j}$ reflects constant expression regardless of the type of transfected miRNAs (Figure S7). After that, I decided to use $x_{\ell_1 i}$ associated with large absolute values of $G(\ell_1, \ell_2 = 1, \ell_3 = 2)$. Because $G(\ell_1 \in (2,3), \ell_2 = 1, \ell_3 = 2)$ have the largest and almost the same absolute values, I decided to employ $x_{\ell_1=2,i}$ and $x_{\ell_1=3,i}$ to select mRNAs.

2.5.8. No. 8: GSE17759

File GSE17759_RAW.tar was downloaded and unpacked. Among these data, six replicates of miR-146a–overexpressing samples (GSM443535 to GSM443540), four replicates of miR-146b–overexpressing samples (GSM443541 to GSM443544), eight replicates of miR-146a knockout samples (GSM443557 to GSM443564), i.e., in total, 18 files were processed. In each file, columns named as gProcessedSignal and rProcessedSignal served as controls and treated samples, respectively. I generated three-mode tensor $x_{ijk}, 1 \leq i \leq 43379, 1 \leq j \leq 18, k = 1,2$, where js are the 18 samples and $k = 1,2$ are control (mock-transfected) and transfected samples. After applying HOSVD to x_{ijk}, I found that $x_{\ell_3=2,k}$ reflects an inverse relation of expression levels between controls and treated samples, while $x_{\ell_2=1,j}$ means constant expression regardless of the type of transfected miRNAs

(Figure S8). Then, I decided to use $x_{\ell_1 i}$ associated with large absolute values of $G(\ell_1, \ell_2 = 1, \ell_3 = 2)$. Given that $G(\ell_1 = 5, \ell_2 = 1, \ell_3 = 2)$ has the largest absolute value, I decided to apply $x_{\ell_1=5,i}$ to select mRNAs.

2.5.9. No. 9: GSE37729

Two files SE37729-GPL6098_series_matrix.txt.gz and GSE37729-GPL6104_series_matrix.txt.gz in the section "Series Matrix File(s)" were downloaded. The two files were merged such that only shared probes were included. Gene expression of HeLa cell lines is considered in Experiment No. 9. GSM926188, GSM926189, GSM926193, GSM926194, GSM926198, and GSM926201 are control samples. GSM926164 and GSM926165 are anti-miR-107-transfected samples. GSM926180, GSM926181, GSM926190, and GSM926191 are miR-107-transfected samples. GSM926162 and GSM926163 are anti-miR-181b-transfected samples. GSM926182, GSM926183, GSM926195, and GSM926196 are miR-181b-transfected samples. Then, I generated three-mode tensor $x_{ijk}, 1 \leq i \leq 9987, 1 \leq j \leq 6, 1 \leq k \leq 3$, where $k = 1,2,3$ correspond to control, miR-7, and miR-181b, respectively. For $k = 2,3$, $1 \leq j \leq 2$ are anti-miR-transfected samples and $3 \leq j \leq 6$ are miR-transfected samples. After applying HOSVD to x_{ijk}, I found that $x_{\ell_3=2,k}$ reflects an inverse relation of expression levels between controls and treated samples, while $x_{\ell_2=1,j}$ indicates constant expression regardless of the type of transfected miRNA (Figure S9). Next, I decided to employ $x_{\ell_1 i}$ associated with large absolute values of $G(\ell_1, \ell_2 = 1, \ell_3 = 2)$. Because $G(\ell_1 = 2, \ell_2 = 1, \ell_3 = 2)$ has the largest absolute value, I decided to use $x_{\ell_1=2,i}$ to select mRNAs.

2.5.10. No. 10: GSE37729

Two files, SE37729-GPL6098_series_matrix.txt.gz and GSE37729-GPL6104_series_matrix.txt.gz, in the section "Series Matrix File(s)" were downloaded. The two files were merged so that only shared probes are included. Gene expression of HEK 293 cell lines was considered in Experiment No. 10. GSM926206, GSM926207, GSM926211, GSM926212, GSM926216, and GSM926217 are control samples. GSM926168 and GSM926169 are anti-miR-107-transfected samples. GSM926184, GSM926185, GSM926208, and GSM926209 are miR-107-transfected samples. GSM926166 and GSM926167 are anti-miR-181b-transfected samples. GSM926186, GSM926187, GSM926213, and GSM926214 are miR-181b-transfected samples. Next, I generated three-mode tensor $x_{ijk}, 1 \leq i \leq 9987, 1 \leq j \leq 6, 1 \leq k \leq 3$. $k = 1,2,3$ corresponding to control, miR-7, and miR-181b, respectively. For $k = 2,3$, $1 \leq j \leq 2$ are anti-miR-transfected samples, and $3 \leq j \leq 6$ are miR-transfected samples. After applying HOSVD to x_{ijk}, I determined that $x_{\ell_3=2,k}$ reflects an inverse relation of expression levels between controls and treated samples, while $x_{\ell_2=1,j}$ reflected constant expression regardless of the type of transfected miRNAs (Figure S10). Thus, I decided to use $x_{\ell_1 i}$ associated with large absolute values of $G(\ell_1, \ell_2 = 1, \ell_3 = 2)$. Because $G(\ell_1 = 2, \ell_2 = 1, \ell_3 = 2)$ has the largest absolute value, I decided to employ $x_{\ell_1=2,i}$ to select mRNAs.

2.5.11. No. 11: GSE37729

Two files, SE37729-GPL6098_series_matrix.txt.gz and GSE37729-GPL6104_series_matrix.txt.gz, in the section "Series Matrix File(s)" were downloaded. The two files were merged such that only shared probes were included. Gene expression of SH-SY5Y cell lines was considered in Experiment No. 11. GSM926170, GSM926171, GSM926178, and GSM926179 are control samples. GSM926176 and GSM926177 are anti-miR-181b-transfected samples. GSM926174 and GSM926175 are miR-181b-transfected samples. After that, I generated three-mode tensor $x_{ijk}, 1 \leq i \leq 9987, 1 \leq j \leq 4, 1 \leq k \leq 2$, where $k = 1,2$ correspond to control and miR-181b, respectively. For $k = 2,3$, $1 \leq j \leq 2$ are anti-miR-transfected samples, whereas $3 \leq j \leq 6$ are miR-transfected samples. After applying HOSVD to x_{ijk}, I found that $x_{\ell_3=2,k}$ reflects an inverse relation of expression levels between controls and treated samples, whereas $x_{\ell_2=1,j}$ reflects constant expression independently of transfected miRNAs (Figure S11). Then, I decided to use $x_{\ell_1 i}$ associated with large absolute values of

$G(\ell_1, \ell_2 = 1, \ell_3 = 2)$. Because $G(\ell_1 = 2, \ell_2 = 1, \ell_3 = 2)$ has the largest absolute value, I decided to use $x_{\ell_1=2,i}$ to select mRNAs.

3. Results

To demonstrate the usefulness of TD-based unsupervised FE, I applied it to artificial data composed of a three-mode tensor, $x_{ijk} \in \mathbb{R}^{N \times M \times 2}$, which is the expression level of the ith gene of the jth sample, where $k = 1$ is control and $k = 2$ is a treated (transfected with distinct miRNAs) sample. Among N genes, N_0 genes are affected by miRNA transfection while the other $N - N_0$ genes are not affected. Among the N_0 genes affected by miRNA transfection, $\frac{N_0}{2}$ genes are supposed to be regulated independently of samples (hence, a sequence-nonspecific off-target effect) while the other $\frac{N_0}{2}$ genes vary from sample to sample (i.e., miRNA-specific regulation). Applying TD-based unsupervised FE to the artificial dataset (averaged across 100 independent trials), I got a result (Figure 1). $x_{\ell_2=1,j}$ (Figure 1A) and $x_{\ell_3=1,k}$ (Figure 1B), which are always associated with core tensor $G(1,1,1)$ with the largest absolute values, represent constant gene expression across M samples and inverted expression levels between control ($k = 1$) and treated ($k = 2$) samples. Accordingly, genes associated with these two are expected to represent sample- or transfected miRNA-independent (thus, sequence-nonspecific off-target) regulation. Given that $x_{\ell_1=1,i}$ is always associated with core tensor $G(1,1,1)$ with the largest absolute values, P-values (Figure 1C) are computed using $x_{\ell_1=1,i}$. It is obvious that there is a sharp peak at the smallest P-values in the histogram of $1 - P$, which presumably does not correspond to the null hypothesis (that $x_{\ell_1=1,i}$ follows the normal distribution). To test whether genes associated with these much smaller P-values include genes $i \leq \frac{N_0}{2}$, the probabilities to be selected by TD-based unsupervised FE are averaged across $i \leq \frac{N_0}{2}$ and $i > \frac{N_0}{2}$, respectively. Then, the former is as large as 0.86, while the latter is as small as 0. (This means that genes $i > \frac{N_0}{2}$ have never been selected by TD-based unsupervised FE.) This observation suggests that TD-based unsupervised FE is effective at sorting out genes—that are expressed independently of samples—from genes expressed only in a limited number of samples and genes not expressed at all. To determine whether TD-based unsupervised FE can outperform the conventional supervised method, the t test and significance analysis of microarrays (SAM) [37] were carried out. For these two methods, P-values obtained were also corrected by means of the BH criterion, and genes associated with adjusted P-values less than 0.01 were selected. Then, the average probability for $i \leq \frac{N_0}{2}$ is 0.43 according to the t test and 0.62 according to SAM. The average probability for $i > \frac{N_0}{2}$ is 2×10^{-4} according to the t test and 3×10^{-5} according to SAM. Therefore, TD-based unsupervised FE is more effective than either the t test or SAM.

To identify genes whose expression alteration is likely to be mediated by sequence-nonspecific off-target regulation caused by miRNA transfection, integrated analysis of gene expression profiles after transfection of various miRNAs was performed. Simultaneous analysis of multiple experiments each of which employs single miRNA transfection will blur sequence-specific regulation of mRNAs while a sequence-nonspecific off-target effect will remain. Furthermore, to avoid biases due to research groups or individual studies, 11 experiments collected from studies involving distinct combinations of transfected miRNAs and cell lines were analyzed simultaneously (Table 1).

Genes—whose expression alteration is likely to be caused by sequence-nonspecific off-target regulation that was induced similarly by various transfected miRNAs—were selected using TD-based unsupervised FE in each of the 11 experiments. The reason why TD-based unsupervised FE was employed is as follows. First, PCA-based unsupervised FE, from which TD-based unsupervised FE was developed, is known to function even when only a small number of samples is available [16,26]. Tensor representation is also more suitable for the present experiments, where multiple genes, multiple miRNAs and controls, or transfected samples are simultaneously considered (they can be represented as a three-mode tensor; see Methods). Second, we can check whether there are miRNAs associated with a common expression pattern among all the miRNA transfection experiments by studying outcomes; we have opportunities to exclude experiments not associated with sequence-nonspecific off-target regulation caused by miRNA transfection.

These 11 analyzed experiments—in which miRNAs associated with sequence-nonspecific off-target regulation caused by miRNA transfection were successfully identified—deal with distinct cell lines into each of which distinct miRNAs were transfected. Nevertheless, gene sets identified in individual experiments not only significantly overlapped with one another but also were associated with a large enough odds ratio (from 300 to 500, Table 2), although the number of genes detected in each experiment varied from ∼100 to ∼800 ("#" in Table 2). This finding suggests that there are some sets of genes whose expression was robustly altered via sequence-nonspecific off-target regulation that was induced similarly by various transfected miRNAs.

Table 2. Fisher's exact tests for coincidence among 11 miRNA transfection experiments. Upper triangle: *P*-value; lower triangle: odds ratio.

Exp.		1	2	3	4	5	6	7	8	9	10	11
	#	232	711	747	441	123	292	246	873	113	104	120
1	232		4.14×10^{19}	6.59×10^{22}	3.96×10^{41}	4.12×10^{71}	9.41×10^{70}	2.90×10^{60}	1.34×10^{17}	1.15×10^{27}	6.84×10^{26}	2.66×10^{7}
2	711	7.68		0.00	1.89×10^{18}	4.93×10^{27}	5.59×10^{20}	2.69×10^{32}	4.62×10^{13}	9.23×10^{16}	8.66×10^{12}	1.37×10^{3}
3	747	8.30	345.52		3.63×10^{20}	7.96×10^{21}	5.70×10^{12}	1.82×10^{27}	9.52×10^{12}	1.18×10^{14}	1.01×10^{12}	3.90×10^{6}
4	441	18.23	5.19	5.34		6.14×10^{41}	1.01×10^{34}	1.44×10^{69}	4.61×10^{11}	2.16×10^{30}	4.09×10^{28}	1.35×10^{10}
5	123	53.86	9.04	7.27	17.48		2.9×10^{179}	1.27×10^{63}	6.24×10^{15}	3.16×10^{25}	2.37×10^{17}	4.69×10^{9}
6	292	61.50	8.15	5.52	17.71	204.39		3.53×10^{53}	2.57×10^{15}	6.65×10^{22}	1.65×10^{12}	5.60×10^{5}
7	246	20.27	5.35	4.67	12.39	20.11	22.03		6.91×10^{42}	1.77×10^{36}	4.50×10^{31}	2.78×10^{14}
8	873	18.61	7.22	6.51	8.29	15.61	18.53	20.73		1.81×10^{7}	1.37×10^{6}	2.76×10^{2}
9	113	39.34	9.87	8.77	25.98	32.44	34.90	21.94	16.02		3.7×10^{125}	9.27×10^{18}
10	104	40.29	8.22	8.27	26.64	23.34	20.86	21.56	15.18	517.87		6.82×10^{16}
11	120	10.15	3.19	4.43	9.19	11.55	8.11	8.28	4.92	19.57	18.70	

#: the number of genes selected for each of 11 experiments via TD- or PCA-based unsupervised FE.

Although this finding itself is remarkable enough to be reported, the observed coincidences may be accidental for some unknown reason and may not be associated with anything biologically valid. To validate biological significance of the identified genes, they were uploaded to Enrichr [38], which is an enrichment analysis server validating various biological terms and concepts. As a result, these genes were found to be enriched with various biological terms and concepts (see below).

First, the identified gene sets mostly included various target genes of transcription factors (TFs) (Table 3). Although the number of TFs detected varied from ∼10 to ∼100 (#2 in Table 3), the detection of multiple instances of enrichment with genes that TFs target may be evidence that these genes cooperatively function in the cell because common TFs' target genes often have shared biological functions [39,40]. In particular, the most frequent cases of enrichment of TFs are common among the 11 experiments analyzed. These include EKLF, MYC, NELFA, and E2F1. These data can be biologically interpreted as follows.

Table 3. In each of 11 experiments, 20 top-ranked significant TFs whose sets of target genes significantly overlap with the set of genes selected for each experiment were identified. Then, EKLF, MYC, NELFA, and E2F1 turned out to be among the 20 top-ranked significant TFs for all 11 experiments.

			EKLF		MYC		NELFA		E2F1	
Exp.	**#1**	**#2**	**OL**	**adj. *P*-Value**	**OL**	**adj. *P*-Value**	**OL**	**adj. *P*-Value**	**OL**	**adj. *P*-Value**
1	232	30	40/1239	2.16×10^{7}	53/1458	6.22×10^{12}	59/2000	8.94×10^{10}	61/1529	1.30×10^{15}
2	711	77	94/1239	1.51×10^{10}	106/1458	1.01×10^{10}	134/2000	3.26×10^{11}	100/1529	6.14×10^{8}
3	747	97	100/1239	2.28×10^{11}	98/1458	2.96×10^{7}	152/2000	1.93×10^{15}	108/1529	3.89×10^{9}
4	441	43	83/1239	4.77×10^{18}	99/1458	2.08×10^{22}	105/2000	1.06×10^{15}	85/1529	9.29×10^{14}
5	123	45	26/1239	2.16×10^{6}	25/1458	9.12×10^{5}	31/2000	4.26×10^{5}	28/1529	7.41×10^{6}
6	292	19	51/1239	2.38×10^{9}	65/1458	5.54×10^{14}	63/2000	2.72×10^{7}	69/1529	8.31×10^{15}
7	246	11	37/1239	5.11×10^{5}	48/1458	5.97×10^{8}	46/2000	1.45×10^{3}	64/1529	7.02×10^{16}
8	873	55	188/1239	8.33×10^{52}	189/1458	8.58×10^{42}	222/2000	3.52×10^{39}	157/1529	8.24×10^{23}
9	113	36	24/1239	4.47×10^{6}	30/1458	2.86×10^{8}	32/2000	2.33×10^{6}	40/1529	6.84×10^{15}
10	104	22	27/1239	1.16×10^{8}	25/1458	4.83×10^{6}	36/2000	1.07×10^{9}	35/1529	3.63×10^{12}
11	120	22	21/1239	8.02×10^{4}	27/1458	2.25×10^{5}	29/2000	4.68×10^{4}	25/1529	3.57×10^{4}

#1: the number of genes selected for each of 11 experiments via TD- or PCA-based unsupervised FE; #2: the number of TFs whose sets of target genes significantly (adjusted P-values < 0.01) overlap with the set of genes selected for each experiment; OL: overlaps, (the number of genes coinciding with the genes selected for each experiment)/(genes listed in Enrichr as TF target genes).

The apparent significant alteration of expression of MYC target genes may be due to miRNAs regulating *MYC* [41–43]. The apparent alteration of expression of E2F1 target genes may be explained similarly because these miRNAs also target *E2F1* [41,43]. In actuality, 1551 genes targeted by only one miRNA but associated with altered gene expression caused by miRNA transfection are significantly targeted by MYC and E2F1. Enrichment with EKLF target genes among these 1551 genes not targeted by more than one miRNA—but showing altered expression caused by transfection with one of miRNAs—is a puzzle, although Yien and Beiker suggested that some miRNAs are likely also regulated by EKLF [44]. Identification of NELFA target genes is explained by the tight interaction between NELFA and c-MYC [45,46]

Additionally, I checked "TargetScan microRNA" in Enrichr to test whether enrichment with genes targeted by the transfected miRNAs would be detected. As a result, only for two of the 11 experiments (Experiments No. 2 and 3), enrichment with genes targeted by nonzero miRNAs was detected. This is possibly because $x_{\ell_2=2,j}$ for these two experiments also detected genes targeted by transfected miRNAs (Figures S2 and S3). In any case, the fact that most of experiments (9 out of 11) did not show enrichment with genes targeted by transfected miRNAs is consistent with the hypothesis that gene expression alteration caused by miRNA transfection is primarily due to the competition for protein machinery between endogenous miRNAs and transfected miRNAs; this pattern can remain unchanged among transfection experiments with different miRNAs.

Next, I checked KEGG pathway enrichment data. Primary enriched KEGG pathways among the identified genes are related to diseases (Table 4). Seven ((ii), (iii), (v), (vi), (vii), (viii), and (x)) out of 10 most frequently enriched KEGG pathways in the 11 experiments are directly related to various diseases. Among the remaining three ((i), (iv), and (ix)), oxidative phosphorylation is a disease-related KEGG pathway because its malfunction causes combined oxidative phosphorylation deficiency (https://www.omim.org/entry/609060). Another one, "endoplasmic reticulum", is also a disease-related pathway because its malfunction is observed in neurological diseases [47]. Therefore, these genes may also contribute to the onset or progression of various diseases and could be therapeutic targets. As a result, sequence-nonspecific off-target regulation caused by miRNA transfection may be a therapeutic method.

Table 4. In each of 11 experiments, 20 top-ranked significant KEGG pathways whose associated genes significantly match some genes selected for each experiment were identified. Thus, the following KEGG pathways are most frequently ranked within the top 20.

Exp.	#	(i)	(ii)	(iii)	(iv)	(v)	(vi)	(vii)	(viii)	(ix)	(x)
1	(232)	31/137	7/168	10/142	6/133	9/55	9/193			7/169	8/203
	[10]	3.69×10^{29}	3.18×10^{2}	1.66×10^{4}	3.45×10^{2}	1.02×10^{6}	6.85×10^{3}			3.18×10^{2}	3.01×10^{2}
2	(711)	36/137	18/168	14/142	12/133	13/55				16/169	18/203
	[12]	3.43×10^{19}	1.48×10^{3}	1.05×10^{2}	3.20×10^{2}	5.92×10^{6}				8.12×10^{3}	8.12×10^{3}
3	(747)	23/137	15/168			14/55				18/169	19/203
	[15]	3.58×10^{7}	1.94×10^{2}			1.20×10^{6}				2.02×10^{3}	4.78×10^{3}
4	(441)	50/137	15/168	19/142	18/133	6/55	19/193	7/78	12/151	9/169	
	[10]	2.92×10^{45}	1.91×10^{4}	3.97×10^{8}	6.42×10^{8}	2.49×10^{2}	3.40×10^{6}	2.74×10^{2}	4.44×10^{3}	1.29×10^{1}	
5	(123)	9/137						8/78		6/169	8/203
	[23]	2.97×10^{6}						6.08×10^{7}		4.29×10^{3}	3.03×10^{4}
6	(292)	45/137	20/168	19/142	18/133	4/55	19/193	11/78	12/151		
	[14]	1.35×10^{46}	3.32×10^{11}	2.27×10^{11}	4.00×10^{11}	7.95×10^{2}	2.24×10^{9}	4.90×10^{7}	4.87×10^{5}		
7	(246)	40/137	9/168	10/142	9/133		11/193	4/78	7/151		6/203
	[6]	5.61×10^{42}	6.60×10^{3}	5.80×10^{4}	1.32×10^{3}		1.16×10^{3}	2.57×10^{1}	6.31×10^{2}		4.52×10^{1}
8	(873)	75/137	30/168	32/142	32/133		36/193	14/78	24/151	25/169	
	[24]	5.59×10^{63}	2.09×10^{9}	9.32×10^{13}	1.89×10^{13}		7.51×10^{12}	1.39×10^{4}	1.62×10^{6}	3.11×10^{6}	
9	(113)	18/137	11/168	12/142	10/133	6/55	12/193	4/78	11/151		
	[20]	8.24×10^{18}	7.10×10^{8}	1.66×10^{9}	8.42×10^{8}	8.85×10^{6}	2.96×10^{8}	6.64×10^{3}	2.96×10^{8}		
10	(104)	11/137	8/168	9/142	8/133	5/55	10/193		8/151		
	[20]	1.98×10^{8}	6.68×10^{5}	3.23×10^{6}	1.71×10^{5}	1.56×10^{4}	3.23×10^{6}		3.60×10^{5}		
11	(120)	6/137		4/142		5/55					5/203
	[3]	9.04×10^{3}		8.49×10^{2}		2.98×10^{3}					6.83×10^{2}

(i) Ribosome: hsa03010; (ii) Alzheimer's disease: hsa05010; (iii) Parkinson's disease: hsa05012; (iv) Oxidative phosphorylation: hsa00190; (v) Pathogenic *Escherichia coli* infection:hsa05130; (vi) Huntington's disease: hsa05016; (vii) Cardiac muscle contraction: hsa04260; (viii) Nonalcoholic fatty liver disease (NAFLD): hsa04932; (ix) Protein processing in endoplasmic reticulum: hsa04141; and (x) Proteoglycans in cancer: hsa05205. (numbers): gene; [numbers]: KEGG pathways. Upper rows in each exp: (the number of genes coinciding with the genes selected for each experiment)/(genes listed in Enrichr in each category). Lower rows in each exp: adjusted P-values provided by Enrichr.

Actually, gene expression alteration mediated by sequence-nonspecific off-target regulation is analogous to treatments with various candidate drugs (Tables 5 and 6). Thus, combinatorial transfection with miRNAs may replace drug treatment in some conditions and can be used for therapeutic purposes, too. For example, LDN-192189 ((ii) in Table 5) is reported to improve neuronal conversion of human fibroblasts [48] and was proposed as a therapy for Alzheimer's disease [49]. GSK-1059615b ((i) and (iii) in Table 5) was once considered a PI3K-kt pathway inhibitor in clinical development for the treatment of cancers [50]. WYE-125132 ((iv) in Table 5) is also known to suppress tumor growth [51] (although the name WYE-125132 does not appear in that article, compound 8a was named WYE-125132 later). Afatinib ((v) in Table 5) is a famous drug for non-small cell lung cancer [52]. PI 103 ((vi) in Table 5) is a drug for acute myeloid leukemia [53]. PD-0325901 was reported to affect heart development [54]. Chelerythrine chloride ((viii) in Table 5) has been reported to affect embryonic chick heart cells [55]. GDC-0980 ((i) in Table 6) is a known anticancer drug [56]. PLX-4720 ((ii) in Table 6) has been considered for both cancer and heart disease treatment [57]. Dinaciclib is another anticancer drug [58]. Because these are related to diseases reported in Table 4, sequence-nonspecific off-target regulation caused by miRNA transfection may be a therapeutic strategy.

Table 5. In each of 11 experiments, 20 top-ranked significant treatments with compounds whose downregulated genes significantly coincide with some genes selected for each experiment were identified. Thus, treatments with the following compounds are most frequently ranked within top 20.

Exp.	#		(i)	(ii)	(iii)	(iv)	(v)	(vi)	(vii)	(viii)
1	(232)	OL	10/85				16/154			
	[129]	adj P-value	6.14×10^5				6.59×10^7			
2	(711)	OL								
	[329]	adj P-value								
3	(747)	OL								
	[417]	adj P-value								
4	(441)	OL	16/85		14/105	15/109		16/144		
	[67]	adj P-value	8.56×10^7		1.15×10^4	5.10×10^5		1.56×10^4		
5	(123)	OL		17/141			12/154			
	[219]	adj P-value		2.87×10^{13}			2.26×10^7			
6	(292)	OL			13/105					19/137
	[132]	adj P-value			9.69×10^6					3.04×10^9
7	(246)	OL	9/85				12/154			
	[61]	adj P-value	1.41×10^3				8.88×10^4			
8	(873)	OL	30/85	46/141					35/162	
	[255]	adj P-value	1.19×10^{15}	1.62×10^{23}					5.21×10^{12}	
9	(119)	OL	9/85	11/141		8/109		8/144	8/162	12/137
	[74]	6.25×10^6	3.48×10^6		3.50×10^4		1.45×10^3	2.33×10^3	2.76×10^7	
10	(104)	OL	9/85		9/105	9/109		10/144		12/137
	[155]	adj P-value	1.87×10^6		4.96×10^6	6.22×10^6		4.96×10^6		7.01×10^8
11	(120)	OL		10/141					9/162	
	[127]	adj P-value		4.74×10^5					5.26×10^4	

(i) LJP005_BT20_24H-GSK-1059615-3.33; (ii) LJP005_HS578T_24H-LDN-193189-10; (iii) LJP005_MCF10A_24H-GSK-1059615-10; (iv) LJP006_BT20_24H-WYE-125132-10; (v) LJP006_HS578T_24H-afatinib-10; (vi) LJP006_MCF10A_24H-PI-103-10; (vii) LJP007_HT29_24H-PD-0325901-0.12; and (viii) LJP009_HEPG2_24H-chelerythrine_chloride-10. #: (numbers): genes; [numbers]: compounds. Upper rows in each exp: OL, overlaps, (the number of genes coinciding with the genes selected for each experiment)/(genes listed in Enrichr in each category). Lower rows in each exp: adjusted P-values provided by Enrichr.

Table 6. In each of the 11 experiments, 20 top-ranked significant treatments with compounds whose sets of upregulated genes significantly overlap with the set of genes selected for each experiment were identified. Therefore, treatment with the following compounds is most frequently ranked within the top 20.

Exp.	#	(i)	(ii)	(iii)
1	(232)			16/166
	[191]			2.16×10^7
2	(711)	23/125	30/163	
	[450]	1.58×10^7	1.11×10^9	
3	(747)		34/163	
	[559]		6.14×10^{12}	
4	(441)	15/125		
	[85]	2.12×10^4		
5	(123)			
	[116]			
6	(292)		18/163	
	[190]		2.46×10^7	
7	(246)			
	[145]			
8	(873)	22/125		31/166
	[69]	6.56×10^5		1.80×10^7
9	(119)			8/166
	[0]			1.95×10^2
10	(104)			
	[0]			
11	(120)			
	[33]			

(i) LJP005_A375_24H-GDC-0980-0.37; (ii) LJP005_HEPG2_24H-PLX-4720-10; and (iii) LJP007_MCF7_24H-dinaciclib-0.12 #: (numbers): genes; [numbers]: compounds. Upper rows in each exp: OL, overlaps (the number of genes coinciding with the genes selected for each experiment)/(genes listed in Enrichr in each category). Lower rows in each exp: adjusted P-values provided by Enrichr.

Next, I studied tissue specificity of the gene expression alteration caused by sequence-nonspecific off-target regulation that was induced by miRNA transfection. Associations with GTEx tissue samples were also observed. For example, in GTEx up (Table 7), enrichment cases were observed in the brain and testes, both of which were reported to be outliers in clustering analysis [59]. Thus, it is not surprising that cases of enrichment in these two tissues were identified primarily. On the other hand, in GTEx down (Table 8), enrichment instances in the skin were mostly observed. Readers may wonder how sequence-nonspecific off-target regulation caused by miRNA transfection can contribute to the tissue specificity of gene expression profiles. Võsa et al. [60] observed that polymorphisms in miRNA response elements (MRE-SNPs) that either disrupt a miRNA-binding site or create a new miRNA-binding site can affect the allele-specific expression of target genes. Therefore, sequence-nonspecific off-target regulation caused by miRNA transfection may have the ability to contribute to tissue specificity of the gene expression profiles through distinct functionality of genetic variations in distinct tissues. Despite these coincidences, how the sequence-nonspecific off-target effect contributes to differentiation is unclear. These coincidences may simply be consequences, not causes. More studies are needed to directly implicate the sequence-nonspecific off-target effect in differentiation itself.

Table 7. In each of the 11 experiments, 20 top-ranked significant tissues whose set of upregulated genes significantly overlapped with the set of genes selected for each experiment were identified.

Exp.	#	(i)	(ii)	(iii)	(iv)	(v)	(vi)	(vii)	(viii)	(ix)	(x)
1	(232) [216]	68/1509 1.11×10^{20}	51/1066 3.65×10^{16}	65/1384 1.28×10^{20}	74/1773 8.31×10^{21}	77/1764 8.66×10^{23}		69/1770 8.65×10^{18}		53/1174 8.12×10^{16}	68/1710 7.14×10^{18}
2	(711) [470]						78/625 3.13×10^{20}				
3	(747) [441]						72/625 1.93×10^{15}				
4	(441) [151]	80/1509 3.20×10^{11}	61/1066 1.01×10^{9}	82/1384 7.82×10^{14}		87/1764 1.08×10^{5}	54/625 6.23×10^{15}		54/525 6.83×10^{18}		
5	(123) [150]	31/1509 2.48×10^{7}	29/1066 9.29×10^{9}	32/1384 1.75×10^{8}	33/1773 5.68×10^{7}			34/1770 1.99×10^{7}		26/1174 9.17×10^{7}	33/1710 2.90×10^{7}
6	(292) [196]	80/1509 7.02×10^{22}	66/1066 3.72×10^{21}	70/1384 4.91×10^{18}	80/1773 4.91×10^{18}	85/1764 4.97×10^{21}	46/625 2.53×10^{17}	78/1770 4.73×10^{17}	50/525 4.99×10^{23}	63/1174 2.53×10^{17}	76/1710 7.45×10^{17}
7	(246) [135]	64/1509 7.77×10^{16}	53/1066 1.13×10^{15}	57/1384 1.02×10^{13}	68/1773 3.42×10^{15}	68/1764 3.28×10^{15}	34/625 5.23×10^{11}	63/1770 1.16×10^{12}	35/525 1.03×10^{13}	52/1174 1.10×10^{13}	63/1710 2.75×10^{13}
8	(873) [178]	164/1509 1.61×10^{25}	131/1066 8.18×10^{25}	156/1384 1.61×10^{25}	155/1773 2.25×10^{15}	168/1764 4.48×10^{20}		161/1770 2.02×10^{17}		128/1174 1.31×10^{19}	157/1710 2.38×10^{17}
9	(119) [225]	39/1509 1.13×10^{13}	28/1066 8.60×10^{10}	40/1384 2.22×10^{15}	38/1773 5.07×10^{11}	37/1764 1.92×10^{10}	24/625 3.31×10^{11}		25/525 1.13×10^{13}		
10	(104) [156]	29/1509 2.83×10^{7}	22/1066 5.82×10^{6}	30/1384 1.93×10^{8}	31/1773 4.52×10^{7}	29/1764 4.31×10^{6}		28/1770 1.19×10^{5}	15/525 1.35×10^{5}	23/1174 5.82×10^{6}	27/1710 1.51×10^{5}
11	(120) [5]						13/625 1.54×10^{2}		12/525 1.38×10^{2}		

(i) GTEX-QDT8-0011-R10A-SM-32PKG_brain_female_30-39_years; (ii) GTEX-QMR6-1426-SM-32PLA_brain_male_50-59_years; (iii) GTEX-TSE9-3026-SM-3DB76_brain_female_60-69_years; (iv) GTEX-PVOW-0011-R3A-SM-32PKX_brain_male_40-49_years; (v) GTEX-PVOW-2526-SM-2XCF7_brain_male_40-49_years; (vi) GTEX-XAJ8-1326-SM-47JYT_testis_male_40-49_years; (vii) GTEX-N7MS-0011-R3a-SM-33HC6_brain_male_60-69_years; (viii) GTEX-OHPM-2126-SM-3LK75_testis_male_50-59_years; (ix) GTEX-PVOW-0011-R5A-SM-32PL7_brain_male_40-49_years; and (x) GTEX-PVOW-2626-SM-32PL8_brain_male_40-49_years. (numbers): gene; [numbers]: tissues. P-values are the ones adjusted by Enrichr.

Table 8. In each of the 11 experiments, 20 top-ranked significant tissues whose set of downregulated genes significantly overlapped with the set of genes selected for each experiment were identified.

Exp.	#	(i)	(ii)	(iii)	(iv)	(v)	(vi)
1	(232) [201]	57/1709 4.88×10^{11}	54/1488 1.40×10^{11}	53/1027 4.65×10^{17}	53/1121 1.13×10^{15}	55/1599 4.65×10^{11}	52/1103 1.96×10^{15}
2	(711) [414]	166/1709 4.72×10^{32}			133/1121 1.27×10^{33}	175/1599 2.48×10^{40}	
3	(747) [365]	174/1709 1.44×10^{33}				165/1599 2.20×10^{32}	
4	(441) [219]	98/1709 2.41×10^{16}	89/1488 8.07×10^{16}	73/1027 2.17×10^{16}		98/1599 5.24×10^{18}	70/1103 1.38×10^{13}
5	(123) [486]			35/1027 9.87×10^{15}	35/1121 1.04×10^{13}		37/1103 2.03×10^{15}
6	(292) [171]		69/1488 2.30×10^{15}	60/1027 3.96×10^{17}	53/1121 5.96×10^{12}		61/1103 9.15×10^{17}
7	(246) [113]		55/1488 2.46×10^{11}	57/1027 7.80×10^{19}	49/1121 1.81×10^{12}	53/1599 3.39×10^{9}	51/1103 1.38×10^{13}
8	(873) [185]	188/1709 1.15×10^{30}		137/1027 8.37×10^{30}	136/1121 1.34×10^{25}	182/1599 3.79×10^{31}	138/1103 3.16×10^{27}
9	(119) [224]	31/1709 4.11×10^{7}	36/1488 6.46×10^{11}				
10	(104) [156]	31/1709 3.24×10^{8}	33/1488 1.64×10^{10}				
11	(120) [20]						

(i) GTEX-Q2AH-0008-SM-48U2J_skin_male_40-49_years; (ii) GTEX-O5YT-0126-SM-48TBW_skin_male_20-29_years; (iii) GTEX-P4PQ-0008-SM-48TDX_skin_male_60-69_years; (iv) GTEX-R55D-0008-SM-48FEV_skin_male_50-59_years; (v) GTEX-R55E-0008-SM-48FCG_skin_male_20-29_years; and (vi) GTEX-RU72-0008-SM-46MV8_skin_female_50-59_years. (numbers): gene; [numbers]: tissues. Upper rows in each exp: OL, overlaps (the number of genes coinciding with the genes selected for each experiment)/(genes listed in Enrichr in each category). Lower rows in each exp: adjusted *P*-values provided by Enrichr.

Genes whose expression alteration is likely to be caused by sequence-nonspecific off-target regulation that miRNA transfection induces are also enriched in pluripotency (Table 9). Because miRNAs are known to mediate reprogramming [61], sequence-nonspecific off-target regulation caused by transfection of multiple miRNAs may contribute to reprogramming too. In actuality, primary binding TFs include MYC ((i) and (iv) in Table 9) and KLF4 ((vi) and (vii) in Table 9), which are two of four Yamanaka factors that mediate pluripotency. Other TFs in Table 9 are DMAP1 (iii) and TIP60 (v). DMAP1 is a member of the TIP60-p400 complex that maintains embryonic stem cell pluripotency [62]. The remaining one, ZFX (x), has been reported to control the self-renewal of embryonic and hematopoietic stem cells [63]. In addition, two gene KO experiments have been conducted ((viii) and (ix)). One of the genes in question, *SUZ12*, encodes a polycomb group protein that mediates differentiation [64], whereas the other, *ZFP281*, is a known pluripotency suppressor [65]. There are no known widely accepted mechanisms by which miRNA transfection can induce pluripotency. Because sequence-nonspecific off-target regulation seems to alter expression of genes critical for pluripotency, it may contribute to the mechanism.

Table 9. In each of the 11 experiments, 20 top-ranked significant terms in the Embryonic Stem Cell Atlas from Pluripotency Evidence (ESCAPE) whose set of associated genes significantly overlapped with the set of genes selected for each experiment were identified.

Exp.	#	(i)	(ii)	(iii)	(iv)	(v)	(vi)	(vii)	(viii)	(ix)	(x)
1	(232) [15]	53/1458 3.85×10^{12}	90/2469 1.63×10^{22}	55/1789 7.36×10^{10}	56/1200 1.35×10^{17}	24/705 7.38×10^{5}	38/1700 1.40×10^{3}	40/1502 2.40×10^{5}	15/315 1.13×10^{4}	17/186 2.84×10^{9}	66/3249 6.10×10^{5}
2	(711) [22]	106/1458 1.29×10^{10}	184/2469 2.83×10^{21}	96/1789 5.46×10^{4}	90/1200 1.09×10^{9}				33/315 1.43×10^{6}		182/3249 4.08×10^{9}
3	(747) [28]	98/1458 4.59×10^{7}	199/2469 7.29×10^{25}	106/1789 2.60×10^{5}	93/1200 1.88×10^{9}	55/705 8.48×10^{6}			32/315 1.09×10^{5}	19/186 1.17×10^{3}	184/3249 9.00×10^{8}
4	(441) [18]	99/1458 1.37×10^{22}	153/2469 1.91×10^{32}	109/1789 1.74×10^{21}	95/1200 2.31×10^{26}	51/705 3.51×10^{12}	64/1700 3.44×10^{4}	60/1502 1.21×10^{4}		23/186 8.71×10^{10}	145/3249 1.39×10^{16}
5	(123) [32]	25/1458 3.65×10^{5}	46/2469 9.91×10^{11}	26/1789 3.38×10^{4}					19/315 1.66×10^{11}		
6	(292) [8]	65/1458 2.14×10^{14}	108/2469 3.68×10^{25}	54/1789 1.03×10^{5}	59/1200 1.26×10^{14}	24/705 3.96×10^{3}	47/1700 6.28×10^{4}	37/1502 2.48×10^{2}	13/315 1.89×10^{2}	19/186 2.10×10^{9}	68/3249 2.34×10^{2}
7	(246) [6]	48/1458 2.16×10^{8}	78/2469 1.87×10^{13}	43/1789 6.95×10^{4}	49/1200 8.53×10^{12}	17/705 1.48×10^{1}	31/1700 2.48×10^{1}	33/1502 2.70×10^{2}	13/315 5.72×10^{3}	15/186 6.01×10^{7}	
8	(873) [28]	189/1458 5.44×10^{42}	303/2469 6.58×10^{67}	184/1789 1.81×10^{27}	159/1200 6.07×10^{36}	89/705 4.59×10^{18}	130/1700 4.78×10^{9}	116/1502 2.74×10^{8}	38/315 3.67×10^{7}	38/186 4.46×10^{14}	247/3249 1.89×10^{18}
9	(119) [14]	30/1458 1.76×10^{8}	54/2469 4.54×10^{18}	37/1789 1.22×10^{10}	28/1200 6.09×10^{9}	16/705 6.01×10^{5}	23/1700 1.46×10^{3}	28/1502 5.60×10^{7}		9/186 3.47×10^{5}	32/3249 1.15×10^{2}
10	(104) [11]	25/1458 3.77×10^{6}	56/2469 3.80×10^{22}	33/1789 4.45×10^{9}	21/1200 2.91×10^{5}	13/705 1.53×10^{3}	23/1700 4.39×10^{4}	19/1502 4.56×10^{3}		10/186 2.87×10^{6}	
11	(120) [14]	27/1458 1.28×10^{5}	44/2469 1.60×10^{9}		17/1200 1.20×10^{2}	13/705 5.75×10^{3}	23/1700 4.08×10^{3}	25/1502 1.97×10^{4}	7/315 3.00×10^{2}		

(i) CHiP_MYC-19079543; (ii) mESC_H3K36me3_18692474; (iii) CHiP_DMAP1-20946988; (iv) CHiP_MYC-18555785; (v) CHiP_TIP60-20946988; (vi) CHiP_KLF4-18358816; (vii) CHiP_KLF4-19030024; (viii) SUZ12-17339329_UP; (ix) ZFP281-21915945_DOWN; and (x) CHiP_ZFX-18555785. (numbers): gene; [numbers]: TF binding, histone modification and a gene KO or overexpression. Upper rows in each exp: OL, overlaps (the number of genes coinciding with the genes selected for each experiment)/(genes listed in Enrichr in each category). Lower rows in each exp: adjusted P-values provided by Enrichr.

Besides, I checked whether protein–protein interactions (PPIs) are enriched in each of the 11 gene sets selected for each of the 11 experiments (Table 10). Gene sets were uploaded to the STRING server [66]. For all the gene sets, instances of PPI enrichment were highly significant. Given that proteins rarely function alone and often function in groups, this is evidence that our analysis is biologically reliable.

Table 10. PPI enrichment by the STRING server. Column "genes" shows numbers of genes recognized by the STRING server.

Exp.	Genes	Edges		P-Values
		Observed	Expected	
1	195	1638	591	0
2	623	4271	2577	0
3	658	4506	2920	0
4	392	3418	1273	0
5	118	539	197	0
6	276	2048	569	0
7	182	1167	303	0
8	640	10176	4342	0
9	112	464	165	0
10	103	323	127	0
11	118	143	104	1.58×10^{4}

I demonstrated enrichment of various biological processes in gene sets. Although there were more cases of enrichment of biological terms in Enrichr, it is impossible to consider and discuss all of them. Instead, I discuss the enrichment cases identified in GeneSigDBs in Enrichr that include various biological properties (Table 11). Numerous GeneSigDBs that reflect a wide range of functional genes were enriched too. This result also suggested that the identified genes may contribute to a wide range of biological activities.

Table 11. In each of the 11 experiments, 20 top-ranked significant terms in GeneSigDB whose set of associated genes significantly overlapped with the set of genes selected for each experiment were identified.

			(i)		(ii)		(iii)		(iv)	
Exp.	#1	#2	OL	adj. *P*-Value	OL	adj. *P*-Value	OL	adj. *P*-Value	OL	adj. *P*-Value
1	232	154	97/1548	7.60×10^{44}	36/663	1.68×10^{12}	77/2585	1.72×10^{13}	36/238	6.63×10^{27}
2	711	194	152/1548	4.03×10^{29}	72/663	3.28×10^{15}	222/2585	8.96×10^{36}	44/238	3.05×10^{17}
3	747	285	149/1548	4.32×10^{25}	87/663	1.58×10^{22}	222/2585	4.01×10^{32}	30/238	2.05×10^{7}
4	441	106	146/1548	7.43×10^{52}	62/663	6.57×10^{20}	145/2585	1.51×10^{25}	46/238	2.15×10^{27}
5	123	183	44/1548	4.46×10^{16}	13/663	1.84×10^{3}	37/2585	1.34×10^{5}	24/238	1.54×10^{19}
6	292	106	103/1548	1.21×10^{38}	36/663	1.79×10^{9}	94/2585	1.51×10^{15}	39/238	5.50×10^{27}
7	246	95	81/1548	5.23×10^{28}	30/663	1.34×10^{7}	66/2585	3.88×10^{7}	30/238	2.86×10^{19}
8	873	269	256/1548	1.82×10^{81}	100/663	8.26×10^{26}	229/2585	4.03×10^{25}	86/238	7.47×10^{53}
9	119	93	60/1548	6.98×10^{34}	30/663	9.55×10^{17}	48/2585	9.73×10^{13}	22/238	4.26×10^{18}
10	104	77	52/1548	1.67×10^{27}	25/663	1.59×10^{12}	45/2585	4.60×10^{12}	15/238	2.49×10^{10}
11	120	68	36/1548	6.11×10^{10}	17/663	4.25×10^{5}	40/2585	1.07×10^{6}	13/238	6.73×10^{7}

(i) A multiclass predictor based on a probabilistic model, i.e., application to gene expression profiling-based diagnosis of thyroid tumors; (ii) A redox signature score identifies diffuse large B-cell lymphoma patients with a poor prognosis; (iii) A comparison of the gene expression profile of undifferentiated human embryonic stem cell lines and differentiating embryoid bodies; and (iv) A gene expression profile of rat left ventricles reveals persisting changes after a chronic mild-exercise protocol: implications for cardioprotection. #1: genes; #2: terms. OL: overlaps (the number of genes coinciding with the genes selected for each experiment)/(genes listed in Enrichr in each category).

As readers can see, top four most frequently significant terms are related to either diseases or differentiation, which are often said to be biological concepts to which miRNAs contribute to. Although canonical target genes of miRNA have mainly been sought to understand biological functions of miRNAs, sequence-nonspecific off-target regulation may be important as well.

Figure 2 shows the summary of results obtained in this section.

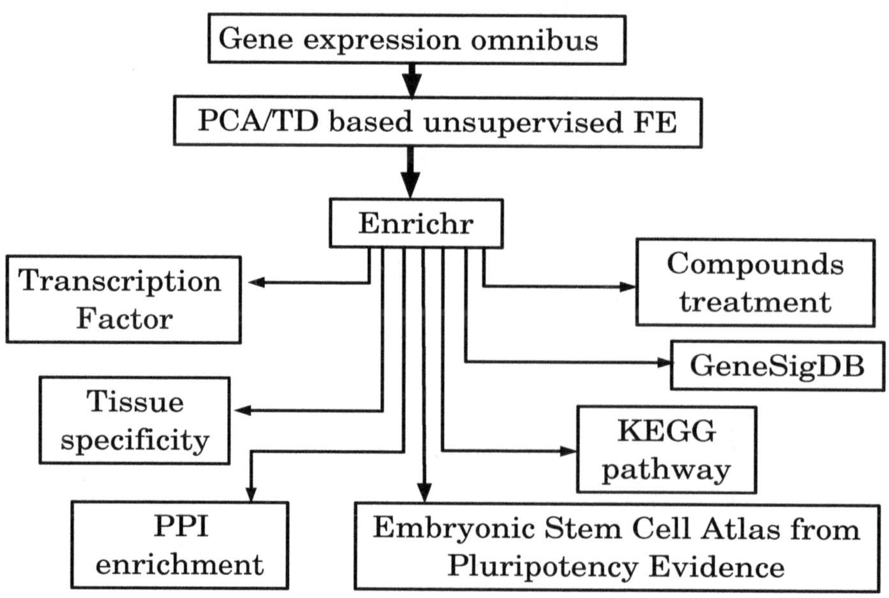

Figure 2. A schematic diagram that summarizes the obtained results.

4. Discussion

Many biological conceptual cases of enrichment were observed in the identified sets of genes across the 11 experiments. Nonetheless, detailed mechanisms by which sequence-nonspecific off-target effects (that transfected miRNAs produce) can regulate expression of these genes are unclear. To identify such a mechanism, enrichment of genes associated with the altered expression pattern caused by a Dicer KO within these 11 gene sets was studied by means of Enrichr. Among 16 experiments included in Enrichr ("Single Gene Perturbations from GEO up" and "Single Gene Perturbations from GEO down"), most of them are associated with enrichment of the identified genes in 11 miRNA transfection experiments (Table 12). In addition, these sets of genes significantly overlap with the set of genes associated with binding to Dicer according to immunoprecipitation (IP) experiments (Table 12). There are also data supporting the hypothesis that sequence-nonspecific off-target regulation is due to the competition for protein machinery between transfected miRNAs and endogenous miRNAs in the cell.

Table 12. GEO DICER KO: The number of experiments among the 16 experiments included in Enrichr whose set of listed genes significantly overlapped with the set of genes identified in each of the 11 experiments. IP: Fisher's exact test for the overlap between the set of genes that bind to Dicer in immunoprecipitation (IP) experiments and the set of genes selected in each of the 11 experiments.

Experiments		1	2	3	4	5	6
GEO DICER KO	up down	12/16 13/16	12/16 12/16	12/16 12/16	12/16 13/16	14/16 14/16	11/16 10/16
IP	*P*-value odds	2.49×10^{23} 47.4	7.22×10^{22} 20.6	1.31×10^{17} 15.9	5.55×10^{29} 38.7	5.21×10^{35} 64.2	1.78×10^{20} 41.2
Experiments			7	8	9	10	11
GEO DICER KO	up down		12/16 12/16	14/16 12/16	12/16 14/16	13/16 14/16	12/16 10/16
IP	*P*-value odds		4.72×10^{32} 37.0	4.29×10^{16} 41.4	2.19×10^{11} 42.6	3.96×10^{10} 39.6	4.64×10^{8} 27.3

On the other hand, the number of miRNAs that target genes whose expression was likely to be altered by miRNA transfection significantly correlated with the number of experiments where individual genes were selected among the 11 conducted experiments (see Figure 3. Pearson's and Spearman's correlation coefficients are 0.13, $P = 3.9 \times 10^{-11}$ and 0.29, $P < 2.2 \times 10^{-16}$, respectively). Genes targeted by a greater number of individual miRNAs are likely to be affected by miRNA transfection, which occupies the protein machinery binding to transcripts of these genes. Therefore, the significant correlation is consistent with the hypothesis that sequence-nonspecific off-target regulation is due to competition for protein machinery between a transfected miRNA and endogenous miRNAs in cells.

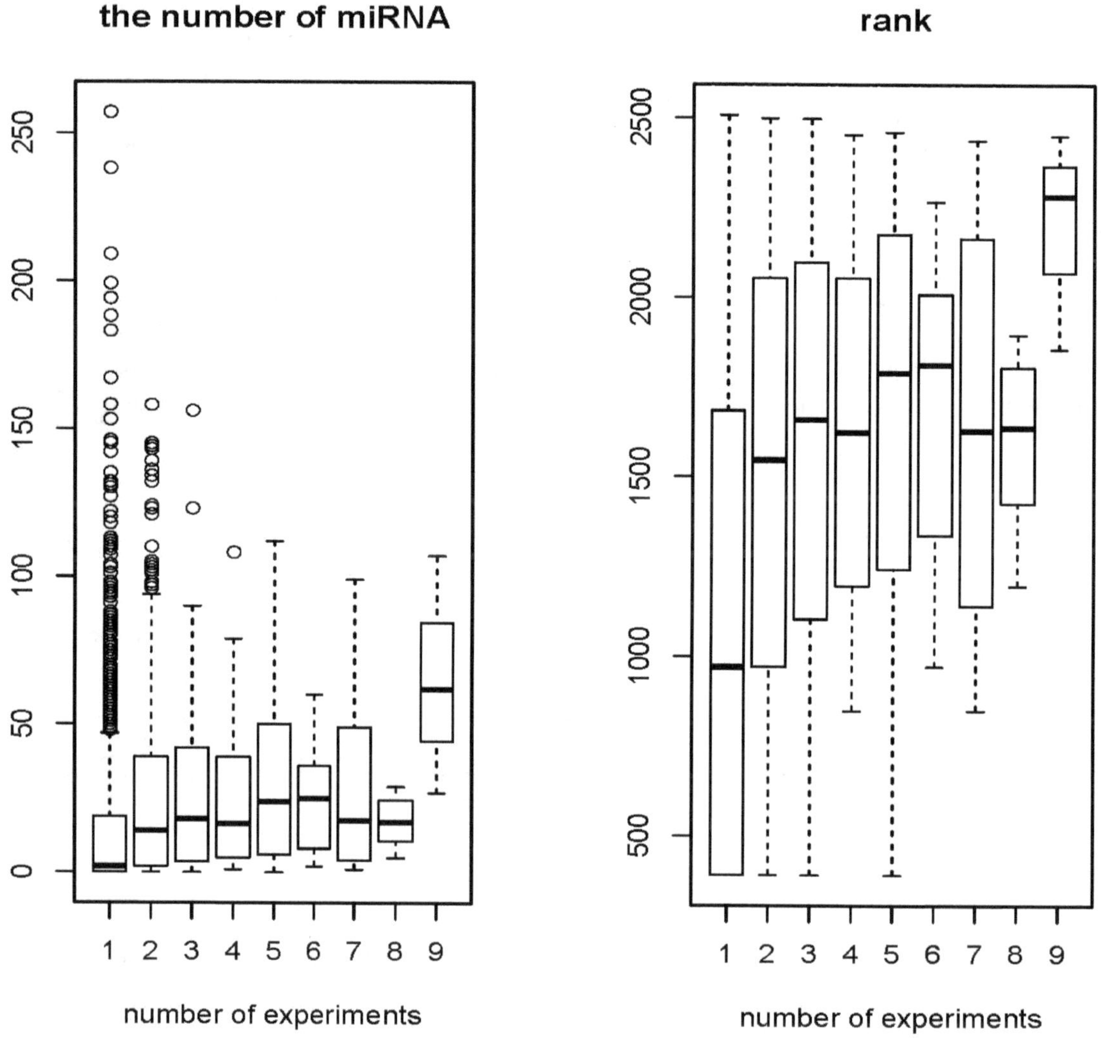

Figure 3. A boxplot of the number of miRNAs that target individual genes as a function of the number of experiments that select individual genes within 11 experiments (most frequently selected genes were selected in nine experiments): (**Left**) raw numbers (Pearson's correlation coefficient $= 0.13$, $P = 3.9 \times 10^{-11}$); and (**Right**) ranks of numbers (Spearman's correlation coefficient $= 0.29$, $P < 2.2 \times 10^{-16}$)

Thus, primarily, gene expression alteration by sequence-nonspecific off-target regulation caused by miRNA transfection is likely due to suppression of miRNA functionality because of a reduction in the amount of available protein machinery owing to occupation by transfected miRNAs.

Despite the above arguments, it cannot be proven that competition for protein machinery is the primary cause of sequence-nonspecific off-target regulation. Upregulation of genes can also be caused by indirect effects, e.g., genes that suppress expression of other genes are repressed by transfected miRNA, although this mechanism is unlikely to cause upregulation of common genes regardless of the transfected miRNAs. Additional experimental validation will clarify which one is the correct scenario.

Furthermore, I compared the performance of TD-based unsupervised FE with the performance of major supervised methods. Previously, when PCA-based unsupervised FE, on which TD-based unsupervised FE is based, has been applied to various problems, PCA-based unsupervised FE often outperformed other (supervised) methods. For example, when PCA-based unsupervised FE was successfully used to identify genes commonly associated with aberrant promoter methylation among three autoimmune diseases [19], no supervised methods—except for PCA-based unsupervised FE—could identify common genes. On the other hand, when PCA-based unsupervised FE was applied to identify common HDACi target genes between two independent HDACis [16], two supervised methods (Limma [67] and categorical regression, i.e. analysis of variance (ANOVA)) identified the same number of common genes as did PCA-based unsupervised FE. Nevertheless, biological validation of the selected genes supported the superiority of PCA-based unsupervised FE. In contrast to the many instances of biological-term enrichment that were observed among genes selected by PCA-based unsupervised FE, such cases of enrichment among the genes selected by a supervised method were not detected.

Although these are only two examples, this kind of advantages of PCA-based unsupervised FE have often been observed. To confirm the superiority of TD-based unsupervised FE too, I consider Experiments No. 9 and No. 10 in Table 1 for the comparisons with other (supervised) methods. The reason for this choice is as follows. At first, these two were taken from the same dataset (GSE37729) and the same miRNAs (miR-107/181b) were transfected; thus, these two experiments are expected to have a greater number of common genes selected than do other pairs of experiments in Table 1. Second, the pair No. 9 and No. 10 has the highest odds ratio in Table 2, as expected. Thus, these data are suitable for the comparison with another method.

When using ANOVA and SAM [37], I found that P-values and the odds ratio computed by Fisher's exact test are 0.09 and 2.9, respectively, for both methods (the number of genes selected is taken to be 103 and 104 for Experiments No. 9 and No. 10, respectively, using ranking based upon P-values assigned to each gene because these numbers are the same as those selected by TD-based unsupervised FE, see "#" in Table 2).

A more sophisticated and advanced supervised method may show somewhat better performance than do SAM and categorical regression. Because TD-based unsupervised FE highly outperformed these two conventional and frequently used supervised methods, it is unlikely that another supervised method can compete with TD-based unsupervised FE (advantages of PCA-based unsupervised FE over various conventional supervised methods have been reported repeatedly too [14–33]). More comprehensive performance comparisons with the t test are provided in the Supplementary Materials.

Readers may also wonder whether the null assumption that $x_{\ell_1,i}$ obeys a normal distribution is appropriate because $x_{\ell_1,i}$ is not proven to follow a normal distribution. This approach is not problematic for the following reasons. First, the null hypothesis that $x_{\ell_1,i}$ obeys a normal distribution is supposed to be rejected later. Thus, even if $x_{\ell_1,i}$ does not follow the normal distribution, this is not a problem. Second, it is reasonable to assume that $x_{\ell_1,i}$ follows the normal distribution under the assumption that x_{ijk} is drawn from a random number (Figure 1C); this statement is suitable as the null hypothesis. Be that as it may, a question may arise whether the null hypothesis that $x_{\ell_1,i}$ obeys the normal distribution is not suitable if this assumption is mostly violated. To evaluate how well the null hypothesis is fulfilled, I demonstrate the result for GSE26996 as a typical example. Figure 4A presents the scatter plot of $x_{\ell_1,i}, \ell_1 = 1, 2$ where $x_{\ell_1=2,i}$ served for selection of genes as mentioned in the Section 2.5.1. Considering the fact that the total number of probes in the microarray is more than 20,000 and the number of probes selected is 379, these 379 probes are obviously outliers along the direction of $x_{\ell_1=2,i}$. Figure 4B depicts the histogram of $1 - P$ under the null hypothesis that $x_{\ell_1=2,i}$ follows the normal distribution. Although smaller $1 - P$ (<0.3) s deviate from constant values that are expected under the null hypothesis, a sharp peak that includes the selected 379 probes is evidently located at the largest $1 - P$. Consequently, it is satisfactory for identifying genes i associated with much larger (that is, invalidating the null hypothesis) absolute $x_{\ell_1=2,i}$ values.

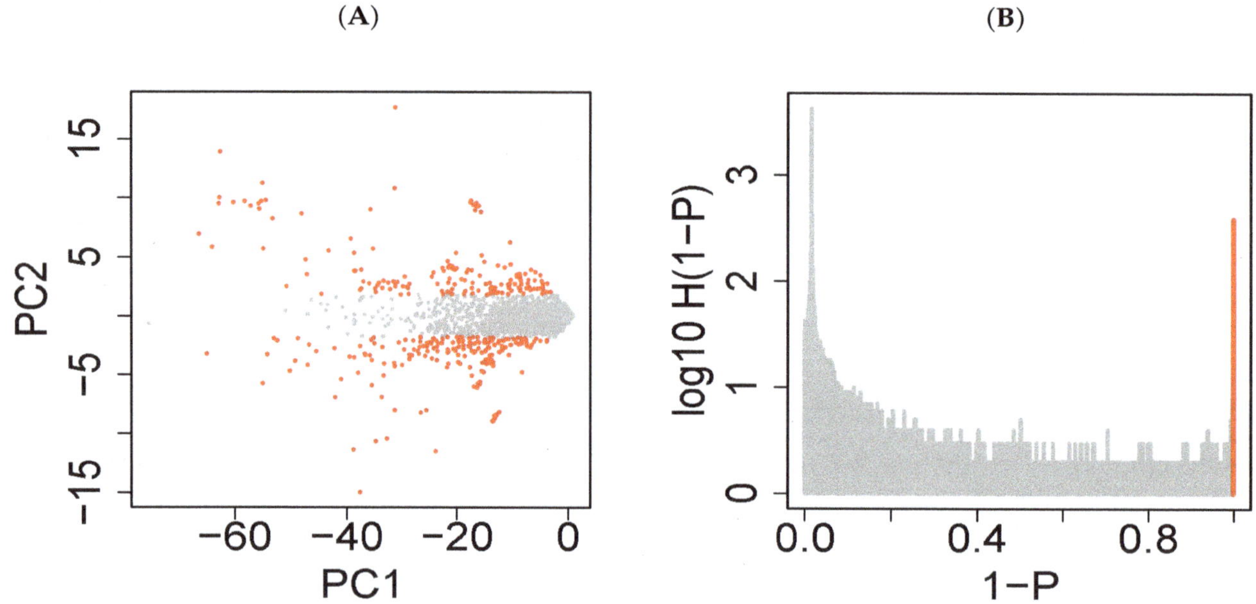

Figure 4. (**A**) The scatter plot of $x_{\ell_1,i}, \ell_1 = 1(\text{horizontal axis}), 2(\text{vertical axis})$ for GSE26996; 379 red dots are selected probes; and (**B**) a semilogarithmic plot of the histogram of $1 - P$ under the null hypothesis that $x_{\ell_1=2,i}$ obeys a normal distribution. A sharp peak is observed in the red bin with the largest $1 - P$, which includes all the probes selected in (**A**).

Finally, I would like to consider possible objections to the hypothesis proposed in this study: sequence-nonspecific off-target regulation of mRNA mediated by miRNA transfection is primarily mediated by competition for the protein machinery. The first possible objection is that some miRNA can bind to a promoter region directly. This process in not mentioned in the above discussion. For example, Kim et al. [68] found that miR-320 can bind to the promoter region of *POLR3D*. Nevertheless, this kind of direct binding to DNA by transfected miRNA cannot be the interpretation of the present findings, because it is still sequence specific. Thus, direct binding to DNA cannot be an alternative interpretation of the sequence-nonspecific regulation presented in Table 2. The second objection is that miRNA can sometimes bind to mRNA with insufficient support by proteins. For example, Lima et al. [69] found that single-stranded siRNAs can bind to mRNA. Given that single-stranded siRNAs do not have to be processed by the DICER that is mentioned in Table 12, this topic apparently seems to be outside the scope of this study. Nonetheless, single-stranded siRNAs that Lima et al. identified still need the AGO2 protein. Thus, the regulation of mRNA expression by single-stranded siRNAs can still be under the control of competition for protein machinery. Therefore, it is hard to say whether the process identified by Lima et al. is independent of protein machinery competition. The third objection to the scenario proposed in this study is that miRNA can often upregulate target mRNAs [70,71]. This means that observation of upregulation caused by miRNA transfection—which was brought up as one of side proofs for protein machinery competition in the text above—does not always have to be mediated by competition for protein machinery. On the other hand, this process is also sequence specific. Thus, direct upregulation by transfected miRNA still cannot explain the sequence-nonspecific regulation of mRNAs presented in Table 2. Therefore, at the moment, protein machinery competition is the only possible explanation of the sequence-nonspecific regulation of mRNAs shown in Table 2.

5. Conclusions

In this study, I applied recently proposed PCA- and TD-based unsupervised FE to mRNA profiles of miRNA-transfected cell lines. mRNAs associated with significant dysregulation turned out to be independent of transfected miRNAs to some extent. This sequence-nonspecific off-target regulation is associated with various biological functions according to enrichment analysis. It is also likely to be caused by protein machinery competition between endogenous miRNAs and transfected miRNAs.

Acknowledgments: The author thanks reviewers for pointing out the references that are useful for the discussion about possible objections.

References

1. Bartel, D.P. MicroRNAs: Target Recognition and Regulatory Functions. *Cell* **2009**, *136*, 215–233. [CrossRef] [PubMed]
2. Olejniczak, M.; Galka, P.; Krzyzosiak, W.J. Sequence-non-specific effects of RNA interference triggers and microRNA regulators. *Nucleic Acids Res.* **2010**, *38*, 1–16. [CrossRef] [PubMed]
3. Khan, A.A.; Betel, D.; Miller, M.L.; Sander, C.; Leslie, C.S.; Marks, D.S. Transfection of small RNAs globally perturbs gene regulation by endogenous microRNAs. *Nat. Biotechnol.* **2009**, *27*, 549–555. [CrossRef] [PubMed]
4. Nagata, Y.; Shimizu, E.; Hibio, N.; Ui-Tei, K. Fluctuation of global gene expression by endogenous miRNA response to the introduction of an exogenous miRNA. *Int. J. Mol. Sci.* **2013**, *14*, 11171–11189. [CrossRef] [PubMed]
5. Carroll, A.P.; Tran, N.; Tooney, P.A.; Cairns, M.J. Alternative mRNA fates identified in microRNA-associated transcriptome analysis. *BMC Genom.* **2012**, *13*, 561. [CrossRef] [PubMed]
6. Taguchi, Y.H. Tensor decomposition-based unsupervised feature extraction identifies candidate genes that induce post-traumatic stress disorder-mediated heart diseases. *BMC Med. Genom.* **2017**, *10*, 67. [CrossRef] [PubMed]
7. Taguchi, Y.H. One-class Differential Expression Analysis using Tensor Decomposition-based Unsupervised Feature Extraction Applied to Integrated Analysis of Multiple Omics Data from 26 Lung Adenocarcinoma Cell Lines. In Proceedings of the 2017 IEEE 17th International Conference on Bioinformatics and Bioengineering (BIBE), Washington, DC, USA, 23–25 October 2017; pp. 131–138.
8. Taguchi, Y.H. Identification of candidate drugs using tensor-decomposition-based unsupervised feature extraction in integrated analysis of gene expression between diseases and DrugMatrix datasets. *Sci. Rep.* **2017**, *7*, 13733. [CrossRef] [PubMed]
9. Taguchi, Y.H. Tensor decomposition-based unsupervised feature extraction applied to matrix products for multi-view data processing. *PLoS ONE* **2017**, *12*, e0183933. [CrossRef] [PubMed]
10. Taguchi, Y.H. Identification of candidate drugs for heart failure using tensor decomposition-based unsupervised feature extraction applied to integrated analysis of gene expression between heart failure and drugmatrix datasets. In *Intelligent Computing Theories and Application*; Huang, D.S., Jo, K.H., Figueroa-García, J.C., Eds.; Springer International Publishing: Cham, Switzerland, 2017; pp. 517–528.
11. Taguchi, Y.H.; Wang, H. Exploring microRNA Biomarker for Amyotrophic Lateral Sclerosis. *Int. J. Mol. Sci.* **2018**, *19*, 1318. [CrossRef] [PubMed]
12. Taguchi, Y.H.; Wang, H. Genetic association between amyotrophic lateral sclerosis and cancer. *Genes* **2017**, *8*, 243. [CrossRef] [PubMed]
13. Taguchi, Y.H.; Iwadate, M.; Umeyama, H.; Murakami, Y. Principal component analysis based unsupervised feature extraction applied to bioinformatics analysis. In *Computational Methods with Applications in Bioinformatics Analysis*; World Scientific: Singapore, 2017; Chapter 8, pp. 153–182.
14. Taguchi, Y.H. microRNA-mRNA Interaction identification in wilms tumor using principal component analysis based unsupervised feature extraction. In Proceedings of the 2016 IEEE 16th International Conference on Bioinformatics and Bioengineering (BIBE), Taichung, Taiwan, 31 October–2 November 2016; pp. 71–78.
15. Taguchi, Y.H. Principal Components Analysis Based Unsupervised Feature Extraction Applied to Gene Expression Analysis of Blood from Dengue Haemorrhagic Fever Patients. *Sci. Rep.* **2017**, *7*, 44016. [CrossRef] [PubMed]
16. Taguchi, Y.H. Principal component analysis based unsupervised feature extraction applied to publicly available gene expression profiles provides new insights into the mechanisms of action of histone deacetylase inhibitors. *Neuroepigenetics* **2016** *8*, 1–18. [CrossRef]

17. Taguchi, Y.H.; Iwadate, M.; Umeyama, H. Principal component analysis-based unsupervised feature extraction applied to in silico drug discovery for posttraumatic stress disorder-mediated heart disease. *BMC Bioinform.* **2015**, *16*, 139. [CrossRef] [PubMed]
18. Taguchi, Y.H.; Okamoto, A. Principal Component Analysis for Bacterial Proteomic Analysis. In *Pattern Recognition in Bioinformatics*; Shibuya, T., Kashima, H., Sese, J., Ahmad, S., Eds.; Springer: Berlin/Heidelberg, Germany, 2012; Volume 7632, *LNCS*, pp. 141–152.
19. Ishida, S.; Umeyama, H.; Iwadate, M.; Taguchi, Y.H. Bioinformatic Screening of Autoimmune Disease Genes and Protein Structure Prediction with FAMS for Drug Discovery. *Protein Pept. Lett.* **2014**, *21*, 828–839. [CrossRef] [PubMed]
20. Kinoshita, R.; Iwadate, M.; Umeyama, H.; Taguchi, Y.H. Genes associated with genotype-specific DNA methylation in squamous cell carcinoma as candidate drug targets. *BMC Syst. Biol.* **2014**, *8*, S4. [CrossRef] [PubMed]
21. Taguchi, Y.H.; Murakami, Y. Principal component analysis based feature extraction approach to identify circulating microRNA biomarkers. *PLoS ONE* **2013**, *8*, e66 714. [CrossRef]
22. Taguchi, Y.H.; Murakami, Y. Universal disease biomarker: Can a fixed set of blood microRNAs diagnose multiple diseases? *BMC Res. Notes* **2014**, *7*, 581. [CrossRef] [PubMed]
23. Murakami, Y.; Toyoda, H.; Tanahashi, T.; Tanaka, J.; Kumada, T.; Yoshioka, Y.; Kosaka, N.; Ochiya, T.; Taguchi, Y.H. Comprehensive miRNA expression analysis in peripheral blood can diagnose liver disease. *PLoS ONE* **2012**, *7*, e48366. [CrossRef] [PubMed]
24. Murakami, Y.; Tanahashi, T.; Okada, R.; Toyoda, H.; Kumada, T.; Enomoto, M.; Tamori, A.; Kawada, N.; Taguchi, Y.H.; Azuma, T. Comparison of Hepatocellular Carcinoma miRNA Expression Profiling as Evaluated by Next Generation Sequencing and Microarray. *PLoS ONE* **2014**, *9*, e106314. [CrossRef] [PubMed]
25. Murakami, Y.; Kubo, S.; Tamori, A.; Itami, S.; Kawamura, E.; Iwaisako, K.; Ikeda, K.; Kawada, N.; Ochiya, T.; Taguchi, Y.H. Comprehensive analysis of transcriptome and metabolome analysis in Intrahepatic Cholangiocarcinoma and Hepatocellular Carcinoma. *Sci. Rep.* **2015**, *5*, 16294. [CrossRef] [PubMed]
26. Umeyama, H.; Iwadate, M.; Taguchi, Y.H. TINAGL1 and B3GALNT1 are potential therapy target genes to suppress metastasis in non-small cell lung cancer. *BMC Genom.* **2014**, *15*, S2. [CrossRef] [PubMed]
27. Taguchi, Y.H.; Iwadate, M.; Umeyama, H. Heuristic principal component analysis-based unsupervised feature extraction and its application to gene expression analysis of amyotrophic lateral sclerosis data sets. In Proceedings of the 2015 IEEE Conference on Computational Intelligence in Bioinformatics and Computational Biology (CIBCB), Niagara Falls, ON, Canada, 12–15 August 2015; pp. 1–10.
28. Taguchi, Y.H.; Iwadate, M.; Umeyama, H.; Murakami, Y.; Okamoto, A. Heuristic principal component analysis-aased unsupervised feature extraction and its application to bioinformatics. In *Big Data Analytics in Bioinformatics and Healthcare*; Wang, B., Li, R., Perrizo, W., Eds.; IGI Global: Hershey, PA, USA, 2015; pp. 138–162.
29. Taguchi, Y.H. Integrative analysis of gene expression and promoter methylation during reprogramming of a non-small-cell lung cancer cell line using principal component analysis-based unsupervised feature extraction. In *Intelligent Computing in Bioinformatics*; Huang, D.S., Han, K., Gromiha, M., Eds.; Springer: Berlin/Heidelberg, Germany, 2014; Volume 8590, *LNCS*, pp. 445–455.
30. Taguchi, Y.H. Identification of aberrant gene expression associated with aberrant promoter methylation in primordial germ cells between E13 and E16 rat F3 generation vinclozolin lineage. *BMC Bioinform.* **2015**, *16*, S16. [CrossRef] [PubMed]
31. Taguchi, Y.H. Identification of More Feasible MicroRNA-mRNA Interactions within Multiple Cancers Using Principal Component Analysis Based Unsupervised Feature Extraction. *Int. J. Mol. Sci.* **2016**, *17*, 696. [CrossRef] [PubMed]
32. Taguchi, Y.H. Principal component analysis based unsupervised feature extraction applied to budding yeast temporally periodic gene expression. *BioData Min.* **2016**, *9*, 22. [CrossRef] [PubMed]
33. Taguchi, Y.H.; Iwadate, M.; Umeyama, H. SFRP1 is a possible candidate for epigenetic therapy in non-small cell lung cancer. *BMC Med. Genom.* **2016**, *9*, 28. [CrossRef] [PubMed]
34. Lathauwer, L.D.; Moor, B.D.; Vandewalle, J. A multilinear singular value decomposition. *SIAM J. Matrix Anal. Appl.* **2000** *21*, 1253–1278. [CrossRef]

35. Benjamini, Y.; Hochberg, Y. Controlling the False Discovery Rate: A Practical and Powerful Approach to Multiple Testing. *J. R. Stat. Soc. Ser. B Methodol.* **1995**, *57*, 289–300.
36. Barrett, T.; Wilhite, S.E.; Ledoux, P.; Evangelista, C.; Kim, I.F.; Tomashevsky, M.; Marshall, K.A.; Phillippy, K.H.; Sherman, P.M.; Holko, M.; et al. NCBI GEO: Archive for functional genomics data sets–update. *Nucleic Acids Res.* **2013**, *41*, D991–D995. [CrossRef] [PubMed]
37. Tusher, V.G.; Tibshirani, R.; Chu, G. Significance analysis of microarrays applied to the ionizing radiation response. *Proc. Natl. Acad. Sci. USA* **2001**, *98*, 5116–5121. [CrossRef] [PubMed]
38. Kuleshov, M.V.; Jones, M.R.; Rouillard, A.D.; Fernandez, N.F.; Duan, Q.; Wang, Z.; Koplev, S.; Jenkins, S.L.; Jagodnik, K.M.; Lachmann, A.; et al. Enrichr: A comprehensive gene set enrichment analysis web server 2016 update. *Nucleic Acids Res.* **2016**, *44*, W90–W97. [CrossRef] [PubMed]
39. Allocco, D.J.; Kohane, I.S.; Butte, A.J. Quantifying the relationship between co-expression, co-regulation and gene function. *BMC Bioinform.* **2004**, *5*, 18. [CrossRef] [PubMed]
40. Karczewski, K.J.; Snyder, M.; Altman, R.B.; Tatonetti, N.P. Coherent functional modules improve transcription factor target identification, cooperativity prediction, and disease association. *PLoS Genet.* **2014**, *10*, e1004122. [CrossRef] [PubMed]
41. Suarez, Y.; Fernandez-Hernando, C.; Yu, J.; Gerber, S.A.; Harrison, K.D.; Pober, J.S.; Iruela-Arispe, M.L.; Merkenschlager, M.; Sessa, W.C. Dicer-dependent endothelial microRNAs are necessary for postnatal angiogenesis. *Proc. Natl. Acad. Sci. USA* **2008**, *105*, 14082–14087. [CrossRef] [PubMed]
42. Dews, M.; Homayouni, A.; Yu, D.; Murphy, D.; Sevignani, C.; Wentzel, E.; Furth, E.E.; Lee, W.M.; Enders, G.H.; Mendell, J.T.; et al. Augmentation of tumor angiogenesis by a Myc-activated microRNA cluster. *Nat. Genet.* **2006**, *38*, 1060–1065. [CrossRef] [PubMed]
43. O'Donnell, K.A.; Wentzel, E.A.; Zeller, K.I.; Dang, C.V.; Mendell, J.T. c-Myc-regulated microRNAs modulate E2F1 expression. *Nature* **2005**, *435*, 839–843. [CrossRef] [PubMed]
44. Yien, Y.Y.; Bieker, J.J. EKLF/KLF1, a tissue-restricted integrator of transcriptional control, chromatin remodeling, and lineage determination. *Mol. Cell. Biol.* **2013**, *33*, 4–13. [CrossRef] [PubMed]
45. Rahl, P.B.; Lin, C.Y.; Seila, A.C.; Flynn, R.A.; McCuine, S.; Burge, C.B.; Sharp, P.A.; Young, R.A. c-Myc regulates transcriptional pause release. *Cell* **2010**, *141*, 432–445. [CrossRef] [PubMed]
46. Das, P.P.; Shao, Z.; Beyaz, S.; Apostolou, E.; Pinello, L.; De Los Angeles, A.; O'Brien, K.; Atsma, J.M.; Fujiwara, Y.; Nguyen, M.; et al. Distinct and combinatorial functions of Jmjd2b/Kdm4b and Jmjd2c/Kdm4c in mouse embryonic stem cell identity. *Mol. Cell* **2014**, *53*, 32–48. [CrossRef] [PubMed]
47. Roussel, B.D.; Kruppa, A.J.; Miranda, E.; Crowther, D.C.; Lomas, D.A.; Marciniak, S.J. Endoplasmic reticulum dysfunction in neurological disease. *Lancet Neurol.* **2013**, *12*, 105–118. [CrossRef]
48. Ladewig, J.; Mertens, J.; Kesavan, J.; Doerr, J.; Poppe, D.; Glaue, F.; Herms, S.; Wernet, P.; Kogler, G.; Muller, F.J.; et al. Small molecules enable highly efficient neuronal conversion of human fibroblasts. *Nat. Methods* **2012**, *9*, 575–578. [CrossRef] [PubMed]
49. Hu, W.; Qiu, B.; Guan, W.; Wang, Q.; Wang, M.; Li, W.; Gao, L.; Shen, L.; Huang, Y.; Xie, G.; et al. Direct Conversion of Normal and Alzheimer's Disease Human Fibroblasts into Neuronal Cells by Small Molecules. *Cell Stem Cell* **2015**, *17*, 204–212. [CrossRef] [PubMed]
50. Engelman, J.A. Targeting PI3K signalling in cancer: Opportunities, challenges and limitations. *Nat. Rev. Cancer* **2009**, *9*, 550–562. [CrossRef] [PubMed]
51. Curran, K.J.; Verheijen, J.C.; Kaplan, J.; Richard, D.J.; Toral-Barza, L.; Hollander, I.; Lucas, J.; Ayral-Kaloustian, S.; Yu, K.; Zask, A. Pyrazolopyrimidines as highly potent and selective, ATP-competitive inhibitors of the mammalian target of rapamycin (mTOR): Optimization of the 1-substituent. *Bioorg. Med. Chem. Lett.* **2010**, *20*, 1440–1444. [CrossRef] [PubMed]
52. Morin-Ben Abdallah, S.; Hirsh, V. Epidermal Growth Factor Receptor Tyrosine Kinase Inhibitors in Treatment of Metastatic Non-Small Cell Lung Cancer, with a Focus on Afatinib. *Front. Oncol.* **2017**, *7*, 97. [CrossRef] [PubMed]
53. Park, S.; Chapuis, N.; Bardet, V.; Tamburini, J.; Gallay, N.; Willems, L.; Knight, Z.A.; Shokat, K.M.; Azar, N.; Viguie, F.; et al. PI-103, a dual inhibitor of Class IA phosphatidylinositide 3-kinase and mTOR, has antileukemic activity in AML. *Leukemia* **2008**, *22*, 1698–1706. [CrossRef] [PubMed]
54. Anastasaki, C.; Rauen, K.A.; Patton, E.E. Continual low-level MEK inhibition ameliorates cardio-facio-cutaneous phenotypes in zebrafish. *Dis. Models Mech.* **2012**, *5*, 546–552. [CrossRef] [PubMed]

55. Wei, H.; Mei, Y.A.; Sun, J.T.; Zhou, H.Q.; Zhang, Z.H. Regulation of swelling-activated chloride channels in embryonic chick heart cells. *Cell Res.* **2003**, *13*, 21–28. [CrossRef] [PubMed]
56. Wallin, J.J.; Edgar, K.A.; Guan, J.; Berry, M.; Prior, W.W.; Lee, L.; Lesnick, J.D.; Lewis, C.; Nonomiya, J.; Pang, J.; et al. GDC-0980 is a novel class I PI3K/mTOR kinase inhibitor with robust activity in cancer models driven by the PI3K pathway. *Mol. Cancer Ther.* **2011**, *10*, 2426–2436. [CrossRef] [PubMed]
57. Bronte, E.; Bronte, G.; Novo, G.; Bronte, F.; Bavetta, M.G.; Lo Re, G.; Brancatelli, G.; Bazan, V.; Natoli, C.; Novo, S.; et al. What links BRAF to the heart function? New insights from the cardiotoxicity of BRAF inhibitors in cancer treatment. *Oncotarget* **2015**, *6*, 35589–35601. [CrossRef] [PubMed]
58. Kumar, S.K.; LaPlant, B.; Chng, W.J.; Zonder, J.; Callander, N.; Fonseca, R.; Fruth, B.; Roy, V.; Erlichman, C.; Stewart, A.K. Dinaciclib, a novel CDK inhibitor, demonstrates encouraging single-agent activity in patients with relapsed multiple myeloma. *Blood* **2015**, *125*, 443–448. [CrossRef] [PubMed]
59. Ardlie, K.G.; Deluca, D.S.; Segrè, A.V.; Sullivan, T.J.; Young, T.R.; Gelfand, E.T.; Trowbridge, C.A.; Maller, J.B.; Tukiainen, T.; Lek, M.; et al. The Genotype-Tissue Expression (GTEx) pilot analysis: Multitissue gene regulation in humans. *Science* **2015**, *348*, 648–660.
60. Vosa, U.; Esko, T.; Kasela, S.; Annilo, T. Altered Gene Expression Associated with microRNA Binding Site Polymorphisms. *PLoS ONE* **2015**, *10*, e0141351. [CrossRef] [PubMed]
61. Anokye-Danso, F.; Trivedi, C.M.; Juhr, D.; Gupta, M.; Cui, Z.; Tian, Y.; Zhang, Y.; Yang, W.; Gruber, P.J.; Epstein, J.A.; et al. Highly efficient miRNA-mediated reprogramming of mouse and human somatic cells to pluripotency. *Cell Stem Cell* **2011**, *8*, 376–388. [CrossRef] [PubMed]
62. Mohan, K.N.; Ding, F.; Chaillet, J.R. Distinct roles of DMAP1 in mouse development. *Mol. Cell. Biol.* **2011**, *31*, 1861–1869. [CrossRef] [PubMed]
63. Galan-Caridad, J.M.; Harel, S.; Arenzana, T.L.; Hou, Z.E.; Doetsch, F.K.; Mirny, L.A.; Reizis, B. Zfx controls the self-renewal of embryonic and hematopoietic stem cells. *Cell* **2007**, *129*, 345–357. [CrossRef] [PubMed]
64. Pasini, D.; Bracken, A.P.; Hansen, J.B.; Capillo, M.; Helin, K. The polycomb group protein Suz12 is required for embryonic stem cell differentiation. *Mol. Cell. Biol.* **2007**, *27*, 3769–3779. [CrossRef] [PubMed]
65. Fidalgo, M.; Shekar, P.C.; Ang, Y.S.; Fujiwara, Y.; Orkin, S.H.; Wang, J. Zfp281 functions as a transcriptional repressor for pluripotency of mouse embryonic stem cells. *Stem Cells* **2011**, *29*, 1705–1716. [CrossRef] [PubMed]
66. Szklarczyk, D.; Franceschini, A.; Wyder, S.; Forslund, K.; Heller, D.; Huerta-Cepas, J.; Simonovic, M.; Roth, A.; Santos, A.; Tsafou, K.P.; et al. STRING v10: Protein-protein interaction networks, integrated over the tree of life. *Nucleic Acids Res.* **2015**, *43*, D447–D452. [CrossRef] [PubMed]
67. Ritchie, M.E.; Phipson, B.; Wu, D.; Hu, Y.; Law, C.W.; Shi, W.; Smyth, G.K. limma powers differential expression analyses for RNA-sequencing and microarray studies. *Nucleic Acids Res.* **2015**, *43*, e47. [CrossRef] [PubMed]
68. Kim, D.H.; Sætrom, P.; Snøve, O.; Rossi, J.J. MicroRNA-Directed Transcriptional Gene Silencing in Mammalian Cells. *Proc. Natl. Acad. Sci. USA* **2008**, *105*, 16230–16235. Available online: http://www.pnas.org/content/105/42/16230.full.pdf (accessed on 1 June 2018). [CrossRef] [PubMed]
69. Lima, W.F.; Prakash, T.P.; Murray, H.M.; Kinberger, G.A.; Li, W.; Chappell, A.E.; Li, C.S.; Murray, S.F.; Gaus, H.; Seth, P.P.; et al. Single-Stranded siRNAs Activate RNAi in Animals. *Cell* **2012**, *150*, 883–894. [CrossRef] [PubMed]
70. Vasudevan, S.; Tong, Y.; Steitz, J.A. Switching from Repression to Activation: MicroRNAs can up-regulate translation. *Science* **2007**, *318*, 1931–1934. [CrossRef] [PubMed]
71. Ghosh, T.; Soni, K.; Scaria, V.; Halimani, M.; Bhattacharjee, C.; Pillai, B. MicroRNA-mediated up-regulation of an alternatively polyadenylated variant of the mouse cytoplasmic β-actin gene. *Nucleic Acids Res.* **2008**, *36*, 6318–6332. [CrossRef] [PubMed]

8

Profiling the Abiotic Stress Responsive microRNA Landscape of *Arabidopsis thaliana*

Joseph L. Pegler, Jackson M. J. Oultram, Christopher P. L. Grof and Andrew L. Eamens *

Centre for Plant Science, School of Environmental and Life Sciences, Faculty of Science, University of Newcastle, Callaghan 2308, Australia; Joseph.Pegler@uon.edu.au (J.L.P.); Jackson.Oultram@uon.edu.au (J.M.J.O.); chris.grof@newcastle.edu.au (C.P.L.G.)

* Correspondence: andy.eamens@newcastle.edu.au;

Abstract: It is well established among interdisciplinary researchers that there is an urgent need to address the negative impacts that accompany climate change. One such negative impact is the increased prevalence of unfavorable environmental conditions that significantly contribute to reduced agricultural yield. Plant microRNAs (miRNAs) are key gene expression regulators that control development, defense against invading pathogens and adaptation to abiotic stress. *Arabidopsis thaliana* (*Arabidopsis*) can be readily molecularly manipulated, therefore offering an excellent experimental system to alter the profile of abiotic stress responsive miRNA/target gene expression modules to determine whether such modification enables *Arabidopsis* to express an altered abiotic stress response phenotype. Towards this goal, high throughput sequencing was used to profile the miRNA landscape of *Arabidopsis* whole seedlings exposed to heat, drought and salt stress, and identified 121, 123 and 118 miRNAs with a greater than 2-fold altered abundance, respectively. Quantitative reverse transcriptase polymerase chain reaction (RT-qPCR) was next employed to experimentally validate miRNA abundance fold changes, and to document reciprocal expression trends for the target genes of miRNAs determined abiotic stress responsive. RT-qPCR also demonstrated that each miRNA/target gene expression module determined to be abiotic stress responsive in *Arabidopsis* whole seedlings was reflective of altered miRNA/target gene abundance in *Arabidopsis* root and shoot tissues post salt stress exposure. Taken together, the data presented here offers an excellent starting platform to identify the miRNA/target gene expression modules for future molecular manipulation to generate plant lines that display an altered response phenotype to abiotic stress.

Keywords: *Arabidopsis thaliana*; abiotic stress; heat stress; drought stress; salt stress; microRNAs (miRNAs); miRNA target gene expression; RT-qPCR

---

1. Introduction

Anthropogenically driven climate change is a rapidly growing concern globally, forcing interdisciplinary research collaborations to provide solutions that address and/or negate the numerous negative consequences of a changing climate, with the provision of sustainable food security the overarching goal of contemporary agricultural research [1–3]. Throughout the last half-century, agriculture has attempted to continue to achieve the food demands of an ever-growing global population via the unsustainable practice of clearing biodiverse terrestrial ecosystems for additional cultivation of traditional cropping species, an alarming practice that further contributes to the global carbon footprint and climate change [4,5]. Considering the capability limitation of the current maximum annual global crop yield to land area ratio, it is obvious that alternate strategies are now required if cropping agriculture is to continue to ensure food security, whilst terminating unsustainable farming practices, and while achieving these goals in an ever increasingly unfavorable environment [4].

Lacking the mobility of metazoa, the sessile nature of a plant requires intricate and interrelated gene expression networks to mediate the plant's ability to physiologically and phenotypically respond to its surrounding environment. Such multilayered molecular networks are especially important to a plant's adaptive and/or defensive response when either abiotic or biotic stress is encountered [6,7]. Elucidating the gene expression cascades that underpin the ability of a plant to adapt to, or mitigate, the negative impact of abiotic stress is the first key step in the development of new plant lines harboring molecular modifications which allow the plant to display an altered response phenotype when exposed to abiotic stress. The genetic model plant, *Arabidopsis thaliana* (*Arabidopsis*), is readily amenable to molecular modification, thereby offering plant biology researchers an excellent experimental system to validate which introduced molecular modifications mediate the expression of abiotic stress tolerance phenotypes.

Since their initial identification in *Arabidopsis* in 2002 [8], plant microRNAs (miRNAs), small non-protein-coding regulatory RNAs, have been repeatedly demonstrated to be key regulators of gene expression across all phases of plant development [9,10], in mediating a defense response against invading viral, bacterial or fungal pathogens [11,12], or to direct a plant's adaptive response to exposure to abiotic stress, including the stresses of heat, drought and salt stress [13,14]. Each *Arabidopsis* miRNA is processed from a stem-loop structured precursor transcript, a non-protein-coding RNA that has folded back upon itself to form this structure, post RNA polymerase II (Pol II)-catalyzed transcription from a unique *MICRORNA* (*MIR*) gene [15–17]. Like protein coding loci, the promoter regions of many *MIR* genes harbor *cis*-elements that contribute to the control of *MIR* gene expression in response to numerous signals external to the cell, including the signals that stem from abiotic stress [18–20]. Altered *MIR* gene expression, and therefore altered mature miRNA abundance, in turn leads to changes in miRNA target gene expression, with each miRNA loaded by the miRNA-induced silencing complex (miRISC) to be used as a sequence specificity guide to modulate target gene expression via either a messenger RNA (mRNA) cleavage or translational repression mechanism of RNA silencing [9,21–23]. To date, in *Arabidopsis*, numerous miRNA/target gene expression modules have been demonstrated to be responsive to abiotic stress, with alteration to the molecular profile of some expression modules further shown to assist the plant to adapt to abiotic stress due to molecular-driven changes to key pathways, such as the photosynthesis, sugar signaling, stomatal control and hormone signaling pathways [6,7]. Of particular interest is the demonstration that considerable numbers of *MIR* gene families identified as abiotic stress responsive in *Arabidopsis*, play a conserved functional role across phylogenetically diverse dicotyledonous and monocotyledonous species, including many of the major monocot grasses (such as *Zea mays*, *Oryza sativa* and *Triticum aestivum*) cultivated to provide much of the daily calorific intake of the world's population [24–28]. This demonstration identifies the use of *Arabidopsis* as an ideal experimental system to molecularly modify the profile of such conserved abiotic stress responsive miRNA expression modules to determine if the introduced modifications enable *Arabidopsis* to display an altered response phenotype to abiotic stress.

High throughput sequencing was therefore employed here to profile the miRNA landscape of wild-type *Arabidopsis* plants exposed to heat, drought and salt stress. Sequencing identified large miRNA cohorts responsive to each applied stress, with 121, 123 and 118 miRNA sRNAs determined to have a greater than 2.0-fold abundance change post heat, drought and salt stress treatment of *Arabidopsis* whole seedlings, respectively. For each assessed stress, a quantitative reverse transcriptase polymerase chain reaction (RT-qPCR)-based approach was used to experimentally confirm the abundance of five miRNA sRNAs. For each miRNA experimentally validated to have altered abundance post stress exposure, RT-qPCR was additionally used to document reciprocal target gene expression profiles to further confirm each miRNA/target gene expression module as abiotic stress responsive. Sequencing, and the initial RT-qPCR profiling of miRNA abundance and miRNA target gene expression was performed on whole seedling samples, therefore RT-qPCR was next employed to confirm the documented whole seedling expression trends in *Arabidopsis* root and shoot tissues post salt stress exposure. This analysis showed that the initial abiotic stress responsive miRNA expression profiles identified in whole plant samples were an accurate representation of the tissue-specific profile of each assessed expression module. Taken together, the data presented here identifies numerous miRNA/target gene expression modules that could be targeted for future molecular modification to determine if such modification allows *Arabidopsis* to display an altered response to abiotic stress. Furthermore, the information gathered in *Arabidopsis* using such a molecular approach could be potentially translated to an agronomically important cropping species for the future generation of plant lines that display an adaptive phenotypic response to abiotic stress.

2. Results

2.1. Response of Wild-type Arabidopsis Seedlings to Heat, Drought and Salt Stress

Post germination and cultivation on standard growth media, 8 day old wild-type *Arabidopsis* seedlings (ecotype Columbia-0 (Col-0)) were exposed to a 7-day period of either heat, drought or salt stress. Figure 1A displays the phenotypes expressed by heat, drought and salt stressed Col-0 plants, compared to that of 15 day old, non-stressed wild-type *Arabidopsis*. The growth of drought (mannitol supplemented media) and salt stressed plants was significantly repressed, to differing degrees, after 7 days of exposure to both stress treatments, as readily demonstrated by a reduction to rosette tissue fresh weight (Figure 1B), and rosette diameter (Figure 1C). Furthermore, salt stress treatment induced the accumulation of anthocyanin (Figure 1D), an antioxidant produced by plants to combat the cellular stress caused by reactive oxygen species [29,30], specifically in the shoot apex of salt stressed *Arabidopsis* whole seedlings (Figure 1A). Unlike the reductions observed to rosette tissue fresh weight and the diameter of the rosette of drought and salt stressed Col-0 plants, the 7-day heat stress treatment resulted in the promotion of both of these phenotypic parameters. Specifically, compared to non-stressed control plants, the rosette leaf fresh weight (Figure 1B), and rosette diameter (Figure 1C), were increased by 75% and 127% respectively, in heat stressed Col-0 plants. Promotion of specific aerial tissue growth parameters, namely hypocotyl and rosette leaf petiole elongation, has been reported previously for *Arabidopsis* plants cultivated under conditions of elevated temperature [31,32]. Although promotion of aerial tissue growth suggested that the 7-day heat stress treatment had a positive influence on *Arabidopsis* development, the 177% increase in anthocyanin accumulation observed in parallel (Figure 1D), alternatively suggested that this treatment actually induced high levels of stress in *Arabidopsis* cells, specifically the cells of the shoot apex and rosette leaf petioles (Figure 1A). Therefore, to additionally demonstrate that each applied stress was influencing *Arabidopsis* at the molecular level, the expression of the well characterized stress-induced gene, *Δ1-PYRROLINE-5-CARBOXYLATE SYNTHETASE1* (*P5CS1*; *AT2G39800*) [33–35] was quantified by RT-qPCR. This analysis revealed that *P5CS1* transcript abundance was upregulated 3.0-, 1.7- and 45-fold in heat, drought and salt stressed *Arabidopsis* whole seedlings, respectively (Figure 1E).

2.2. Profiling of the microRNA Landscape of Heat, Drought and Salt Stressed Arabidopsis Whole Seedlings

Demonstrated induction of *P5CS1* expression (Figure 1E), a well characterized [33–35] stress-induced gene in *Arabidopsis*, post exposure to heat, drought and salt stress, suggested that all three applied stresses were inducing molecular responses in *Arabidopsis*. Therefore, total RNA was extracted from non-stressed plants, and from heat, drought and salt stressed *Arabidopsis* whole seedlings, and the sRNA fraction of each analyzed via high throughput sequencing to profile the respective miRNA landscapes (Figure 2A). In total, 333 miRNA sRNAs were identified by sequencing across the control and stress treatments (see Supplemental Table S1). Sequencing further revealed a greater than 2-fold abundance change for 121, 123 and 118 mature miRNA sRNAs for heat, drought and salt stressed *Arabidopsis* whole seedlings, respectively (Figure 2B). More specifically, heat stress promoted the accumulation of 90 miRNAs (miR395a abundance was increased to the greatest degree at 89.4-fold) and repressed the abundance of 17 miRNAs (miR3932b levels showed the greatest degree of reduction at -19.6-fold). Exposure of wild-type *Arabidopsis* whole seedlings to drought stress enhanced the abundance of 111 miRNAs and reduced the levels of 2 miRNAs, with the accumulation of miRNAs, miR851 (31.1-fold) and miR397b (-7.8-fold), determined to be influenced to the greatest degree by drought stress treatment. Post salt stress exposure, 86 miRNAs were determined to have a greater than 2-fold elevated abundance (miR778 was upregulated to the greatest degree at 34.0-fold), and further, 22 miRNAs were determined to be reduced in their abundance (miR169g showed the greatest degree of reduction at -8.7-fold). It was also of interest to document reciprocal abundance trends for an additional 5, 5 and 1 miRNA sRNAs upon comparison of each of the applied stresses, namely the heat/drought, heat/salt and drought/salt stress comparisons, respectively. In addition, a further 4 miRNA sRNAs, including miR169f, miR169h, miR397b and miR857, were determined to have an opposing change in abundance upon exposure to each of the three abiotic stresses assessed (Figure 2B). These findings indicate that the promoter regions of the encoding loci of these miRNAs harbor multiple *cis*-elements that direct the changes in *MIR* gene expression which would be required to result in the observed changes in the abundance of this miRNA cohort post exposure to different abiotic stresses [18–20].

A modified RT-qPCR approach [36] was next employed to experimentally validate the sequencing determined abundance of five miRNAs for each stress treatment. For heat stressed *Arabidopsis* whole seedlings, sequencing determined that the abundance of miRNAs, miR169, miR395 and miR396, was altered -7.4, 37.8 and 2.9-fold, respectively, compared to their abundance in non-stressed plants. The altered abundance trend of all three miRNAs was confirmed by RT-qPCR with quantified fold changes of -3.2-, 2.2- and 2.9-fold for the miR169, miR395 and miR396 sRNAs, respectively (Figure 3A–C). Although the abundance changes determined by RT-qPCR were not as dramatic as those determined via sequencing for miRNAs miR169 and miR395 (especially for miR395), the obtainment of a matching abundance trend for each quantified miRNA post heat stress exposure was highly encouraging. Therefore, RT-qPCR was again applied to confirm the sequencing identified abundance fold changes of -2.7-, 4.0- and 2.7-fold for miRNAs, miR857, miR156 and miR399, respectively (Figure 3D–F). Fold changes of -4.3-, 3.0- and 3.2-fold were determined for the miR857, miR156 and miR399 sRNAs respectively, post the 7-day drought stress treatment of *Arabidopsis* whole seedlings by RT-qPCR. RT-qPCR also confirmed the sequencing identified miRNA abundance trends for salt stressed *Arabidopsis* whole seedlings. Fold changes of -2.5-, 2.9- and 3.9-fold were determined by RT-qPCR for miRNAs, miR169, miR399 and miR778 respectively (Figure 3G–I), compared to the abundance fold changes of -4.8-, 4.0- and 34.0-fold determined via sequencing for these three miRNAs in response to salt stress treatment. In addition, miR839 and miR855 abundance in heat, drought and salt stressed plants was also quantified via RT-qPCR due to sequencing indicating that the level of both of these miRNAs did not vary significantly post application of each stress (Figure 3J,K). RT-qPCR confirmed that the levels of these two miRNAs varied less than 0.5-fold post stress exposure, a finding that suggests that neither miRNA is abiotic stress responsive. Taken together, the data presented in Figure 3 demonstrated that the high throughput sequencing employed here (Figure 2) was a reliable tool for profiling miRNA abundance changes in abiotic stressed *Arabidopsis* whole seedlings, and that once identified, RT-qPCR quantification provides a more biologically accurate reflection of the changes in sRNA abundance post exposure to each stress treatment.

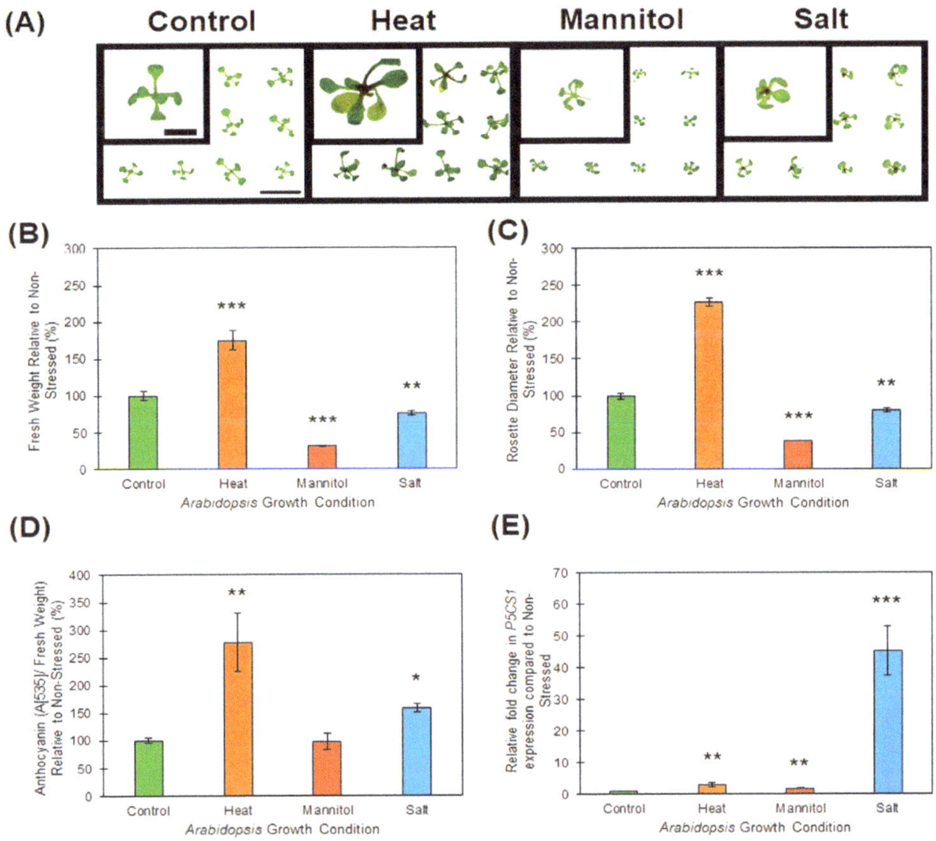

Figure 1. Phenotypic and physiological consequence of heat, drought and salt stress treatment of 15-day-old wild-type *Arabidopsis* whole seedlings. (**A**) Phenotypes displayed by 15-day old wild-type *Arabidopsis* whole seedlings post a 7-day treatment regime with heat, drought or salt stress, compared to non-stressed seedlings of the same age (left panel). Scale bar = 1.0 centimeter (cm) on larger sized panels and 0.5 cm on the superimposed images of a single representative seedling. (**B**) Whole seedling fresh weight of heat, drought (mannitol) and salt stressed *Arabidopsis* compared to their non-stressed counterparts of the same age. (**C**) Rosette diameter of 15-day-old *Arabidopsis* whole seedlings post 7-day exposure to heat, drought (mannitol) and salt stress compared to the non-stressed control. (**D**) Anthocyanin accumulation in heat, drought (mannitol) and salt stressed *Arabidopsis* whole seedlings compared to non-stressed whole seedlings of the same age (15 days). (**E**) RT-qPCR assessment of the expression of the stress induced gene, *Δ1-PYRROLINE-5-CARBOXYLATE SYNTHETASE1* (*P5CS1*; *AT2G39800*) expression in 15-day-old *Arabidopsis* whole seedlings post a 7-day heat, drought (mannitol) and salt stress treatment regime compared to the abundance of the *P5CS1* transcript in non-stressed *Arabidopsis* whole seedlings of the same age. (**B–E**) Error bars represent the standard deviation of four biological replicates and each biological replicate consisted of a pool of six individual plants. The presence of an asterisk above a column represents a statistically significantly difference between the stress treated sample and the non-stressed control sample (p-value: * < 0.05; ** < 0.005; *** < 0.001).

2.3. Assessment of microRNA Target Gene Expression in Heat, Drought and Salt Stressed Arabidopsis Whole Seedlings

It has been extensively documented in *Arabidopsis* that miRNA sRNAs direct expression regulation of their targeted gene(s) via either a mRNA cleavage or translational repression mode of miRNA-directed RNA silencing [9,21–23]. Therefore, to identify the mode of target gene expression regulation directed by the miRNAs experimentally validated here to be responsive to heat, drought or salt stress, RT-qPCR was next employed to reveal the changes in miRNA target gene transcript abundance post exposure of *Arabidopsis* whole seedlings to these three abiotic stresses. For miRNAs, miR169, miR395 and miR396, the three miRNAs determined to be responsive to heat stress treatment via their RT-qPCR-determined, -3.2-, 2.2- and 2.9-fold change in abundance, the transcript level of a single target gene for each miRNA was demonstrated to have a reciprocally altered trend in abundance to that of their targeting miRNA (Figure 4A–C). Specifically, the miR169 target, *NUCLEAR FACTOR Y, SUBUNIT A5* (*NFYA5; AT1G54160*) was determined to have a 32.9-fold elevation in expression in heat stressed wild-type *Arabidopsis* compared to its levels in non-stressed whole seedlings (Figure 4A). In addition, *ATP SULFURYLASE1* (*ATPS1; AT3G22890*) and *GROWTH REGULATING FACTOR7* (*GRF7; AT5G53660*), the target genes of the heat-induced miRNAs, miR395 and miR396, respectively, were determined to have 2.4- (Figure 4B) and 2.2-fold (Figure 4C) reduced expression. RT-qPCR identified similar trends in altered expression for *LACCASE7* (*LAC7; AT3G09220*), *SQUAMOSA PROMOTER BINDING PROTEIN-LIKE9* (*SPL9; AT2G42200*) and *PHOSPHATE2* (*PHO2; AT2G33770*), the target genes of drought responsive miRNAs, miR857, miR156 and miR399, respectively. That is, *LAC7* expression was elevated 2.5-fold (Figure 4D) in response to the 2.7-fold reduction in miR857 levels (Figure 3D), and *SPL9* (Figure 4E) and *PHO2* (Figure 4F) expression was repressed by 2.8- and 4.5-fold respectively, in accordance with the documented 4.0- and 2.7-fold elevated abundance of the targeting miRNAs, miR156 (Figure 3E) and miR399 (Figure 3F), post drought stress treatment of *Arabidopsis* whole seedlings.

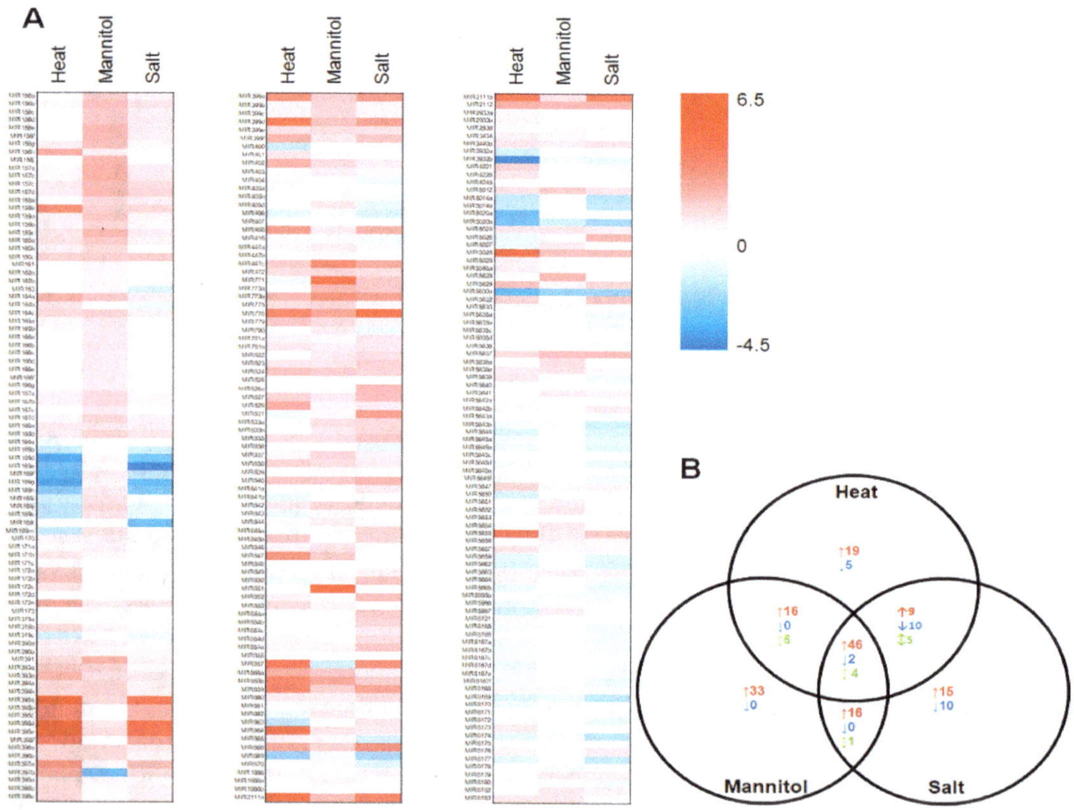

Figure 2. Profiling of the miRNA landscape of heat, drought and salt stressed 15-day-old wild-type *Arabidopsis* whole seedlings. (**A**) Red (up) and blue (down) shaded tiles represent a Log2 fold change in abundance of the *Arabidopsis* miRNA sRNAs detected via high throughput sequencing (see Supplemental Table S1 for the normalized read numbers used to determine fold change values). (**B**) The number of miRNA sRNAs determined to have a greater than 2-fold change in abundance in heat, drought and salt stressed 15-day-old wild-type *Arabidopsis* whole seedlings compared to the abundance of each detected miRNA sRNA in non-stressed control plants of the same age. Red colored up arrows indicate the number of miRNAs with elevated abundance under each assessed stress, blue colored down arrows represent the number of miRNAs with reduced abundance post stress treatment and green colored up/down arrows state the number of miRNA sRNAs with a differing abundance trend between the individual stress treatments.

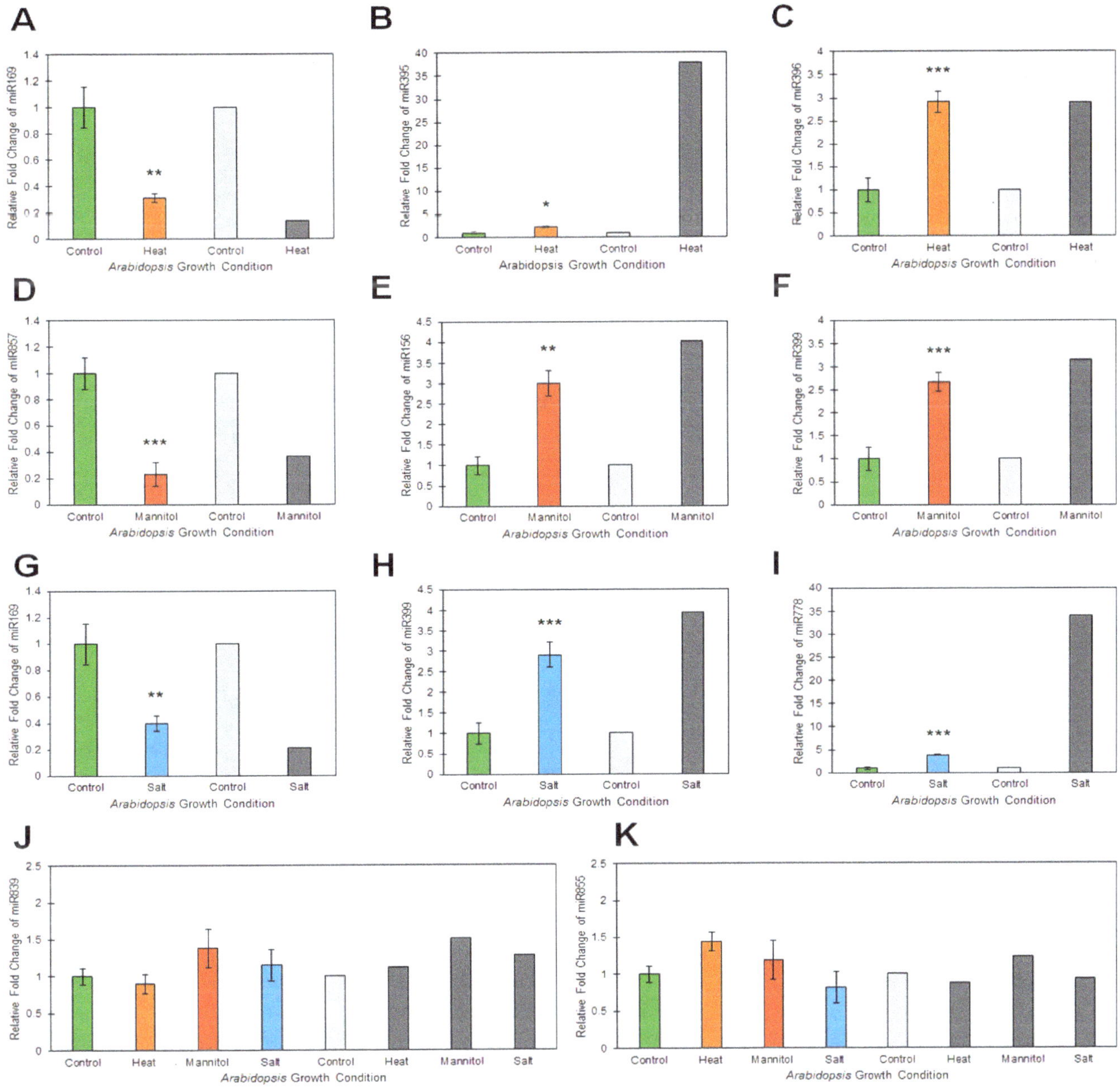

Figure 3. Quantification of miRNA abundance via RT-qPCR analysis of 15-day-old wild-type *Arabidopsis* whole seedlings post exposure to heat, drought and salt stress treatment. (**A**–**C**) RT-qPCR assessment of miR169 (**A**), miR395 (**B**) and miR396 (**C**) abundance in heat stressed *Arabidopsis* whole seedlings. (**D**–**F**) RT-qPCR assessment of miR857 (**D**), miR156 (**E**) and miR399 (**F**) abundance in drought (mannitol) stressed *Arabidopsis* whole seedlings. (**G**–**I**) RT-qPCR assessment of miR169 (**G**), miR399 (**H**) and miR778 (**I**) abundance in salt stressed *Arabidopsis* whole seedlings. (**J**,**K**) RT-qPCR assessment of miR839 (**J**) and miR855 (**K**) abundance across heat, drought (mannitol) and salt stressed *Arabidopsis* whole seedlings. (**A**–**K**) Colored columns (green = non-stressed control; orange = heat stress; red = drought stress, and; blue = salt stress) represent RT-qPCR determined abundance of each quantified miRNA sRNA and the light (control) and dark grey (stress) shaded columns present the fold changes in miRNA abundance as determined via high throughput sequencing. Error bars represent the standard deviation of four biological replicates and each biological replicate consisted of a pool of six individual plants. The presence of an asterisk above a column represents a statistically significantly difference between the stress treated sample and the non-stressed control sample (p-value: * < 0.05; ** < 0.005; *** < 0.001).

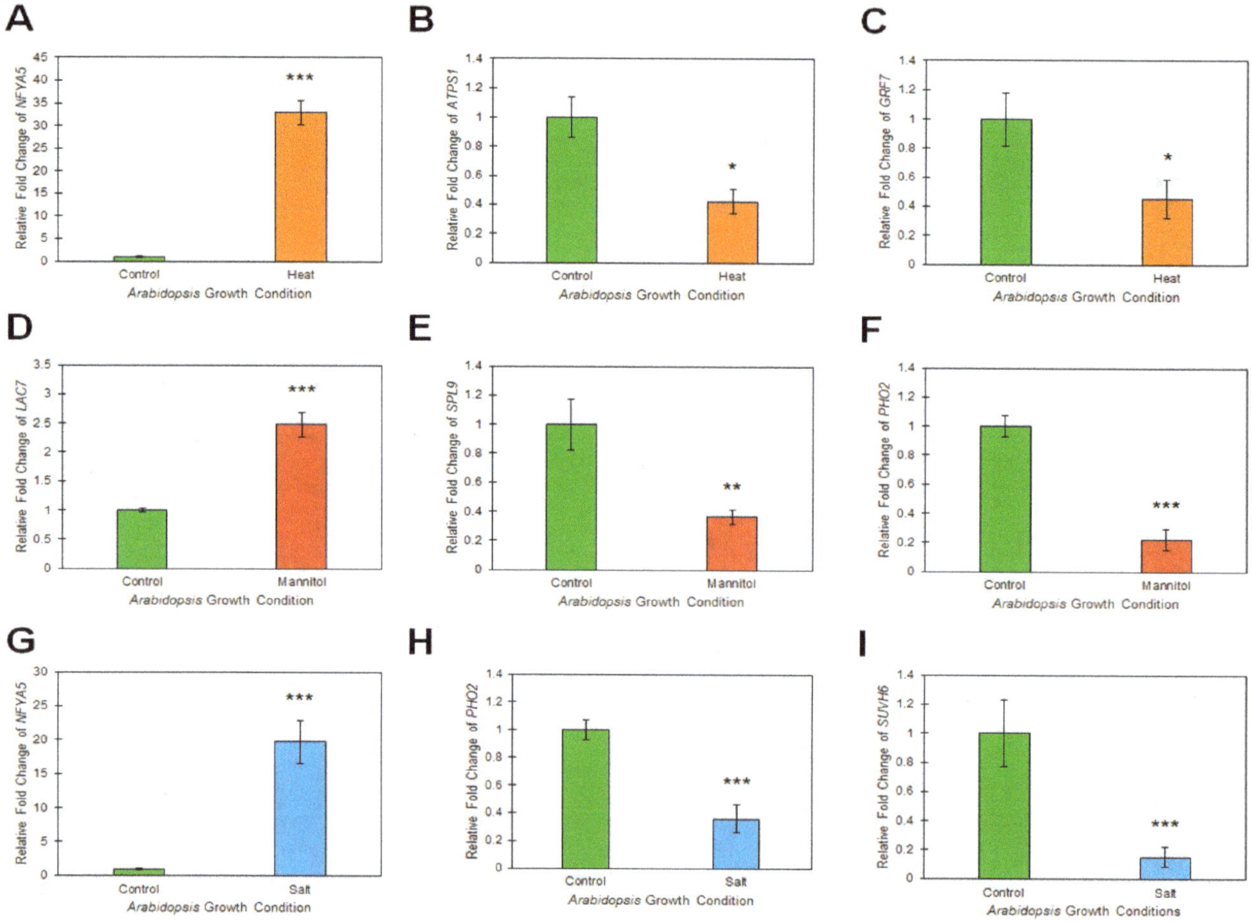

Figure 4. Determination of miRNA target gene expression via RT-qPCR analysis of 15-day-old wild-type *Arabidopsis* whole seedlings post exposure to heat, drought and salt stress treatment. **(A–C)** RT-qPCR assessment of *NFYA5* (**A**), *ATPS1* (**B**) and *GRF7* (**C**) miRNA target gene expression in heat stressed *Arabidopsis* whole seedlings. **(D–F)** RT-qPCR assessment of *LAC7* (**D**), *SPL9* (**E**) and *PHO2* (**F**) miRNA target gene expression in drought (mannitol) stressed *Arabidopsis* whole seedlings. **(G–I)** RT-qPCR assessment of *NFYA5* (**G**), *PHO2* (**H**) and *SUVH6* (**I**) miRNA target gene expression in salt stressed *Arabidopsis* whole seedlings. **(A–I)** Colored columns (green = non-stressed control; orange = heat stress; red = drought stress, and; blue = salt stress) represent RT-qPCR quantified expression of a single target gene for each miRNA assessed via RT-qPCR analysis in Figure 3. Error bars represent the standard deviation of four biological replicates and each biological replicate consisted of a pool of six individual plants. The presence of an asterisk above a column represents a statistically significantly difference between the stress treated sample and the non-stressed control sample (p-value: * < 0.05; ** < 0.005; *** < 0.001).

Reciprocal expression trends were again observed post RT-qPCR assessment of target gene expression in salt stressed samples. Namely, *NFYA5* transcript abundance was significantly elevated 19.7-fold (Figure 4G) in response to the RT-qPCR documented 2.5-fold reduction in miR169 levels (Figure 3G). Further, the transcript abundance of *PHO2* and *SU(VAR)3-9 HOMOLOG6* (*SUVH6; AT2G22740*) was reduced by -2.8- (Figure 4H) and -6.6-fold (Figure 4I), respectively. Reduced *PHO2* and *SUVH6* expression in salt stressed *Arabidopsis* whole seedlings was not a surprising observation considering that the abundance of their targeting miRNAs, miR399 and miR778, was determined to be elevated by 2.9- and 3.9-fold (Figure 3H,I), respectively. Taken together, the target gene expression data presented in Figure 4 indeed suggested that altered miRNA abundance in response to each assessed stress was in turn leading to changes in miRNA target gene transcript abundance. In addition, demonstration of reciprocal trends in abundance for the miRNAs determined to be abiotic stress responsive in Figure 3 compared to the miRNA target gene expression profiles presented in Figure 4, strongly suggested that each abiotic stress responsive miRNA was regulating the expression of its assessed target gene via a mRNA cleavage mode of miRNA-directed RNA silencing.

2.4. Profiling of Salt Responsive microRNA Expression Modules in Arabidopsis Root and Shoot Tissues

To determine whether the documented alterations to abiotic stress responsive miRNA expression modules identified in *Arabidopsis* whole seedlings was an accurate indication of the changes occurring in specific and developmentally distinct *Arabidopsis* tissues, the three miRNA expression modules determined to be salt responsive in *Arabidopsis* whole seedlings (the miR169/*NFYA5*, miR399/*PHO2* and miR778/*SUVH6* expression modules), were profiled in *Arabidopsis* root and shoot tissue by RT-qPCR post exposure to salt stress. Prior to performing this molecular analysis however, the root architecture of wild-type *Arabidopsis* plants cultivated on vertically orientated control and salt stress growth media was assessed. It has been demonstrated previously that the major phenotypic response of the *Arabidopsis* root system to exposure to salt stress is reduced expansion of the primary root [37]. Figure 5A clearly shows that compared to non-stressed wild-type *Arabidopsis* seedlings, the primary phenotypic response of the root system of Col-0 plants exposed to the 7-day salt stress regime was inhibition of primary root elongation, with primary root length reduced by ~60% in salt stressed plants compared to the primary root length of non-stressed control plants (Figure 5A,B).

Inhibition of primary root elongation, coupled with the vertically cultivated salt stressed plants again displaying reductions to the overall size of aerial tissue (i.e., rosette size; Figure 5A), as demonstrated in Figure 1A for *Arabidopsis* plants cultivated on horizontally orientated growth media, led us to next assess *P5SC1* expression in the vertically cultivated salt stressed *Arabidopsis* root and shoot tissue (Figure 5C). Compared to its levels in non-stressed roots and shoots, RT-qPCR revealed *P5CS1* expression to be significantly induced with transcript abundance elevated by 9.1- and 44.0-fold respectively, in salt stressed *Arabidopsis* roots and shoots (Figure 5C). This finding strongly suggested that both tissues types were indeed *'stressed'* by the 7 days of vertical cultivation on plant growth media supplemented with 150 mM sodium chloride. RT-qPCR was therefore next employed to profile the miR169/*NFYA5*, miR399/*PHO2* and miR778/*SUVH6* expression modules in salt stressed root and shoot samples and revealed an opposing trend in abundance for each profiled expression module across both assessed tissues. For example, miR169 abundance was determined to be reduced by 1.6- and 2.2-fold in salt stressed roots and shoots respectively (Figure 5D), while *NFYA5* expression was elevated 4.6- and 16.5-fold in these two tissues (Figure 5E). Similar altered trends in abundance for the miR399 sRNA and its targeted transcript, the *PHO2* mRNA, were also revealed by RT-qPCR, namely; miR399 abundance was elevated by 3.7- and 3.0-fold in salt stressed roots and shoots (Figure 5F), and *PHO2* target gene expression was repressed accordingly in the corresponding tissues by 3.6- and 2.3-fold (Figure 5G), respectively. In addition, RT-qPCR revealed that the abundance of the *SUVH6*-targeting miRNA, miR778, was elevated in both salt stressed root and shoot tissue (Figure 5H). The 3.9- and 2.1-fold elevated abundance of the miR778 sRNA in *Arabidopsis* roots and shoots following the salt stress treatment was determined to result in repressed target gene expression, with the abundance of the *SUVH6* transcript reduced by ~2.0-fold in both assessed tissues (Figure 5I). Taken together, the data presented in Figure 5 confirmed that for the three miRNAs determined to be responsive to salt stress, via their profiling in *Arabidopsis* whole seedlings using a high throughput sequencing approach, provided an accurate reflection of the altered abundance of these three miRNAs in developmentally distinct tissues.

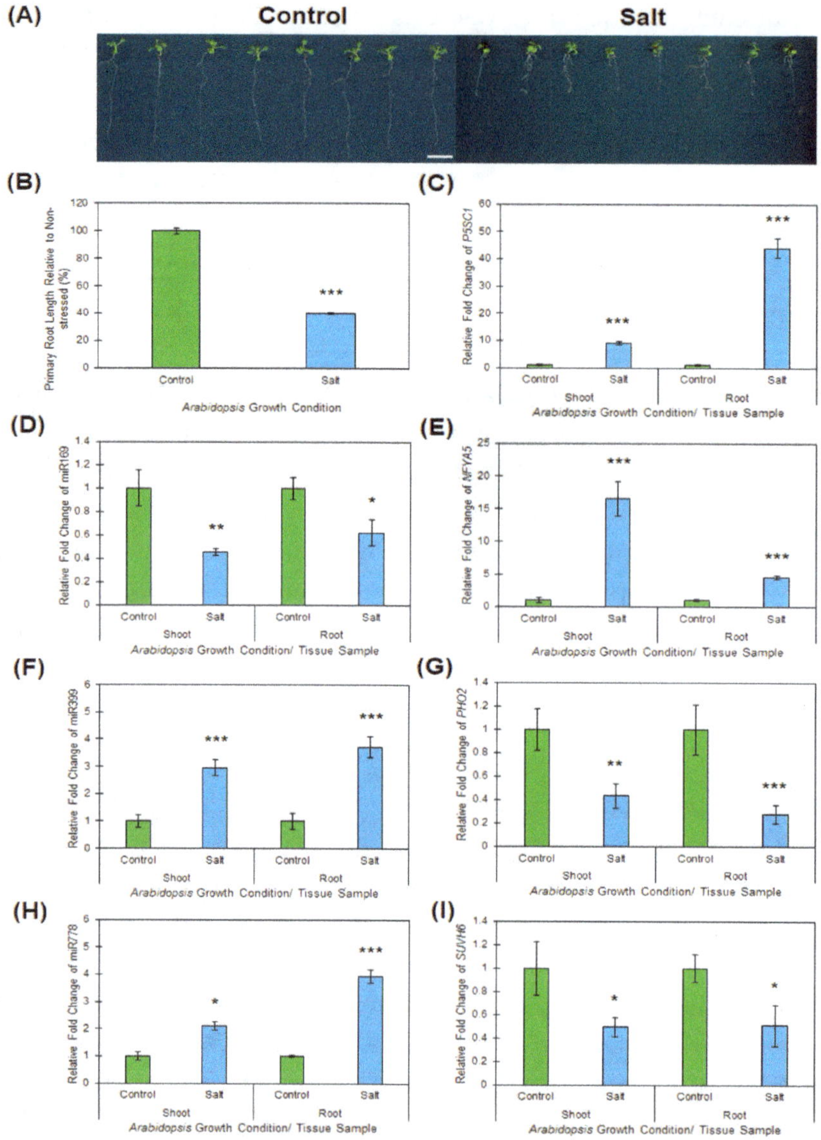

Figure 5. Phenotypic and molecular assessment of the root and shoot tissues of 15-day-old wild-type *Arabidopsis* plants post the 7-day salt stress treatment regime. (**A**) Root and shoot architecture of 15-day-old wild-type *Arabidopsis* seedlings post a 7-day salt stress treatment (right panel) during which the growth media plates were orientated for vertical growth. Scale bar = 1.0 cm. (**B**) Primary root length of 15-day-old *Arabidopsis* whole seedlings cultivated on vertically oriented media growth plates that contained either standard *Arabidopsis* growth media (non-stressed control) or growth media that had been supplemented with 150 mM sodium chloride (stress treatment). (**C**) RT-qPCR assessment of the expression of the stress induced gene, *P5CS1*, expression in 15-day-old *Arabidopsis* root and shoot material post 7-day salt stress treatment compared to the abundance of the *P5CS1* transcript in non-stress control plants of the same age. (**D,E**) RT-qPCR quantification of miR169 abundance (**D**) and *NFYA5* target gene expression (**E**) in salt stressed *Arabidopsis* root and shoot tissues. (**F,G**) RT-qPCR quantification of miR399 abundance (**F**) and *PHO2* target gene expression (**G**) in salt stressed *Arabidopsis* root and shoot tissues. (**H,I**) RT-qPCR quantification of miR778 abundance (**H**) and *SUVH6* target gene expression (**I**) in salt stressed *Arabidopsis* root and shoot tissues. (**B–I**) Colored columns represent the values obtained for non-stressed control plants (green colored columns) and the salt stressed samples (blue colored columns). Error bars represent the standard deviation of four biological replicates and each biological replicate consisted of a pool of six individual plants. The presence of an asterisk above a column represents a statistically significantly difference between the salt stress sample and the non-stressed controls (p-value: * < 0.05; ** < 0.005; *** < 0.001).

3. Discussion

In an attempt to provide sustainable food security into the future, it is essential that the complex, fundamental molecular networks that underpin the ability of a plant to maintain yield, particularly during extended periods of abiotic stress, are elucidated. This would provide the foundation for plant biology researchers to use a molecular approach to develop new plant lines that are readily able to adapt to, or mitigate the negative impacts that result from exposure to abiotic stress. With the miRNA class of sRNA demonstrated to be a key regulator of all aspects of plant development, as well as playing a central role in the ability of a plant to mount a defensive response against invading pathogens, or to mediate an adaptive response to abiotic stress, there currently remains a significant lack of resource datasets available for *Arabidopsis* to allow researchers to identify candidate miRNA expression modules for molecular modification as part of the future development of new plant lines that display adaptive phenotypes to abiotic stress. Towards this goal, here the genetic model plant species, *Arabidopsis thaliana*, was used to profile the miRNA landscape that potentially underpins, in part, the physiological and phenotypic responses of *Arabidopsis* to exposure to the abiotic stresses, heat, drought and salt stress. Most notably, sRNA sequencing revealed that the abundance of 121, 123 and 118 mature miRNA sRNAs was significantly (>2.0-fold) up- or down-regulated in response to heat, drought and salt stress treatment of *Arabidopsis* whole seedlings, respectively. The subsequent experimental validation of the miRNA abundance changes identified via high throughput sequencing by RT-qPCR, in combination with the additional use of RT-qPCR to document reciprocal trends in transcript abundance for each assessed miRNA target gene, was essential to confidently identify the miRNA expression modules responsive to each assessed stress.

In response to heat stress, a significant reduction to miR169 abundance (-3.2-fold) was observed in *Arabidopsis* whole seedlings (Figure 3A). Reduced miR169 levels have been reported previously for *Arabidopsis* post exposure to either drought stress or nitrogen starvation [38,39]. Further, [39] went on to demonstrate that reduced miR169 abundance in drought stressed *Arabidopsis* resulted in deregulated *NFYA5* expression, a target gene expression trend also observed here for heat stressed *Arabidopsis* (Figure 4A). The authors also revealed that overexpression of *NFYA5* in *Arabidopsis* resulted in these molecularly modified plant lines displaying reduced leaf water loss, due to reduced stomatal aperture, and drought stress tolerance [39]. Documentation of similar alterations to the miR169/*NFYA5* expression module in this study post heat stress treatment of *Arabidopsis* whole seedlings (Figure 3A, Figure 4A), to those previously reported for drought stressed *Arabidopsis* [39], suggests that these molecular changes are potentially in part driving the physical adaptation of *Arabidopsis* to both stresses, namely, alteration of stomatal aperture to promote water retention during exposure to such stress. It is also important to note here that the expression of several of the *MIR169* gene loci from which the miR169 precursor transcripts are transcribed were demonstrated to be induced in *Arabidopsis* by heat stress [40]. Induction of *MIR169* gene expression would be expected to result in elevated mature miR169 sRNA accumulation, and not reduced miR169 abundance, as observed here (Figure 3A). Curiously however, [40] did not report on whether *MIR169* gene expression induction actually resulted in elevated miR169 abundance in heat stressed *Arabidopsis* plants. The noted reduction to miR169 abundance reported here for heat stressed *Arabidopsis* plants, suggests that in *Arabidopsis*, heat stress represses *MIR169* gene expression, rather than promote transcription from these loci as reported by [40]. The opposing effect of heat stress exposure on miR169 accumulation in *Arabidopsis* reported here in Figure 3A, to that previously reported [40], could potentially be the result of differences in the application of the stress. Specifically, in [40], two-week-old *Arabidopsis* plants were transferred to moistened filter paper and exposed to the 40 °C heat stress treatment for a duration of either 3 or 6 h, whereas here, 8-day-old *Arabidopsis* seedlings were exposed to a prolonged 7-day heat stress treatment of elevated day/night (16/8 h) temperatures of 32 °C/28 °C. Nonetheless, the detection of elevated target gene (*NFYA5*) expression (Figure 4A), in accordance with the documented reduction in the abundance of the miR169 sRNA (Figure 3A), indicates that the alterations to the miR169/*NFYA5* expression module observed here in heat stressed *Arabidopsis*, are biologically relevant.

Elevated abundance has previously been reported for miRNAs, miR395 and miR396, post exposure of *Arabidopsis* to heat stress [41,42]. Similar abundance changes for the miR395 (Figure 3B) and miR396 (Figure 3C) sRNAs were observed here for heat stressed wild-type *Arabidopsis* whole seedlings. In addition, elevated miRNA abundance was further demonstrated to direct enhanced miRNA-directed target gene expression repression, with both the miR395 and miR396 target genes, *ATPS1* (Figure 4B) and *GRF7* (Figure 4C) respectively, determined to have reduced transcript abundance in heat stressed *Arabidopsis*. Taken together, comparison of the findings reported here, to those reported previously for miR395 and miR396 [41,42], strongly suggest that these two miRNAs are indeed heat stress responsive miRNAs, and further, that enhanced miR395- and miR396-directed expression repression of *ATPS1* and *GRF7*, respectively, potentially forms part of the adaptive response of *Arabidopsis* to elevated temperature.

Here, mannitol was used as an osmoticum to stimulate osmotic stress in *Arabidopsis* whole seedlings in an attempt to replicate drought stress conditions in a tightly controlled growth environment (i.e., sealed plant tissue culture plates). Post stress treatment, miR857 abundance was revealed to be reduced in *Arabidopsis* whole seedlings via both high throughput sequencing and RT-qPCR (Figure 2A, Figure 3D). The miR857 sRNA has previously been demonstrated to post-transcriptionally regulate the expression of *LAC7*, a laccase enzyme involved in mediating lignin deposition in the secondary xylem [43]. In addition to revealing reduced miR857 abundance, RT-qPCR showed that *LAC7* expression was elevated in drought stressed *Arabidopsis* whole seedlings (Figure 4D). Considering its documented role in secondary xylem development, the observed alterations to the miR857/*LAC7* expression module may potentially mediate an adaptive response to osmotic stress in *Arabidopsis*, potentially directing a change to tissue architecture in response to drought stress. Unlike miR857, miR156 and miR399 abundance was elevated by the mannitol-induced drought stress treatment (Figure 3E,F). In accordance, RT-qPCR showed that the transcript abundance of *SPL9* (Figure 4E) and *PHO2* (Figure 4F), the target genes of miR156 and miR399, respectively, was reduced in response to the elevated abundance of their targeting miRNA sRNAs. Interestingly, the miR156/*SPL9* expression module, together with the downstream gene, *DIHYDROFLAVONOL-4-REDUCTASE* (*DFR*; *AT5G42800*) have been demonstrated previously to play a role in anthocyanin metabolism [44], and Figure 1D shows that anthocyanin accumulation remained at its non-stressed levels in *Arabidopsis* plants cultivated on standard growth media supplemented with 200 mM mannitol, in spite of these plants displaying reductions to their fresh weight and rosette diameter, in addition to elevated expression of the stress induced gene, *P5CS1* (Figure 1E). Furthermore, in *Arabidopsis*, both of these miRNAs (miR156 and miR399) have been previously demonstrated to be responsive to mannitol-induced drought stress [41,44], findings that when taken together with those presented here in Figures 3 and 4, strongly suggest that the miR156/*SPL9* and miR399/*PHO2* expression modules are indeed responsive to mannitol-induced drought stress.

High throughput sRNA sequencing and RT-qPCR revealed miR169 abundance to be reduced by -4.8 and -2.5-fold respectively, post exposure to salt stress. This abundance change opposes that reported previously for the miR169 sRNA in rice and cotton [45,46], where miR169 accumulation was demonstrated to be induced by salt stress. However, the observed differences in miR169 abundance post salt stress exposure in rice [45], cotton [46] and *Arabidopsis* (Figure 3G), is most likely the result of unique *cis*-element landscapes of the promoter regions of *MIR169* loci across these three species [19]. In *Arabidopsis*, miR169 abundance has been previously demonstrated to be reduced by drought stress [40], conditions of limited phosphate [47], and nitrogen starvation [38]. The findings of these reports [38,40,47], together with those presented here, namely deregulated *NFYA5* target gene expression in *Arabidopsis* whole seedlings (Figure 4G), roots and shoots (Figure 5E), due to loss of miR169-directed *NFYA5* expression repression in these tissues, indicates that the miR169/*NFYA5* expression module potentially plays a central role in mediating the response of *Arabidopsis* to a range of abiotic stresses, potentially even forming a *'crosstalk junction'* to link the highly complicated molecular networks that are required to directed the physiological and phenotypic responses of *Arabidopsis* to abiotic stress.

Salt stress treatment was shown to enhance miR399 sRNA abundance in *Arabidopsis* whole seedlings (Figure 3H) as well as in root and shoot tissue (Figure 5F), a previously reported finding [41]. Furthermore, and using a molecular approach in *Arabidopsis*, [48] revealed *MIR399F* gene expression to be induced by salt stress and that *Arabidopsis* plants modified to constitutively overexpress the *MIR399F* gene were more tolerant to salt stress than unmodified wild-type plants. Given *PHO2* targeting by miR399, and the previously documented role for phosphate in modulating root system architecture alterations under salt stress conditions in *Arabidopsis*, the elevated abundance of miR399 shown here, in conjunction with the demonstrated reductions to the level of the *PHO2* target transcript (Figure 4H, Figure 5G), are consistent with the proposed role of the miR399/*PHO2* expression module in the complex phosphate-salt regulatory network in *Arabidopsis* tissues [49,50]. Like miR399, miR778 has previously been classed as a phosphate responsive miRNA in *Arabidopsis* [47,51,52]. Here we demonstrate that miR778 abundance is also elevated in response to salt stress in *Arabidopsis* whole seedlings (Figure 3I), roots and shoots (Figure 5H). Accordingly, via a RT-qPCR approach, we further revealed that elevated miR778 abundance resulted in enhanced expression repression of the miR778 target gene, *SUVH6*, in salt stressed *Arabidopsis* tissues (Figure 4I, Figure 5I). Interestingly, the miR778 target, SUVH6, is involved in directing methylation of the lysine 9 residue of histone H3 (H3K9 methylation), and further, *SUVH6* expression repression via the constitutive overexpression of the *MIR778* precursor transcript resulted in the modified *Arabidopsis* plants displaying moderately enhanced primary and lateral root growth, and elevated levels of free phosphate and anthocyanin accumulation in the aerial tissues of these plants when cultivated in a phosphate deficient growth environment [52]. These findings, together with the alterations to both the miR399/*PHO2* and miR778/*SUVH6* expression module reported here for salt stressed *Arabidopsis* (Figures 3 and 5), add further weight to the importance of phosphate-mediated responses in *Arabidopsis* tissues as part of the adaptive response of *Arabidopsis* to salt stress.

Altered miRNA abundance, and miRNA target gene expression, have been identified as key molecular responses to an array of abiotic stresses across an evolutionary diverse range of plant species [6,14,24,41,48,49]. Here we have specifically assessed alterations to the miRNA landscapes of heat, drought and salt stressed wild-type *Arabidopsis* whole seedlings and identified large miRNA cohorts responsive to each stress. Alteration to a select number of miRNA/target gene expression modules for the heat, drought and salt stress treatments were experimentally validated via an RT-qPCR approach. Considering that many abiotic responsive miRNAs have been demonstrated to play a conserved functional role across phylogenetically diverse plant species [14,24,25], it is envisaged that the dataset generated in this study forms a valuable resource for the wider plant biology research community; a resource that can be used as the starting point to identify the specific miRNA expression modules to be molecularly manipulated in plant species amenable to genetic modification as part of the future development of plant lines with an altered miRNA and/or miRNA target gene abundance that display a tolerance phenotype to either heat, drought or salt stress. Alternatively, for plant species that are not readily amenable to genetic modification, this dataset can additionally be used to identify the specific miRNA expression modules to be targeted for rapid high throughput screening (via RT-qPCR) across diverse germplasm of a specific species to select those genotypes that harbor natural alterations to the molecular profile of the miRNA expression module of interest.

4. Materials and Methods

4.1. Plant Material

The seeds of wild-type *Arabidopsis thaliana* (*Arabidopsis*), ecotype Columbia-0 (Col-0), were surface sterilized using chlorine gas and post sterilization, seeds were plated out onto standard *Arabidopsis* plant growth media (half strength Murashige and Skoog (MS) salts) and stratified in the dark at 4°C for 48 h. Post stratification, sealed plates containing the surface sterilized seeds were transferred to a temperature-controlled growth cabinet (A1000 Growth Chamber, Conviron® Australia) and cultivated for 8 days under a standard growth regime of 16 h light/8 h dark and a 22 °C/18 °C day/night temperature. Following this 8-day cultivation period, equal numbers of Col-0 seedlings were transferred under sterile conditions to either fresh standard *Arabidopsis* plant growth media (control treatment), or to plant growth media that had been supplemented with 200 millimolar (mM) mannitol (drought stress treatment) or 150 mM of sodium chloride (salt stress treatment). Post seedling transfer, the non-stressed control and the drought and salt stress treatment plates were returned to the growth cabinet and cultivated for an additional 7-day period under standard growth conditions. For the heat stress treatment, 8-day-old seedlings were also transferred under sterile conditions to standard growth media, however the 16/8 h day/night temperature was elevated to 32 °C/28 °C for the duration of the 7-day stress treatment period. At the end of the 7-day treatment period, all of the phenotypic and molecular assessments reported here were conducted on 15 day old *Arabidopsis* whole seedlings. For the tissue specific analyses reported in Figure 5, plants were treated exactly as outlined above, except for the 7-day treatment period, when 8 day old seedlings were transferred and cultivated on control and salt stress media plates that were orientated for vertical growth.

4.2. *Phenotypic and Physiological Assessments*

All phenotypic assessments reported here were conducted on 15 day old *Arabidopsis* seedlings. The performance of wild-type *Arabidopsis* plants exposed to each assessed stress is therefore presented relative to non-stressed control plants. More specifically, each phenotypic measurement collected for *Arabidopsis* seedlings exposed to each assessed abiotic stress regime is presented as a percentage of the corresponding measurement determined for non-stressed control seedlings cultivated under standard growth conditions for the duration of the 7-day stress treatment period. Rosette diameter and primary root length analysis was determined via assessment of photographic images using the ImageJ software. A standard 99:1 (v/v) methanol:HCl extraction protocol was used to extract anthocyanin from control and stress treated Col-0 plants. Post extraction, anthocyanin content was determined using a spectrophotometer (Thermo Scientific, Australia) at an absorbance wavelength of 535 nanometers (A_{535}) and using the 99:1 (v/v) methanol:HCl solution as the blank.

4.3. *Total RNA Extraction and High Throughput Sequencing of the small RNA Fraction*

Total RNA was isolated from four biological replicates (each biological replicate contained 6 individual plants) of 15 day old Col-0 whole seedlings cultivated under normal growth conditions for the duration of the experimental period, or post 7-days of heat, drought or salt stress treatment, using TRIzol™ Reagent (Invitrogen™) according to the manufacturer's instructions. The quality and quantity of the isolated total RNA was assessed using a Nanodrop spectrophotometer (NanoDrop® ND-1000, Thermo Scientific, Australia) and via standard electrophoresis on a 1.2% (w/v) ethidium bromide-stained agarose gel to allow for RNA visualization. Next, 5.0 micrograms (μg) of each of the four biological replicates for each treatment, were pooled together and diluted in RNase-free water to obtain a final preparation of 25 microliters (μL) of total RNA at a concentration of 800 nanograms (ng) per μL. Samples were shipped to the Australian Genome Research Facility (AGRF; Melbourne node, Australia) with the AGRF performing all subsequent preparatory steps prior to sequencing the small RNA fraction of each sample on an Illumina HiSeq 2500 platform.

4.4. *Bioinformatic Assessment of the microRNA Landscape of Arabidopsis Whole Seedlings*

Using the Qiagen CLC Genomics Workbench (11) software, next-generation sequencing adapter sequences were removed prior to performing sequence quality trimming to remove any sRNA reads that were either shorter than 15 nucleotides (nts), or longer than 35 nts in length. Additionally, parameters within the CLC Genomic Workbench were applied to remove any ambiguous nucleotides at either the 5′ or 3′ terminus of each sequencing read (i.e., the removal of any 'N' nucleotides on sequence ends), or to 'trim' low quality sequences using a modified *'Mott trimming'* algorithm. The remaining sequences that aligned perfectly (i.e., zero mismatches) to known *Arabidopsis* miRNAs listed in miRBase 22 were then annotated. The values determined for the; (1) raw read count of each detected miRNA sRNA across the four treatments (control, heat, drought and salt); (2) Log2 fold change in abundance for each miRNA sRNA per stress treatment, compared to the non-stressed control values; (3) total number of high quality raw reads per library; (4) total number of miRNA sRNA raw reads per library, and; (5) percentage of the total library size that the miRNA class of sRNA represents, is presented in Supplemental Table S1.

4.5. Quantitative Reverse Transcriptase Polymerase Chain Reaction Analyses

Quantitative reverse transcriptase polymerase chain reaction (RT-qPCR) assessment of miRNA sRNA and miRNA target gene transcript abundance was conducted on 4 biological replicates: the same four biological replicates that were pooled together to perform the high throughput sequencing analysis of the sRNA fraction of each sample. The synthesis of miRNA-specific complementary DNA (cDNA) was conducted using 200 ng of DNase I treated (New England BioLabs, Australia) total RNA as template and 1.0 unit (U) of ProtoScript® II Reverse Transcriptase (New England BioLabs, Australia) according to manufacturer's instructions. The cycling conditions for miRNA-specific cDNA synthesis were: 1 cycle at 16°C for 30 min; 60 cycles of 30°C for 30 s, 42°C for 30 s 50°C for 2 s, and; 1 cycle of 85°C for 5 min. To generate a global high molecular weight cDNA library for the quantification of miRNA target gene expression, 5.0 µg of total RNA was treated with 5.0 U of DNase I (New England BioLabs, Australia) according to manufacturer's instructions. Post DNase I treatment, the total RNA was purified using an RNeasy Mini Kit according to the manufacturer's instructions (Qiagen, Australia), and then 1.0 µg of this preparation was used as template for cDNA synthesis with 1.0 U of ProtoScript® II Reverse Transcriptase according to the manufacturer's instructions (New England Biolabs, Australia) along with 2.5 µM of oligo $dT_{(18)}$. All single stranded cDNA preparations were next diluted to 50 ng/µL in RNase-free water prior to RT-qPCR quantification of miRNA sRNA abundance or miRNA target gene expression. The GoTaq® qPCR Master Mix (Promega, Australia) was used as the fluorescent reagent for all performed RT-qPCRs, and all RT-qPCRs had the same cycling conditions of: 1 cycle of 95°C for 10 min, followed by 45 cycles of 95°C for 10 s and 60°C for 15 s. The abundance of each assessed miRNA sRNA and the expression of each examined miRNA target gene was determined using the $2^{-\Delta\Delta CT}$ method with the small nucleolar RNA, *snoR101*, and *UBIQUITIN10* (*UBI10*; *AT4G05320*) used as the respective internal controls to normalize the relative abundance of each assessed transcript. All DNA oligonucleotides used for either miRNA-specific cDNA synthesis or the quantification of miRNA target gene expression are provided in Supplemental Table S2. For the synthesis of miRNA-specific cDNA of a miRNA sRNA that belongs to a multimember family, and where multiple family members were detected via the high throughput sequencing approach, a miRNA family consensus sequence was determined and the primer designed to hybridize will all detected family members.

Supplementary Materials:
Table S1: Raw miRNA reads and the Log2 fold change in abundance of each *Arabidopsis thaliana* miRNA sRNA detected via high throughput sRNA sequencing, Table S2: Sequences of the DNA oligonucleotides used in this study for the synthesis of miRNA-specific cDNAs and the RT-qPCR based quantification of miRNA abundance or miRNA target gene expression.

Author Contributions: C.P.L.G. and A.L.E. conceived and designed the research. J.L.P. and J.M.J.O. performed the experiments and analyzed the data. J.L.P., C.P.L.G., J.M.J.O. and A.L.E. authored the manuscript and, J.L.P., C.P.L.G., J.M.J.O. and A.L.E. have read and approved the final version of the manuscript.

Acknowledgments: The authors would like to thank fellow members of the Centre for Plant Science for their guidance with plant growth care and RT-qPCR experiments.

References

1. Erb, K.H.; Haberl, H.; Plutzar, C. Dependency of global primary bioenergy crop potentials in 2050 on food systems, yields, biodiversity conservation and political stability. *Energy Policy* **2012**, *47*, 260–269. [CrossRef] [PubMed]
2. Godfray, H.C.J.; Beddington, J.R.; Crute, I.R.; Haddad, L.; Lawrence, D.; Muir, J.F.; Pretty, J.; Robinson, S.; Thomas, S.M.; Toulmin, C. Food security: The challenge of feeding 9 billion people. *Science* **2010**, *327*, 812–818. [CrossRef] [PubMed]
3. Tilman, D.; Balzar, C.; Hill, J.; Befort, B.L. Global food demand and the sustainable intensification of agriculture. *Proc. Natl. Acad. Sci. USA* **2011**, *108*, 20260–20264. [CrossRef] [PubMed]
4. Foley, J.A.; Ramankutty, N.; Brauman, K.A.; Cassidy, E.S.; Gerber, J.S.; Johnston, M.; Mueller, N.D.; O'Connell, C.; Ray, D.K.; West, P.C.; et al. Solutions for a cultivated planet. *Nature* **2011**, *478*, 337–342. [CrossRef] [PubMed]
5. Ramankutty, N.; Evan, A.T.; Monfreda, C.; Foley, J.A. Farming the plant: 1. Geographic distribution of global agricultural lands in the year 2000. *Glob. Biogeochem. Cycles* **2008**, 22. [CrossRef]
6. Khraiwesh, B.; Zhu, J.K.; Zhu, J. Role of miRNAs and siRNAs in biotic and abiotic stress responses of plants. *Biochim. Biophys. Acta* **2012**, *1819*, 137–148. [CrossRef] [PubMed]
7. Sunkar, R.; Chinnusamy, V.; Zhu, J.; Zhu, J.K. Small RNAs as big players in plant abiotic stress responses and nutrient deprivation. *Trends Plant Sci.* **2007**, *12*, 301–309. [CrossRef] [PubMed]
8. Reinhart, B.J.; Weinstein, E.G.; Rhoades, M.W.; Bartel, B.; Bartel, D.P. MicroRNAs in plants. *Genes Dev.* **2002**, *16*, 1616–1626. [CrossRef] [PubMed]
9. Chen, X. A microRNA as a translational repressor of *APETALA2* in *Arabidopsis* flower development. *Science* **2004**, *303*, 2022–2025. [CrossRef] [PubMed]
10. Palatnik, J.F.; Allen, E.; Wu, X.; Schommer, C.; Schwab, R.; Carrington, J.C.; Weigel, D. Control of leaf morphogenesis by microRNAs. *Nature* **2003**, *425*, 257–263. [CrossRef] [PubMed]
11. Kasschau, K.D.; Xie, Z.; Allen, E.; Llave, C.; Chapman, E.J.; Krizan, K.A.; Carrington, J.C. P1/HC-Pro, a viral suppressor of RNA silencing, interferes with *Arabidopsis* development and miRNA function. *Dev. Cell* **2003**, *4*, 205–217. [CrossRef]
12. Li, Y.; Zhang, Q.; Zhang, J.; Wu, L.; Qi, Y.; Zhou, J.M. Identification of microRNAs involved in pathogen-associated molecular pattern-triggered plant innate immunity. *Plant Physiol.* **2010**, *152*, 2222–2231. [CrossRef] [PubMed]
13. Liu, H.H.; Tian, X.; Li, Y.J.; Wu, C.A.; Zheng, C.C. Microarray-based analysis of stress-regulated microRNAs in *Arabidopsis thaliana*. *RNA* **2008**, *14*, 836–843. [CrossRef] [PubMed]
14. Sunkar, R.; Zhu, J.K. Novel and stress-regulated microRNAs and other small RNAs from *Arabidopsis*. *Plant Cell* **2004**, *16*, 2001–2019. [CrossRef] [PubMed]
15. Bartel, D.P. MicroRNAs: Genomics, biogenesis, mechanism, and function. *Cell* **2004**, *116*, 281–297. [CrossRef]
16. Kurihara, Y.; Takashi, Y.; Watanabe, Y. The interaction between DCL1 and HYL1 is important for efficient and precise processing of pri-miRNA in plant microRNA biogenesis. *RNA* **2006**, *12*, 206–212. [CrossRef] [PubMed]
17. Vazquez, F.; Gasciolli, V.; Crété, P.; Vaucheret, H. The nuclear dsRNA binding protein HYL1 is required for microRNA accumulation and plant development, but not posttranscriptional transgene silencing. *Curr. Biol.* **2004**, *14*, 346–351. [CrossRef] [PubMed]
18. Megraw, M.; Baev, V.; Rusinov, V.; Jensen, S.T.; Kalantidis, K.; Hatzigeorgiou, A.G. MicroRNA promoter element discovery in *Arabidopsis*. *RNA* **2006**, *21*, 1612–1619. [CrossRef] [PubMed]
19. Pegler, J.L.; Grof, C.P.L.; Eamens, A.L. Profiling of the differential abundance of drought and salt stress-responsive microRNAs across grass crop and genetic model plant species. *Agronomy* **2018**, *8*, 118. [CrossRef]
20. Zhao, X.; Li, L. Comparative analysis of microRNA promoters in *Arabidopsis* and rice. *Genom. Proteom. Bioinform.* **2013**, *11*, 56–60. [CrossRef] [PubMed]
21. Baulcombe, D.C. RNA silencing in plants. *Nature* **2004**, *431*, 356–363. [CrossRef] [PubMed]
22. Baumberger, N.; Baulcombe, D.C. *Arabidopsis* ARGONAUTE1 is an RNA Slicer that selectively recruits microRNAs and short interfering RNAs. *Proc. Natl. Acad. Sci. USA* **2005**, *102*, 11928–11933. [CrossRef] [PubMed]

23. Reis, R.S.; Hart-Smith, G.; Eamens, A.L.; Wilkins, M.R.; Waterhouse, P.M. Gene regulation by translational inhibition is determined by Dicer partnering proteins. *Nat. Plants* **2015**, *1*, 14027. [CrossRef] [PubMed]
24. Noman, A.; Fahad, S.; Aqeel, M.; Ali, U.; Amanullah; Anwar, S.; Baloch, S.K.; Zainab, M. miRNAs: Major modulators for crop growth and development under abiotic stresses. *Biotechnol. Lett.* **2017**, *39*, 685–700. [CrossRef] [PubMed]
25. Shriram, V.; Kumar, V.; Devarumath, R.M.; Khare, T.S.; Wani, S.H. MicroRNAs as potential targets for abiotic stress tolerance in plants. *Front. Plant Sci.* **2016**, *14*, 817. [CrossRef] [PubMed]
26. Zhang, B. MicroRNA: A new target for improving plant tolerance to abiotic stress. *J. Exp. Bot.* **2015**, *66*, 1749–1761. [CrossRef] [PubMed]
27. Zhang, B.; Unver, T. A critical and speculative review on microRNA technology in crop improvement: Current challenges and future directions. *Plant Sci.* **2018**, *274*, 193–200. [CrossRef] [PubMed]
28. Liu, H.; Able, A.J.; Able, J.A. SMARTER De-Stressed Cereal Breeding. *Trends Plant Sci.* **2016**, *21*, 909–925. [CrossRef] [PubMed]
29. Chalker-Scott, L. Environmental significance of anthocyanins in plant stress responses. *Photochem. Photobiol.* **1999**, *70*, 1–9. [CrossRef]
30. Lotkowska, M.E.; Tohge, T.; Fernie, A.R.; Xue, G.P.; Balazadeh, S.; Mueller-Roeber, B. The *Arabidopsis* transcription factor MYB112 promotes anthocyanin formation during salinity and under high light stress. *Plant Physiol.* **2015**, *169*, 1862–1880. [CrossRef] [PubMed]
31. Gray, W.M.; Östin, A.; Sandberg, G.; Romano, C.P.; Estelle, M. High temperature promotes auxin-mediated hypocotyl elongation in *Arabidopsis*. *Proc. Natl. Acad. Sci. USA* **1998**, *95*, 7197–7202. [CrossRef] [PubMed]
32. Koini, M.A.; Alvey, L.; Allen, T.; Tilley, C.A.; Harberd, N.P.; Whitelam, G.C.; Franklin, K.A. High temperature-mediated adaptations in plant architecture require the bHLH transcription factor PIF4. *Curr. Biol.* **2009**, *19*, 408–413. [CrossRef] [PubMed]
33. Lv, W.T.; Lin, B.; Zhang, M.; Hua, X.J. Proline accumulation is inhibitory to *Arabidopsis* seedlings during heat stress. *Plant Physiol.* **2011**, *156*, 1921–1933. [CrossRef] [PubMed]
34. Urano, K.; Maruyama, K.; Ogata, Y.; Morishita, Y.; Takeda, M.; Sakurai, N.; Suzuki, H.; Saito, K.; Shibata, D.; Kobayashi, M.; et al. Characterization of the ABA-regulated global responses to dehydration in *Arabidopsis* by metabolomics. *Plant J.* **2009**, *56*, 106–1078. [CrossRef] [PubMed]
35. Strizhov, N.; Abrahám, E.; Okrész, L.; Blickling, S.; Zilberstein, A.; Schell, J.; Koncz, C.; Szabados, L. Differential expression of two P5CS genes controlling proline accumulation during salt-stress requires ABA and is regulated by ABA1, ABI1 and AXR2 in *Arabidopsis*. *Plant J.* **1997**, *12*, 557–569. [CrossRef] [PubMed]
36. Mestdagh, P.; Feys, T.; Bernard, N.; Guenther, S.; Chen, C.; Speleman, F.; Vandesompele, J. High-throughput stem-loop RT-qPCR miRNA expression profiling using minute amounts of input RNA. *Nucleic Acids Res.* **2008**, *36*, e143. [CrossRef] [PubMed]
37. Potters, G.; Pasternak, T.P.; Guisez, Y.; Palme, K.J.; Jansen, M.A.K. Stress-induced morphogenic responses: Growing out of trouble? *Trends Plant Sci.* **2007**, *12*, 98–105. [CrossRef] [PubMed]
38. Zhao, M.; Ding, H.; Zhu, J.K.; Zhang, F.; Li, W.X. Involvement of miR169 in the nitrogen-starvation resposnses in *Arabidopsis*. *New Phytol.* **2011**, *190*, 906–915. [CrossRef] [PubMed]
39. Li, W.X.; Oono, Y.; Zhu, J.; He, X.J.; Wu, J.M.; Iida, K.; Lu, X.Y.; Cui, X.; Jin, H.; Zhu, J.K. The *Arabidopsis* NFYA5 transcription factor is regulated transcriptionally and posttranscriptionally to promote drought resistance. *Plant Cell* **2008**, *20*, 2238–2251. [CrossRef] [PubMed]
40. Li, Y.; Fu, Y.; Ji, L.; Wu, C.; Zheng, C. Characterization and expression analysis of the *Arabidopsis* miR169 family. *Plant Sci.* **2010**, *178*, 271–280. [CrossRef]
41. Barciszewska-Pacak, M.; Milanowska, K.; Knop, K.; Bielewicz, D.; Nuc, P.; Plewka, P.; Pacak, A.M.; Vazquez, F.; Karlowski, W.; Jarmolowski, A.; et al. *Arabidopsis* microRNA expression regulation in a wide range of abiotic stress responses. *Front. Plant Sci.* **2015**, *6*, 410. [CrossRef] [PubMed]
42. Zhong, S.H.; Liu, J.Z.; Jin, H.; Lin, L.; Li, Q.; Chen, Y.; Yuan, Y.X.; Wang, Z.Y.; Huang, H.; Qi, Y.J.; et al. Warm temperatures induce transgenerational epigenetic release of RNA silencing by inhibiting siRNA biogenesis in *Arabidopsis*. *Proc. Natl. Acad. Sci. USA* **2013**, *110*, 9171–9176. [CrossRef] [PubMed]
43. Zhao, Y.; Lin, S.; Qiu, Z.; Cao, D.; Wen, J.; Deng, X.; Wang, X.; Lin, J.; Li, X. MicroRNA857 is involved in the regulation of secondary growth of vascular tissues in *Arabidopsis*. *Plant Physiol.* **2015**, *169*, 2539–2552. [CrossRef] [PubMed]

44. Cui, L.G.; Shan, J.X.; Shi, M.; Gao, J.P.; Lin, H.X. The *miR156-SPL9-DFR* pathway coordinates the relationship between development and abiotic stress tolerance in plants. *Plant J.* **2014**, *80*, 1108–1117. [CrossRef] [PubMed]
45. Zhao, B.; Ge, L.; Liang, R.; Li, W.; Ruan, K.; Lin, H.; Jin, Y. Members of the miR-169 family are induced by high salinity and transiently inhibit the NF-YA transcription factor. *BMC Mol. Biol.* **2009**, *10*, 29. [CrossRef] [PubMed]
46. Yin, Z.; Li, Y.; Yu, J.; Liu, Y.; Li, C.; Han, X.; Shen, F. Difference in miRNA expression profiles between two cotton cultivars with distinct salt sensitivity. *Mol. Biol. Rep.* **2012**, *39*, 4961–4970. [CrossRef] [PubMed]
47. Hsieh, L.C.; Lin, S.I.; Shih, A.C.; Chen, J.W.; Lin, W.Y.; Tseng, C.Y.; Li, W.H.; Chiou, T.J. Uncovering small RNA-mediated responses to phosphate deficiency in *Arabidopsis* by deep sequencing. *Plant Physiol.* **2009**, *151*, 2120–2132. [CrossRef] [PubMed]
48. Baek, D.; Chun, H.J.; Kang, S.; Shin, G.; Park, S.J.; Hong, H.; Kim, C.; Kim, D.H.; Lee, S.Y.; Kim, M.C.; et al. A role for *Arabidopsis miR399f* in salt, drought, and ABA signaling. *Mol. Cell* **2015**, *39*, 111–118.
49. Bari, R.; Pant, B.D.; Stitt, M.; Scheible, W.-R. PHO_2, microRNA399, and PHR1 define a phosphate-signaling pathway in plants. *Plant Physiol.* **2006**, *141*, 988–999. [CrossRef] [PubMed]
50. Kawa, D.; Julkowska, M.M.; Sommerfield, H.M.; Ter Horst, A.; Haring, M.A.; Testerink, C. Phosphate-dependent root system architecture responses to salt stress. *Plant Physiol.* **2016**, *172*, 690–706. [CrossRef] [PubMed]
51. Pant, B.D.; Musialak-Lange, M.; Nuc, P.; May, P.; Buhtz, A.; Kehr, J.; Walther, D.; Scheible, W.R. Identification of nutrient-responsive *Arabidopsis* and rapeseed microRNAs by comprehensive real-time polymerase chain reaction profiling and small RNA sequencing. *Plant Physiol.* **2009**, *150*, 1541–1555. [CrossRef] [PubMed]
52. Wang, L.; Zeng, J.H.Q.; Song, J.; Feng, S.J.; Yang, Z.M. miRNA778 and *SUVH6* are involved in phosphate homeostasis in *Arabidopsis*. *Plant Sci.* **2015**, *238*, 273–285. [CrossRef] [PubMed]

9

Circular RNA circHIPK3 Promotes the Proliferation and Differentiation of Chicken Myoblast Cells by Sponging miR-30a-3p

Biao Chen [1,2,3], Jiao Yu [1,2,3], Lijin Guo [1,2,3], Mary Shannon Byers [4], Zhijun Wang [1,2,3], Xiaolan Chen [1,2,3], Haiping Xu [1,2,3] and Qinghua Nie [1,2,3,*]

1 Department of Animal Genetics, Breeding and Reproduction, College of Animal Science, South China Agricultural University, Guangzhou 510642, China; biaochen@stu.scau.edu.cn (B.C.); 13539763630@163.com (J.Y.); guolijin2016@163.com (L.G.); zhijunwang@stu.scau.edu.cn (Z.W.); xiaolanchen@stu.scau.edu.cn (X.C.); music-three@163.com (H.X.)

2 National-Local Joint Engineering Research Center for Livestock Breeding, Guangzhou 510642, China

3 Guangdong Provincial Key Lab of Agro-Animal Genomics and Molecular Breeding, and Key Laboratory of Chicken Genetics, Breeding and Reproduction, Ministry of Agriculture, Guangzhou 510642, China

4 Department of Biological Sciences, College of Life and Physical Sciences, Tennessee State University, Nashville, TN 37209, USA; maryshannonbyers@yahoo.com

* Correspondence: nqinghua@scau.edu.cn;

Abstract: Circular RNAs and microRNAs widely exist in various species and play crucial roles in multiple biological processes. It is essential to study their roles in myogenesis. In our previous sequencing data, both miR-30a-3p and circular HIPK3 (circHIPK3) RNA, which are produced by the third exon of the *HIPK3* gene, were differentially expressed among chicken skeletal muscles at 11 embryo age (E11), 16 embryo age (E16), and 1-day post-hatch (P1). Here, we investigated their potential roles in myogenesis. Proliferation experiment showed that miR-30a-3p could inhibit the proliferation of myoblast. Through dual-luciferase assay and Myosin heavy chain (MYHC) immunofluorescence, we found that miR-30a-3p could inhibit the differentiation of myoblast by binding to Myocyte Enhancer Factor 2 C (*MEF2C*), which could promote the differentiation of myoblast. Then, we found that circHIPK3 could act as a sponge of miR-30a-3p and exerted a counteractive effect of miR-30a-3p by promoting the proliferation and differentiation of myoblasts. Taking together, our data suggested that circHIPK3 could promote the chicken embryonic skeletal muscle development by sponging miR-30a-3p.

Keywords: circular RNA; circHIPK3; microRNA; miR-30a-3p; skeletal muscle; proliferation; differentiation

1. Introduction

Skeletal muscles are important components in animals. Chicken skeletal muscle, which can provide high quality protein, is one of the most important meat source for humans. The development of skeletal muscle is regulated by multiple factors, including genetics, nutrition, disease, environment and so on [1,2]. Heritability estimates showed that chicken growth could be enhanced by genetic improvement [3]. The genetic factors which control skeletal muscle development include genes and non-coding RNAs.

MicroRNAs (miRNAs) have been shown to be involved in many biological processes, including muscle development [4]. Some myogenic miRNAs, including the miR-1 family, miR-206 and miR-133 family, regulate muscle development by targeting and inhibiting the expression of muscle-related gene [5,6]. Previous studies in our group showed that miR-203, miR-16, miR-29, and miR-1611 all played crucial roles in myoblast proliferation and differentiation [7–10]. Two other studies found that miR-30a-3p could suppress tumor growth [11,12]. However, the molecular function of chicken miR-30a-3p has not yet been reported.

Circular RNAs, which widely exist in the transcriptomes of different species and tissues, were previously considered as a kind of non-coding RNA, but they have now been demonstrated to have both coding and regulating functions [13–15]. Circular RNA, formed by the covalently joined 5′ and 3′ ends of linear RNA, possess a more stable structure than linear RNA. The functions of circular RNA include, acting as a miRNA sponge, participating in regulating the expression of its own linear RNA in different ways, coding protein, and deriving pseudogenes [16,17]. Previous studies found that circular RNAs were abundantly expressed in skeletal muscle tissue in many species [14,18,19]. Circular RNAs in chicken skeletal muscle could act as an miRNA sponge and regulate chicken muscle development [19]. A circular RNA, produced by *SVIL* could promote the proliferation and differentiation of myoblast cells by sponging miR-203 [20]. Circular RNA circFGFR2, generated by the *FGFR2* gene, could interact with miR-133a-5p and miR-29b-1-5p to regulate myoblast cells development [7].

A circular RNA produced by the third exon of the chicken *HIPK3* gene (circHIPK3—01, we named it as circHIPK3, hereinafter) has the highest expression level compare to other circular RNAs generated from *HIPK3* gene. It was also differentially expressed in different stages of skeletal development. We predicted it has three potential binding sites for miR-30a-3p. In this study, we aimed to examine the interaction of circHIPK3 and miR-30a-3p and their functions on myoblast proliferation and differentiation.

2. Materials and Methods

2.1. Ethics Statement

All animal experiments performed in this study met the requirements of the Institutional Animal Care and Use Committee at the South China Agricultural University (approval ID: SCAU#0014). All efforts were made to minimize the suffering of animals.

2.2. Primers

All primers that were used in this study were synthesized by Sangon (Sangon Biotech, Shanghai, China). The primers listed in Table 1 were designed by Premier Primer 5.0 software (Premier Bio-soft International, Palo Alto, CA, USA). Information of the qRT-PCR primers for *MYOD, MYOG, MYHC* and *GAPDH* were shown in our previous study [21].

2.3. Cell Culture and Transfection

Chicken primary myoblasts (CPMs) were isolated from the leg muscle of 10-day Yuhe chicken embryos (E10; Zhuhai Yuhe Company Ltd., Zhuhai, China), as described in our previous study [7]. Briefly, the legs of E10 chickens were collected, and the skin and bones were removed. Then, leg muscles were minced with scissors and trypsinized (Gibco, Grand Island, NY, USA) at 37 °C for 20 min). Digestion was done with complete 1640 medium-(RPMI), containing 20% fetal bovine serum (FBS), 1% nonessential amino acids, and 0.2% penicillin/streptomycin (Invitrogen, Carlsbad, CA, USA). The mixture was filtered and centrifuged at 500 g for 5 min. Following the serial plating, the cells were cultured in complete medium and incubated at 37 °C, in a 5% CO_2 humidified atmosphere. Chicken fibroblast DF-1 cells were cultured in Dulbecco's modified Eagle medium (DMEM) (Gibco, Grand Island, NY, USA) supplemented with 10% FBS and 0.2% penicillin/streptomycin (Invitrogen, Carlsbad, CA, USA), then incubated with 5% CO_2 at 37 °C humidity. DNA plasmids, miRNA mimic, mimic negative control (mimic NC), miRNA inhibitors, inhibitor negative control (inhibitor NC), small interfering RNA (siRNA), and siRNA negative control (siRNA NC) were transiently transfected into cells using Lipofectamine 3000 reagent (Invitrogen, Carlsbad, CA, USA).

2.4. RNA Exaction, cDNA Synthesis and Quantitative Real-Time PCR (qRT-PCR)

All RNAs were exacted using Trizol reagent (TaKaRa, Otsu, Japan) according to the manufacturer's instructions. The quality and concentrations of the RNA samples were detected by 1.5% agarose gel electrophoresis. Total RNA was employed to synthesize cDNA, using a Primescript RT Reagent Kit with gDNA Eraser (Perfect Real Time) (TaKaRa, Otsu, Japan). Synthesized cDNA libraries were diluted with RNase-free water at a ratio of 1:3 before real-time PCR. Relative mRNA expression levels were detected by qRT-PCR using SsoFast Eva Green Supermix (Bio-Rad, Hercules, CA, USA). *GAPDH* was used as an internal control. Reverse transcription for miRNA was conducted using ReverTra Ace qPCR RT Kit (Toyobo, Osaka, Japan). The specific bulge-loop miRNA qRT-PCR primer for miR-30a-3p and U6 were designed by RiboBio (RiboBio, Guangzhou, China). All qRT-PCR reactions were conducted with a CFX96 system (Bio-Rad, Hercules, CA, USA). All reactions were run in triplicates and fold expression changes were calculated using the comparative $2^{-\Delta\Delta Ct}$ method.

2.5. Validation of circHIPK3

Based on the NCBI reference sequences of *HIPK3* (NCBI accession number: NM_001199411.1), convergent and divergent primers were designed to validate the existence of circHIPK3. To confirm the cirHIPK3 junction, genomic DNA, and cDNA from CPMs were used for PCR reaction. All PCR products were sequenced by Sangon Biotech Co Ltd. Sequence analysis was conducted using DNASTAR software (DNASTAR 7.1, http://www.dnastar.com). For RNase R treatment, 2 mg of total RNA was incubated 20 min at 37 °C with RNase R (Epicentre Technologies, Madison, WI, USA), and employed to synthesize cDNA for qPCR. For the control group, the same amount of RNA was incubated 20 min at 37 °C and subsequently used to synthesize cDNA.

2.6. Plasmids Construction and RNA Oligonucleotides

For the construction of the circHIPK3 over-expression vector, exon 3 of *HIPK3* was amplified using cDNA, produced from CPMs and cloned into a pCD5ciR vector (Geneseed Biotech, Guangzhou, China) between *EcoRI* and *BamHI* restriction sites. The siRNAs to circHIPK3, which especially target the circHIPK3 rather than the linear HIPK3, were designed and synthesized by Geneseed using the sequence shown in Table 1. The gga-miR-30a-3p mimic, mimic NC, the gga-miR-30a-3p inhibitor and inhibitor NC were synthesized by RiboBio (Guangzhou, China). For the construction of pmirGLO Dual-Luciferase reporter vector, wild-type and mutated sequences in the 3′UTR region of *MEF2C* and the partial region of circHIPK3, which include the predicted binding sites of miR-30a-3p, were synthesized and inserted into pmirGLO vectors (Promega, Madison, WI, USA), according to instructions, using *NheI* and *XhoI* restriction sites. The gga-miR-30a sequence was also synthesized and inserted into pmirGLO vectors.

2.7. 5-Ethynyl-2′-Deoxyuridine (EdU) Assay

After 48 h of transfection, the treated CPMs and negative control groups in 24-well plates were incubated with 50 μM 5-ethynyl-20-deoxyuridine (RiboBio, Guangzhou, China) for 2 h at 37 °C. After washing twice, the cells were stained with C10310 EdU Apollo. EdU-stained cells were counted using a Leica DMi8 fluorescent microscope (400× magnification) (Leica, Wetzlar, Germany). The ratio of EDU-stained cells to Hoechst 33342-stained cells was calculated and represented the CPM proliferation rate. Detailed protocols were described in the manufacturer's instruction.

2.8. Flow Cytometry of the Cell Cycle

After 48 h of transient transfection with the over-expression plasmid (blank vector) and siRNA (siRNA NC), CPMs were collected from the 12-well plates and kept overnight in 70% ethanol at −20 °C. The cells were then incubated with 50 μg/mL PI (propidium iodide) (Sigma, Louis, MO, USA), 10 μg/mL RNase A (Takara, Otsu, Japan) and 0.2% (*v*/*v*) Triton X-100 (Sigma, Louis, MO, USA) at 4 °C for 30 min. Lastly, cells were detected with a BD AccuriC6 flow cytometer (BD Biosciences, San Jose, CA, USA), and the results were analyzed by FlowJo7.6 software.

2.9. Cell Counting Kit 8 (CCK-8) Assay

CPMs were seeded in a 48-well plate and cultured in complete medium. After transfection, cell proliferation was detected at 12, 24, 36, and 48 h using the TransDetect CCK Kit (TransGen Biotech, Beijing, China), following the manufacturer's protocol. Cells were added in 25 uL CCK solution to each well and incubated for 2 h at 37 °C in a 5% CO2 cell incubator. Then absorbance of treated and control groups were measured with a Fluorescence/Multi-Detection Microplate Reader (BioTek, Winooski, VT, USA) by optical density at a wavelength of 450 nm.

2.10. Immunofluorescence

For immunofluorescence, after transfection, cells in 12-well plates were fixed for 30 min with 4% formaldehyde. Cells were then permeabilized by adding 0.1% Triton X-100 for 5 min and blocked for 30 min with goat serum. Following overnight incubation at 4 °C with anti-MYHC (B103; DHSB, Iowa City, IA, USA; 0.5 μg/mL), Fluorescein (FITC)-conjugated AffiniPure Goat Anti-Mouse IgG (H + L) (Bioworld, Minneapolis, MN, USA; 1:200) was added to the plate and incubated at room temperature for 1 h. Cell nuclei were stained with DAPI (1:50, Beyotime, Shanghai, China) for 5 min. The images were captured with fluorescence microscopy (Leica, Wetzlar, Germany). The area of cells labeled with anti-MYHC was measured using Photoshop software (Adobe Photoshop CC 2018, Adobe, San Jose, CA, USA), and the total myotube area was calculated as a percentage of the total image area covered by myotubes.

2.11. Binding Relationship Prediction and Dual-Luciferase Reporter Assay

To predict the relationship between target genes and miR-30a-3p, miRDB (http://mirdb.org/miRDB/) and RNAhybrid (http://bibiserv2.cebitec.uni-bielefeld.de/rnahybrid) were employed. After seeding DF-1 cells in the 96-well plate and culturing for 24 h, four groups (wild type pmirGLO plasmids and mimic as the treatment, mutated pmirGLO plasmids and mimic, wild type pmirGLO plasmids and mimic NC, mutated type pmirGLO plasmids and mimic NC) were set and co-transfected. For the confirmation of the target relationship between circHIPK3 and miR-30a-3p, another method of Dual-Luciferase reporter assay was employed. Three groups (circHIPK3 over-expression plasmid and miR-30a-3p mimic, pCD5ciR and miR-30a-3p mimic, pCD5ciR and mimic NC) were set and co-transfected with a pmirGLO vector containing the miR-30a sequence. After 48 h, Dual-GLO Luciferase Assay System kit (Promega, Madison, WI, USA) was employed to detect luminescent signals of firefly and Renilla Luciferase with a Fluorescence/Multi-Detection Microplate Reader (BioTek, Winooski, VT, USA). Firefly luciferase activities were normalized to Renilla luminescence in each well. Detailed protocols were described in the manufacturer's instruction.

2.12. Western Blotting

Briefly, cells were lysed in the radio immune precipitation assay (RIPA) buffer (Beyotime, Shanghai, China) containing phenylmethane sulfonyl fluoride (PMSF) protease inhibitor (Beyotime, Shanghai, China). After incubation on ice for 30 min, the samples were centrifuged at 10,000 g for 10 min at 4 °C, and the supernatant was collected. Proteins were separated by SDS-PAGE and blotted onto nitrocellulose membranes (Whatman, Maidstone, UK), then membranes were probed with primary and secondary antibodies. The primary antibodies used were anti-MYHC (1:1000, B103; DHSB, Iowa City, IA, USA), anti-GAPDH (1:1500, AB-P-R 001, Hangzhou Goodhere Biotech, Hangzhou, China), and anti-Tubulin (1:1000, Beyotime, Shanghai, China). The secondary antibodies used were goat anti-rabbit IgG-HRP (1:5000, BA1054, Boster, Wuhan, China) and peroxidase-goat anti-mouse IgG (1:2500, BA1051, Boster, Wuhan, China). Image J software (d1.47, National Institutes of Health, Bethesda, MD, USA) was used to quantify the band intensity.

2.13. Statistical Analysis

All results were presented as a mean ± SEM and were subjected to statistical analysis by two-tailed t-test. The level of significance was presented as * ($p < 0.05$), ** ($p < 0.01$) and *** ($p < 0.001$).

Table 1. Primers and RNA oligos used in this study.

Name	Nucleotide Sequences (5′→3′)	Tm. (°C)	Product Size (bp)	Application
QcircHIPK3	F: GTTTAATCCACGCTGACCTCA R: GACTTGTGAGGCCATACCTATA	61.3	130	qPCR for circHIPK3
QHIPK3	F: GGGGTATGTCCCGGAG R: CTTCGCTAATGGAACAACAC	61.3	261	qPCR for HIPK3
QMEF2C	F: AGGGTGTATGTGCAGGAACG R: AGCAATCTCGCAGTCACACA	60	288	qPCR for MEF2C
Convergent primers	F: TGGTACAAGCGGAGATGG R: TTGAGGTCAGCGTGGATTA	55	450	Amplification of partial sequence of exon 3 of HIPK3
Divergent primers	F: GCACGCCAAGGACAAATA R: TACGCTTCAATCCACATCG	58	782	Amplification of partial sequence of circHIPK3 which contain the joint site
β-actin	F: CTCCCCCATGCCATCCTCCGTCTG R: GCTGTGGCCATCTCCTGCTC	52–65	179	qPCR forβ-actin
si-circHIPK3-001	CCCGGTATTATAGGTATGG	-	-	-
si-circHIPK3-002	GGTATTATAGGTATGGCCT	-	-	-
si-circHIPK3-003	ATTATAGGTATGGCCTCAC	-	-	-

Note: The nucleotide sequences of si-circHIPK3 represent the target sequences of each siRNA.

3. Results

3.1. circHIPK3 Differentially Expressed during Embryonic Leg Muscle Development

Previous circular RNA sequencing data from our lab revealed 11 circular RNAs were generated by the *HIPK3* gene (available in the Gene Expression Omnibus with accession number GSE89355). The genomic structure of chicken *HIPK3* and the regions, in which all the circular HIPK3 (circHIPK3) RNA were derived, are shown in Figure 1A. Interestingly, circHIPK3 (referred as circHIPK3—01 in Figure 1A), which was derived from exon3 of *HIPK3*, was the only exonic circular RNA. Compared with other circular RNAs derived from *HIPK3*, circHIPK3 had the highest expression level. Its expression level in E16 was significantly higher than in E11 and P1 (Figure 1B). The expression levels of circHIPK3 and *HIPK3* mRNA in E11, E12, E16, and E18 were detected by qRT-PCR (Figure 1C). The trend of the expression level of circHIPK3 was consistent with the result from the sequencing data. However, the expression patterns of circHIPK3 and *HIPK3* mRNA were not identical, which indicated that they might have different functions during leg muscle development. To confirm the sequence and the junction of circHIPK3, genomic DNA and cDNA were used for the PCR reaction, with convergent and divergent primers. The result of the PCR product electrophoresis showed the expectants of convergent primers were amplified with both templates. However, there was no PCR product of divergent primers with the genomic DNA template (Figure 1D). PCR products of divergent primers were analyzed by Sanger sequencing (Figure 1E). Sequencing results showed that circHIPK3 was generated from the third exon of *HIPK3*. The circHIPK3 was also validated by RNase R digestion. The result of qRT-PCR showed that RNase R had no impact on circHIPK3, whereas the levels of linear RNA, HIPK3 and β-actin, were significantly decreased (Figure 1F). These results validated the existence and differential expression of circHIPK3 during skeletal muscle development of chicken.

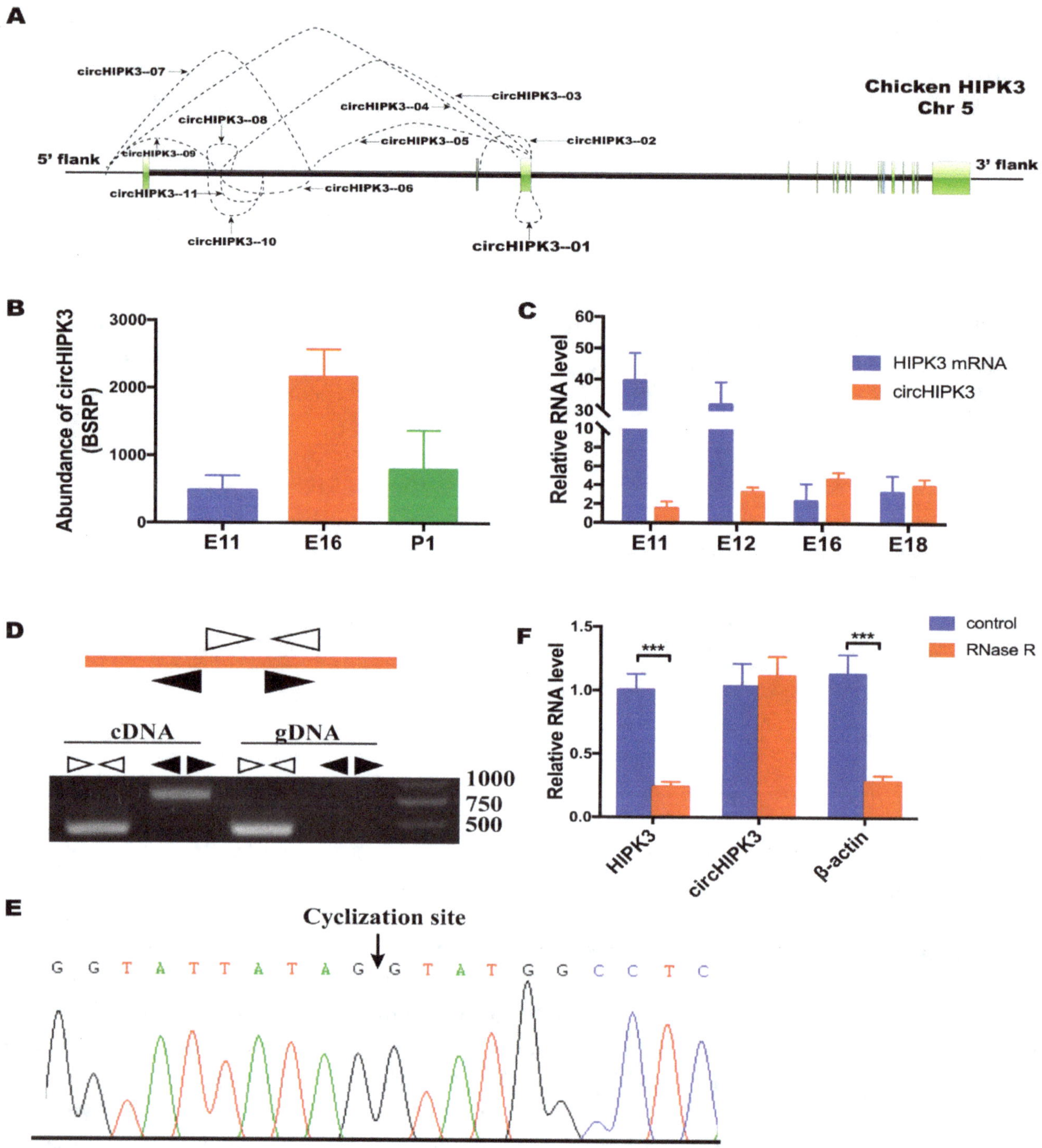

Figure 1. The differential expression and validation of circular HIPK3 (circHIPK3). (**A**) The schema of all circular RNA derived from *HIPK3*. The green rectangles represent the exons of *HIPK3*. (**B**) The RNA-Seq result showed that circHIPK3 was differentially expressed in E11, E16, and P1 of leg muscle. The expressed abundances were normalized as the number of back-spliced reads per million mapped reads (BSRP). (**C**) The expression profiles of circHIPK3 and *HIPK3* mRNA in E11, E12, E16 and E18. (**D**) Divergent primers amplified circHIPK3 in cDNA but not genomic DNA (gDNA). White triangles represent convergent primers and black triangles represent divergent primers. (**E**) Sanger sequencing confirmed the junction sequence of circHIPK3. (**F**) Quantitative real-time PCR (qRT-PCR) showed the resistance of circHIPK3 to RNase R digestion. In all panels, values represent mean ± SEM from three independent experiments. * $p < 0.05$; ** $p < 0.01$; *** $p < 0.001$.

3.2. *circHIPK3 Interacts with miR-30a-3p*

Many studies showed that circular RNAs exerted their functions by acting as the miRNA sponge. CircHIPK3 was predicted by miRDB and RNAhybrid to be a target of multiple miRNAs. Among these miRNAs, ggs-miR-30a-3p was chosen as a candidate because there were three potential binding sites in circHIPK3 (Figure 2A). The seed sequence of miR-30a-3p matched with three sites in circHIPK3 (Figure 2B). Besides, the prediction results from RNAhybrid indicated that the binding site 2 was the most stable format (Figure 2C). To identify the interactions between circHIPK3 and miR-30a-3p, the over-expression vector of circHIPK3 was constructed and transfected into DF-1 cells. The expression efficiency of the over-expression vector was detected by qPCR. Compared with the group transfected with pCD5ciR, the circHIPK3 over-expression vector expressed a higher level of circHIPK3 (Figure 2D). Then, circHIPK3 over-expression vector and miR-30a-3p mimic were co-transfected into DF-1 cells with a pmirGLO vector, containing the miR-30a sequence. Meanwhile, as the control group, the pCD5ciR and miR-30a-3p mimic (or mimic NC) were co-transfected with the pmirGLO vector, containing the miR-30a sequence. The results showed that the relative luminescence activity of the group with circHIPK3 over-expression vector and miR-30a-3p mimic was significantly higher than the group with pCD5ciR and miR-30a-3p mimic, but had no difference compared with the group pCD5ciR and mimic NC (Figure 2E). These results suggest that circHIPK3 could bind with miR-30a-3p mimic. In addition, sequences which contained binding site 2 or the mutated sequence were inserted into pmirGLO vector. Recombinant vectors with the wild type sequence was then co-transfected into DF-1 cells with miR-30a-3p mimic, meantime, three control groups were set (pmirGLO vector with mutated sequence and miR-30a-3p mimic, pmirGLO vector with wild type sequence and mimic NC, pmirGLO vector with mutated sequence and mimic NC). The results showed that the relative luminescence activity of the group with a wild-type plasmid and mimic was significantly decreased compared to the group transfected with mutated plasmid and mimic, and the group with wild type plasmids and mimic NC (Figure 2F). Moreover, the RNA level of circHIPK3 was significantly down-regulated after over-expression of miR-30a-3p mimic, compared to the group transfected with mimic NC (Figure 2G). Subsequently, the result of flow cytometry analysis showed that miR-30a-3p could reverse the effect of circHIPK3 on a cell cycle (Figure 2H). Altogether, these results indicated that miR-30a-3p could interact with circHIPK3.

3.3. *miR-30a-3p Inhibits Myoblast Proliferation*

To explore the function of miR-30a-3p on the proliferation of CPMs, miR-30a-3p mimic and inhibitor were transfected into CPMs with 100 nM to detect an over-expression effect and an inhibitory effect. The results showed that the two oligos of miR-30a-3p had the effect as expected compared with the mimic NC group, and inhibitor NC group, respectively, and could be used in the subsequent experiments (Figure 3A,B). After being transfected with miR-30a-3p mimic/mimic NC and miR-30a-3p inhibitor/inhibitor NC, flow cytometry analysis was performed in CPMs and the results showed that ectopic expression of miR-30a-3p suppressed the cell cycle markedly, while knock-down of miR-30a-3p significantly promoted the cell cycle (Figure 3C,D). Besides, CCK-8 assay was conducted to detect of proliferation vitality in CPMs. The results showed that the group which transfected with miR-30a-3p mimic had a lower proliferation vitality than mimic NC. In contrast, the group which transfected with miR-30a-3p inhibitor had a higher proliferation vitality than inhibitor NC group (Figure 3E,F). Furthermore, the EdU assay demonstrated that the rate of the cells, which were in the cell division in the ectopic expression group, was significantly less than in the mimic NC group, and the statistics of the cell proliferation rate of the miR-30a-3p over-expression group, were markedly lower than the control group (Figure 3G). Conversely, knock-down of miR-30a-3p dramatically increased the numbers of EdU strained cells compare with the inhibitor NC group (Figure 3H). Altogether, these results indicated that miR-30a-3p could suppress myoblast proliferation.

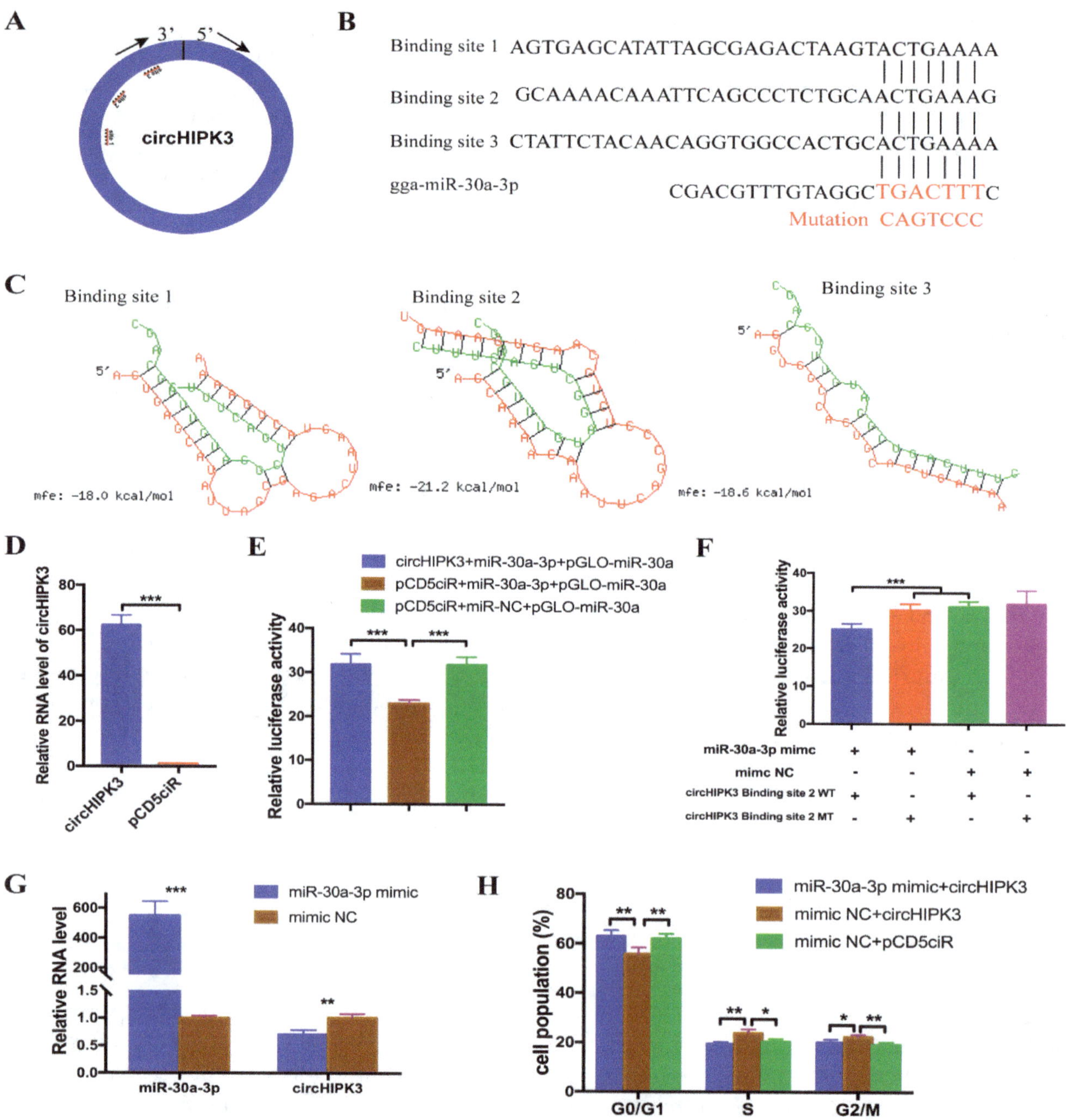

Figure 2. CircHIPK3 interacts with miR-30a-3p. (**A**) A schematic illustration showing the putative binding sites of miR-30a-3p on circHIPK3. (**B**) The potential binding site sequence of miR-30a-3p on circHIPK3. The seed sequences and mutant sequences were highlighted in red. (**C**) The potential interaction model between miR-30a-3p and circHIPK3 from RNAhybrid. (**D**) The expression efficiency of circHIPK3 over-expression vector in DF-1 cells. (**E**) Luminescence was measured after co-transfected with the luciferase reporter and miR-30a-3p mimic (or mimic NC) and circHIPK3 over-expression vector (or pCD5ciR). The relative levels of firefly luminescence normalized to Renilla luminescence are plotted. (n = 6). (**F**) Luminescence was measured after co-transfecting wild type or mutant linear sequence of circHIPK3 with miR-30a-3p mimic (or mimic NC) in DF-1 cells. (n = 6). (**G**) The RNA levels of miR-30a-3p and circHIPK3 from miR-30a-3p mimic transfected DF-1 cells. (**H**) The effect of co-transfected with miR-30a-3p mimic (or mimic NC) and circHIPK3 over-expression vector (or pCD5ciR) on cell-cycle progression of DF-1 cells. The plot of cell-cycle analysis in different cell-cycle phases was compared. In all panels, values represent mean ± SEM from three independent experiments. * $p < 0.05$; ** $p < 0.01$; *** $p < 0.001$.

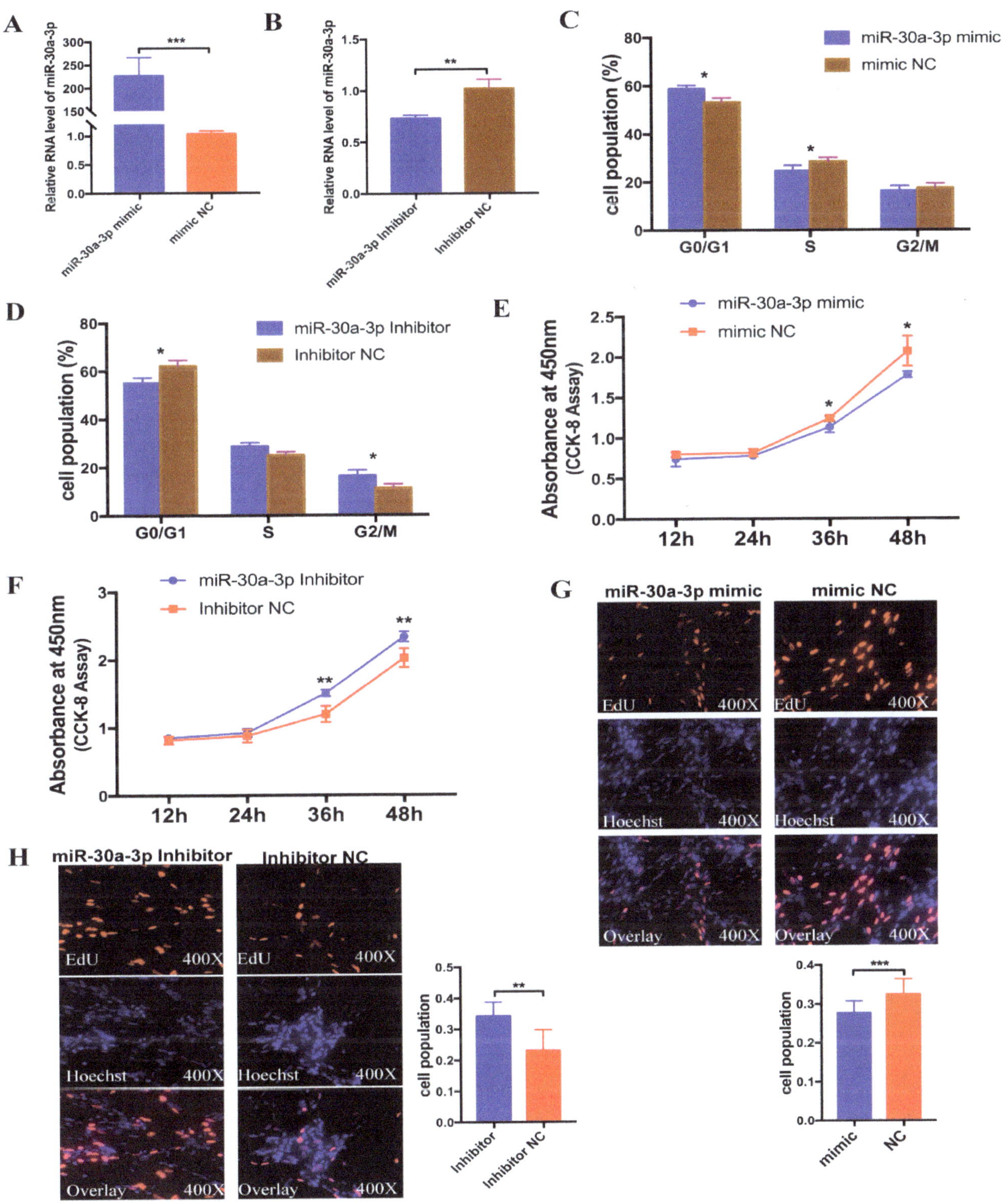

Figure 3. miR-30a-3p inhibits myoblast proliferation. (**A**,**B**) The over-expression and inhibitory effects of miR-30a-3p mimic and inhibitor in CPMs. (**C**,**D**) Effect of miR-30a-3p mimic and inhibitor on cell-cycle progression of chicken primary myoblasts (CPMs). The plot of cell-cycle analysis in different cell-cycle phases was compared. (**E**,**F**) The growth curves of CPMs were measured after the transfection of miR-30a-3p mimic and inhibitor. (**G**,**H**) 5-Ethynyl-2′-Deoxyuridine (EdU) assays for CPMs with over-expression and inhibition of miR-30a-3p. In all panels, values represent mean ± SEM from three independent experiments. * $p < 0.05$; ** $p < 0.01$; *** $p < 0.001$.

3.4. *MEF2C* Is a Target Gene of miR-30a-3p

To investigate the potential function of miR-30a-3p on CPM differentiation, we try to find the differentiation-related genes among the targets of miR-30a-3p. Interestingly, there are three potential binding sites in the 3′UTR region of *MEF2C* (Figure 4A). The potential interaction model from RNAhybrid showed that the binding site 3 was the most stable format (Figure 4B). To validate the target relationship between *MEF2C* and miR-30a-3p, wild-type and mutated-type sequences containing three binding sites, separately, were inserted into the pmirGLO vector for construction of pmirGLO dual-luciferase miRNA target expression vector. After co-transfection of vectors and mimics (or mimic NC), relative luminescence activities were detected by a Fluorescence/Multi-Detection Microplate Reader. For the Binding site 1, the relative luminescence activity of the group, with wild type reporter and mimic, was significantly lower compared with the control groups, which transfected with mutated reporters and mimic, wild type reporter and mimic NC, separately (Figure 4C). For the Binding site 2 and Binding site 3, the luminescence activity of the groups with wild type plasmids and mimic were all dramatically lower than the three control groups (mutated reporters and mimic, wild type reporter and mimic NC, mutated reporters and mimic NC) (Figure 4D,E). Particularly, the binding site 3 exerted the most significant interaction with miR-30a-3p compare with the control groups. The RNA expression level of *MEF2C* was significantly decreased after ectopic expression of miR-30a-3p mimic compared with the group transfected with mimic NC (Figure 4F).

3.5. miR-30a-3p Represses CPM Differentiation

After the confirmation of the target relationship between miR-30a-3p and *MEF2C*, we try to investigate the potential role of miR-30a-3p on CPM differentiation. First, the expression profile of miR-30a-3p was detected in the process of differentiation in CPMs. Interestingly, the expression level of miR-30a-3p was decreased in the first two days of differentiation medium (DM) compare to GM (growth medium), then increased in DM3 and DM4 (Figure 5A). Then, the expression of the myoblast differentiation marker genes, including *MYOD*, *MYOG*, and *MYHC* were evaluated by qPCR in myoblast transfected with miR-30a-3p mimic (or mimic NC) and inhibitor (or inhibitor NC). Over-expression of miR-30a-3p notably inhibited the expression of *MYOD*, *MYOG*, and *MYHC* compared with the groups transfected with mimic NC. Conversely, knock-down of miR-30a-3p promoted the expression of *MYOD*, *MYOG* and *MYHC* relative to the inhibitor NC group (Figure 5B,C). The protein level of *MYHC* was also detected by western blotting. Ectopic expression of miR-30a-3p inhibited the expression *MYHC*, conversely, knock-down of miR-30a-3p promoted the protein expression of *MYHC* (Figure 5D). MYHC immunofluorescence staining was employed on those transfected differentiated myoblasts at DM5. The results showed that the total area of myotubes of miR-30a-3p mimic transfected group was markedly less than that of the group transfected with mimic NC (Figure 5E). On the contrary, the areas of myotubes in miR-30a-3p inhibitor transfected group was more than that of control group (Figure 5F). To sum up, these results revealed that miR-30a-3p could repress CPM differentiation.

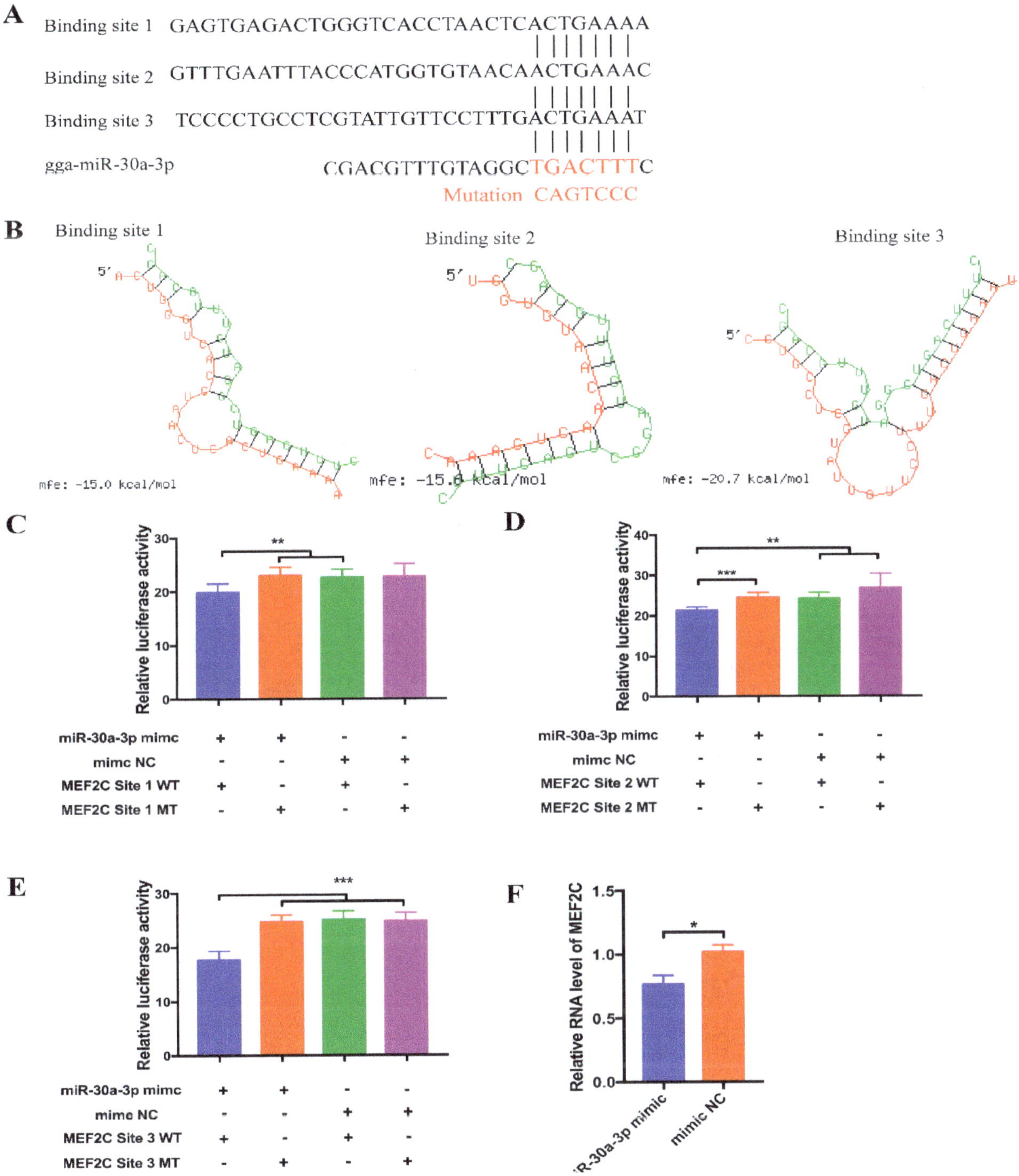

Figure 4. *MEF2C* is a target gene of miR-30a-3p. (**A**) The potential binding site sequence of miR-30a-3p on MEF2C. The seed sequences and mutant sequences were highlighted in red. (**B**) The potential interaction model between miR-30a-3p and MEF2C from RNAhybrid. (**C–E**) Luminescence was measured after co-transfecting wild type or mutant sequence of *MEF2C* with miR-30a-3p mimic (or mimic NC) in DF-1 cells. (n = 6). (**F**) The RNA level of *MEF2C* from miR-30a-3p mimic transfected CPMs. In all panels, values represent mean ± SEM from three independent experiments. * $p < 0.05$; ** $p < 0.01$; *** $p < 0.001$.

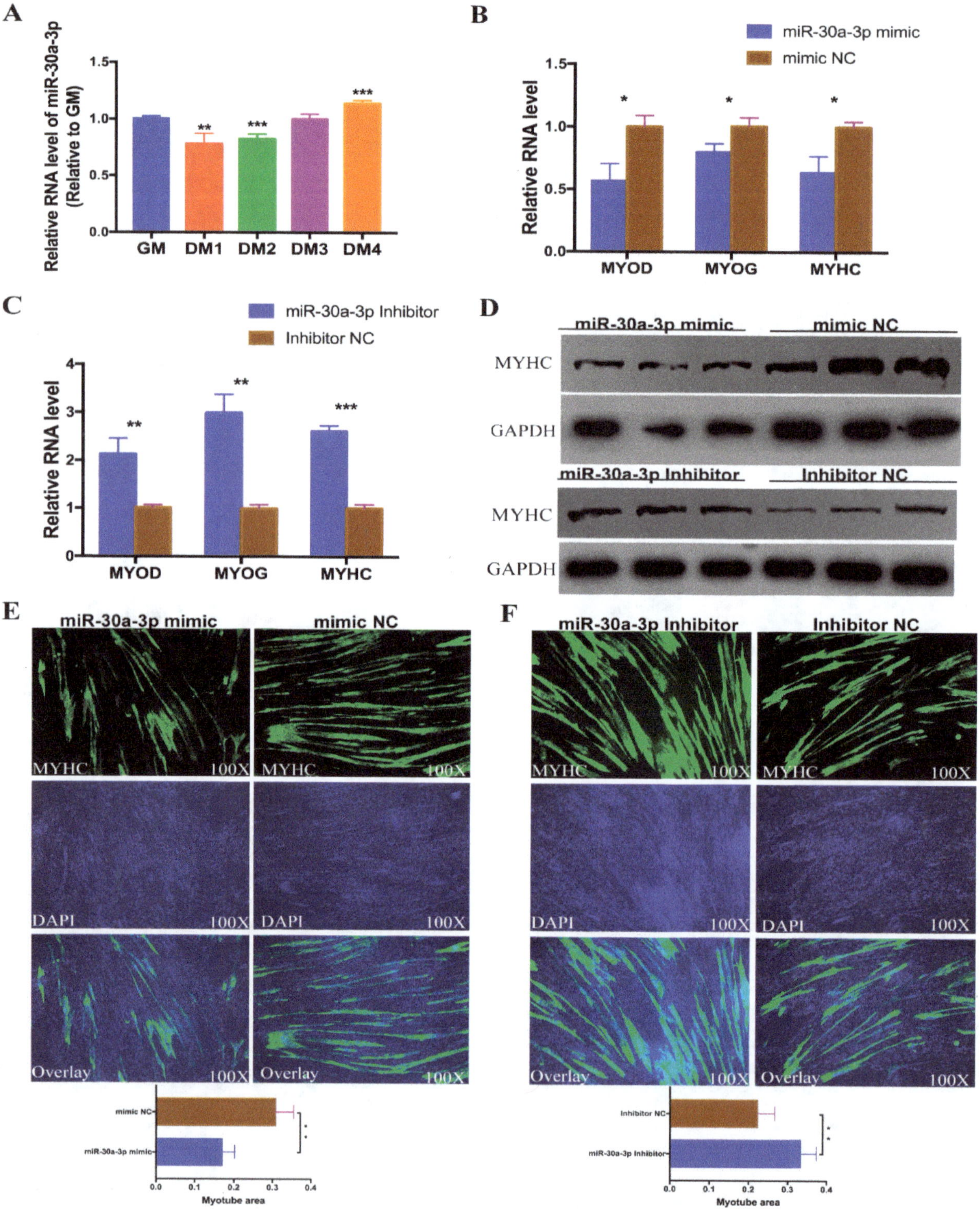

Figure 5. miR-30a-3p represses CPM differentiation. (**A**) The expression profile of miR-30a-3p in the process of CPMs induced differentiation. (**B**) The expression of *MYOD*, *MYOG* and *MYHC* in CPMs after over-expression of miR-30a-3p. (**C**) The expression of *MYOD*, *MYOG* and *MYHC* in CPMs after knock-down of miR-30a-3p. (**D**) The protein level of *MYHC* in CPMs after over-expression and knock-down of miR-30a-3p. (**E**) Immunofluorescence analysis of MYHC-staining cells after over-expression miR-30a-3p in CPMs. (**F**) Immunofluorescence analysis of MYHC-staining cells after knock-down of miR-30a-3p in CPMs. * $p < 0.05$; ** $p < 0.01$; *** $p < 0.001$.

3.6. circHIPK3 Promotes the Proliferation of CPMs

Given that circHIPK3 could act as a sponge of miR-30a-3p, and miR-30a-3p could repress the proliferation of CPMs, we hypothesized that circHIPK3 played an opposite role on the proliferation of CPMs. To investigate the role of circHIPK3 on skeletal muscle cell proliferation, the over-expression vector and siRNAs were transfected into CPMs. The over-expression vectors expressed high levels of circHIPK3 compared with the group transfected with pCD5ciR (Figure 6A), and all three siRNAs could significantly knock down the level of circHIPK3 relative to siRNA NC group (Figure 6B). Among the three siRNAs, si-circHIPK3-003 had the highest efficiency of interference effect and was chosen for the following experiments. After being transfected with circHIPK3 over-expression vector/pCD5ciR and si-circHIPK3-003/siRNA NC, flow cytometry analysis in CPMs was performed and the results showed that ectopic expression of circHIPK3 notably promoted the cell cycle, conversely, knock-down of circHIPK3 significantly retarded the cell cycle (Figure 6C,D). Moreover, CCK-8 assay was used to detect the proliferation vitality in CPMs. The results showed that the groups which transfected with circHIPK3 over-expression vectors, had higher proliferation vitality than the negative control group. In contrast, the groups which transfected with si-circHIPK3-003, had lower proliferation vitality than siRNA NC group (Figure 6E,F). The EdU assay indicated that the rate of the cells which were in the cell division in circHIPK3 over-expression group was significantly more than pCD5ciR group, the statistics of the cell proliferation rate of the circHIPK3 over-expression group were markedly higher (Figure 6G). Conversely, knock-down of circHIPK3 decreased the numbers of EdU strained cells dramatically (Figure 6H). In summary, these results demonstrated that circHIPK3 promoted the proliferation of CPMs.

3.7. circHIPK3 Promotes the Differentiation of CPMs

As the miR-30a-3p had the effect on CPM differentiation, and miR-30a-3p could interact with circHIPK3, we speculated that circHIPK3 might have the opposite effect on CPM differentiation. To confirm our hypothesis, the expression profile of circHIPK3 was detected in the process of differentiation in CPMs. Interestingly, the expression trend of circHIPK3 coincided with miR-30a-3p, and it dramatically decreased in the DM phase compare to GM but raised gradually in DM4 (Figure 7A), indicating that circHIPK3 may be related to the differentiation of CPMs. Then, the expression level of the myoblast differentiation marker genes including *MYOD*, *MYOG* and *MYHC* were evaluated by qPCR in CPMs transfected with circHIPK3 over-expression vector/pCD5ciR and siRNA/siRNA NC. The expression levels of *MYOG* and *MYHC* but not *MYOD* were significantly increased in CPMs transfected with the over-expression vector compare with the group transfected with pCD5ciR (Figure 7B). On the contrary, knock-down of circHIPK3 significantly inhibited the expression of *MYOD*, *MYOG* and *MYHC* compare with the siRNA NC group (Figure 7C). The relative protein level of *MYHC* was also decreased after knock-down of circHIPK3 (Figure 7D). Moreover, MYHC immunofluorescence staining was conducted on transfected differentiated CPMs at DM5. The results showed that the total area of myotubes of si-circHIPK3-003 transfected group was notably less than that of siRNA NC group, and the statistics of the myotube area rate of the si-circHIPK3-003 group were markedly lower than that of siRNA NC group (Figure 7E). These results suggested that circHIPK3 had counteractive effect of miR-30a-3p on CPM differentiation and knock-down of circHIPK3 suppressed the differentiation of CPMs.

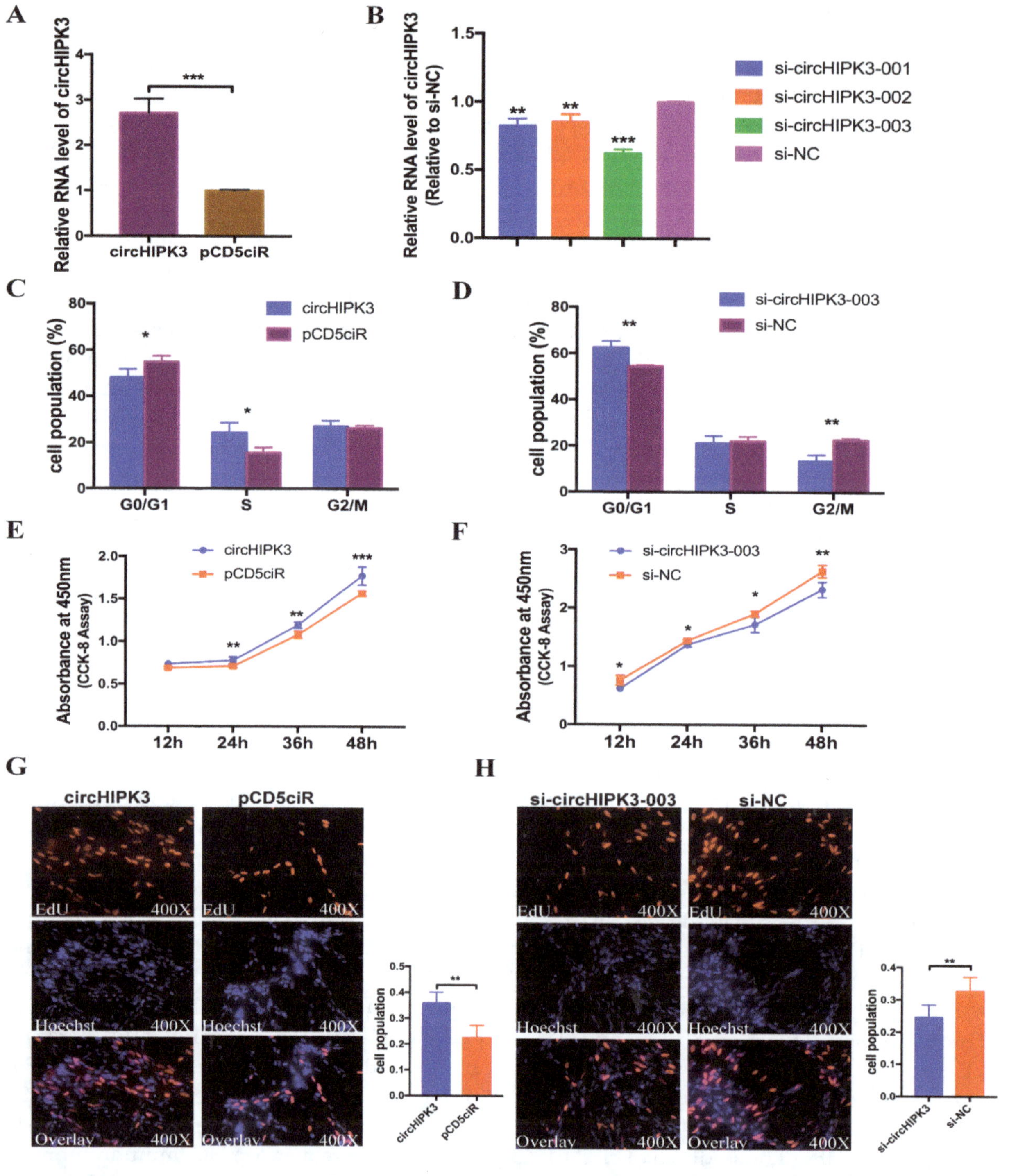

Figure 6. circHIPK3 promotes the proliferation of CPMs. (**A**) The over-expression effect of circHIPK3 over-expression vector in CPMs. (**B**) The interference effects of three siRNAs of circHIPK3 in CPMs. (**C**,**D**) Effect of circHIPK3 on cell-cycle progression of CPMs. The plot of cell-cycle analysis in different cell-cycle phases was compared. (**E**,**F**) The growth curves of CPMs were measured after the transfection of over-expression vector and siRNA of circHIPK3. (**G**,**H**) EdU assays for CPMs with over-expression and inhibition of circHIPK3. In all panels, the values represent mean ± SEM from three independent experiments. * $p < 0.05$; ** $p < 0.01$; *** $p < 0.001$.

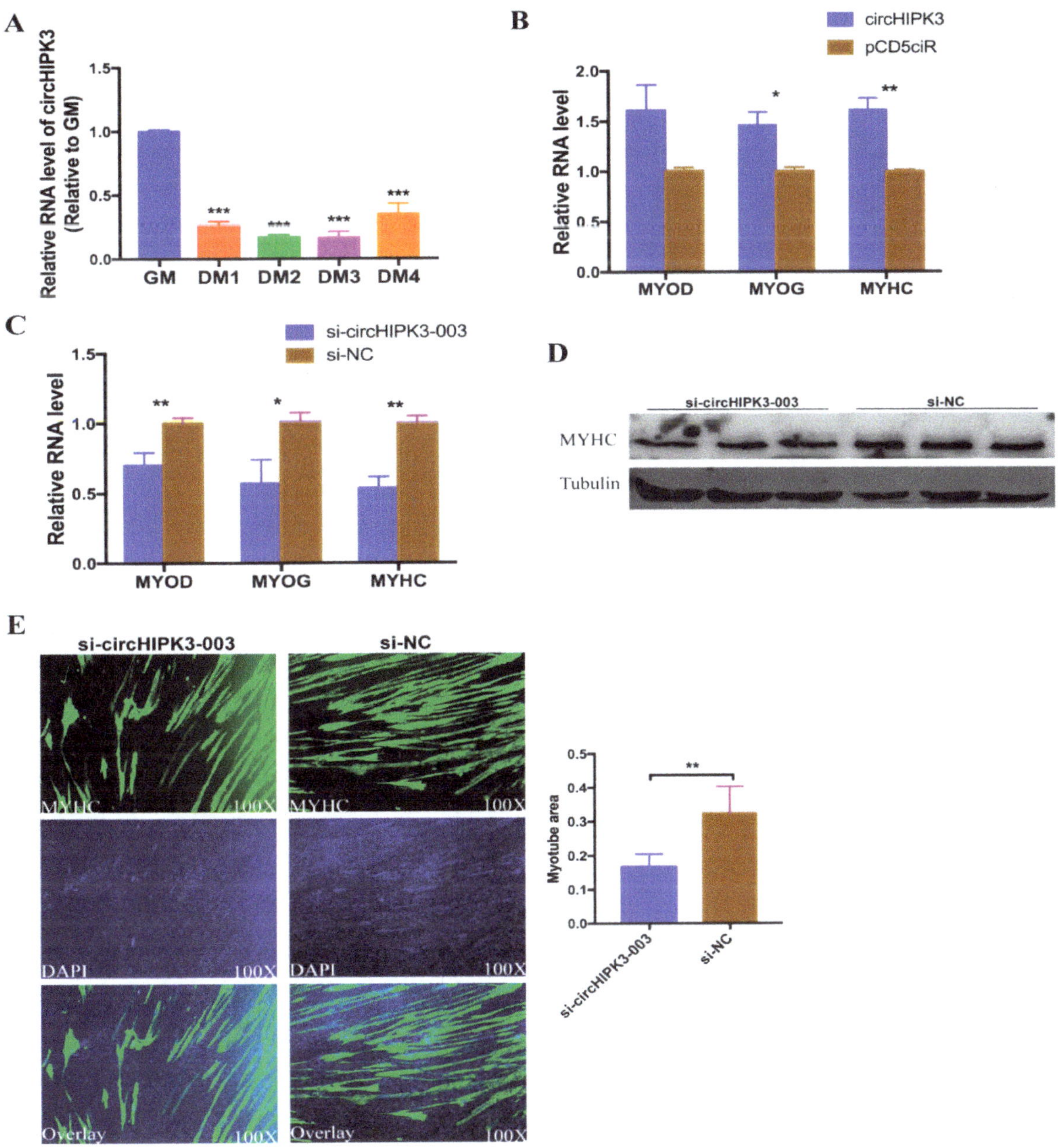

Figure 7. CircHIPK3 promotes the differentiation of CPMs. (**A**) The expression profile of circHIPK3 in the process of CPMs induced differentiation. (**B**) The expression level of *MYOD*, *MYOG* and *MYHC* in CPMs after over-expression of circHIPK3. (**C**) The expression of *MYOD*, *MYOG* and *MYHC* in CPMs after knock-down of circHIPK3. (**D**) The protein level of *MYHC* in CPMs after knock-down of circHIPK3. (**E**) Immunofluorescence analysis of MYHC-staining cells after knock-down of circHIPK3 in CPMs, * $p < 0.05$; ** $p < 0.01$; *** $p < 0.001$.

4. Discussion

Recently, circular RNAs have been identified as a new regulatory factor in multiple biological processes in different kinds of species [18,22–24]. Generally, the expression pattern of circular RNAs were corresponding to their linear parental transcript, the expression levels of circular RNAs were lower and had less functions compare to the corresponding mRNA [25,26]. However, some circular RNAs expressed independently high levels and exerted crucial roles in some special cell lines or tissues [14,27]. Many studies proved that circular RNAs were differentially expressed in different muscle developmental stages and played crucial roles in muscle development [28]. Circular RNA sequencing analysis, in different kinds of muscle cells or tissues, among many species, showed that most of the circular RNAs were exonic circular RNAs and differentially expressed during aging [29,30]. Functional circular RNAs in myoblast differentiation play a role by acting as an miRNA sponge or translating micropeptide [15,29,31,32]. Previous circular RNA sequencing data (GSE89355) showed that circHIPK3 was differentially expressed in three stages of chicken muscle development and expressed the highest level, while compared with the other ten circular RNAs, which was produced by chicken *HIPK3*. In this study, we found that circHIPK3 maintained a high expression level in myoblast and decreased sharply from DM1 to DM4, then the expression trend of circHIPK3 increased. This unique expression pattern suggested that circHIPK3 may have an important impact on CPM proliferation and differentiation.

MicroRNAs have been found to be widespread in various cell types or tissues and exerted a regulatory role in biological development [33–35]. It exerted important functions in muscle development and was associated with phenotypic changes in skeletal muscle [36,37]. miR-30a-3p is an inhibitor of proliferation in cancer cells [11,12]. In chicken skeletal muscle development, miR-30a-3p differentially expressed in three stages [38]. In this study, we demonstrated that miR-30a-3p inhibited the proliferation of CPMs and suppressed the differentiation by inhibiting the expression of *MEF2C*. According to the resources of microRNA viewer (http://people.csail.mit.edu/akiezun/microRNAviewer/index.html) (Supplementary Figure), miR-30a-3p was conserved in multiple species. The current study provided information to similar studies in other species.

Circular RNA that possesses miRNA response elements (MREs) is known to be a sponge for miRNAs [39]. Previous studies confirmed that the miR-30a family could bind to circHIPK3 in mice [40]. Given the homology of circHIPK3 and miR-30a family, we found that circHIPK3 possessed three potential binding sites for miR-30a-3p, through bioinformatic analyses by miRDB and RNAhybrid. Then two methods were conducted and we confirmed there are interactions between circHIPK3 and miR-30a-3p. Therefore, we hypothesized that circHIPK3 might play an opposite role of miR-30a-3p on skeletal muscle development.

CircHIPK3 was found to be involved in cell proliferation by acting as a sponge of multiple miRNAs, and *HIPK3* gene could produce several circular RNAs with different types in previous study [40,41]. miR-30a inhibited the proliferation by involving in different pathways in different kinds of cells [11,42,43]. In this study, we found that circHIPK3 promoted the proliferation and differentiation of CPMs by combing with miR-30a-3p. However, the underlying mechanism of miR-30a-3p retarding the myoblast proliferation needs further investigation.

In conclusion, our results suggested that miR-30a-3p could inhibit the proliferation of CPMs and repress the differentiation of CPMs by decreasing the expression of *MEF2C*, while circHIPK3 could promote the proliferation and differentiation of CPMs by sponging miR-30a-3p.

Author Contributions: B.C. carried out all the experiments, analyzed the data and wrote the paper. J.Y., L.G., Z.W. and X.C. participated in partial experiments. M.S.B. and H.X. joined the revision of the manuscript. Q.N. designed the experiments, analyzed data, and participated in its revision.

Acknowledgments: We would like to thank Jing Liu and Zhuo Chen for their support.

References

1. Scanes, C.G.; Harvey, S.; Marsh, J.A.; King, D.B. Hormones and Growth in Poultry. *Poult. Sci.* **1984**, *63*, 2062–2074. [CrossRef] [PubMed]
2. Güller, I.; Russell, A.P. MicroRNAs in skeletal muscle: Their role and regulation in development, disease and function. *J. Physiol.* **2010** *588*, 4075–4087. [CrossRef] [PubMed]

3. Flisar, T.; Malovrh, Š.; Terčič, D.; Holcman, A.; Kovač, M. Thirty-four generations of divergent selection for 8-week body weight in chickens. *Poult. Sci.* **2014**, *93*, 16–23. [CrossRef] [PubMed]
4. Bartel, D.P. MicroRNAs: Genomics, Biogenesis, Mechanism, and Function. *Cell* **2004**, *116*, 281–297. [CrossRef]
5. Chen, J.-F.; Mandel, E.M.; Thomson, J.M.; Wu, Q.; Callis, T.E.; Hammond, S.M.; Conlon, F.L.; Wang, D.-Z. The role of microRNA-1 and microRNA-133 in skeletal muscle proliferation and differentiation. *Nat. Genet.* **2006**, *38*, 228–233. [CrossRef] [PubMed]
6. Van Rooij, E.; Liu, N.; Olson, E.N. MicroRNAs flex their muscles. *Trends Genet.* **2008**, *24*, 159–166. [CrossRef] [PubMed]
7. Chen, X.; Ouyang, H.; Wang, Z.; Chen, B.; Nie, Q. A Novel Circular RNA Generated by FGFR2 Gene Promotes Myoblast Proliferation and Differentiation by Sponging miR-133a-5p and miR-29b-1-5p. *Cells* **2018**, *7*, 199. [CrossRef] [PubMed]
8. Luo, W.; Wu, H.; Ye, Y.; Li, Z.; Hao, S.; Kong, L.; Zheng, X.; Lin, S.; Nie, Q.; Zhang, X. The transient expression of miR-203 and its inhibiting effects on skeletal muscle cell proliferation and differentiation. *Cell Death Dis.* **2014**, *5*, e1347. [CrossRef] [PubMed]
9. Jia, X.; Ouyang, H.; Abdalla, B.A.; Xu, H.; Nie, Q.; Zhang, X. miR-16 controls myoblast proliferation and apoptosis through directly suppressing Bcl2 and FOXO1 activities. *Biochim. Biophys. Acta* **2017**, *1860*, 674–684. [CrossRef] [PubMed]
10. Ma, M.; Cai, B.; Jiang, L.; Abdalla, B.A.; Li, Z.; Nie, Q.; Zhang, X. lncRNA-Six1 Is a Target of miR-1611 that Functions as a ceRNA to Regulate Six1 Protein Expression and Fiber Type Switching in Chicken Myogenesis. *Cells* **2018**, *7*, 243. [CrossRef]
11. Qi, B.; Wang, Y.; Chen, Z.-J.; Li, X.-N.; Qi, Y.; Yang, Y.; Cui, G.-H.; Guo, H.-Z.; Li, W.-H.; Zhao, S. Down-regulation of miR-30a-3p/5p promotes esophageal squamous cell carcinoma cell proliferation by activating the Wnt signaling pathway. *World J. Gastroenterol.* **2017**, *23*, 7965–7977. [CrossRef]
12. Wang, W.; Lin, H.; Zhou, L.; Zhu, Q.; Gao, S.; Xie, H.; Liu, Z.; Xu, Z.; Wei, J.; Huang, X.; et al. MicroRNA-30a-3p inhibits tumor proliferation, invasiveness and metastasis and is downregulated in hepatocellular carcinoma. *Eur. J. Surg. Oncol.* **2014**, *40*, 1586–1594. [CrossRef] [PubMed]
13. Memczak, S.; Jens, M.; Elefsinioti, A.; Torti, F.; Krueger, J.; Rybak, A.; Maier, L.; Mackowiak, S.D.; Gregersen, L.H.; Munschauer, M.; et al. Circular RNAs are a large class of animal RNAs with regulatory potency. *Nature* **2013**, *495*, 333–338. [CrossRef]
14. Rybak-Wolf, A.; Stottmeister, C.; Glažar, P.; Jens, M.; Pino, N.; Giusti, S.; Hanan, M.; Behm, M.; Bartok, O.; Ashwal-Fluss, R.; et al. Circular RNAs in the Mammalian Brain Are Highly Abundant, Conserved, and Dynamically Expressed. *Mol. Cell* **2015**, *58*, 870–885. [CrossRef] [PubMed]
15. Legnini, I.; Di Timoteo, G.; Rossi, F.; Morlando, M.; Briganti, F.; Sthandier, O.; Fatica, A.; Santini, T.; Andronache, A.; Wade, M.; et al. Circ-ZNF609 Is a Circular RNA that Can Be Translated and Functions in Myogenesis. *Mole. Cell* **2017**, *66*, 22–37. [CrossRef] [PubMed]
16. Chen, L.-L. The biogenesis and emerging roles of circular RNAs. *Nat. Rev. Mol. Cell Biol.* **2016**, *17*, 205–211. [CrossRef] [PubMed]
17. Dong, R.; Zhang, X.-O.; Zhang, Y.; Ma, X.-K.; Chen, L.-L.; Yang, L. CircRNA-derived pseudogenes. *Cell Res.* **2016**, *26*, 747–750. [CrossRef]
18. Abdelmohsen, K.; Panda, A.C.; De, S.; Grammatikakis, I.; Kim, J.; Ding, J.; Noh, J.H.; Kim, K.M.; Mattison, J.A.; de Cabo, R.; et al. Circular RNAs in monkey muscle: Age-dependent changes. *Aging* **2015**, *7*, 903–910. [CrossRef] [PubMed]
19. Ouyang, H.; Chen, X.; Wang, Z.; Yu, J.; Jia, X.; Li, Z.; Luo, W.; Abdalla, B.A.; Jebessa, E.; Nie, Q.; et al. Circular RNAs are abundant and dynamically expressed during embryonic muscle development in chickens. *DNA Res.* **2017**, *3*. [CrossRef] [PubMed]
20. Ouyang, H.; Chen, X.; Li, W.; Li, Z.; Nie, Q.; Zhang, X. Circular RNA circSVIL Promotes Myoblast Proliferation and Differentiation by Sponging miR-203 in Chicken. *Front. Genet.* **2018**, *9*. [CrossRef]
21. Guo, L.; Huang, W.; Chen, B.; Jebessa Bekele, E.; Chen, X.; Cai, B.; Nie, Q. gga-mir-133a-3p Regulates Myoblasts Proliferation and Differentiation by Targeting PRRX1. *Front. Genet.* **2018**, *9*. [CrossRef] [PubMed]
22. Li, C.; Li, X.; Ma, Q.; Zhang, X.; Cao, Y.; Yao, Y.; You, S.; Wang, D.; Quan, R.; Hou, X.; et al. Genome-wide analysis of circular RNAs in prenatal and postnatal pituitary glands of sheep. *Sci. Rep.* **2017**, *7*, 97165–97177. [CrossRef] [PubMed]

23. Liang, G.; Yang, Y.; Niu, G.; Tang, Z.; Li, K. Genome-wide profiling of Sus scrofa circular RNAs across nine organs and three developmental stages. *DNA Res.* **2017**, *24*, 523–535. [CrossRef] [PubMed]
24. Piwecka, M.; Glažar, P.; Hernandez-Miranda, L.R.; Memczak, S.; Wolf, S.A.; Rybak-Wolf, A.; Filipchyk, A.; Klironomos, F.; Cerda Jara, C.A.; Fenske, P.; et al. Loss of a mammalian circular RNA locus causes miRNA deregulation and affects brain function. *Science* **2017**, *22*, eaam8526. [CrossRef] [PubMed]
25. Salzman, J.; Chen, R.E.; Olsen, M.N.; Wang, P.L.; Brown, P.O. Cell-Type Specific Features of Circular RNA Expression. *PLoS Genet.* **2013**, *9*. [CrossRef]
26. Guo, J.U.; Agarwal, V.; Guo, H.; Bartel, D.P. Expanded identification and characterization of mammalian circular RNAs. *Genome Biol.* **2014**, *15*. [CrossRef] [PubMed]
27. Conn, S.J.; Pillman, K.A.; Toubia, J.; Conn, V.M.; Salmanidis, M.; Phillips, C.A.; Roslan, S.; Schreiber, A.W.; Gregory, P.A.; Goodall, G.J. The RNA Binding Protein Quaking Regulates Formation of circRNAs. *Cell* **2015**, *160*, 1125–1134. [CrossRef]
28. Greco, S.; Cardinali, B.; Falcone, G.; Martelli, F. Circular RNAs in Muscle Function and Disease. *Int. J. Mol. Sci.* **2018**, *19*. [CrossRef]
29. Wei, X.; Li, H.; Yang, J.; Hao, D.; Dong, D.; Huang, Y.; Lan, X.; Plath, M.; Lei, C.; Lin, F.; et al. Circular RNA profiling reveals an abundant circLMO7 that regulates myoblasts differentiation and survival by sponging miR-378a-3p. *Cell Death Dis.* **2017**, *8*, e3153. [CrossRef]
30. Chen, J.; Zou, Q.; Lv, D.; Wei, Y.; Raza, M.A.; Chen, Y.; Li, P.; Xi, X.; Xu, H.; Wen, A.; et al. Comprehensive transcriptional landscape of porcine cardiac and skeletal muscles reveals differences of aging. *Oncotarget* **2017**, *9*, 1524–1541. [CrossRef]
31. Li, H.; Yang, J.; Wei, X.; Song, C.; Dong, D.; Huang, Y.; Lan, X.; Plath, M.; Lei, C.; Ma, Y.; et al. CircFUT10 reduces proliferation and facilitates differentiation of myoblasts by sponging miR-133a. *J. Cell. Physiol.* **2018**, *233*, 4643–4651. [CrossRef] [PubMed]
32. Li, H.; Wei, X.; Yang, J.; Dong, D.; Hao, D.; Huang, Y.; Lan, X.; Plath, M.; Lei, C.; Ma, Y.; et al. circFGFR4 Promotes Differentiation of Myoblasts via Binding miR-107 to Relieve Its Inhibition of Wnt3a. *Mol. Ther. Nucleic Acids* **2018**, *11*, 272–283. [CrossRef] [PubMed]
33. Rogg, E.M.; Abplanalp, W.T.; Bischof, C.; John, D.; Schulz, M.H.; Krishnan, J.; Fischer, A.; Poluzzi, C.; Schaefer, L.; Bonauer, A.; et al. Analysis of Cell Type-Specific Effects of MicroRNA-92a Provides Novel Insights Into Target Regulation and Mechanism of Action. *Circulation* **2018**, *138*, 2545–2558. [CrossRef]
34. Yasmeen, S.; Kaur, S.; Mirza, A.H.; Brodin, B.; Pociot, F.; Kruuse, C. miRNA-27a-3p and miRNA-222-3p as Novel Modulators of Phosphodiesterase 3a (PDE3A) in Cerebral Microvascular Endothelial Cells. *Mol. Neurobiol.* **2019**, *2*. [CrossRef] [PubMed]
35. Treiber, T.; Treiber, N.; Meister, G. Regulation of microRNA biogenesis and its crosstalk with other cellular pathways. *Nat. Rev. Mol. Cell Biol.* **2019**, *20*, 5–20. [CrossRef]
36. Wei, W.; He, H.-B.; Zhang, W.-Y.; Zhang, H.-X.; Bai, J.-B.; Liu, H.-Z.; Cao, J.-H.; Chang, K.-C.; Li, X.-Y.; Zhao, S.-H. miR-29 targets Akt3 to reduce proliferation and facilitate differentiation of myoblasts in skeletal muscle development. *Cell Death Dis.* **2013**, *4*, e668. [CrossRef]
37. Li, G.; Luo, W.; Abdalla, B.A.; Ouyang, H.; Yu, J.; Hu, F.; Nie, Q.; Zhang, X. miRNA-223 upregulated by MYOD inhibits myoblast proliferation by repressing IGF2 and facilitates myoblast differentiation by inhibiting ZEB1. *Cell Death Dis.* **2017**, *8*, e3094. [CrossRef]
38. Jebessa, E.; Ouyang, H.; Abdalla, B.A.; Li, Z.; Abdullahi, A.Y.; Liu, Q.; Nie, Q.; Zhang, X. Characterization of miRNA and their target gene during chicken embryo skeletal muscle development. *Oncotarget* **2018**, *9*. [CrossRef]
39. Hansen, T.B.; Jensen, T.I.; Clausen, B.H.; Bramsen, J.B.; Finsen, B.; Damgaard, C.K.; Kjems, J. Natural RNA circles function as efficient microRNA sponges. *Nature* **2013**, *495*, 384–388. [CrossRef]
40. Shan, K.; Liu, C.; Liu, B.H.; Chen, X.; Dong, R.; Liu, X.; Zhang, Y.Y.; Liu, B.; Zhang, S.J.; Wang, J.J.; et al. Circular Noncoding RNA HIPK3 Mediates Retinal Vascular Dysfunction in Diabetes Mellitus. *Circulation* **2017**, *136*, 1629–1642.
41. Zheng, Q.; Bao, C.; Guo, W.; Li, S.; Chen, J.; Chen, B.; Luo, Y.; Lyu, D.; Li, Y.; Shi, G.; et al. Circular RNA profiling reveals an abundant circHIPK3 that regulates cell growth by sponging multiple miRNAs. *Nat. Commun.* **2016**, *7*, 11215. [CrossRef]
42. Cheng, C.-W.; Wang, H.-W.; Chang, C.-W.; Chu, H.-W.; Chen, C.-Y.; Yu, J.-C.; Chao, J.-I.; Liu, H.-F.; Ding, S.; Shen, C.-Y. MicroRNA-30a inhibits cell migration and invasion by downregulating vimentin expression and

is a potential prognostic marker in breast cancer. *Breast Cancer Res. Treat.* **2012**, *134*, 1081–1093. [CrossRef] [PubMed]

43. Peng, R.; Zhou, L.; Zhou, Y.; Zhao, Y.; Li, Q.; Ni, D.; Hu, Y.; Long, Y.; Liu, J.; Lyu, Z.; et al. MiR-30a Inhibits the Epithelial—Mesenchymal Transition of Podocytes through Downregulation of NFATc3. *Int. J. Mol. Sci.* **2015**, *16*, 24032–24047. [CrossRef] [PubMed]

10

Conserved Cu-MicroRNAs in *Arabidopsis thaliana* Function in Copper Economy under Deficiency

Muhammad Shahbaz and Marinus Pilon *

Biology Department, Colorado State University, Fort Collins, CO 80523-1878, USA; ms@sppg.ca
* Correspondence: pilon@colostate.edu;

Abstract: Copper (Cu) is a micronutrient for plants. Three small RNAs, which are up-regulated by Cu deficiency and target transcripts for Cu proteins, are among the most conserved microRNAs in plants. It was hypothesized that these Cu-microRNAs help save Cu for the most essential Cu-proteins under deficiency. Testing this hypothesis has been a challenge due to the redundancy of the Cu microRNAs and the properties of the regulatory circuits that control Cu homeostasis. In order to investigate the role of Cu-microRNAs in Cu homeostasis during vegetative growth, we used a tandem target mimicry strategy to simultaneously inhibit the function of three conserved Cu-microRNAs in *Arabidopsis thaliana*. When compared to wild-type, transgenic lines that express the tandem target mimicry construct showed reduced Cu-microRNA accumulation and increased accumulation of transcripts that encode Cu proteins. As a result, these mimicry lines showed impaired photosynthesis and growth compared to wild type on low Cu, which could be ascribed to a defect in accumulation of plastocyanin, a Cu-containing photosynthetic electron carrier, which is itself not a Cu-microRNA target. These data provide experimental support for a Cu economy model where the Cu-microRNAs together function to allow maturation of essential Cu proteins under impending deficiency.

Keywords: plastocyanin; photosynthesis; copper deficiency; Cu-microRNA; copper protein; target mimicry

---

1. Introduction

Copper deficiency in plants leads to defects in photosynthesis, chlorosis, reduced respiration, and wilting of leaves. Copper is not a "mobile" element. This means that under impending deficiency, Cu is not efficiently transported from older tissues to newly developing leaves, which become chlorotic [1]. In green tissue, the majority of Cu is found in the chloroplast [2,3]. Plastocyanin (PC) is a blue Cu protein that mediates electron transfer from the cytochrome-b_6f complex to photosystem I (PSI) in the thylakoid lumen of oxygenic photosynthetic organisms [4]. *Arabidopsis* has two genes for PC, *PC1 (PETE1)* and *PC2 (PETE2)*. The PC2 protein is more abundant and responsive to Cu [5,6]. PC2 accumulation is a consequence of protein stability due to cofactor presence and is not due to transcript abundance changes [6–8]. In higher plants, PC is the only protein that can accept electrons from the cytochrome-b_6f complex and *Arabidopsis* mutants with insertions in both PC genes are seedling-lethal on soil [9,10]. Cu is also a cofactor of Cu/Zn superoxide dismutase (CSD) proteins that function in the metabolism of reactive oxygen radicals. In *Arabidopsis*, CSD2 is active in the plastids, and CSD1 is active in the cytosol [11]. Both CSD1 and CSD2 receive the Cu cofactor from a copper chaperone called copper chaperone for superoxide dismutase, CCS [12,13]. In mitochondria, copper is required for the function of cytochrome-*c* oxidase (COX), the proton-pumping terminal oxidase in the respiratory electron transport chain in the inner membrane [14]. Three Cu atoms are bound by the core COX subunits I, II, and III, which are encoded in the mitochondrial genome. The ethylene receptors in the endomembrane system are copper-binding proteins [15,16]. All remaining Cu proteins are most likely apoplastic. These include the plant-specific blue copper proteins called phytocyanins, which includes plantacyanin, apoplastic ascorbate oxidases, amine oxidases, and laccases [17–24]. The laccase family has 17 members in *Arabidopsis* [22], and for several of these, a role in lignification and secondary growth of the vasculature has now been established [25,26].

MicroRNAs that are regulated by Cu availability and that target mRNAs encoding for Cu proteins are called Cu-microRNAs [27]. The four Cu-microRNAs of *Arabidopsis*—*miR397*, *miR398*, *miR408*, and *miR857*—were first discovered in deep sequencing projects [28] and regulation of CSD1 and CSD2 by *miR398* was shown by Sunkar et al. [29]. It was later found that *miR398*, and thus CSD1, CSD2, and CCS, are regulated primarily by Cu levels [13,30]. In addition, *miR397*, *miR408*, *and miR857* regulate the abundance of other Cu proteins in *Arabidopsis*, specifically laccases and the secreted protein plantacyanin in response to Cu availability [31]. The Cu-microRNAs are in turn regulated via a Cu-responsive transcription factor called SPL7 (squamosa promotor binding protein-like7), which also regulates Cu assimilation [8,32–34]. Three of the four *Arabidopsis* Cu microRNAs (*miR397*, *miR398*, and *miR408*) are among the most highly conserved microRNAs in plants [35]. This conservation suggests an important function. Because of the strong link with Cu, it was hypothesized that the Cu-microRNAs function in the Cu economy. According to this idea, in plants, which have symplasmic connections that allow both the sharing of nutrients and communication via small RNAs, the regulation via microRNAs provides a mechanism to save Cu and to allow essential Cu protein maturation, such as PC and COX, in actively growing cells of a tissue during impending deficiency [27]. The hypothesis rests on three legs. Two well-supported tenets are the very tight bonding of Cu atoms to its ligands [36] and the presence of symplasmic connections between cells in plants. The third leg of the hypothesis is that effective signaling in and between cells via microRNAs works to signal Cu status and that it can indeed help to tune Cu protein expression. This third leg needs further experimental support. The possible role of Cu-microRNAs in the Cu economy is based mostly on correlative evidence with some more direct support in *Arabidopsis* and in poplar [30,31,33,34,37,38]. In fact, the idea that Cu delivery to PC is a priority is not supported by the observation that PC2 protein levels are strongly affected by Cu deficiency even if mRNA levels were not affected [6]. Do Cu microRNAs actually make a difference for Cu economy? To test this, we aimed to inactivate the conserved Cu-microRNAs (*miR397*, *miR398*, and *miR408*) and to analyze, especially under impending deficiency, the effects on growth and Cu allocation to abundant and essential Cu proteins, such as PC1 and PC2, which are not down-regulated via a microRNA. Because Cu deficiency affects flower and pollen development with several compounding effects, we limited our study to the vegetative shoot before flowering [31,39].

The Cu-microRNAs are under control of SPL7 [8,32]. Because in a *spl7*-loss-of-function mutant, Cu uptake is also defective, a good test of Cu microRNA function requires that its regulation is uncoupled from SPL7. We have tested single microRNA loss-of-function mutants for *miR398a/c* and for *miR408* and found no discernable phenotypes and certainly no defect in PC maturation [3,13]. Similarly, *miR398* overexpression caused no severe phenotypes on soil or on agar media [13,40]. However, Zhang et al. have reported a small effect on PC accumulation due to altered miR408 expression in a hy5/spl7 background [34]. Due to properties of the SPL7-mediated system, changes in single Cu-microRNAs can be predicted to show attenuated effects on Cu homeostasis because the addition or elimination of Cu-binding targets will affect the Cu pool sensed by SPL7, which will reset the expression of Cu-microRNAs and Cu uptake systems [27]. Due to the "dampening" properties of the system, we predict that disruption of just one Cu-microRNA will have very small effects. Thus, perhaps all or the majority of the Cu-microRNAs must be perturbed simultaneously before strong effects on Cu homeostasis can be expected. We focused here on the conserved Cu-microRNAs. Each of *miR397* and *miR398* have multiple loci and thus combinations (crosses) of knock-out (KO) lines to inactivate all Cu-microRNAs are virtually impossible. We therefore used a target mimicry construct to simultaneously inhibit all the conserved Cu-microRNAs [41]. Target mimicry relies on the production of a "designed" RNA transcript, which is modified from INDUCED BY PHOSPHATE STARVATION1 (*IPS1*), a non-coding RNA in *Arabidopsis* that functions to sequester miR399, a microRNA that functions in the regulation of phosphate homeostasis [41]. The native *IPS1* transcript contains a microRNA target site for *miR399* that cannot be cleaved and that should sequester and inactivate the microRNA. The original mimicry target site in *miR399* can be modified to inhibit other microRNAs [41].

2. Results

In order to simultaneously inhibit all three Cu-microRNAs, we designed a tandem target mimicry strategy (see Supplementary Figure S1). In this approach we modified the native IPS1 sequence [41] by replacing the miR399 target sequence with three, in tandem, target sites for miR397, miR398, and miR408, respectively. The construct was placed under the control of a constitutive 35S-CaMV promoter. The three conserved Cu-microRNAs should be sequestered by the modified IPS1-Cu-microRNA tandem mimicry construct, which is not a target for microRNA-directed cleavage due to inserted mismatches at the predicted target cleavage sites.

Three transgenic *Arabidopsis* lines with high expression of the tandem mimicry construct were selected and compared to wild-type plants. To control Cu status, we germinated plants on Cu replete agar media for 10 days before transplanting the seedlings to hydroponics. We first verified that the expression in the shoot of the tandem mimicry construct was not affected by Cu-feeding status after 3.5 weeks in hydroponics (5 weeks after the start of germination) (Figure 1). We used three Cu regimes; no Cu addition for mild deficiency after 3–4 weeks in this growth condition; 5 nM $CuSO_4$ for Cu-sufficient; and 50 nM $CuSO_4$ for Cu-replete, a condition where all Cu proteins are expected to acquire this cofactor [31]. As expected, the target mimicry construct expression was not affected by Cu availability (Figure 1).

The tandem target mimicry construct was designed to affect the function of the three conserved Cu-microRNAs and prevent strong down-regulation of the transcripts for Cu proteins normally targeting the Cu-microRNAs, especially in low Cu conditions. We investigated, using quantitative reverse transcription polymerase chain reaction (qRT-PCR), the effect of the tandem target mimicry construct on the accumulation of the three conserved Cu-microRNAs at the three Cu concentrations (Figure 1). For both miR397 and miR398, three loci are present, whereas for miR408, a single locus exists. For miR397 and miR398, we utilized primers optimized to amplify the more abundant and Cu induced isoforms [8]. As expected, all three conserved Cu-microRNAs were strongly regulated by Cu availability in the wild-type, with highest expression seen when Cu was omitted from the growth medium (Figure 1). Conversely, in plants grown on 50 nM $CuSO_4$, the expression of all three Cu-microRNAs was strongly repressed. Expression of the tandem target mimicry construct caused a notable reduction of all three Cu microRNAs. The effect on Cu-microRNA expression was more pronounced for miR398 and miR408 compared to miR397 (Figure 1). The effect of target tandem mimicry was most pronounced at 0 nM and 5 nM $CuSO_4$, where in the wild-type, the Cu-microRNAs accumulated. Higher expression of the mimicry construct in lines M5 and M17 correlated with a stronger effect on Cu-microRNA accumulation.

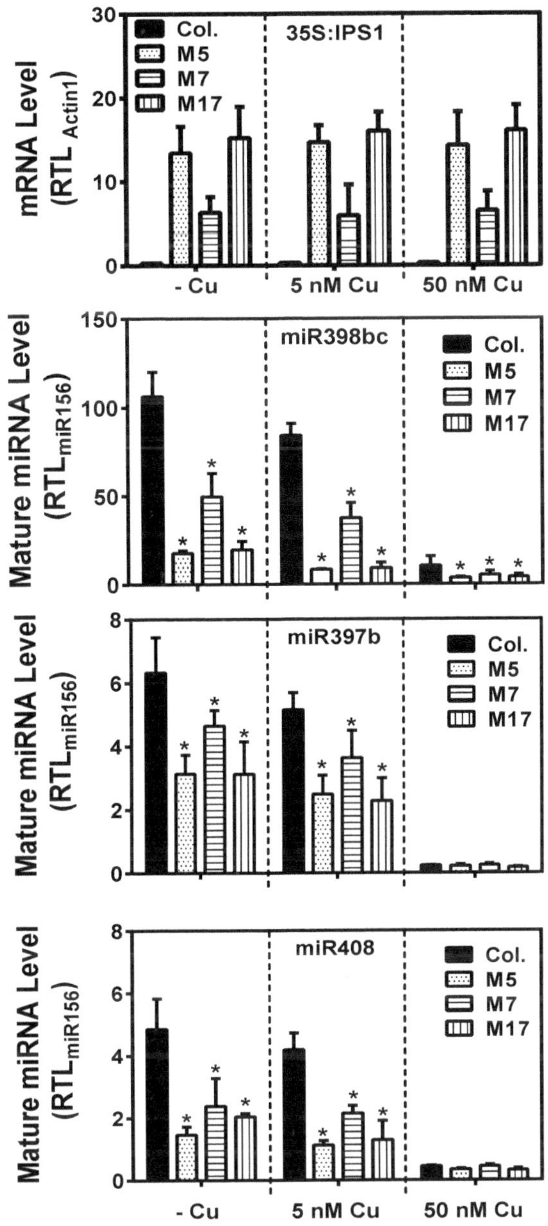

Figure 1. Expression of the IPS1 tandem target mimicry construct and three conserved Cu-microRNAs in wild type (Col) and three transgenic lines (M5, M7, and M17). 35S:IPS1 transgene relative transcript levels (RTL) were determined using qRT-PCR. Values are normalized relative to actin 1 expression. Mature microRNA levels were measured by qRT-PCR using primers designed for miR398bc, miR397b, and miR408. Relative microRNA levels were normalized to miR156 expression. All data and are given as averages ± SD ($n = 3$). * indicate significant differences from the control ($p < 0.05$, Student's t-test).

We next analyzed the effect of tandem target mimicry on the transcript levels of selected confirmed Cu-microRNA targets (Figure 2). As expected, a strong effect of tandem target mimicry was seen for all target transcripts, this effect was most pronounced under Cu deficiency (- Cu). The deregulation by tandem target mimicry on Cu deficiency was strong for all the tested targets of the three conserved microRNAs (Figure 2). We conclude that accumulation of all the three conserved Cu-microRNAs was disrupted in the shoot by tandem target mimicry, and in turn this caused aberrant accumulation of transcripts encoding Cu proteins under Cu deficiency.

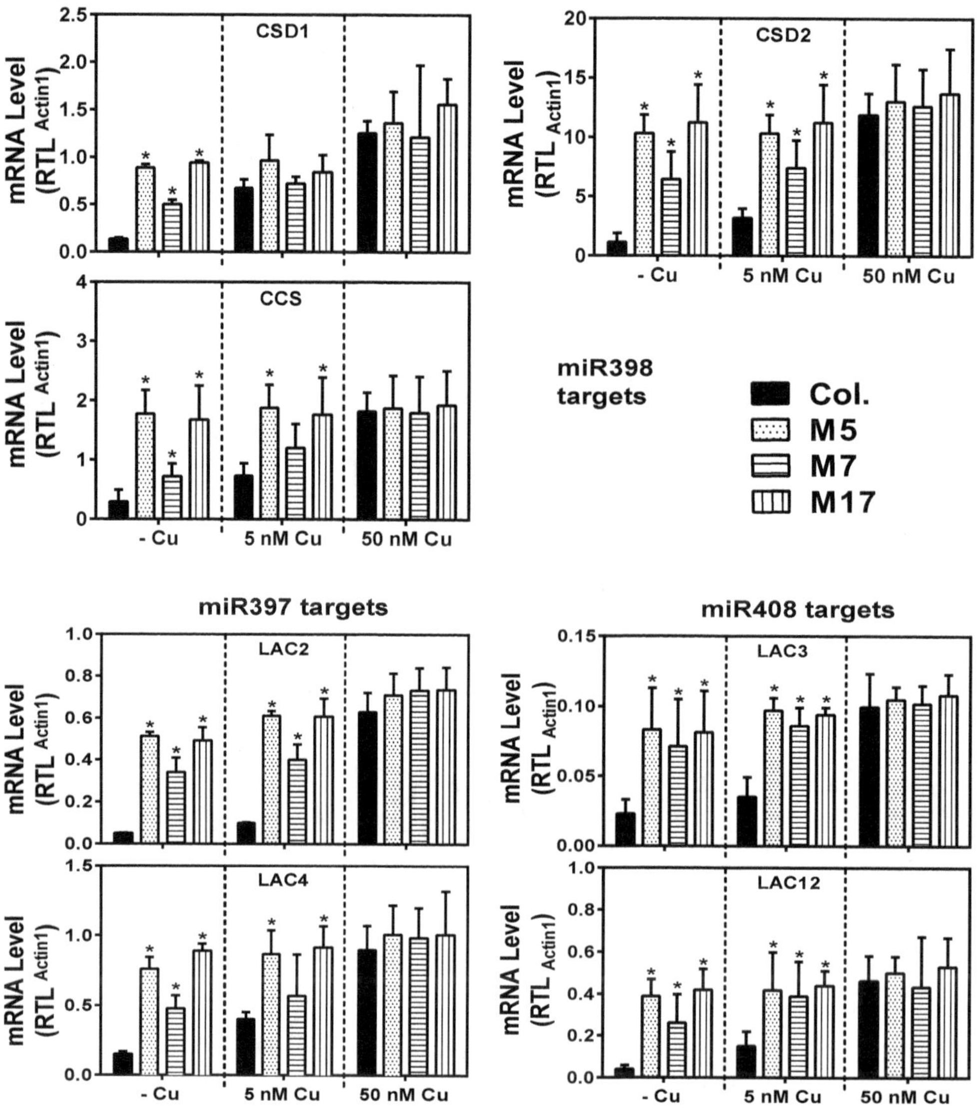

Figure 2. Expression analysis of selected Cu-microRNA target transcripts using qRT-PCR. mRNA-relative transcript levels (RTL) were normalized relative to actin 1 expression and given as averages ± SD (n = 3). * indicate significant differences from the control ($p < 0.05$, Student's t-test).

We next analyzed vegetative growth. Because the Cu-microRNAs are highly expressed in plants grown on low Cu and repressed in Cu-replete conditions, phenotypes due to target mimicry were mainly expected on low Cu conditions. Because we were especially interested in the effects on photosynthesis during vegetative growth, we compared plants at 5 weeks of age, which in our short-day growth condition is about a week before bolting, with the onset of symptoms of deficiency showing in the wild-type, without irreversible secondary effects. Because there is Cu in the seeds and in the germination medium, Cu depletion treatment in hydroponics takes several weeks to show visible deficiency symptoms. The plants of all genotypes showed overall comparable morphology on all three conditions after 3.5 weeks in hydroponics (Figure 3A). However, when the plant biomass was measured, a clear effect was seen for Cu omission compared to 5 nM and 50 nM $CuSO_4$ (Figure 3B). Interestingly, in the deficient condition, a significant further reduction in biomass was seen for the three tandem target mimicry lines. We measured the elemental composition of the shoots. As expected, $CuSO_4$ in the medium strongly affected Cu concentrations in the shoots. There was, however, no difference between the transgenics and wild-type, indicating that tandem target mimicry did not affect Cu levels (Figure 3C). Other elements did not show line-specific differences (See Table S1). In summary, tandem target mimicry for Cu-microRNAs resulted in a mild but significant effect on biomass accumulation, but only under Cu deficiency.

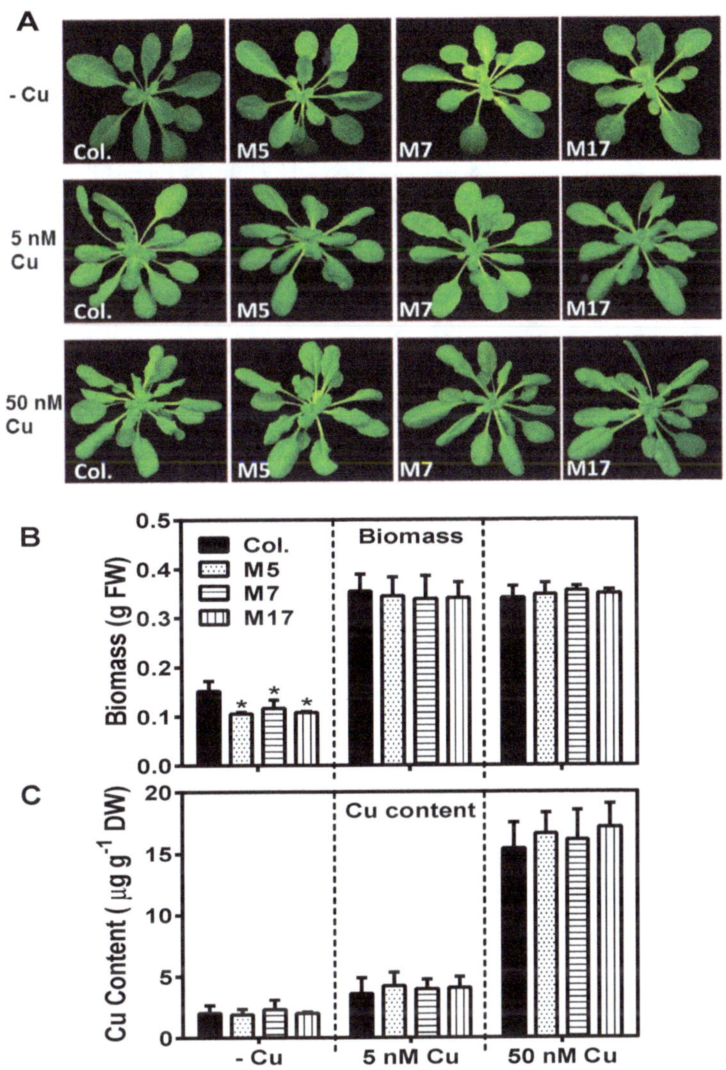

Figure 3. Effect of Cu-microRNA target mimicry on plant growth in hydroponics. (**A**) Representative images of 5-week-old WT (Col.); M5, M7, and M17 are All3-Cu miRNA mimicry lines. Plants were grown on 1/10th strength Hoagland solution containing 0, 5, or 50 nM $CuSO_4$. (**B**) Fresh weight and (**C**) Cu content (μg g^{-1} DW) of 5-week grown plants. Values are given as averages ± SD (n = 6). * indicate significant differences ($p < 0.05$, Student's t-test).

We hypothesized that tandem target mimicry, because of the misregulated and higher expression of Cu-microRNA targets, affected growth by limiting the pool of Cu available for plastocyanin, which is not a microRNA target. A lack of plastocyanin function should result in a decreased photosynthetic electron transport and a more reduced plastoquinone pool. Measurement of chlorophyll fluorescence parameters (Figure 4) indicated that Cu depletion in the wild-type caused a mild decrease in the flux through PSII (ΦPSII), which is an estimate of photosynthetic electron transport activity. However, there was a significantly larger decrease in ΦPSII for the three tandem target mimicry lines compared to the wild-type, but only under Cu deficiency. Similarly, the parameter 1-qP, which indicates the redox state of the plastoquinone pool, was affected by Cu depletion with a strong effect seen in tandem target mimicry. These results are strongly indicative of a defect in plastocyanin function in plants grown on low Cu, which is exacerbated by tandem target mimicry.

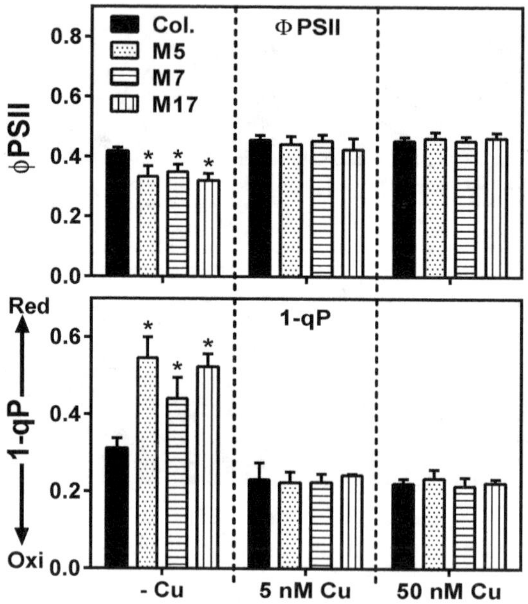

Figure 4. Photosynthetic electron transport parameters as measured by chlorophyll fluorescence. FluorCam measurement done on dark-adopted rosette leaves from WT (Col.) and transgenic lines (M5, M7, M17) growing in hydroponic culture at 0, 5, or 50 nM $CuSO_4$, where values are given as averages ± SD ($n = 6$). Top panel: ΦPSII (estimate of electron transport). Bottom panel: 1-qP (indicative of the redox state of the plastoquinone pool). * indicate significant differences between lines within a growth condition ($p < 0.05$, Student's t-test).

We next analyzed Cu protein accumulation using immunoblotting levels (Figure 5A). In *Arabidopsis*, two plastocyanin isoforms are expressed, PC1 and PC2 [10]. The more abundant PC2 isoform is known to be strongly affected at the protein level by Cu depletion, most likely due to a post-translational process as neither PC transcript is a target of a microRNA [6]. Indeed, Cu depletion in the wild-type resulted in a marked decrease in PC2 accumulation, while PC1 was relatively less affected (Figure 5A). Remarkably, there was a much stronger reduction in both PC isoforms in the three tandem target mimicry lines on low Cu. The severity of the effect on PC correlated strongly with the tandem target mimicry expression level (Figures 1 and 5A). At 5 and 50 nM $CuSO_4$, there was however no noticeable difference between the wild-type and transgenics. Cytochrome-c oxidase (COX) is a major Cu protein in the mitochondria where it functions as the terminal oxidase. We used antibodies specific to the Cu-binding COX core subunit II (COXII) as a proxy for COX accumulation (Figure 5A). COXII was affected by Cu depletion in the wild-type, albeit that the effect seemed not as strong as for PC in the chloroplast. There was a mild but noticeable larger effect of Cu depletion on COXII accumulation in the three target mimicry lines (Figure 5A).

We also verified accumulation of the miR398 targets: CSD1, CSD2, and CCS. As expected, accumulation of these proteins was strongly affected by Cu levels (Figure 5A). However, tandem target mimicry attenuated the effects of lowering Cu on the accumulation of these proteins. For CSD1 and CSD2 protein accumulation, the strongest effect of target mimicry was seen at 5 nM $CuSO_4$, whereas for CCS protein accumulation, a strong effect was also seen at 0 nM $CuSO_4$ (Figure 5A). The superoxide dismutase isozyme activity was analyzed using native gel assays (Figure 5B). As expected, the decrease in CSD1 and CSD2 protein levels resulted in a reduced CSD activity. FeSOD expression and activity, which is known to be highly Cu responsive and regulated via SPL7 [7,8,32], was not affected by tandem-target mimicry. In conclusion, lines that express the tandem target mimicry construct show a stronger reduction in plastocyanin and COXII levels on low Cu, which is accompanied by an elevated accumulation and activity of the miR398 targets.

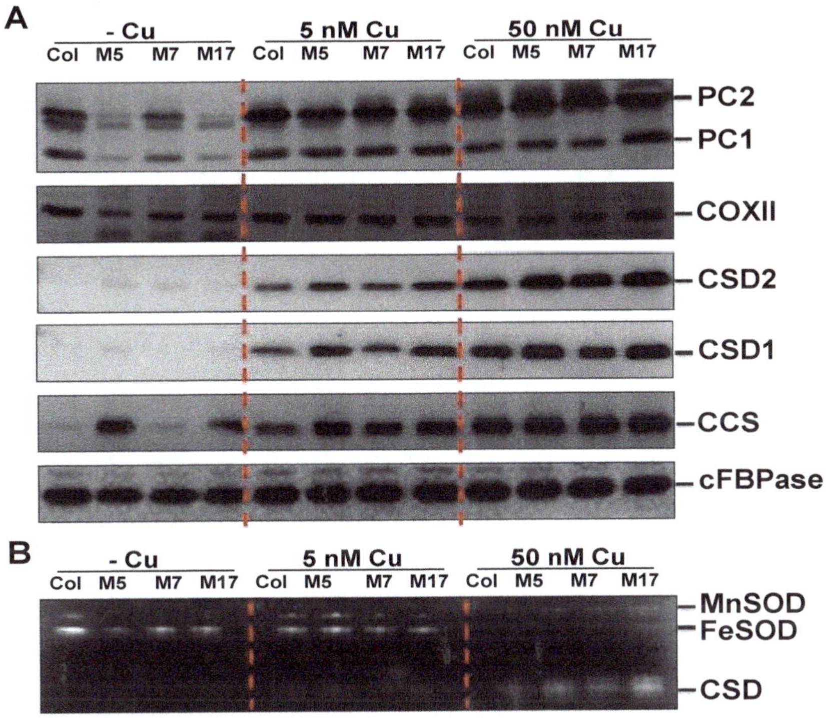

Figure 5. Comparison of Cu protein accumulation in wild-type (Col) and three transgenics (M5, M7, M17). (**A**) Western-blot analysis using the following antibodies: PC, plastocyanin two isoforms (PC2 and PC1 are detected); COXII, cytochrome-c oxidase subunit II; CSD2 and CSD1 (chloroplast and cytosolic isoforms of Cu/Z superoxide dismutase; and CCS, copper chaperone for Cu/ZnSOD. cFBPase was used as a loading control. Total soluble proteins (30 μg) were fractioned using SDS-PAGE (15% gel) and blotted onto nitrocellulose membranes. (**B**) For SOD isoform activity, total soluble proteins (30 μg) were fractioned on a non-denaturing 15% acryl amide gel and stained for SOD activity. The indicated detected isoforms were identified based on known relative mobility.

3. Discussion

The conservation of the Cu microRNA sequences, their target transcripts encoding Cu proteins, and their regulation by Cu via SPL7 suggests an important function related to copper directly. Neither plastocyanin nor cytochrome-c oxidase core subunits are targets of a microRNA in plants, presumably because these proteins are indispensible. Could a role of the Cu-microRNAs be to maintain a pool of Cu accessible for plastocyanin and cytochrome-c oxidase? While such a role in the Cu economy has been proposed before, the direct evidence has been limited. We aimed to test a role of conserved Cu-microRNAs in the Cu economy in *Arabidopsis* under conditions that require photoautotrophic growth. In order to characterize a function of the conserved Cu microRNAs, we employed a tandem target mimicry strategy. Expression of target mimicry constructs caused a decrease in the microRNAs that were targeted (Figure 1). Such a decrease has been reported before for other microRNAs and indicates that target mimicry not only leads to target sequestration, but also causes increased microRNA instability. The Cu-microRNA deregulation was well correlated with the extent of mimicry construct expression. The tandem target mimicry approach clearly caused a deregulation of a large set of transcripts but (as expected) only on low Cu, a condition where Cu-microRNAs are highly expressed in the wild-type (Figures 1 and 2). As a consequence of tandem target mimicry, the plants accumulated lower amounts of the photosynthetic electron carrier plastocyanin, one of the most abundant Cu proteins in plants. The exacerbated reduction in plastocyanin content in tandem target mimicry lines was strongly correlated with a defect in photosynthetic electron transfer and reduced biomass on low Cu (Figures 3–5). These data provide strong and direct support for a role of the Cu-microRNAs in the Cu economy in *Arabidopsis*, requiring optimization of vegetative growth under low-Cu conditions.

While the effect on growth was relatively small and required low Cu to be noted, we think it is likely a strong enough phenotype to explain the conservation of the Cu-microRNAs. In this context it should be noted that the non-conserved and fourth Cu-microRNA (miR857) was not targeted here [31]. Perhaps stronger phenotypes could be observed if this fourth target was also de-regulated.

Because of the presence of Cu in seeds and in the germination medium, which is needed to ensure comparable development, there is a significant lag time of about two and a half to three weeks before symptoms of Cu deficiency manifest themselves [31]. We wanted to avoid secondary effects, and therefore grew all plants for less than four weeks in hydroponics without Cu. We did this to avoid the compounding effects of bolting and flowering, which are strongly affected by Cu deficiency treatment, but only after strong and irreversible symptoms of Cu deficiency, such as leaf curling and browning, are evident in the vegetative shoot [31]. Therefore, in this study, we focused on vegetative growth and on shoots because this is where the Cu demand for plastocyanin is high. It seems likely, though, that the role of Cu-microRNAs, which are expressed in roots, stems, leaves, and flowers, is not limited to the vegetative shoot and indeed Cu feeding status is known to affect the timing of flowering, with deficient plants showing a delay [31]. Furthermore, Cu is also important for pollen development, presumably to allow for sufficient mitochondrial cytochrome-c oxidase activity [39]. We did not note any defect in fertility or seed set for our three target mimicry lines compared to wild-type. This was probably due to the sufficiency of Cu in the soil. It is possible that if the plants are to be subjected to Cu deficiency after the onset of flowering that a defect in pollen maturation would cause a decrease in fertility. However, such an experiment is difficult to control, and effects of both Cu and development would have to be considered simultaneously. Nevertheless, accumulation of mitochondrial COXII protein, one of three mitochondrial-encoded Cu-binding core subunits of cytochrome-c oxidase in the respiratory electron transport chain [14], was clearly negatively affected by tandem target mimicry in the shoots. This observation indicates that optimal cytochrome-c oxidase activity under Cu deficiency requires functional Cu-microRNAs. A role in regulating grain yield was also reported for miR408 in rice [42]. In this same study, overexpression of the rice miR408 target UCL8 (uclacyanin 8, a phytocyanin), showed a negative effect on plastocyanin and chloroplastic Cu accumulation in leaves [42].

In response to Cu availability, the Cu-microRNAs are regulated directly by SPL7, which is the plant homolog of the chlamydomonas (CRR1) copper response regulator [43]. SPL7 and CRR1 regulation requires *cis* elements called Cu-response elements (CuRe) with a GTAC core motif [43]. Thus, a strong mechanistic link exists for Cu and Cu-microRNA regulation. Another identified transacting factor that directly mediates miR408 Cu-microRNA expression via promoter area binding is HY5 (elongated hypocotyl 5). Co-regulation was shown for *miR408* by both SPL7 and HY5 [33,34]. This regulation makes sense since HY5 mediates gene expression in the light required for photosynthetic growth, which has a high demand for Cu. Plants that lack SPL7 or HY5 function had only a small defect in PC [34]. Overexpression of miR408 in the *spl7* and *hy5* mutants could, to some extent, alleviate this defect in PC maturation, albeit that plants were grown in vitro on agar media to see this effect [34]. Interestingly, miR408 overexpression could also rescue some developmental defects in *hy5/spl7* lines [34]. Therefore, a feedback loop was proposed as a mechanism to allow better PC maturation in miR408-overproducing lines, but it can also be argued that an improved Cu economy in such lines allows for better Cu cofactor availability to PC [34]. Consistent with this idea are the observations reported in several plant species where miR408 expression causes higher chloroplast Cu content and improved photosynthesis, growth, and seed yield in diverse plant species [42,44,45]. On the other hand, Carrió-Seguí et al. have reported that both overexpression and loss of miR408 caused a diminished plant performance, especially in low Fe conditions [46].

Besides Cu availability, other environmental conditions have been reported to affect Cu-microRNA expression [27]. This was convincingly shown in *Arabidopsis* for cold [47–49], Fe availability [46,50], or stresses that are predicted to induce reactive oxygen species accumulation such as high light, the herbicide methyl-viologen, and excessive toxic metal [29]. In our setup, we controlled growth conditions and attempted to avoid any additional stressors. It seems likely that Cu-microRNA expression is further tuned by environmental conditions apart from Cu and light. However, unlike for Cu and light, where *cis*-regions and interacting *trans*-acting factors (SPL7 and HY5) are identified, it is presently unclear at a mechanistic level how these conditions could be linked to gene expression of Cu-microRNAs. Therefore, the effects of other stresses on Cu microRNAs could be direct or indirect. For Fe, for instance, it is well-reported in the literature that lower Cu levels increases Fe uptake and vice versa. A mechanism for this interaction can be proposed based on the observation that both Fe and Cu uptake at the root require a reductase activity of the ferric reductase oxidase (FRO) family. A cross reactivity for the low Cu-induced FRO4/FRO5 with Fe, or for the low-Fe induced FRO2 with Cu, would lead to increased Cu uptake under Fe deficiency, thus resetting the SPL7 regulation of Cu microRNAs, making it appear as if Fe regulates Cu-microRNAs [8,51]. Therefore, to more directly link Cu-microRNAs to other abiotic stresses, it will be important to uncover how potential *cis*-regions in the promoters and transacting factors interact to mediate stress response.

4. Materials and Methods

4.1. Plant Material and Growth Conditions

Lines were propagated on PRO-MIX HP soil that was fertilized with Miracle-Gro Liquid All Purpose Plant Food (Scotts Company, Marysville, OH, USA). Wild type (WT; Col.) and All3-mimicry transgenic lines seeds were surface-sterilized and germinated on agar plates containing one-half strength Murashige and Skoog (MS) medium [52] supplemented with 1% sucrose. For hydroponics, 7–10 days old seedlings were placed on a one-tenth-strength Hoagland's solution prepared with deionized water [53]. The nutrient solution was aerated and was replaced each week. For copper depletion (deficiency), Cu was omitted from the nutrient solution, while $CuSO_4$ was added to 5 nM for low but sufficient Cu (sufficient for high PC activity, but with low CSD activity in the wild-type) and to 50 nM for the Cu-replete conditions (where all Cu proteins accumulate to maximum levels in the wild-type) [31]. To minimize Cu contamination, all containers and buckets used for Cu-deprivation had never been in contact with Cu. The plants were grown in a light intensity of 150 μmol m^{-2} s^{-1}, with a 10-h-light/14-h-dark cycle and the temperature was maintained at 25 °C ± 2 °C in a climate-controlled room. After 4-weeks, plant material was harvested and stored frozen at −80 °C before analysis.

4.2. Transgenic Lines

Artificial target mimicry constructs were generated by modifying the sequence of the IPS1 gene [41]. The sequence of the construct, which was ordered as a synthesized piece of DNA (GenScript USA Inc. Piscataway, NJ, USA), is given in the Supplementary Materials. The All3-Cu miRNA target mimic construct was placed behind the CaMV 35S promoter in the pGWB41 vector, conferring resistance to kanamycin and hygromycin [54]. The construct was introduced into *A. thaliana* (accession Col-0) plants via a *Agrobacterium tumefaciens*-mediated transformation followed by selection on kanamycin containing agar media [55].

4.3. Elemental Analysis

For mineral content analysis, the plant tissue was placed in a drying oven at 55 °C for 48–72 h. One hundred milligrams of the dried material was digested in 1 mL of trace element grade nitric acid and heated at 60 °C for 2 h, followed by 130 °C for 6 h. The digests were subsequently diluted to 10 mL with double-distilled water before analysis using inductively coupled plasma – atomic emission spectrometry (ICP-AES) as described [56].

4.4. Chlorophyll Fluorescence Measurements

For chlorophyll fluorescence assays, whole rosettes of intact plants were used and dark-adapted for 30 minutes prior to analysis. Chlorophyll fluorescence imaging was done using a FluoroCam 701 MF

(Photon Systems International, Brno, Czech Republic) at an actinic light intensity of 150 micro-Einsteins as described [13]. The parameters ΦPSII (flux through PSII, an estimate of photosynthetic electron the transport rate) and 1-qP (an estimate of the redox state of plastoquinone pool) were calculated as described [57].

4.5. Protein Accumulation

Soluble proteins were extracted as described [38]. Protein concentration was determined using the Pierce BCA protein assay kit (Thermo Scientific, Waltham, MA, USA) using bovine serum albumin as a standard. For western blotting, 20 μg of total protein was separated using 15% SDS-PAGE and then transferred onto a nitrocellulose membrane. Antibodies used for immunodetection of PC, CSD1, CSD2, and CCS have been described [13,37]. Antisera for COXII and cFBPase were obtained from Agrisera (Vannas, Sweden). All protein detection experiments were done at least in biological triplicate with comparable results, and representative gels are shown.

4.6. Quantitative Reverse Transcription-PCR

RNA from plants sampled in biological triplicate was isolated using the Trizol reagent following the manufacturer's recommended protocol (Life Technologies, Carlsbad, CA, USA). After determination of total RNA concentrations, equal amounts per sample were reverse-transcribed using a First Strand cDNA Synthesis Kit (Life Technologies) and random hexamer primers. Quantitative RT-PCR was performed using the Light Cycler SYBR Green 1 master mix (Life Technologies) using gene-specific primer pairs as described previously for *Populus trichocarpa* using actin 1 as a housekeeping reference gene [37]. Samples without a template were used as negative controls. All primers are listed in Supplementary Table S2. qRT-PCR results and quality controls were analyzed using Light-Cycler 480 data-analysis software (version 1.5.1, Roche, Basel, Switzerland). The ΔΔCt method was used to calculated relative transcript expression levels.

4.7. Mature miRNA Stem-Loop qRT-PCR

For the quantification of mature microRNAs, a stem-loop pulsed RT was used [37]. The RNA was extracted using the Trizol method as described above; however, ethanol washes were avoided and nucleic acid precipitation steps were carried out after addition of 1/10th volume of sodium acetate (3 M; pH 5.2), followed by an equal volume of isopropanol [37]. The stem-loop pulsed RT and miRNA qRT-PCR were performed as described previously [37,58]. miRNA397b, 398b/c, and 408 abundance were analyzed using gene-specific primers, as described previously for *P. trichocarpa* [37]; see Supplementary Table S2). Relative mature miRNA abundance was standardized using miR156 expression [37]. Each sample was analyzed in biological triplicate.

4.8. Statistical Analysis

For statistical analyses the JMP software package (version 9.0.2; SAS Institute, Cary, NC, USA) was used. All results represent the averages and SD from at least three independent biological replicates. A Student's *t*-test was used to calculate significant differences ($p < 0.05$).

5. Conclusions

Simultaneous inhibition of the function of three conserved Cu-microRNAs caused a mild but significant growth phenotype on low-Cu media, which can be ascribed to a lack of function of plastocyanin, which is essential for photosynthesis. These observations provide support for a function of Cu-microRNas is the Cu economy.

Author Contributions: Authors' contributions: M.S and M.P. conceived the research plans; M.S. did the experiments; M.S. and M.P. analyzed the data; M.S. drafted the first manuscript; M.P. supervised the overall project and finalized the writing.

Acknowledgments: This project was supported by the Agriculture and Food Research Initiative competitive grant 2012-67-13-19416 of the USDA National Institute of Food and Agriculture to M.P.

Abbreviations

CSD, Cu/Zn-superoxide dismutase; COX, cytochrome-c oxidase; PSII, flux through PSII; FeSOD, iron superoxide dismutase; 1-qP, redox state of plastoquinone pool; LAC, laccase; PC, Plastocyanin.

References

1. Marschner, H. *Mineral Nutrition of Higher Plants*; Academic Press: London. UK, 1995.
2. Shikanai, T.; Muller-Moule, P.; Munekage, Y.; Niyogi, K.K.; Pilon, M. PAA1, a P-type ATPase of arabidopsis, functions in copper transport in chloroplasts. *Plant Cell* **2003**, *15*, 1333–1346. [CrossRef]
3. Tapken, W.; Ravet, K.; Pilon, M. Plastocyanin controls the stabilization of the thylakoid Cu-transporting P-type ATPase PAA2/HMA8 in response to low copper in Arabidopsis. *J. Biol. Chem.* **2012**, *287*, 18544–18550. [CrossRef] [PubMed]
4. Raven, J.A.; Evans, M.C.; Korb, R.E. The role of trace metals in photosynthetic electron transport in O_2-evolving organisms. *Photosynth. Res.* **1999**, *60*, 111–149. [CrossRef]
5. Pesaresi, P.; Scharfenberg, M.; Weigel, M.; Granlund, I.; Schroeder, W.P.; Finazzi, G.; Rappaport, F.; Masiero, S.; Furini, A.; Jahns, P.; et al. Mutants, Overexpressors, and Interactors of *Arabidopsis* Plastocyanin Isoforms: Revised Roles of Plastocyanin in Photosynthetic Electron Flow and Thylakoid Redox State. *Mol. Plant* **2009**, *2*, 236–248. [CrossRef] [PubMed]
6. Abdel-Ghany, S.E. Contribution of plastocyanin isoforms to photosynthesis and copper homeostasis in *Arabidopsis thaliana* grown at different copper regimes. *Planta* **2009**, *229*, 767–779. [CrossRef]
7. Abdel-Ghany, S.E.; Muller-Moule, P.; Niyogi, K.K.; Pilon, M.; Shikanai, T. Two P-type ATPases are required for copper delivery in *Arabidopsis thaliana* chloroplasts. *Plant Cell* **2005**, *17*, 1233–1251. [CrossRef]
8. Bernal, M.; Casero, D.; Singh, V.; Wilson, G.T.; Grande, A.; Yang, H.; Dodani, S.C.; Pellegrini, M.; Huijser, P.; Connolly, E.L.; et al. Transcriptome sequencing identifies SPL7-regulated copper acquisition genes FRO4/FRO5 and the copper dependence of iron homeostasis in Arabidopsis. *Plant Cell* **2012**, *24*, 738–761. [CrossRef] [PubMed]
9. Molina-Heredia, F.P.; Wastl, J.; Navarro, J.A.; Bendall, D.S.; Hervás, M.; Howe, C.J.; De La Rosa, M.A. Photosynthesis: a new function for an old cytochrome? *Nature* **2003**, *424*, 33–34. [CrossRef]
10. Weigel, M.; Varotto, C.; Pesaresi, P.; Finazzi, G.; Rappaport, F.; Salamini, F.; Leister, D. Plastocyanin Is Indispensable for Photosynthetic Electron Flow in *Arabidopsis thaliana*. *J. Biol. Chem.* **2003**, *278*, 31286–31289. [CrossRef] [PubMed]
11. Kliebenstein, D.J.; Monde, R.-A.; Last, R.L. Superoxide dismutase in Arabidopsis: An eclecticenzyme family with disparate regulation and protein localization. *Plant Physiol.* **1998**, *118*, 637–650. [CrossRef]
12. Chu, C.C.; Lee, W.C.; Guo, W.Y.; Pan, S.M.; Chen, L.J.; Li, H.M.; Jinn, T.L. A copper chaperone for superoxide dismutase that confers three types of copper/zinc superoxide dismutase activity in Arabidopsis. *Plant Physiol.* **2005**, *139*, 425–436. [CrossRef]
13. Cohu, C.M.; Abdel-Ghany, S.E.; Gogolin Reynolds, K.A.; Onofrio, A.M.; Bodecker, J.R.; Kimbrel, J.A.; Niyogi, K.K.; Pilon, M. Copper delivery by the copper chaperone for chloroplast and cytosolic copper/zinc-superoxide dismutases: regulation and unexpected phenotypes in an Arabidopsis mutant. *Mol. Plant* **2009**, *2*, 1336–1350. [CrossRef]
14. Carr, H.S.; Winge, D.R. Assembly of Cytochrome *c* Oxidase within the Mitochondrion. *Acc. Chem. Res.* **2003**, *36*, 309–316. [CrossRef]
15. Rodriguez, F.I.; Esch, J.J.; Hall, A.E.; Binder, B.M.; Schaller, G.E.; Bleecker, A.B. A copper cofactor for the ethylene receptor ETR1 from Arabidopsis. *Science* **1999**, *283*, 996–998. [CrossRef]
16. Alonso, J.M.; Stepanova, A.N. The Ethylene Signaling Pathway. *Science* **2004**, *306*, 1513–1515. [CrossRef]
17. Dong, J.; Kim, S.T.; Lord, E.M. Plantacyanin plays a role in reproduction in Arabidopsis. *Plant Physiol.* **2005**, *138*, 778–789. [CrossRef]

18. Pignocchi, C.; Fletcher, J.M.; Wilkinson, J.E.; Barnes, J.D.; Foyer, C.H. The function of ascorbate oxidase in tobacco. *Plant Physiol.* **2003**, *132*, 1631–1641. [CrossRef]
19. Yamamoto, A.; Bhuiyan, M.N.; Waditee, R.; Tanaka, Y.; Esaka, M.; Oba, K.; Jagendorf, A.T.; Takabe, T. Suppressed expression of the apoplastic ascorbate oxidase gene increases salt tolerance in tobacco and Arabidopsis plants. *J. Exp. Bot.* **2005**, *56*, 1785–1796. [CrossRef]
20. Frébort, I.; Sebela, M.; Svendsen, I.; Hirota, S.; Endo, M.; Yamauchi, O.; Bellelli, A.; Lemr, K.; Pec, P. Molecular mode of interaction of plant amine oxidase with the mechanism-based inhibitor 2-butyne-1,4-diamine. *Eur. J. Biochem.* **2000**, *267*, 1423–1433.
21. Groß, F.; Rudolf, E.E.; Thiele, B.; Durner, J.; Astier, J. Copper amine oxidase 8 regulates arginine-dependent nitric oxide production in Arabidopsis thaliana. *J. Exp. Bot.* **2017**, *68*, 2149–2162. [CrossRef]
22. McCaig, B.C.; Meagher, R.B.; Dean, J.F. Gene structure and molecular analysis of the laccase-like multicopper oxidase (LMCO) gene family in Arabidopsis thaliana. *Planta* **2005**, *221*, 619–636. [CrossRef]
23. Turlapati, P.V.; Kim, K.W.; Davin, L.B.; Lewis, N.G. The laccase multigene family in Arabidopsis thaliana: towards addressing the mystery of their gene function(s). *Planta* **2001**, *233*, 439–470. [CrossRef] [PubMed]
24. Cai, X.; Davis, E.J.; Ballif, J.; Liang, M.; Bushman, E.; Haroldsen, V.; Torabinejad, J.; Wu, Y. Mutant identification and characterization of the laccase gene family in Arabidopsis. *J. Exp. Bot.* **2006**, *57*, 2563–2569. [CrossRef] [PubMed]
25. Berthet, S.; Demont-Caulet, N.; Pollet, B.; Bidzinski, P.; Cézard, L.; Le Bris, P.; Borrega, N.; Hervé, J.; Blondet, E.; Balzergue, S.; et al. Disruption of LACCASE4 and 17 Results in Tissue-Specific Alterations to Lignification of Arabidopsis thaliana Stems. *Plant Cell* **2011**, *23*, 1124–1137. [CrossRef]
26. Zhao, Y.; Lin, S.; Qiu, Z.; Cao, D.; Wen, J.; Deng, X.; Wang, X.; Lin, J.; Li, X. MicroRNA857 is Involved in the Regulation of Secondary Growth of Vascular Tissues in Arabidopsis. *Plant Physiol.* **2015**, *169*, 2539–2552. [CrossRef] [PubMed]
27. Pilon, M. The copper microRNAs. *New Phytol.* **2017**, *213*, 1030–1035. [CrossRef]
28. Jones-Rhoades, M.W.; Bartel, D.P.; Bartel, B. MicroRNAs and their regulatory roles in plants. *Ann. Rev. Plant Biol.* **2006**, *57*, 19–53. [CrossRef]
29. Sunkar, R.; Kapoor, A.; Zhu, J.-K. Posttranscriptional induction of two Cu/Zn superoxide dismutase genes in Arabidopsis is mediated by downregulation of miR398 and important for oxidative stress tolerance. *Plant Cell* **2006**, *18*, 2051–2065. [CrossRef]
30. Yamasaki, H.; Abdel-Ghany, S.E.; Cohu, C.M.; Kobayashi, Y.; Shikanai, T.; Pilon, M. Regulation of Copper Homeostasis by Micro-RNA in Arabidopsis. *J. Biol. Chem.* **2007**, *282*, 16369–16378. [CrossRef]
31. Abdel-Ghany, S.E.; Pilon, M. MicroRNA-mediated Systemic Down-regulation of Copper Protein Expression in Response to Low Copper Availability in Arabidopsis. *J. Biol. Chem.* **2008**, *283*, 15932–15945. [CrossRef]
32. Yamasaki, H.; Hayashi, M.; Fukazawa, M.; Kobayashi, Y.; Shikanai, T. SQUAMOSA Promoter Binding Protein-Like7 Is a Central Regulator for Copper Homeostasis in *Arabidopsis*. *Plant Cell.* **2009**, *21*, 347–361. [CrossRef] [PubMed]
33. Zhang, H.; Li, L. SQUAMOSA promoter binding protein-like7 regulated microRNA408 is required for vegetative development in Arabidopsis. *Plant J.* **2013**, *74*, 98–109. [CrossRef] [PubMed]
34. Zhang, H.; Zhao, X.; Li, J.; Cai, H.; Deng, X.W.; Li, L. MicroRNA408 is critical for the HY5-SPL7 gene network that mediates the coordinated respose to light and copper. *Plant Cell* **2014**, *26*, 4933–4953. [CrossRef]
35. Cuperus, J.T.; Fahlgren, N.; Carrington, J.C. Evolution and functional diversification of MIRNA genes. *Plant Cell* **2011**, *23*, 431–442. [CrossRef] [PubMed]
36. Waldron, K.J.; Rutherford, J.C.; Ford, D.; Robinson, N.J. Metalloproteins and metal sensing. *Nature* **2009**, *460*, 823–830. [CrossRef]
37. Ravet, K.; Danford, F.L.; Dihle, A.; Pittarello, M.; Pilon, M. Spatial-temporal analysis of Cu homeostasis in Populus trichocarpa reveals an integrated molecular remodeling for a preferential allocation of copper to plastocyanin in the chloroplasts of developing leaves. *Plant Physiol.* **2011**, *157*, 1300–1312. [CrossRef]
38. Shahbaz, M.; Ravet, K.; Peers, G.; Pilon, M. Prioritization of copper for the use in photosynthetic electron transport in developing leaves of hybrid poplar. *Front. Plant Sci.* **2015**, *6*, 407. [CrossRef]
39. Yan, J.; Chia, J.C.; Sheng, H.; Jung, H.I.; Zavodna, T.O.; Zhang, L.; Huang, R.; Jiao, C.; Craft, E.J.; Fei, Z.; et al. Arabidopsis Pollen Fertility Requires the Transcription Factors CITF1 and SPL7 That Regulate Copper Delivery to Anthers and Jasmonic Acid Synthesis. *Plant Cell* **2017**, *29*, 3012–3029. [CrossRef]

40. Dugas, D.V.; Bartel, B. Sucrose induction of Arabidopsis miR398 represses two Cu/Zn superoxide dismutases. *Plant Mol. Biol.* **2008**, *67*, 403–417. [CrossRef]
41. Franco-Zorrilla, J.M.; Valli, A.; Todesco, M.; Mateos, I.; Puga, M.I.; Rubio-Somoza, I.; Leyva, A.; Weigel, D.; García, J.A.; Paz-Ares, J. Target mimicry provides a new mechanism for regulation of microRNA activity. *Nat. Genet.* **2007**, *39*, 1033–1037. [CrossRef]
42. Zhang, J.P.; Yu, Y.; Feng, Y.Z.; Zhou, Y.F.; Zhang, F.; Yang, Y.W.; Lei, M.Q.; Zhang, Y.C.; Chen, Y.Q. MiR408 Regulates Grain Yield and Photosynthesis via a Phytocyanin Protein. *Plant Physiol.* **2017**, *175*, 1175–1185. [CrossRef] [PubMed]
43. Kropat, J.; Tottey, S.; Birkenbihl, R.P.; Depege, N.; Huijser, P.; Merchant, S. A regulator of nutritional copper signaling in Chlamydomonas is an SBP domain protein that recognizes the GTAC core of copper response element. *Proc. Natl. Acad. Sci. USA* **2005**, *102*, 18730–18735. [CrossRef] [PubMed]
44. Pan, J.; Huang, D.; Guo, Z.; Kuang, Z.; Zhang, H.; Xie, X.; Ma, Z.; Gao, S.; Lerdau, M.T.; Chu, C.; et al. Overexpression of microRNA408 enhances photosynthesis, growth, and seed yield in diverse plants. *J. Integr. Plant Biol.* **2018**, *60*, 323–340. [CrossRef] [PubMed]
45. Song, Z.; Zhang, L.; Wang, Y.; Li, H.; Li, S.; Zhao, H.; Zhang, H. Constitutive Expression of miR408 Improves Biomass and Seed Yield in Arabidopsis. *Front. Plant Sci.* **2018**, *8*, 2114. [CrossRef] [PubMed]
46. Carrió-Seguí, À.; Ruiz-Rivero, O.; Villamayor-Belinchón, L.; Puig, S.; Perea-García, A.; Peñarrubia, L. The Altered Expression of microRNA408 Influences the Arabidopsis Response to Iron Deficiency. *Front. Plant Sci.* **2019**, *10*, 324. [CrossRef]
47. Sunkar, R.; Zhu, J.K. Novel and stress-regulated micro-RNAs and other small RNAs from Arabidopsis. *Plant Cell* **2004**, *16*, 2001–2019. [CrossRef]
48. Chen, Y.; Jiang, J.; Song, A.; Chen, S.; Shan, H.; Luo, H.; Gu, C.; Sun, J.; Zhu, L.; Fang, W.; et al. Ambient temperature enhanced freezing tolerance of Chrysanthemum dichrum CdICE1 Arabidopsis via miR398. *BMC Biol.* **2013**, *11*, 121. [CrossRef]
49. Guan, Q.; Lu, X.; Zeng, H.; Zhang, Y.; Zhu, J. Heat stress induction of miR398 triggers a regulatory loop that is critical for thermotolerance in Arabidopsis. *Plant J.* **2013**, *74*, 840–851. [CrossRef]
50. Waters, B.M.; McInturf, S.A.; Stein, R.J. Rosette iron deficiency transcript and microRNA profiling reveals links between copper and iron homeostasis in Arabidopsis thaliana. *J. Exp. Bot.* **2012**, *63*, 5903–5918. [CrossRef]
51. Robinson, N.J.; Procter, C.M.; Connolly, E.L.; Guerinot, M.L. A ferric-chelate reductase for iron uptake from soils. *Nature* **1999**, *397*, 694–697. [CrossRef]
52. Murashige, T.; Skoog, F. A revised medium for rapid growth and bio assays with tobacco tissue cultures. *Physiol. Plant.* **1962**, *15*, 473–497. [CrossRef]
53. Epstein, E.; Bloom, A. *Mineral Nutrition in Plants: Principles and Perspectives*, 2nd ed.; Sinauer Associates: Sunderland, MA, USA, 2005.
54. Nakagawa, T.; Kurose, T.; Hino, T.; Tanaka, K.; Kawamukai, M.; Niwa, Y.; Toyooka, K.; Matsuoka, K.; Jinbo, T.; Kimura, T. Development of series of gateway binary vectors, pGWBs, for realizing efficient construction of fusion genes for plant transformation. *J. Biosci. Bioeng.* **2007**, *104*, 34–41. [CrossRef] [PubMed]
55. Clough, S.J.; Bent, A.F. Floral dip: a simplified method for Agrobacterium-mediated transformation of Arabidopsis thaliana. *Plant J.* **1998**, *16*, 735–743. [CrossRef] [PubMed]
56. Pilon-Smits, E.A.; Hwang, S.; Lytle, C.M.; Zhu, Y.; Tai, J.C.; Bravo, R.C.; Chen, Y.; Leustek, T.; Terry, N. Overexpression of ATP sulfurylase in Indian mustard leads to increased selenite uptake, reduction and tolerance. *Plant Physiol.* **1999**, *119*, 123–132. [CrossRef]
57. Maxwell, K.; Johnson, G.N. Chlorophyll fluorescence—a practical guide. *J. Exp. Bot.* **2000**, *51*, 659–668. [CrossRef]
58. Varkonyi-Gasic, E.; Hellens, R.P. Quantitative stem-loop RT-PCR for detection of microRNAs. *Methods Mol. Biol.* **2007**, *744*, 145–157.

11

Exploring MicroRNA Biomarkers for Parkinson's Disease from mRNA Expression Profiles

Y-h. Taguchi [1] and Hsiuying Wang [2,*]

[1] Department of Physics, Chuo University, 1-13-27 Kasuga, Bunky-ku, Tokyo 112-8551, Japan; tag@granular.com
[2] Institute of Statistics, National Chiao Tung University, Hsinchu 30010, Taiwan
[*] Correspondence: wang@stat.nctu.edu.tw;

Abstract: Parkinson's disease (PD) is a chronic, progressive neurodegenerative disease characterized by both motor and nonmotor features. The diagnose of PD is based on a review of patients' signs and symptoms, and neurological and physical examinations. So far, no tests have been devised that can conclusively diagnose PD. In this study, we explore both microRNA and gene biomarkers for PD. Microarray gene expression profiles for PD patients and healthy control are analyzed using a principal component analysis (PCA)-based unsupervised feature extraction (FE). 244 genes are selected to be potential gene biomarkers for PD. In addition, we implement these genes into Kyoto Encyclopedia of Genes and Genomes (KEGG) pathways, and find that the 15 microRNAs (miRNAs), hsa-miR-92a-3p, 16-5p, 615-3p, 877-3p, 100-5p, 320a, 877-5p, 23a-3p, 484, 23b-3p, 15a-5p, 324-3p, 19b-3p, 7b-5p and 505-3p, significantly target these 244 genes. These miRNAs are shown to be significantly related to PD. This reveals that both selected genes and miRNAs are potential biomarkers for PD.

Keywords: biomarker; gene; microRNA; parkinson's disease

---

1. Introduction

Parkinson's disease (PD), first described by Dr. James Parkinson in 1817, is a chronic, progressive neurodegenerative disease characterized by both motor and nonmotor features [1]. PD motor symptoms such as shaking, rigidity, and slowness of movement are caused by the loss of striatal dopaminergic neurons [2]. The nonmotor symptoms of PD include sleep disorders, depression, and cognitive changes [3–5].

The incidence and prevalence of PD increase with age. So far, it is problematic to conclusively diagnose PD because of the lack of a reference standard test [6]. The diagnose of PD is based on a review of patients' signs and symptoms, and neurological and physical examinations. Resting tremor, cogwheel rigidity, and bradykinesia are three "cardinal signs" of PD, and postural instability, a late finding in PD, is the fourth cardinal sign of PD [6].

Owing to the lack of a standard test for diagnosing PD, genetic testing of mutations in disease-causing genes may be one of a helpful way to diagnose familial Parkinson's disease (fPD) and sporadic Parkinson's disease (sPD). A number of PD disease-causing genes have been discovered and debated for both physicians and patients regarding diagnostic and presymptomatic genetic testing of PD in the clinic [7].

A common form of monogenic PD with dominant inheritance is caused by mutations in the gene for leucine-rich repeat kinase 2 (LRRK2) [8,9]; H-Synuclein (SNCA), which is a presynaptic neuronal protein, is linked genetically and neuropathologically to PD [10]; mutations in Parkin are the second most common known cause of PD [11]; mutations in DJ-1 and PTEN Kinase 1 (PINK1) can cause PD [11]; there is a significant inverse association of the ubiquitin carboxy-terminal hydrolase L1

(UCHL1) S18Y variant with PD [12]; and variants in glucocerebrosidase (GBA) and SNCA influence PD risk [13].

In addition to neurological and physical examinations and genetic testing, the gene expression differences between PD and healthy controls can be used as a potential prognosis of PD. Several studies have explored gene biomarkers for PD [14,15]. In this study, we explore gene and microRNA (miRNA) biomarkers for PD.

2. Materials and Methods

2.1. Gene Expression

We downloaded three mRNA expression profiles of PD and healthy control from Gene Expression Omnibus (GEO) under the GEO ID, GSE20295, GSE20163, and GSE20164. For all three profiles, raw data files (CEL files) (Supplementary file S1) were read with the function ReadAffy of the package Affy (Affymetrix; Thermo Fisher Scientific, Inc., Waltham, MA, USA) in R. Loaded CEL files were treated by mas5 function in the affy package. Then, write.exprs function was used to output normalized mRNA expression profiles. The tissue of the data was substantia nigra (see Supplementary file S1). Integrating the three datasets, we had 35 normal control and 25 PD patients' substantia nigra mRNA expression profiles.

2.2. Principal Component Analysis Based Unsupervised Feature Extraction

We applied principal component analysis (PCA)-based unsupervised feature extraction (FE) to mRNA expression profiles in order to select mRNAs that were expressed distinctly between controls and PD patients (see Supplementary file S2 for more details). This method has been successful in identifying potential biomarkers for other neurological disorders [16,17]. PCA was applied to only mRNAs identified by PCA-based unsupervised feature extraction (FE). PC loadings and PC scores were attributed to samples and genes, respectively. Furthermore, PC loadings are associated with the distinction between controls and PD patients; PD patients and controls were differentiated using linear discriminant analysis (LDA) with these selected PC loadings (see Supplementary file S2 for more details).

2.3. Validation of Obtained mRNAs

In order to validate the obtained mRNAs, we computed the area under the curve (AUC) of the receiver operating characteristic curve (ROC). Gene symbols associated with mRNAs selected by PCA-based unsupervised FE were uploaded to Enrichr [18,19], and various enriched biological terms were identified (see Supplementary file S4 for more details).

3. Results

255 probes (Supplementary file S3) were identified using PCA-based unsupervised FE. PCA was applied to 255 probes, and PC loadings attributed to samples were computed. The fourth PC loadings turned out to be associated with distinction between PD patients and controls. Table 1 shows the confusion table obtained by the linear discriminant analysis (LDA) using the fourth PC loading. Sensitivity of PD is 0.88, precision of PD is 0.73, F1 score (F-measure) is 0.8, and accuracy is 0.80. AUC is 0.95. Odds ratio is 20.5. p value computed by Fisher's exact test is 2.5×10^{-6}. All of these evaluations suggest that the fourth PC loadings can significantly discriminate controls and PD patients.

Although we found 255 probes that successfully discriminate PD patients from controls, biological evaluation of obtained probes is important. Two hundred and forty four gene symbols corresponding to these 255 probes (See Supplementary Materials) were uploaded to Enrichr for biological evaluation. There turned out to be many biological terms enriched in these gene symbols. Table 2 shows the top-ranked five categories in "Disease Perturbations from GEO down" of Enrichr (full list is available in Supplementary Material S4). The first four among these five categories are the PD. Since the GEO datasets used in Enrichr are not the same datasets used in our study, PCA based unsupervised FE successfully identified genes downregulated in PD patients.

Table 1. Confusion table obtained by linear discriminant analysis (LDA) using the fourth principal component (PC) loading obtained by principal component analysis (PCA) using 255 mRNAs identified by PCA-based unsupervised feature extraction (FE). Rows: true, columns: prediction.

	Control	**PD**
Control	24	8
PD	3	22

Table 2. Top ranked five categories in "Disease Perturbations from Gene Expression Omnibus (GEO) down" of Enrichr.

Term	**Overlap**	***p*-Value**	**Adjusted *p*-Value**
Parkinson's disease DOID-14330 human GSE19587 sample 740	65/207	5.02×10^{-83}	4.18×10^{-80}
Parkinson's disease DOID-14330 human GSE19587 sample 1080	56/167	5.88×10^{-73}	1.60×10^{-70}
Parkinson's disease DOID-14330 human GSE19587 sample 496	73/361	3.90×10^{-78}	1.59×10^{-75}
Parkinson's disease DOID-14330 human GSE7621 sample 940	67/365	2.96×10^{-68}	6.06×10^{-66}
Dystonia C0393593 human GSE3064 sample 329	62/317	1.06×10^{-64}	1.74×10^{-62}

As for "Disease Perturbations from GEO up" of Enrichr, PD is less enriched, but there are still seven PD experiments with a significant *p*-value (Table 3, full list in Supplementary Material S4).

Table 3. Seven PD expression profiles in "Disease Perturbations from GEO up" where 244 gene symbols are enriched.

Term	**Overlap**	***p*-Value**	**Adjusted *p*-value**
Parkinson's disease DOID-14330 human GSE19587 sample 741	33/158	5.14×10^{-35}	1.22×10^{-33}
Parkinson's disease DOID-14330 human GSE7621 sample 940	35/235	1.05×10^{-31}	1.74×10^{-30}
Parkinson's disease DOID-14330 human GSE7621 sample 941	38/342	1.55×10^{-29}	2.19×10^{-28}
Parkinson's disease DOID-14330 human GSE19587 sample 1080	37/433	1.17×10^{-24}	8.78×10^{-24}
Parkinson's disease DOID-14330 human GSE6613 sample 788	26/274	1.24×10^{-18}	5.50×10^{-18}
Parkinson's disease DOID-14330 human GSE19587 sample 496	15/239	1.03×10^{-8}	2.15×10^{-8}

The next biological term investigated is KEGG pathway (Table 4). Although PD was not top ranked, it is the eighth most significant enriched KEGG pathway. Tables 1–4 suggest that the identified 244 genes are significantly related to PD.

Table 4. Top ranked 10 KEGG pathways enriched in 244 identified gene symbols.

Term	**Overlap**	***p*-Value**	**Adjusted *p*-Value**
Ribosome_Homo sapiens_hsa03010	28/137	1.68×10^{-29}	2.92×10^{-27}
Phagosome_Homo sapiens_hsa04145	16/154	1.72×10^{-12}	1.49×10^{-10}
Synaptic vesicle cycle_Homo sapiens_hsa04721	10/63	4.11×10^{-10}	2.38×10^{-8}
Pathogenic Escherichia coli infection_Homo sapiens_hsa05130	9/55	2.38×10^{-9}	1.04×10^{-7}
Gap junction_Homo sapiens_hsa04540	10/88	1.18×10^{-8}	4.11×10^{-7}
Mineral absorption_Homo sapiens_hsa04978	8/51	2.68×10^{-8}	7.76×10^{-7}
Oxidative phosphorylation_Homo sapiens_hsa00190	10/133	6.09×10^{-7}	1.51×10^{-5}
Parkinson's disease_Homo sapiens_hsa05012	10/142	1.11×10^{-6}	2.42×10^{-5}
Vibrio cholerae infection_Homo sapiens_hsa05110	6/51	8.84×10^{-6}	1.71×10^{-4}
GABAergic synapse_Homo sapiens_hsa04727	7/88	2.15×10^{-5}	3.73×10^{-4}

In this regard, we found that 15 miRNAs significantly target 244 gene symbols by checking "miRTarBase 2017" in Enrichr (Table 5); 113 gene symbols out of 244 gene symbols were targeted by either of 15 miRNAs. These 15 miRNAs were reported to be related to PD (Table 5). Thus, we should consider these 15 miRNAs as key regulators of PD instead of 244 gene symbols. In reality, miRNAs are generally suggested as key regulator of PD [20].

Table 5. "miRTarBase 2017" in Enrichr when 244 gene symbols were uploaded.

Term	Overlap	*p*-Value	Adjusted *p*-Value	Reference
hsa-miR-92a-3p	37/1404	1.41×10^{-8}	2.71×10^{-5}	[21]
hsa-miR-16-5p	37/1555	1.93×10^{-7}	1.85×10^{-4}	[22]
hsa-miR-615-3p	25/891	1.38×10^{-6}	8.85×10^{-4}	[23]
hsa-miR-877-3p	19/606	5.92×10^{-6}	2.28×10^{-3}	[24]
hsa-miR-100-5p	12/250	5.37×10^{-6}	2.28×10^{-3}	[25]
hsa-miR-320a	18/584	1.33×10^{-5}	4.25×10^{-3}	[26]
hsa-miR-877-5p	11/235	1.68×10^{-5}	4.63×10^{-3}	[24]
hsa-miR-23a-3p	11/249	2.88×10^{-5}	6.91×10^{-3}	[25]
hsa-miR-484	22/890	4.37×10^{-5}	9.33×10^{-3}	[25]
hsa-miR-23b-3p	12/322	6.55×10^{-5}	1.26×10^{-2}	[27]
mmu-miR-15a-5p	15/499	9.42×10^{-5}	1.65×10^{-2}	[25]
hsa-miR-324-3p	12/338	1.04×10^{-4}	1.66×10^{-2}	[28]
mmu-miR-19b-3p	11/310	2.03×10^{-4}	3.00×10^{-2}	[20]
mmu-miR-7b-5p	13/438	3.13×10^{-4}	4.02×10^{-2}	[20]
hsa-miR-505-3p	9/222	2.93×10^{-4}	4.02×10^{-2}	[29]

4. Discussion

More Brain Synapse-Related Biological Terms Are Enriched

Although in the previous section, we notice that downregulated genes and upregulated genes in PD GEO datasets and in PD KEGG pathway are enriched in the 244 identified gene symbols, more detailed analyses give us more convincing insight about the relationship between these 244 gene symbols and PD. Other than PD, some KEGG pathways in Table 4 are also related to PD. For example, as for ribosome that is top-ranked in Table 4, ribosomal protein s15 phosphorylation was reported to mediate LRRK2 neurodegeneration in Parkinson's disease [30] As for phagosome that was the second top-ranked, LRRK2 was reported to be a negative regulator of Mycobacterium tuberculosis phagosome maturation in macrophages [31]. As for synaptic vesicle cycle that was the third top-ranked, synaptic vesicle trafficking was reported to be related to Parkinson's disease [32]. These results suggest that 244 gene symbols are expected to be closely related to PD progression mechanisms.

Although LRRK2 itself was not included in 244 gene symbols, significant number of genes obtained from "Single Gene Perturbations from GEO up/down" affected by LRKK2 KO/KI (knocked out/knocked in) were included in 244 gene symbols (Tables 6 and 7). This also supports that 244 gene symbols are related to mechanisms of PD.

Other than PD specificity, it is important to check if 244 genes are specifically upregulated in brain/synapse, since, otherwise, genes identified are not convincing. Tables 8 and 9 show that 244 genes are highly brain tissue-specific, while Table 10 shows that 244 genes are overlapped with genes downregulated in other tissues than brain (full list is in Supporting Materials).

Table 6. Top ranked four LRRK2 KO/KI experiments in "Single Gene Perturbations from GEO up".

Name	Overlap	p-Value	Adjusted p-Value
LRRK2 Gly2019Ser (G2019S) mutation knockin human GSE36321 sample 1688	21/335	1.03×10^{-11}	4.82×10^{-11}
LRRK2 mutant human GSE33298 sample 2039	16/309	4.94×10^{-8}	1.55×10^{-7}
LRRK2 dominant negative mutation-G2019S homozygous human GSE33298 sample 1743	12/280	1.68×10^{-5}	4.14×10^{-5}
LRRK2 dominant negative mutation-G2019S homozygous human GSE33298 sample 1741	12/337	1.01×10^{-4}	2.33×10^{-4}

Table 7. Top ranked two LRRK2 KO/KI experiments in "Single Gene Perturbations from GEO down".

Name	Overlap	p-Value	Adjusted p-Value
LRRK2 Gly2019Ser (G2019S) mutation knockin human GSE36321 sample 1688	24/265	8.38×10^{-17}	9.78×10^{-16}
LRRK2 dominant negative mutation-G2019S heterozygous human GSE33298 sample 1739	9/282	1.61×10^{-3}	3.56×10^{-3}

Table 8. Top ranked five gene expression profiles in drug matrix whose altered genes are associated with 244 gene symbols.

Term	Overlap	p-Value	Adjusted p-Value
Oxcarbazepine-1600-mg/kg-in_CMC-Rat-Brain-3d-dn	31/369	1.66×10^{-20}	1.31×10^{-16}
Carbachol-15-mg/kg_in_Water-Rat-Brain-3d-up	26/318	4.96×10^{-17}	7.82×10^{-14}
Piracetam-2500_mg/kg_in_CMC-Rat-Brain-5d-up	27/325	7.89×10^{-18}	2.56×10^{-14}
Theophylline-225_mg/kg_in_Water-Rat-Brain-3d-dn	25/314	3.87×10^{-16}	5.08×10^{-13}
Tramadol-114_mg/kg_in_Water-Rat-Brain-5d-dn	26/315	3.93×10^{-17}	7.75×10^{-14}

Table 9. Top ranked five gene expression profiles in "Genotype-Tissue Expression (GTEx) Tissue Sample Gene Expression Profiles up" whose altered genes are associated with 244 gene symbols.

Term	Overlap	p-Value	Adjusted p-Value
GTEX-X585-0011-R2B-SM-46MVF_brain_male_50-59_years	81/1895	1.63×10^{-33}	4.72×10^{-30}
GTEX-WHSE-0011-R2A-SM-3P5ZL_brain_male_20-29_years	71/1660	7.67×10^{-29}	1.11×10^{-25}
GTEX-X261-0011-R8A-SM-4E3I5_brain_male_50-59_years	70/1878	8.35×10^{-25}	8.08×10^{-22}
GTEX-N7MT-0011-R10A-SM-2I3E1_brain_female_60-69_years	70/1918	2.88×10^{-24}	2.09×10^{-21}
GTEX-TSE9-0011-R8A-SM-3DB7R_brain_female_60-69_years	62/1548	2.52×10^{-23}	1.47×10^{-20}

Table 10. Top ranked five gene expression profiles in "GTEx Tissue Sample Gene Expression Profiles down" whose altered genes are associated with 244 gene symbols.

Term	Overlap	p-Value	Adjusted p-Value
GTEX-S4Q7-1226-SM-4AD5I_testis_male_20-29_years	22/329	8.84×10^{-13}	2.36×10^{-9}
GTEX-U4B1-1526-SM-4DXSL_testis_male_40-49_years	20/282	3.48×10^{-12}	3.09×10^{-9}
GTEX-UPK5-1426-SM-4JBHH_liver_male_40-49_years	79/3879	2.06×10^{-12}	2.74×10^{-9}
GTEX-OHPM-2126-SM-3LK75_testis_male_50-59_years	26/525	6.48×10^{-12}	4.07×10^{-9}
GTEX-S7PM-0626-SM-4AD4Q_testis_male_60-69_years	34/911	7.63×10^{-12}	4.07×10^{-9}

Thus, we can conclude that 244 gene symbols are not only PD-specific but also brain tissue-specific. This also suggests that we successfully identified gene-related PD mechanisms. In spite of successfully identifications, it is not very easy to investigate as many as 244 gene symbols one by one. It is better to find a more limited number of factors that regulate 244 gene symbols.

For the selected 15 miRNAs in Table 5, we confirmed our results by comparing other studies. hsa-miR-92a appeared as novel hub miR in both regulatory and co-expression network, indicating its strong functional role in PD; GBA deficiency is associated with PD, and miR-16-5p has been shown to correspond to enhanced GBA protein levels [22,33]; and both PD and HD (Huntington's disease) are neurodegenerative and caused by protein inclusions. miR-615-3p was identified as differentially expressed in HD prefrontal cortex compared to non-neurological disease controls, and hsa-miR-615-3p was identified up-regulated in HD [23]; miR-100, miR-23a, and miR484 were identified to be PD-related miRNAs [25]; miR-320a was identified as a PD-related miRNA [26]; tumor necrosis factor-alpha (TNF-α), a pro-inflammatory cytokine, was elevated in blood, CSF, and striatum regions of the brain in PD patients, and hsa-miR-23a, hsa-miR-23b, and hsa-miR-320a significantly decreased in the presence of TNF-α [27]; mmu-miR-15a-5p, mmu-miR-19b-3p, and mmu-miR-7b-5p miR-7 were shown to downregulate the inflammatory response in cellular in vitro and/or in vivo PD neurotoxic models, and miR-19b was shown to be downregulated in the prodromal stage of alpha-synucleinopathies and pinpointed this miRNA as a potential biomarker also for PD and DLB [20]; and miR-505-3p was identified as a PD-predictive miRNA by microarrays [29].

5. Conclusions

PD is a long-term degenerative disorder that usually affects elderly people. Since there are no standard tests that can conclusively diagnose PD, biological biomarkers can help in early diagnosis. In this study, PD-related mRNAs are selected using a PCA-based, unsupervised FE method, and miRNA biomarkers of PD are explored based on these selected mRNAs. Two hundred and forty-four genes and 15 miRNAs are identified to be related to PD. Biological evidence shows the selected genes and miRNAs are potential PD biomarkers. However, since the tissues used in this study are substantia nigra from postmortem brain, it needs to be further verified whether these selected biomarkers can be used as blood/serum PD biomarkers.

Author Contributions: Conceptualization, H.W.; Methodology, Y-h.T.; Formal Analysis, Y-h.T.; Writing, Y-h.T. and H.W.

References

1. DeMaagd, G.; Philip, A. Parkinson's disease and its management: Part 1: Disease entity, risk factors, pathophysiology, clinical presentation, and diagnosis. *Pharm. Ther.* **2015**, *40*, 504.
2. Drui, G.; Carnicella, S.; Carcenac, C.; Favier, M.; Bertrand, A.; Boulet, S.; Savasta, M. Loss of dopaminergic nigrostriatal neurons accounts for the motivational and affective deficits in Parkinson's disease. *Mol. Psychiatry* **2014**, *19*, 358–367. [CrossRef] [PubMed]
3. Neikrug, A.B.; Maglione, J.E.; Liu, L.; Natarajan, L.; Avanzino, J.A.; Corey-Bloom, J.; Palmer, B.W.; Loredo, J.S.; Ancoli-Israel, S. Effects of sleep disorders on the non-motor symptoms of Parkinson disease. *J. Clin. Sleep Med.* **2013**, *9*, 1119–1129. [CrossRef] [PubMed]
4. Marsh, L. Depression and Parkinson's disease: Current knowledge. *Curr. Neurol. Neurosci. Rep.* **2013**, *13*, 409. [CrossRef] [PubMed]
5. Watson, G.S.; Leverenz, J.B. Profile of cognitive impairment in Parkinson's disease. *Brain Pathol.* **2010**, *20*, 640–645. [CrossRef] [PubMed]
6. Levine, C.B.; Fahrbach, K.R.; Siderowf, A.D.; Estok, R.P.; Ludensky, V.M.; Ross, S.D. Diagnosis and treatment of Parkinson's disease: A systematic review of the literature. *Evid. Rep. Technol. Assess.* **2003**, *57*, 1–4.
7. Tan, E.-K.; Jankovic, J. Genetic testing in Parkinson disease: Promises and pitfalls. *Arch. Neurol.* **2006**, *63*, 1232–1237. [CrossRef]
8. Zimprich, A.; Biskup, S.; Leitner, P.; Lichtner, P.; Farrer, M.; Lincoln, S.; Kachergus, J.; Hulihan, M.; Uitti, R.J.; Calne, D.B.; et al. Mutations in LRRK2 cause autosomal-dominant parkinsonism with pleomorphic pathology. *Neuron* **2004**, *44*, 601–607. [CrossRef]

9. Gasser, T. Usefulness of genetic testing in PD and PD trials: A balanced review. *J. Park. Dis.* **2015**, *5*, 209–215. [CrossRef]
10. Stefanis, L. α-Synuclein in Parkinson's disease. *Cold Spring Harbor Perspect. Med.* **2012**, *2*, a009399. [CrossRef]
11. Dawson, T.M.; Dawson, V.L. The role of parkin in familial and sporadic Parkinson's disease. *Mov. Disord.* **2010**, *25*, S32–S39. [CrossRef] [PubMed]
12. Tan, E.K.; Puong, K.Y.; Fook-Chong, S.; Chua, E.; Shen, H.; Yuen, Y.; Pavanni, R.; Wong, M.-C.; Puvan, K.; Zhao, Y. Case-control study of UCHL1 S18Y variant in Parkinson's disease. *Mov. Disord.* **2010**, *25*, S32–S39. [CrossRef] [PubMed]
13. Davis, A.A.; Andruska, K.M.; Benitez, B.A.; Racette, B.A.; Perlmutter, J.S.; Cruchaga, C. Variants in GBA, SNCA, and MAPT influence Parkinson disease risk, age at onset, and progression. *Neurobiol. Aging* **2016**, *37*, 209. e1–209. e7. [CrossRef] [PubMed]
14. Pinho, R.; Guedes, L.C.; Soreq, L.; Lobo, P.P.; Mestre, T.; Coelho, M.; Rosa, M.M.; Gonçalves, N.; Wales, P.; Mendes, T. Gene expression differences in peripheral blood of Parkinson's disease patients with distinct progression profiles. *PLoS ONE* **2016**, *11*, e0157852. [CrossRef]
15. Aguiar, P.M.C.; Severino, P. Biomarkers in Parkinson Disease: global gene expression analysis in peripheral blood from patients with and without mutations in PARK2 and PARK8. *Einstein* **2010**, *8*, 291–297. [CrossRef] [PubMed]
16. Taguchi, Y.H.; Wang, H. Exploring microRNA Biomarker for Amyotrophic Lateral Sclerosis. *Int. J. Mol. Sci.* **2018**, *19*, 1318. [CrossRef] [PubMed]
17. Taguchi, Y.H.; Wang, H. Genetic association between amyotrophic lateral sclerosis and cancer. *Genes* **2017**, *8*, 243. [CrossRef]
18. Kuleshov, M.V.; Jones, M.R.; Rouillard, A.D.; Fernandez, N.F.; Duan, Q.; Wang, Z.; Koplev, S.; Jenkins, S.L.; Jagodnik, K.M.; Lachmann, A.; et al. Enrichr: A comprehensive gene set enrichment analysis web server 2016 update. *Nucleic Acids Res.* **2016**, *44*, W90–W97. [CrossRef]
19. Chen, E.Y.; Tan, C.M.; Kou, Y.; Duan, Q.; Wang, Z.; Meirelles, G.V.; Clark, N.R.; Ma'ayan, A. Enrichr: Interactive and collaborative HTML5 gene list enrichment analysis tool. *BMC Bioinf.* **2013**, *14*, 128. [CrossRef]
20. Leggio, L.; Vivarelli, S.; L'Episcopo, F.; Tirolo, C.; Caniglia, S.; Testa, N.; Marchetti, B.; Iraci, N. microRNAs in Parkinson's disease: From pathogenesis to novel diagnostic and therapeutic approaches. *Int. J. Mol. Sci.* **2017**, *18*, 2698. [CrossRef]
21. Chatterjee, P.; Bhattacharyya, M.; Bandyopadhyay, S.; Roy, D. Studying the system-level involvement of microRNAs in Parkinson's disease. *PLoS ONE* **2014**, *9*, e93751. [CrossRef] [PubMed]
22. Hoss, A.G.; Labadorf, A.; Beach, T.G.; Latourelle, J.C.; Myers, R.H. microRNA profiles in Parkinson's disease prefrontal cortex. *Front. Aging Neurosci.* **2016**, *8*, 36. [CrossRef] [PubMed]
23. Hoss, A.G.; Kartha, V.K.; Dong, X.; Latourelle, J.C.; Dumitriu, A.; Hadzi, T.C.; Macdonald, M.E.; Gusella, J.F.; Akbarian, S.; Chen, J.F.; et al. MicroRNAs located in the Hox gene clusters are implicated in huntington's disease pathogenesis. *PLoS Genet.* **2014**, *10*, e1004188. [CrossRef] [PubMed]
24. Sibley, C.R.; Seow, Y.; Curtis, H.; Weinberg, M.S.; Wood, M.J. Silencing of Parkinson's disease-associated genes with artificial mirtron mimics of miR-1224. *Nucleic Acids Res.* **2012**, *40*, 9863–9875. [CrossRef] [PubMed]
25. Chen, L.; Yang, J.; Lu, J.; Cao, S.; Zhao, Q.; Yu, Z. Identification of aberrant circulating miRNAs in Parkinson's disease plasma samples. *Brain Behav.* **2018**, *8*, e00941. [CrossRef]
26. Heman-Ackah, S.M.; Hallegger, M.; Rao, M.S.; Wood, M.J. RISC in PD: The impact of microRNAs in Parkinson's disease cellular and molecular pathogenesis. *Front. Mol. Neurosci.* **2013**, *6*, 40. [CrossRef]
27. Prajapati, P.; Sripada, L.; Singh, K.; Bhatelia, K.; Singh, R.; Singh, R. TNF-α regulates miRNA targeting mitochondrial complex-I and induces cell death in dopaminergic cells. *Biochim. Biophys. Acta Mol. Basis Dis.* **2015**, *1852*, 451–461. [CrossRef]
28. Vallelunga, A.; Ragusa, M.; Di Mauro, S.; Iannitti, T.; Pilleri, M.; Biundo, R.; Weis, L.; Di Pietro, C.; De Iuliis, A.; Nicoletti, A. Identification of circulating microRNAs for the differential diagnosis of Parkinson's disease and Multiple System Atrophy. *Front. Cell. Neurosci.* **2014**, *8*, 156. [CrossRef]
29. Khoo, S.K.; Petillo, D.; Kang, U.J.; Resau, J.H.; Berryhill, B.; Linder, J.; Forsgren, L.; Neuman, L.A.; Tan, A.C. Plasma-based circulating MicroRNA biomarkers for Parkinson's disease. *J. Park. Dis.* **2012**, *2*, 321–331.
30. Martin, I.; Kim, J.W.; Lee, B.D.; Kang, H.C.; Xu, J.C.; Jia, H.; Stankowski, J.; Kim, M.S.; Zhong, J.; Kumar, M.; et al. Ribosomal protein s15 phosphorylation mediates LRRK2 neurodegeneration in Parkinson's disease. *Cell* **2014** *157*, 472–485. [CrossRef]

31. Härtlova, A.; Herbst, S.; Peltier, J.; Rodgers, A.; Bilkei-Gorzo, O.; Fearns, A.; Dill, B.D.; Lee, H.; Flynn, R.; Cowley, S.A. LRRK2 is a negative regulator of Mycobacterium tuberculosis phagosome maturation in macrophages. *EMBO J.* **2018**, *37*, e98694. [CrossRef] [PubMed]
32. Esposito, G.; Clara, F.A.; Verstreken, P. Synaptic vesicle trafficking and Parkinson's disease. *Dev. Neurobiol.* **2012**, *72*, 134–144. [CrossRef] [PubMed]
33. Siebert, M.; Westbroek, W.; Chen, Y.-C.; Moaven, N.; Li, Y.; Velayati, A.; Saraiva-Pereira, M.L.; Martin, S.E.; Sidransky, E. Identification of miRNAs that modulate glucocerebrosidase activity in Gaucher disease cells. *RNA Biol.* **2014**, *11*, 1291–1300. [CrossRef] [PubMed]

12

Circular and microRNAs from *Arabidopsis thaliana* Flowers are Simultaneously Isolated from AGO-IP Libraries

Érika Frydrych Capelari [1,2,†], Guilherme Cordenonsi da Fonseca [1,†], Frank Guzman [1] and Rogerio Margis [1,2,3,*]

[1] Programa de Pós-graduação em Biologia Celular e Molecular (PPGBCM), Centro de Biotecnologia, Universidade Federal do Rio Grande do Sul, Porto Alegre 91501-970, Brazil
[2] Programa de Pós-graduação em Genética e Biologia Molecular (PPGBM), Universidade Federal do Rio Grande do Sul, Porto Alegre 91501-970, Brazil
[3] Centro de Biotecnologia, Laboratório de Genomas e Populações de Plantas (LGPP), Universidade Federal do Rio Grande do Sul, Av. Bento Gonçalves, 9500—Laboratório 206 Prédio 43422, Porto Alegre 91501-970, Brazil
* Correspondence: rogerio.margis@ufrgs.br
† These authors contributed equally to this work.

Abstract: Competing endogenous RNAs (ceRNAs) are natural transcripts that can act as endogenous sponges of microRNAs (miRNAs), modulating miRNA action upon target mRNAs. Circular RNAs (circRNAs) are one among the various classes of ceRNAs. They are produced from a process called back-splicing and have been identified in many eukaryotes. In plants, their effective action as a miRNA sponge was not yet demonstrated. To address this question, public mRNAseq data from Argonaute-immunoprecipitation libraries (AGO-IP) of *Arabidopsis thaliana* flowers were used in association with a bioinformatics comparative multi-method to identify putative circular RNAs. A total of 27,812 circRNAs, with at least two reads at the back-splicing junction, were identified. Further analyses were used to select those circRNAs with potential miRNAs binding sites. As AGO forms a ternary complex with miRNA and target mRNA, targets count in AGO-IP and input libraries were compared, demonstrating that mRNA targets of these miRNAs are enriched in AGO-IP libraries. Through this work, five circRNAs that may function as miRNA sponges were identified and one of them were validated by PCR and sequencing. Our findings indicate that this post-transcriptional regulation can also occur in plants.

Keywords: circRNA; microRNA; non-coding RNA; argonaute; immunoprecipitation; plant

---

1. Introduction

The advancements in high throughput sequencing technologies and the development of new bioinformatics tools expanded the knowledge about non-coding RNAs (ncRNAs) and their functions as regulators of gene expression. The ncRNAs can be subdivided into two major classes: (i) small non-coding RNAs (sncRNAs) and (ii) long non-coding RNAs (lncRNAs) [1]. The lncRNAs are usually more than 300 nucleotides in length and can be regulated by microRNAs (miRNAs) [2,3]. miRNAs represent small RNAs, with approximately 19–24 nucleotides, and their main function is to act as a post-transcriptional gene regulator, through the RNA Induced Silencing Complex (RISC). Argonaute (AGO) is the main protein involved in this regulatory complex. It harbors small RNAs in its active site and promotes the interaction between the miRNA sequence and the target messenger RNA (mRNA), forming a ternary miRNA:AGO:mRNA complex. It leads to a repression in gene expression either by the mRNA cleavage or by translational repression [4]. The regulation mediated by miRNA occurs through the base pairing of complementary sequences, known as miRNA response elements (MREs), between mRNA and miRNA [5].

Competing endogenous RNAs (ceRNAs) are transcripts of coding or non-coding genes that have MREs and can compete with mRNA targets for miRNAs binding. They can promote a reduction in miRNA action by decreasing their availability in the cytoplasm [6–9]. A typical example of this mechanism is represented by ncRNA IPS1, which interacts with miR-399 by mimicking the MRE of its target mRNA PHO2, in a mechanism called target mimicry [10]. It has been suggested that an interaction network exists among ceRNAs, which communicate and co-regulate themselves through competition for a limited set of miRNA [7]. Therefore, all transcripts that share similar MREs can potentially compete for a specific miRNA.

A distinct class of newly discovered endogenous non-coding RNA was denominated as circular RNA (circRNA) [11,12]. In the early 1990s, due to their low levels of expression, circRNAs were considered as being splicing artifacts, corresponding to transcripts with scrambled exon order and splicing errors [13–15]. circRNAs were also associated with pathologic agents like hepatitis delta virus (HDV) [16] and plant viroids [17]. With the advent of next-generation sequencing technology and bioinformatic tools, the identification, biogenesis, and functions of circRNAs have been described, allowing a better understanding of these molecules [18,19]. Thus, many circRNAs were shown to be expressed as abundant and stable molecules [18] in different organisms, like humans [20,21], animals [12,22], yeast [23], bacteria [24], and plants [19,23,25–29]. In addition, circRNAs exhibited development-specific, tissue-specific and cell type-specific expression in animals, suggesting a regulatory role [18,30,31].

CircRNAs are characterized by the lack of 5′ caps and 3′ poly-A tails. Instead, they form a covalently closed loop structure originated by back-splicing circularization in a mechanism mediated by the spliceosomes. In this process, the 3′ region of a downstream exon of a given gene is linked to the 5′ region of an upstream exon of that same gene. The circularization enhances the RNA stability, making circRNAs resistant to RNase R, an exonuclease that degrades linear RNAs [32]. Due to this stability, some exonic circRNAs have been shown to be at higher concentrations than their linear counterparts [18,33,34]. circRNAs can be originated from exons [15,34], introns [12,35] or both [36]. However, most of the circRNAs are originated from exons of protein-coding genes [37]. Thus, circRNAs may comprise a single or multiples exons.

Another feature of circRNAs that has aroused great interest is its multi-functionality. circRNAs have been implicated in: (i) regulation of RNA processing [22,38], (ii) transcription regulation [39], (iii) interaction with RNA binding proteins and ribonucleoproteins complexes [40,41], and (iv) acting as microRNA sponges, preventing miRNAs to bind their target mRNAs [12]. Furthermore, miRNA binding sites in circRNAs are less likely to have polymorphisms than flanking sequences or random sites, suggesting an important role of circRNAs in the regulation of miRNA activities [42]. Up to now, the study of circRNAs in plants has received much less attention, compared to the wide comprehensive knowledge of circRNAs in mammals, in which a large number of circRNAs have been identified and characterized [43].

Recent studies have shown that circRNA are present in many species of plants [19,23,25–27,29]. However, it was not yet demonstrated whether they could effectively act as miRNA sponges. To address this question, we used a publicly available sequencing data from an Argonaute-immunoprecipitation experiment (AGO-IP) from *Arabidopsis thaliana* flowers followed by sequencing of the associated RNAs [44] to screen for circRNAs with miRNA binding sites. In the present work, five putative circRNAs that may function as miRNA sponges were found, with one of them being validated by PCR and sequencing. Our findings suggest the existence of AGO-miRNA-circRNAs complexes, and contribute another step in the understanding of post-transcriptional regulation mechanisms in plants.

2. Results

2.1. Identification of circRNAs in AGO-IP Libraries

Circular RNAs with potential to act as sponges for miRNAs were identified in RNAseq data from libraries prepared from total RNA extracted from flowers A. thaliana. In a previous study, Carbonel and coworkers produced three independent libraries corresponding to Argonaute immunoprecipitation (AGO-IP) libraries [44]. These libraries were used in our analyses. Two lines of A. thaliana overexpressing the Argonaute wild type (DDH) and another overexpressing a mutant line with no ability to slice (DAH). Specific AGO-IP was carried using monoclonal antibodies directed against the human influenza hemagglutinin (HA) sequence tag present in the recombinant AGO (Figure 1).

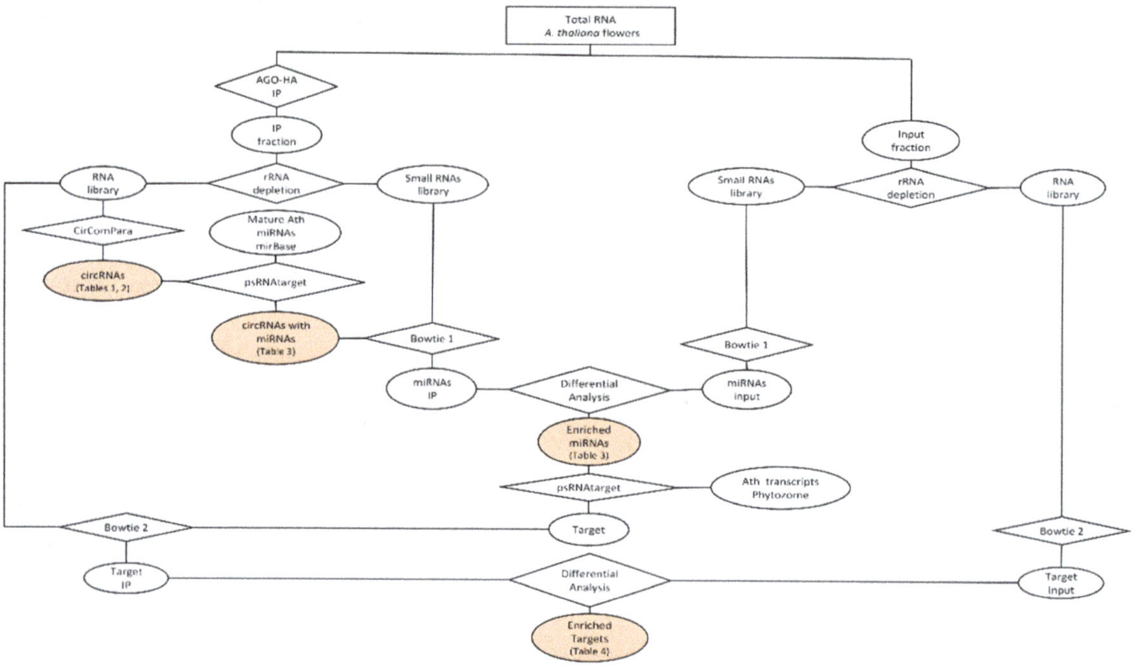

Figure 1. Flowchart for identification of circRNAs, miRNAs and target mRNAs in AGO-IP and control libraries. The total RNA from *A. thaliana* flowers was divided in two fractions. One of them went through Argonaute immunoprecipitation (IP fraction) and the other was used as control (Input fraction). Different methodologies are represented by rhombus, while the outputs are represented by ellipses. Filled ellipses correspond to results also presented in tables.

The use of CirComPara allowed the identification of up to 29.358 circRNAs in AGO-IP RNAseq libraries (Table 1). Using the CircExplorer2 with the Segemehl anchor 86 putative circRNAs were identified, while using the Star anchor 15 and with TopHat, 23. The number of predicted circRNAs identified by FindCirc algorithm was 1422 and by the TestRealign was 27.812. The number of circRNAs hits is reduced to only three when certain methods that are more stringent are used (Table 1).

Table 1. Number of circRNAs identified in AGO-IP libraries by 5 different methods.

Identification Method		CircExplorer2			FindCirc	TestRealign
		Segemehl	Star	Tophat	-	-
CircExplorer2	Segemehl	86	9	12	10	26
	Star	-	15	7	3	3
	Tophat	-	-	23	7	7
FindCirc		-	-	-	1422	198
TestRealign		-	-	-	-	27,812

So far, we decided to focus on those circRNAs identified by at least three different methods. The description of these 12 circRNAs, including the library from which they were identified, the locus and function of parental gene, their origin and length are described in Table 2. The coordinates of the 12 circRNAs in the A. thaliana genome is listed in Supplementary Table S3. The majority of circRNAs was originated from perfect exon back-splicing, while two were produced from introns and another resulted from an imperfect exon back-splicing. The number of exons that form the chosen circRNAs varied from one to four. Their sizes ranged from 49 nt (At5g16880) to 1063 nt (At2g42170).

Table 2. Description of 12 putative circRNAs predicted by at least three methods.

Library	Gene_ID	Circ_ID	Parental Gene Function	Origin	Exons	Length (nt) ***	Methods
DDH-IP	At1g02560	circ_At1g02560	Nuclear encoded CLP protease 5	exonic	2	123	5
DDH-IP	At1g12080 **	circ_At1g12080	Vacuolar calcium-binding protein-related	exonic *	1	95	4
DDH-IP	At1g31810	circ_At1g31810	Formin Homology 14	exonic	1	50	3
DDH-IP	At1g52360	circ_At1g52360	Coatomer beta subunit	intronic	1	224	3
DDH-IP	At2g02410	circ_At2g02410	K06962—uncharacterized protein (K06962)	exonic	1	71	4
DDH-IP	At2g35940 **	circ_At2g35940	BEL1-like homeodomain 1	exonic	1	930	4
DDH-IP	At2g42170 **	circ_At2g42170	Actin family protein	exonic	4	1063	5
DDH-IP	At5g16880	circ_At5g16880	Target of Myb protein 1	exonic *	1	49	4
DDH-IP	At5g56950	circ_At5g56950	NAP-1 Nucleosome assembly protein	intronic *	1	118	4
DAH-IP	At3g01800	circ_At3g01800	Ribosome recycling factor	exonic	1	68	4
DAH-IP	At3g13990 **	circ_At3g13990	Kinase-related protein (DUF1296)	exonic	3	349	3
DAH-IP	At5g27720 **	circ_At5g27720	Small nuclear ribonucleoprotein family protein	exonic	4	321	5

* circRNA originated from an imperfect back-splicing; ** circRNAs with miRNA binding site; *** only exons considered.

2.2. circRNAs with miRNA Binding Sites

In order to identify those plants circRNAs that can function as miRNA sponges, only the five circRNAs that have binding sites to miRNAs were selected, among the 12 previous circRNAs (Table 3). The read count of each of the 12 circRNA, matching the back-splicing junction, was analyzed in both AGO-IP and control libraries, in order to identify the enrichment in the AGO-IP (Table 3). In total, AGO-IP libraries had 284,490,887 reads, while the control library had 594,458,195. From the 428 mature A. thaliana miRNAs, 14 miRNAs were predicted as having at least one of the five circRNAs as targets. 10 from these miRNAs were more abundant in AGO-IP libraries (highlighted with an *) in comparison to input library (Table 3). All the miRNAs that have predicted sites of translational inhibition were enriched. Those with cleavage sites were poorly represented or not detected at any library.

Table 3. Read counts of circRNA and microRNAs that are potentially associated.

circRNA	circRNA Read Counts ***		miRNA	miRNA Read Counts			Inhibiton By
	AGO-IP	Total RNA		AGO1-DDH	AGO1-DAH	Empty Vector	
circ_At1g12080 **	175	13	miR4221-5p *	265	171	7	Cleavage
			miR838-3p *	226	59	41	Translation
-	-	-	miR397a-5p *	1014	520	29	Translation
-	-	-	miR5654-3p	2	0	0	Cleavage
circ_At2g35940	8	0	miR8182-5p *	2	9	0	Translation

Table 3. *Cont.*

circRNA	circRNA Read Counts ***		miRNA	miRNA Read Counts			Inhibiton By
	AGO-IP	Total RNA		AGO1-DDH	AGO1-DAH	Empty Vector	
-	-	-	miR830-3p *	29	13	0	Cleavage
-	-	-	miR833a-5p *	50	35	17	Translation
-	-	-	miR8174-3p	-	-	-	Cleavage
circ_At2g42170	8	0	miR831-3p *	81	11	3	Translation
-	-	-	miR838-3p *	226	59	41	Cleavage
-	-	-	miR4239-5p *	17	7	0	Translation
circ_At3g13990 **	17	0	miR5637-5p	-	-	-	Cleavage
-	-	-	miR780.2-3p	-	-	-	Cleavage
circ_At5g27720	17	0	miR838-3p *	226	59	41	Cleavage

* miRNA considered enriched in AGO-IP libraries; ** circRNAs validated; *** Read count normalized by the library size and with difference between assembled AGO-IPs (AGO-DAH/DDH) and Input libraries ($p < 0.05$); Expectation value ≤5.

2.3. The circRNAs Harbor Reverse Complementary Sequences of miRNAs which Targeted mRNAs Present in AGO-IP Libraries

The enriched miRNAs were selected to evaluate if their predicted target mRNAs were also present and enriched in AGO-IP libraries. In total, 260 mRNA targets were identified with an expectation range from 0.5 to 3 (Supplementary Table S2). Six out of the 10 enriched miRNAs presented mRNA targets with reads that were significantly more frequent in AGO-IP libraries than in the control input, reducing the number of predicted targets to 64 (Table 4).

Table 4. mRNAs targeted by miRNAs with circRNAs and enriched in AgoIP libraries.

Target_Access	miRNA	Expectation	Inhibition By	Lenght	Target Counts *		Function
					AgoIP	Input	
At2g38080.1		1	Cleavage	2021	58	21	Laccase/Diphenol oxidase
At5g60020.1	miR397a-5p	1	Cleavage	2049	33	18	Laccase 17
At3g06040.1		3	Cleavage	864	29	14	Ribosomal protein L12
At3g06470.1		3	Cleavage	1092	75	4	GNS1/SUR4 membrane protein
At3g54170.1		2.5	Cleavage	1262	22	10	FKBP12 interacting protein 37
At4g13070.1		2.5	Cleavage	1775	8	2	RNA-binding CRS1
At5g60040.1		2.5	Cleavage	4582	62	22	Nuclear RNA polymerase C1
At1g13350.1	miR4221-5p	3	Cleavage	2454	142	24	Protein kinase
At1g77660.1		3	Cleavage	1765	22	12	H3K4-specific methyltransferase
At2g33240.1		3	Cleavage	5313	36	12	Myosin XI D
At3g02170.1		3	Cleavage	3300	319	155	Longifolia2
At4g14510.1		3	Cleavage	2940	57	22	CRM family member 3B
At1g31650.1		3	Translation	2255	164	28	RHO guanyl-exchange factor 14
At2g38610.1		3	Translation	1452	56	26	RNA-binding KH protein
At2g35160.1	miR8182-5p	3	Cleavage	2798	20	9	SU(VAR)3-9 homolog 5
At4g22580.1		3	Cleavage	1628	39	10	Exostosin family protein
At1g23400.1		3	Cleavage	1822	81	24	RNA-binding CRS1
At1g49880.1	miR831-3p	2.5	Translation	803	50	2	FAD-linked sulfhydryl oxidase
At3g46060.1		3	Translation	1132	75	41	RAS-related protein RABE1C

Table 4. *Cont.*

Target_Access	miRNA	Expectation	Inhibition By	Lenght	Target Counts *		Function
					AgoIP	Input	
At2g36890.1	miR833a-5p	2.5	Cleavage	971	6	1	Myb-like DNA-binding domain
At3g12380.1		2.5	Cleavage	2323	33	16	Actin-related protein 5
At1g21740.1		3	Cleavage	2862	63	22	Protein of unknown function
At1g64180.1		3	Cleavage	2072	13	3	Intracellular transport protein
At1g70470.1		3	Cleavage	765	17	4	No annotated domains
At4g01080.1	miR838-3p	3	Cleavage	1583	98	33	Trichome-birefringence like 26
At5g09460.1		3	Cleavage	2546	124	41	Transcription Factor SAC51
At5g09461.1		3	Cleavage	2546	124	41	Conserved peptide upstream ORF
At5g20110.1		3	Cleavage	778	28	2	Dynein light chain type 1
At5g46030.1		2	Translation	732	26	12	Elongation factor Elf1 like
At2g44430.1		2.5	Translation	2196	98	19	DNA-binding protein
At5g22640.1		2.5	Translation	2814	247	115	MORN repeat-containing protein
At5g40340.1		2.5	Translation	3096	624	75	Tudor/PWWP/MBT protein
At5g56210.1		2.5	Translation	2004	22	5	WPP domain interacting protein 2
At5g62390.1		2.5	Translation	1859	349	152	BCL-2-associated athanogene 7
At5g17910.1		3	Translation	4532	178	77	No annotated domains
At5g41960.1		3	Translation	874	9	4	No annotated domains
At5g57790.1		3	Translation	1407	29	12	No annotated domains

* Read count normalized by the library size and with difference between assembled AGO-IPs (AGO-DAH/DDH) and Input libraries ($p < 0.05$).

2.4. circRNAs Validation by RT-PCR and Sequencing

PCR reactions with divergent primers were used in order to validate the back-splicing site of the five circRNAs presenting miRNA binding sites, all with more than two reads in AGO-IP libraries. Only one of the five circRNAs predicted by bioinformatics was amplified by RT-PCR using total RNAs extracted from A. thaliana flowers followed by RNase treatment and divergent primers (Figure 2). The circ_At3g13990 showed the expected electrophoretic band profile of 312 bp (Table 2). PCR negative and positive controls were done using genomic DNA (gDNA) and cDNAs from the parental gene with divergent and convergent primers, respectively. These amplification products were not detected in RNA samples from leaf, silique and steam (data not shown).

The total RT-PCR product from circ_At3g1399080 was purified and submitted to Sanger sequencing. The sequence resulted from back-splicing of At3g13990 exon 4 (E4) and exon 2 (E2) was obtained using the Primer circular Forward (PcF) (Figure 3). This result was also corroborated by 34 reads, present in AGO-IP libraries, that overhang with 3 or 4 nucleotides over the back-splicing site.

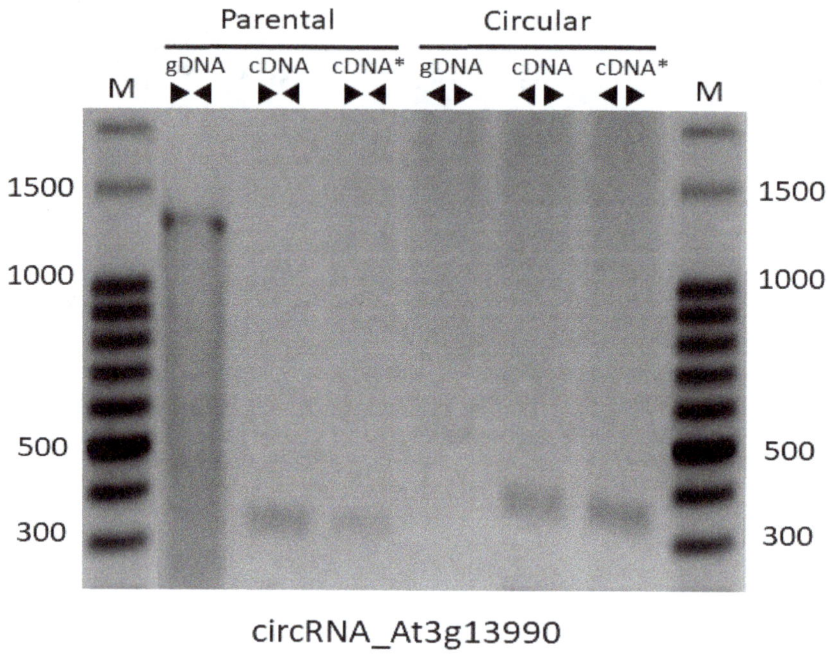

Figure 2. Validation of circRNA by RT-PCR. PCR reactions were performed using divergent primers (◀▶) to amplify the circRNA_At3g13990. Convergent primers (▶◀) were used to amplify parental mRNA. Genomic DNA (gDNA) was used as control. Samples were analyzed on 1,5% agarose gel. (M) DNA size marker of 100 bp; cDNA: complementary DNA; cDNA*: complementary DNA produced from total RNA treated with RNase R previously to reverse transcription. bp: base pairs.

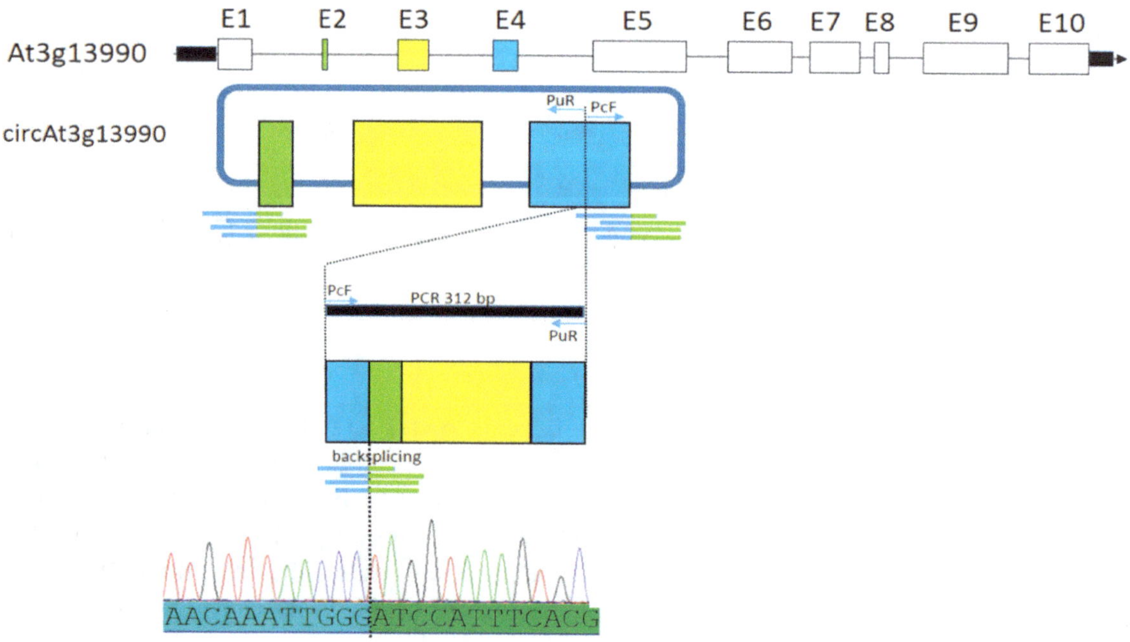

Figure 3. circRNA_At3g13990 back-splicing validation by sequencing. The parental gene structure is represented by exons (boxes), introns (black lines) and 5′ and 3′ untranslated regions (black rectangles). Filled boxes represent exons encompassing the circRNA. Sequencing reactions were performed using PcF and PuR primer. Lines indicated below the colored boxes represent reads matching the back-splicing junction. The nucleotide sequence flanking the back-splicing is represented as an electropherogram. Primer universal Reverse (PuR), Primer circular Forward (PcF) and base pairs (bp).

3. Discussion

At present, the role of circular RNA (circRNA) as one of the several classes of competing endogenous RNA (ceRNA) was only demonstrated in animals. They can act as sponges of miRNAs, modulating miRNA action upon target mRNAs. Nevertheless, circRNAs have been identified all across the eukaryotic tree of life [23]. Argonaute (AGO) is an important regulatory protein, with nuclease activity, that is involved in the pathway of RNA-induced silencing. AGO harbors a small RNA in its active site and places it in the correct sequence position in relation to the RNA target in the silencing complex (RISC). The interaction between miRNA-AGO and mRNA target forms a ternary complex and leads to transcripts regulation either by the mRNA cleavage or by translational repression [4]. Another possible molecular component in this ternary complex would be miRNA-AGO and circRNAs. Considering that circRNAs can act as miRNA sponges in mammals [12], publicly available sequencing data from AGO-IP RNAseq libraries were used to screen for circRNAs with this same function in *A. thaliana*. There are studies using the AGO-IP protocol [2,45]. However, because there are no other experiments available with AGO-IP followed by both small RNAseq and RNAseq, we used the data from Carbonell et al. [44], to develop our work.

There are several algorithms available to identify circRNAs. In this work, we used the CirComPara pipeline to detect, quantify and annotate circRNAs from RNA-seq data. This software comprises four different methods for back-splice identification. Each of them has its own features and requirements for the identification. This is the reason why we observed a such variability in the number of identified circRNAs. In order to increase the detection reliability of circRNAs, only those identified by at least 3 methods were selected. These circRNAs present a wide chromosomal distribution, since their loci are on chromosomes 1, 2, 3 and 5 of *A. thaliana*. In addition, their parental genes presented a considerable diversity of functions [46,47]. Using the PlantcircBase, which is a database for plant circular RNAs [48], only circRNA_At1g31810, circRNA_At2g35940, circRNA_At5g16880, circRNA_At3g13990 and circRNA_At5g27720 were previously identified. Besides, they are not conserved between others plant species. Until the present work, none of them had been validated. However, we show the circ_At3g13990 validation, by quantitative PCR and by sequencing. From the 12 circRNAs, only 5 have miRNA binding sites for miRNAs with read counts that are higher in AGO-IP libraries than in the empty vector library.

Interestingly, the majority of enriched miRNAs (6 out 10) have mismatches at the central region of hybridization with their target circRNAs and were predicted as having translation inhibition, which should avoid the degradation of the circRNAs, as observed in mammals' miRNAs sponges [49]. In this scenario, circRNAs would be able to capture miRNAs for longer periods and increase their sponge activity efficiency, avoiding the negative regulation of miRNAs on their target transcripts. At the same time, the miRNAs: mRNA targets found in AGO-IP libraries were predicted as being inhibited by cleavage (Table 4).

One circRNAs with miRNA binding sites (circ_At3g13990 and circ_At1g12080) was validated by both RT PCR and sequencing. Except in flower, no amplification products were detected in the other tissues analyzed. This indicates that these circRNAs are tissue specific, a feature also observed in other works [27,50,51]. Besides that, circ_At1g12080 presented more than one amplification product in the electrophoresis analysis. This indicates that different circular isoforms can be produced from a given gene and that their expression can be specific to cell type, tissue, and developmental stage.

The circ_At3g13990 is a perfect case study. It was predicted by 3 methods; the amplified PCR product had the expected size of 312 bp and contains perfect back-splicing site confirmed by Sanger sequencing. RT-PCR performed in RNA samples treated with RNAse R produce the expected 312 bp amplification, with the same intensity as in untreated samples, thereby demonstrating the circular nature of this RNA molecule. The PCR product from the parental cDNA, which originated from RNA samples previously treated with RNase R, revealed a weak amplification. It could indicate that not all RNA was degraded. The circ_At3g13990 contains three predicted miRNA binding sites, but just one of them, miR4239-5p, was significantly more frequently found in AGO-IP libraries. It indicates that this circRNA may be acting as miRNA sponge, blocking the action of miRNA upon its target. Curiously, no mRNA target for miR4239-5p was identified among the reads enriched in the AGO-IP libraries. It suggests that the majority of miR4239-5p molecules are associated to the miRNA:AGO:circRNAs ternary complex. Thus, not allowing the formation of the alternative miRNA:AGO:mRNA complex that would downregulate the target gene expression. It is noteworthy that miR4239-5p has as predicted targets: the small RNA degrading nuclease 3-SDN3 (At5g67240), the UBP1-associated protein 2A (At3g56860) and the gamma tubulin complex protein (At3g43610). All these three gene present a higher expression in flowers and carpels compared to other tissues, according to the BAR eFP Browser from the TAIR database (www.arabidopsis.org).

Our data contributed to the knowledge about the role of circRNAs in plants, since no work until now had demonstrated the existence of a ternary complex formed by AGO:miRNA:circRNA. These findings allow us to propose that plants circRNAs are also able to act as miRNA sponges and modulate the mRNA target regulation by using miRNA.

4. Materials and Methods

4.1. mRNAseq and Small RNAs Libraries

The RNAseq and AGO-IP small RNAs libraries [44] were downloaded from Gene Expression Omnibus (GEO, accession number GSM989339—GSM989346 and GSM989350—GSM989352) of NCBI. Quality and the presence of adapters in these libraries was visualized using FastQC software (http://www.bioinformatics.babraham.ac.uk/projects/fastqc/). Next, quality trimming and adaptor removal in the small RNAs and RNAseq libraries were carried out using Cutadapt/Sickle (https://github.com/najoshi/sickle) and Trimmomatic [52], respectively.

4.2. circRNAs Identification in mRNAseq Libraries from AGO-IP

Clean data from the AGO-IP RNAseq libraries SRR546147, SRR546148, SRR546149 and SRR546150 were used to identify, quantify and annotate potential circRNAs using the CirComPara pipeline [53], which uses five different methods in parallel: FindCirc, TestRealign and CircExplorer2, that works with three different aligners (Segemhel, Star and Tophat). All methods realized back-splice identification in each library with a minimum of 2 reads. *A. thaliana* genome and annotation files obtained from Ensembl Plants (https://plants.ensembl.org/index.html) were used as references. Only circRNAs identified by at least 3 methods were selected for the subsequent analyses.

4.3. Analysis of Target mRNAs and miRNAs Counts in AGO-IP and Control Libraries

The psRNATarget tool [54] was used to identify potential miRNAs that could interact with the circRNAs identified in the AGO-IP libraries. All mature miRNAs of *A. thaliana* from miRBase release 21 [55] were used in this analysis. Those miRNAs with an Expectation value (number of mismatches allowed) of 5 or less were selected to subsequent analysis. The Bowtie algorithm [56] was used to align the small RNAs sequences from each library to the miRNA sequences of the selected miRNAs to obtain read count values. The default parameters were used for the alignment and no mismatch was allowed. The miRNAs read counts were normalized according to the size of the libraries. A miRNA was considered to be enriched in the AGO-IP libraries if the normalized read count values of the miRNA were higher in the two AGO-IP libraries compared to the control (input).

The data from the same libraries used for the identification of the circRNAs was used to evaluate if the mRNAs targets of the selected miRNAs were also enriched in the AGO-IP libraries. The putative target mRNAs were selected using the psRNATarget tool. The enriched miRNAs and the transcriptome from *A. thaliana* TAIR version 10 obtained from Phytozome database (https://phytozome.jgi.doe.gov/pz/portal.html) were selected for this analysis using a maximum expectation value of 3.

To obtain the read count value, the reads from each library were mapped against the *A. thaliana* transcriptome using the Bowtie2 algorithm [57] with the default parameters. The DESeq package from the R software [49] was used to identify the target mRNAs significantly more frequent (maximum adjusted *p*-value of 0.05) in the four AGO-IP libraries compared to the input total mRNA controls.

4.4. Plant Material and Growth Condition

A. thaliana plants of ecotype Columbia were used. After incubation in the dark at 4 °C for 3 days, seeds were cultivated in soil for six weeks, at a temperature of 22 °C and a photoperiod of 16 h of light. Samples of leaves, flowers, axis and siliques were collected and stored at liquid nitrogen for subsequent storage at −80 °C.

4.5. RNA Extraction, RNase R Treatment and cDNA Synthesis

The RNA was extracted using the Trizol (Invitrogen) reagent, according to the manufacturer's instructions. The RNA integrity was performed using 1% agarose gel electrophoresis, where it was visualized under UV light and a digital image generated by the Gel-Doc (Bio-Rad) system. Prior to cDNA synthesis, samples containing 1 g of total RNA were treated with 2 units of RNase R (Lucigen) for 60 min at 37 °C. For cDNA synthesis were used the reverse primer of each analyzed circRNA and the M-MLV Reverse Transcriptase (Promega), according to the manufacturer's instructions.

4.6. Primers Design

Primers were projected using the Primer3 tool [58]. To validate the circRNAs identified by bioinformatic, divergent primers were projected [59]. In order to amplify part of the circRNA we used the primer combination PcF/PuR and to amplify all circRNA sequence we used the primer combination PcF/PuRi. As control, a set of convergent primers were designed for the parental mRNA detection The Reverse universal primer was the same for both circRNA and mRNA detection (Supplementary Table S1).

4.7. circRNAs and Parental mRNAs Amplification and Sequencing

The expression of the five circRNAs was evaluated by RT-PCR, using divergent primers. Samples were analyzed in technical triplicates and biological quadruplicates. The Polymerase Chain Reactions (PCR) reactions were realized using the Platinum Taq DNA polymerase (Invitrogen) enzyme. All RT-PCR reactions were performed on the Applied Biosystems Veriti apparatus. PCR conditions were conducted in a volume of 20 µL containing 10 µL of the diluted cDNA (1:100), 0.4 mM dNTPs, 10× Buffer, 3 mM MgCl 2, 0.25 U Platinum Taq DNA polymerase (Invitrogen) and 0.1 µM of each oligonucleotide. PCR conditions were: an initial 2 min step at 95 °C followed by 40 cycles of 10 s denaturing at 95 °C, 15 s annealing at 60 °C and 15 s extension at 72 °C. Confirmation of the fragments was performed by 3% agarose gel electrophoresis. The circRNAs predicted by bioinformatics and confirmed by PCR were purified using the Wizard SV gel PCR clean-up system (Ludwig Biotecnologia) according to the manufacturer's recommendations. Sanger sequencing reactions were performed with purified PCR products at a final concentration of the reaction of 4.5 pmol/µL, using the PcF, PuR, or PuRi primers (Supplementary Table S1).

Author Contributions: Conceptualization, E.F.C., G.C.F. and R.M.; Methodology, E.F.C.; Software, E.F.C., G.C.F. and F.G.; Validation, E.F.C. and G.C.F.; Formal Analysis, G.C.F.; Investigation, E.F.C., G.C.F. and R.M.; Data Curation, E.F.C., G.C.F., F.G. and R.M.; Writing—Original Draft Preparation, E.F.C.; Writing—Review & Editing, G.C.F. and R.M.; Supervision, R.M.; Project Administration, R.M.; Funding Acquisition, R.M.

Acknowledgments: RM is the recipient of a research fellowship 309030/2015-3 and EF is recipient of a PhD fellowship from Conselho Nacional de Pesquisa e Desenvolvimento Científico e Tecnológico. EF is recipient of M.Sc. fellowship and GC and FG of Post-Doctoral fellowships from Coordenação de Aperfeiçoamento de Pessoal de Nível Superior CAPES.

References

1. Tay, Y.; Rinn, J.; Pandolfi, P.P. The multilayered complexity of ceRNA crosstalk and competition. *Nature* **2014**, *505*, 344–352. [CrossRef]
2. Chi, S.W.; Zang, J.B.; Mele, A.; Darnell, R.B. Argonaute HITS-CLIP decodes microRNA-mRNA interaction maps. *Nature* **2009**, *460*, 479–486. [CrossRef]
3. Licatalosi, D.D.; Mele, A.; Fak, J.J.; Ule, J.; Kayikci, M.; Chi, S.W.; Clark, T.A.; Schweitzer, A.C.; Blume, J.E.; Wang, X.; et al. HITS-CLIP yields genome-wide insights into brain alternative RNA processing. *Nature* **2008**, *456*, 464–469. [CrossRef]
4. Huntzinger, E.; Izaurralde, E. Gene silencing by microRNAs: Contributions of translational repression and mRNA decay. *Nat. Rev. Genet.* **2011**, *12*, 99–110. [CrossRef]
5. Pasquinelli, A.E. MicroRNAs and their targets: Recognition, regulation and an emerging reciprocal relationship. *Nat. Rev. Genet.* **2012**, *13*, 271–282. [CrossRef]
6. Poliseno, L.; Salmena, L.; Zhang, J.; Carver, B.; Haveman, W.J.; Pandolfi, P.P. A coding-independent function of gene and pseudogene mRNAs regulates tumour biology. *Nature* **2010**, *465*, 1033–1038. [CrossRef]
7. Salmena, L.; Poliseno, L.; Tay, Y.; Kats, L.; Pandolfi, P.P. A ceRNA Hypothesis: The Rosetta Stone of a Hidden RNA Language? *Cell* **2011**, *146*, 353–358. [CrossRef]
8. Seitz, H. Redefining MicroRNA Targets. *Curr. Biol.* **2009**, *19*, 870–873. [CrossRef]
9. Tay, Y.; Kats, L.; Salmena, L.; Weiss, D.; Tan, S.M.; Ala, U.; Karreth, F.; Poliseno, L.; Provero, P.; Di Cunto, F.; et al. Coding-Independent Regulation of the Tumor Suppressor PTEN by Competing Endogenous mRNAs. *Cell* **2011**, *147*, 344–357. [CrossRef]
10. Franco-Zorrilla, J.M.; Valli, A.; Todesco, M.; Mateos, I.; Puga, M.I.; Rubio-Somoza, I.; Leyva, A.; Weigel, D.; García, J.A.; Paz-Ares, J. Target mimicry provides a new mechanism for regulation of microRNA activity. *Nat. Genet.* **2007**, *39*, 1033–1037. [CrossRef]
11. Hansen, T.B.; Wiklund, E.D.; Bramsen, J.B.; Villadsen, S.B.; Statham, A.L.; Clark, S.J.; Kjems, J. miRNA-dependent gene silencing involving Ago2-mediated cleavage of a circular antisense RNA. *EMBO J.* **2011**, *30*, 4414–4422. [CrossRef]
12. Memczak, S.; Jens, M.; Elefsinioti, A.; Torti, F.; Krueger, J.; Rybak, A.; Maier, L.; Mackowiak, S.D.; Gregersen, L.H.; Munschauer, M.; et al. Circular RNAs are a large class of animal RNAs with regulatory potency. *Nature* **2013**, *495*, 333–338. [CrossRef]
13. Capel, B.; Swain, A.; Nicolis, S.; Hacker, A.; Walter, M.; Koopman, P.; Goodfellow, P.; Lovell-Badge, R. Circular transcripts of the testis-determining gene Sry in adult mouse testis. *Cell* **1993**, *73*, 1019–1030. [CrossRef]
14. Cocquerelle, C.; Daubersies, P.; Majérus, M.A.; Kerckaert, J.P.; Bailleul, B. Splicing with inverted order of exons occurs proximal to large introns. *EMBO J.* **1992**, *11*, 1095–1098. [CrossRef]
15. Nigro, J.M.; Cho, K.R.; Fearon, E.R.; Kern, S.E.; Ruppert, J.M.; Oliner, J.D.; Kinzler, K.W.; Vogelstein, B. Scrambled exons. *Cell* **1991**, *64*, 607–613. [CrossRef]
16. Kos, A.; Dijkema, R.; Arnberg, A.C.; van der Meide, P.H.; Schellekens, H. The hepatitis delta (delta) virus possesses a circular RNA. *Nature* **1986**, *323*, 558–560. [CrossRef]
17. Sanger, H.L.; Klotz, G.; Riesner, D.; Gross, H.J.; Kleinschmidt, A.K. Viroids are single-stranded covalently closed circular RNA molecules existing as highly base-paired rod-like structures. *Proc. Natl. Acad. Sci. USA* **1976**, *73*, 3852–3856. [CrossRef]
18. Salzman, J.; Chen, R.E.; Olsen, M.N.; Wang, P.L.; Brown, P.O. Cell-Type Specific Features of Circular RNA Expression. *PLoS Genet.* **2013**, *9*, e1003777. [CrossRef]
19. Ye, C.-Y.; Chen, L.; Liu, C.; Zhu, Q.-H.; Fan, L. Widespread noncoding circular RNAs in plants. *New Phytol.* **2015**, *208*, 88–95. [CrossRef]
20. Salzman, J.; Gawad, C.; Wang, P.L.; Lacayo, N.; Brown, P.O. Circular RNAs Are the Predominant Transcript Isoform from Hundreds of Human Genes in Diverse Cell Types. *PLoS ONE* **2012** *7*, e30733. [CrossRef]
21. Zhang, Y.; Zhang, X.-O.; Chen, T.; Xiang, J.-F.; Yin, Q.-F.; Xing, Y.-H.; Zhu, S.; Yang, L.; Chen, L.-L. Circular Intronic Long Noncoding RNAs. *Mol. Cell* **2013** *51*, 792–806. [CrossRef]

22. Zhang, X.-O.; Wang, H.-B.; Zhang, Y.; Lu, X.; Chen, L.-L.; Yang, L. Complementary Sequence-Mediated Exon Circularization. *Cell* **2014**, *159*, 134–147. [CrossRef]
23. Wang, P.L.; Bao, Y.; Yee, M.-C.; Barrett, S.P.; Hogan, G.J.; Olsen, M.N.; Dinneny, J.R.; Brown, P.O.; Salzman, J. Circular RNA Is Expressed across the Eukaryotic Tree of Life. *PLoS ONE* **2014**, *9*, e90859. [CrossRef]
24. Danan, M.; Schwartz, S.; Edelheit, S.; Sorek, R. Transcriptome-wide discovery of circular RNAs in Archaea. *Nucleic Acids Res.* **2012**, *40*, 3131–3142. [CrossRef]
25. Conn, V.M.; Hugouvieux, V.; Nayak, A.; Conos, S.A.; Capovilla, G.; Cildir, G.; Jourdain, A.; Tergaonkar, V.; Schmid, M.; Zubieta, C.; et al. A circRNA from SEPALLATA3 regulates splicing of its cognate mRNA through R-loop formation. Nature Plants, v. 3, n. 5, p. 17053, 18 abr. 2017. cognate mRNA through R-loop formation. *Nat. Plants* **2017**, *3*, 17053. [CrossRef]
26. Darbani, B.; Noeparvar, S.; Borg, S. Identification of Circular RNAs from the Parental Genes Involved in Multiple Aspects of Cellular Metabolism in Barley. *Front. Plant Sci.* **2016**, *7*, 776. [CrossRef]
27. Lu, T.; Cui, L.; Zhou, Y.; Zhu, C.; Fan, D.; Gong, H.; Zhao, Q.; Zhou, C.; Zhao, Y.; Lu, D.; et al. Transcriptome-wide investigation of circular RNAs in rice. *RNA* **2015**, *21*, 2076–2087. [CrossRef]
28. Wang, Y.; Yang, M.; Wei, S.; Qin, F.; Zhao, H.; Suo, B. Identification of Circular RNAs and Their Targets in Leaves of *Triticum aestivum* L. under Dehydration Stress. *Front. Plant Sci.* **2017**, *7*, 2024. [CrossRef]
29. Zuo, J.; Wang, Q.; Zhu, B.; Luo, Y.; Gao, L. Deciphering the roles of circRNAs on chilling injury in tomato. *Biochem. Biophys. Res. Commun.* **2016**, *479*, 132–138. [CrossRef]
30. Rybak-Wolf, A.; Stottmeister, C.; Glažar, P.; Jens, M.; Pino, N.; Giusti, S.; Hanan, M.; Behm, M.; Bartok, O.; Ashwal-Fluss, R.; et al. Circular RNAs in the Mammalian Brain Are Highly Abundant, Conserved, and Dynamically Expressed. *Mol. Cell* **2015**, *58*, 870–885. [CrossRef]
31. Westholm, J.O.; Miura, P.; Olson, S.; Shenker, S.; Joseph, B.; Sanfilippo, P.; Celniker, S.E.; Graveley, B.R.; Lai, E.C. Genome-wide Analysis of Drosophila Circular RNAs Reveals Their Structural and Sequence Properties and Age-Dependent Neural Accumulation. *Cell Rep.* **2014**, *9*, 1966–1980. [CrossRef]
32. Suzuki, H.; Tsukahara, T. A view of pre-mRNA splicing from RNase R resistant RNAs. *Int. J. Mol. Sci.* **2014**, *15*, 9331–9342. [CrossRef]
33. Burd, C.E.; Jeck, W.R.; Liu, Y.; Sanoff, H.K.; Wang, Z.; Sharpless, N.E. Expression of Linear and Novel Circular Forms of an INK4/ARF-Associated Non-Coding RNA Correlates with Atherosclerosis Risk. *PLoS Genet.* **2010**, *6*, e1001233. [CrossRef]
34. Jeck, W.R.; Sorrentino, J.A.; Wang, K.; Slevin, M.K.; Burd, C.E.; Liu, J.; Marzluff, W.F.; Sharpless, N.E. Circular RNAs are abundant, conserved, and associated with ALU repeats. *RNA* **2013**, *19*, 141–157. [CrossRef]
35. Talhouarne, G.J.S.; Gall, J.G. Lariat intronic RNAs in the cytoplasm of Xenopus tropicalis oocytes. *RNA* **2014**, *20*, 1476–1487. [CrossRef]
36. Li, Z.; Huang, C.; Bao, C.; Chen, L.; Lin, M.; Wang, X.; Zhong, G.; Yu, B.; Hu, W.; Dai, L.; et al. Exon-intron circular RNAs regulate transcription in the nucleus. *Nat. Struct. Mol. Biol.* **2015**, *22*, 256–264. [CrossRef]
37. Suzuki, H.; Zuo, Y.; Wang, J.; Zhang, M.Q.; Malhotra, A.; Mayeda, A. Characterization of RNase R-digested cellular RNA source that consists of lariat and circular RNAs from pre-mRNA splicing. *Nucleic Acids Res.* **2006**, *34*, e63. [CrossRef]
38. Ashwal-Fluss, R.; Meyer, M.; Pamudurti, N.R.; Ivanov, A.; Bartok, O.; Hanan, M.; Evantal, N.; Memczak, S.; Rajewsky, N.; Kadener, S. circRNA Biogenesis Competes with Pre-mRNA Splicing. *Mol. Cell* **2014**, *56*, 55–66. [CrossRef]
39. Chao, C.W.; Chan, D.C.; Kuo, A.; Leder, P. The mouse formin (Fmn) gene: Abundant circular RNA transcripts and gene-targeted deletion analysis. *Mol. Med.* **1998**, *4*, 614–628. [CrossRef]
40. Hentze, M.W.; Preiss, T. Circular RNAs: splicing's enigma variations. *EMBO J.* **2013**, *32*, 923–925. [CrossRef]
41. Romeo, T. Global regulation by the small RNA-binding protein CsrA and the non-coding RNA molecule CsrB. *Mol. Microbiol.* **1998**, *29*, 1321–1330. [CrossRef]
42. Thomas, L.F.; Sætrom, P. Circular RNAs are depleted of polymorphisms at microRNA binding sites. *Bioinformatics* **2014**, *30*, 2243–2246. [CrossRef]
43. Sablok, G.; Zhao, H.; Sun, X. Plant Circular RNAs (circRNAs): Transcriptional Regulation Beyond miRNAs in Plants. *Mol. Plant* **2016**, *9*, 192–194. [CrossRef]
44. Carbonell, A.; Fahlgren, N.; Garcia-Ruiz, H.; Gilbert, K.B.; Montgomery, T.A.; Nguyen, T.; Cuperus, J.T.; Carrington, J.C. Functional Analysis of Three Arabidopsis ARGONAUTES Using Slicer-Defective Mutants. *Plant Cell* **2012** *24*, 3613–3629. [CrossRef]

45. Voinnet, O.; Ponce, M.R.; Vaucheret, H.; Baumberger, N.; Sarazin, A.; Clavel, M.; Micol, J.L.; Ziegler-Graff, V.; Genschik, P.; Derrien, B.; et al. A Suppressor Screen for AGO1 Degradation by the Viral F-Box P0 Protein Uncovers a Role for AGO DUF1785 in sRNA Duplex Unwinding. *Plant Cell* **2018**, *30*, 1353–1374.
46. Lasda, E.; Parker, R. Circular RNAs: Diversity of form and function. *RNA* **2014**, *20*, 1829–1842. [CrossRef]
47. Li, X.; Yang, L.; Chen, L.-L. The Biogenesis, Functions, and Challenges of Circular RNAs. *Mol. Cell* **2018**, *71*, 428–442. [CrossRef]
48. Chu, Q.; Zhang, X.; Zhu, X.; Liu, C.; Mao, L.; Ye, C.; Zhu, Q.; Fan, L. PlantcircBase: A database for plant circular RNAs. *Mol. Plant.* **2017**, *10*, 1126–1128. [CrossRef]
49. Anders, S.; Huber, W. Differential expression analysis for sequence count data. *Genome Biol.* **2010**, *11*, R106. [CrossRef]
50. Ye, J.; Wang, L.; Li, S.; Zhang, Q.; Zhang, Q.; Tang, W.; Wang, K.; Song, K.; Sablok, G.; Sun, X.; et al. AtCircDB: A tissue-specific database for Arabidopsis circular RNAs. *Brief. Bioinform.* **2017**, *20*, 58–65. [CrossRef]
51. Zhao, W.; Cheng, Y.; Zhang, C.; You, Q.; Shen, X.; Guo, W.; Jiao, Y. Genome-wide identification and characterization of circular RNAs by high throughput sequencing in soybean. *Sci. Rep.* **2017**, *7*, 5636. [CrossRef]
52. Bolger, A.M.; Lohse, M.; Usadel, B. Trimmomatic: A flexible trimmer for Illumina sequence data. *Bioinformatics* **2014**, *30*, 2114–2120. [CrossRef]
53. Gaffo, E.; Bonizzato, A.; Kronnie, G.; Bortoluzzi, S. CirComPara: A Multi-Method Comparative Bioinformatics Pipeline to Detect and Study circRNAs from RNA-seq Data. *Non Coding RNA* **2017**, *3*, 8. [CrossRef]
54. Dai, X.; Zhao, P.X. psRNATarget: A plant small RNA target analysis server. *Nucleic Acids Res.* **2011**, *39*, W155–W159. [CrossRef]
55. Kozomara, A.; Griffiths-Jones, S. miRBase: Annotating high confidence microRNAs using deep sequencing data. *Nucleic Acids Res.* **2014**, *42*, D68–D73. [CrossRef]
56. Langmead, B.; Trapnell, C.; Pop, M.; Salzberg, S.L. Ultrafast and memory-efficient alignment of short DNA sequences to the human genome. *Genome Biol.* **2009**, *10*, R25. [CrossRef]
57. Langmead, B.; Salzberg, S.L. Fast gapped-read alignment with Bowtie 2. *Nat. Methods* **2012**, *9*, 357–359. [CrossRef]
58. Rozen, S.; Skaletsky, H. Primer3 on the WWW for general users and for biologist programmers. *Methods Mol. Biol.* **2000**, *132*, 365–386.
59. Kulcheski, F.R.; Christoff, A.P.; Margis, R. Circular RNAs are miRNA sponges and can be used as a new class of biomarker. *J. Biotechnol.* **2016**, *238*, 42–51. [CrossRef]

13

Substantial Dysregulation of miRNA Passenger Strands Underlies the Vascular Response to Injury

Karine Pinel [†,*], **Louise A. Diver, Katie White, Robert A. McDonald and Andrew H. Baker** [‡,*]

Institute of Cardiovascular and Medical Sciences, BHF Glasgow Cardiovascular Research Centre, University of Glasgow, Glasgow G12 8TA, UK; drlouisediver@gmail.com (L.A.D.); kt_m_white@hotmail.com (K.W.); robertamcdonald8@gmail.com (R.A.M.)

* Correspondence: pinel.karine.m@gmail.com (K.P.); Andy.Baker@ed.ac.uk (A.H.B.),

† Current address: INRA, UMR 1419 NuMeA (Nutrition Metabolism Aquaculture), Aquapôle, 64310 St-Pee-sur-Nivelle, France.

‡ Current address: BHF/University Centre for Cardiovascular Science, University of Edinburgh, Edinburgh EH16 4UU, Scotland, UK.

Abstract: Vascular smooth muscle cell (VSMC) dedifferentiation is a common feature of vascular disorders leading to pro-migratory and proliferative phenotypes, a process induced through growth factor and cytokine signaling cascades. Recently, many studies have demonstrated that small non-coding RNAs (miRNAs) can induce phenotypic effects on VSMCs in response to vessel injury. However, most studies have focused on the contribution of individual miRNAs. Our study aimed to conduct a detailed and unbiased analysis of both guide and passenger miRNA expression in vascular cells in vitro and disease models in vivo. We analyzed 100 miRNA stem loops by TaqMan Low Density Array (TLDA) from primary VSMCs in vitro. Intriguingly, we found that a larger proportion of the passenger strands was significantly dysregulated compared to the guide strands after exposure to pathological stimuli, such as platelet-derived growth factor (PDGF) and IL-1α. Similar findings were observed in response to injury in porcine vein grafts and stent models in vivo. In these studies, we reveal that the miRNA passenger strands are predominantly dysregulated in response to vascular injury.

Keywords: miRNA expression and regulation; passenger miRNA; biomarker; vascular injury; smooth muscle cells; porcine vein graft and stent models

1. Introduction

MicroRNAs (miRNAs) are a class of non-coding RNAs known to play a prominent role in gene regulation at a post-transcriptional level [1–3]. Pri-miRNAs are transcribed by RNA polymerase II and are processed successively by two RNase III enzymes, Drosha and Dicer, to a ~22 nt miRNA:miRNA * duplex [4], which, on processing, generates two single-stranded RNA molecules. miRNAs are loaded onto Argonaute to produce a RNA-induced silencing complex (RISC), which exerts translational repression via an imperfect binding to the mRNA target, generally localized in the 3′UTR. An equal amount of the two strands from the miRNA:miRNA * duplex is produced by the transcription but their accumulation becomes asymmetric [5]. Current miRNA nomenclature provides information on the direction of the mature miRNA strand using the -3p or -5p suffixes and, normally, the miRNA guide strand is thought to be the strand that preferentially accumulates in the RISC and the miRNA * or passenger strand (its partner) is degraded [6]. However, recent studies suggest that these passenger strands can accumulate in a number of disease pathologies and mediate strand-specific roles based on their distinct seed sequences and targets. Indeed, the mature sequences of the two strand forms can regulate distinct sets of mRNA transcripts or common targets through different hybridization sites on the mRNA targeted.

miRNAs have been shown to induce phenotypic effects on smooth muscle cells (SMCs) in vascular disorders. Vascular SMC (VSMCs) within the vessel wall perform both contractile and proliferative functions in response to diverse cellular stimuli [7]. Under normal physiological conditions, VSMCs are relatively quiescent, harboring a contractile phenotype responding to changes in vascular flow to mediate relaxation and contraction. However, vascular injury or insult results in the initiation of phenotypic switching to a dedifferentiated phenotype. Indeed, the transition of VSMCs from a differentiated to a dedifferentiated phenotype is a common feature of vascular disorders where VSMCs harbor pro-proliferative and pro-migratory status. The VSMC phenotype can be modulated by many environmental cues and triggers, including a number of cytokines and growth factors released in response to injury. Indeed, platelet-derived growth factor (PDGF) promotes VSMC proliferation and migration resulting in neointimal formation after artery injury [8,9].

Since there is some recent evidence of the importance of the miRNA passenger strands [10–12], their individual functions have been studied in settings such as vascular cells with miR-126-5p and miR-10A*, [13,14] and cancer [15,16]. However, a more detailed or global evaluation of the importance of the miRNA passenger strands in vascular remodeling is still lacking. As miRNAs emerge as potential prognostic biomarkers and as attractive targets for disease intervention, we utilized array technology to evaluate the global expression of miRNA guide and passenger strands from hairpin precursors in pre-clinical porcine models of vascular injury and human VSMCs.

2. Materials and Methods

2.1. Animals and Tissues

All the procedures for surgery, for the pig vein graft model and the stent model, were performed in reference [17] and [18], respectively, and were performed in accordance with the UK Home Office Guidance on the Operation of the Animals (Scientific Procedures) Act 1986 and with the institutional ethical approval (PPL60/4114 and PPL 60/4429). Male white Landrace pigs (SAC Commercial Ltd, Edinburgh, UK) were maintained on a 12 h light/dark cycle with free access to food and water at a designated Biological Procedures Unit. Briefly, the animals (weighing 19–26 kg) were pre-dosed orally with aspirin and clopidogrel 24 h prior to surgery. The animals were sedated by intramuscular (IM) injection of Tiletamine/Zolazepam (Zoletil; Virbac, France); general anesthesia was conducted using intravenous (IV) Propofol (Rapinovet; Schering-Plough, Welwyn Garden City, UK) and maintained with isoflurane (Abbott Laboratories Ltd, Berkshire, UK). Animals received 100 IU/kg of IV heparin (Leo Laboratories, London, UK) before surgery. After vascular access, coronary angiography was performed prior to the deployment of either bare metal stents (BMS) (Gazelle™, Biosensors, Morges, Switzerland) or drug eluting stents (DES) (biolimus A9 eluting stents; Biomatrix Flex™, Biosensors, Hägglingen, Switzerland). The unstented animals were used as controls. The pigs received 0.15 mg of buprenorphine IM (Vetergesic: Alstore Ltd, Sheriff Hutton, York, UK) and 350 mg of ampicillin IM (Amfipen LA, Intervet, UK) immediately following the procedure and were euthanized after 7 or 28 days by an intravenous overdose of euthatal, and their coronary vessels carefully removed. For the porcine vein graft surgery, following the induction with Ketaset (100 mg/mL ketamine hydrochloride), the animals were intubated and anesthetized with halothane and allowed to spontaneously ventilate. The method for the saphenous vein-carotid interpositional graft model has been described previously [19]. The saphenous vein was surgically exposed and harvested from the hind leg by a 'no-touch' technique. The vein was cannulated and gently irrigated with heparinized iso-osmotic sodium chloride (NaCl) solution (0.9 g/L). Each carotid was exposed via a longitudinal neck incision. The animal was heparinized by the IV administration of 100 IU/kg of heparin (CP Parmaceuticals Ltd, Wrexham, UK). A 4–5 cm segment of the carotid artery was isolated between the vascular clamps and 10 mm was then removed. The residual ends were beveled at 45°

and a small longitudinal incision was made to lengthen the anastomotic area. The ends of the vein were similarly beveled and anastomosed as an interposition graft under optical magnification using continuous 7/0 Surgipro (auto Suture, Dagford, UK) sutures. The saphenous veins from the ungrafted animals were used as controls. The animals were recovered, returned to their pen and maintained on a normal chow diet for the duration of the experiment.

Surplus human saphenous vein tissue was obtained with informed consent from patients undergoing CABG (Coronary Artery Bypass Grafting). All procedures received local ethical approval (Research Ethical Committee number: 06/S0703/110) and the experimental procedures conformed to the principles outlined in the Declaration of Helsinki.

2.2. *Vessel Storage and RNA Isolation from the Pig Vein Graft Model and Stent Model*

All the procedures for RNA extraction from the pig vein graft model and stent model are described in detail in [17,18], respectively. Briefly, the harvested vessels were placed in RNAlater®-ice (Invitrogen, Paisley, UK) and stored at −80 °C until the day of isolation. RNAs from the veins or arteries were isolated following disruption of the vessels under liquid nitrogen using a pestle and mortar, and these vessel fragments were placed in Qiazol (Qiagen, Hilden, Germany) and homogenized using a tissue homogenizer (Polytron, Switzerland). The RNAs were processed through miRNEasy Mini Kit (Qiagen) following the manufacturer's instructions, treated with DNAse 1 (amplification grade; Sigma, St. Louis, MO, USA) in order to eliminate genomic DNA contamination and quantified using a NanoDrop ND-1000 Spectrophotometer (Nano-Drop Technologies, Wilmington, DE, USA). RNA integrity was assessed using the RNA 6000 Nano LabChip kit (Agilent Technologies, Santa Clara, CA, USA). Only The RNAs with an RNA integrity number value >7 were used for the TLDA experiments.

2.3. *Cell Culture*

VSMCs were isolated from human saphenous vein segments within 24 h of surgery using the explant technique as previously described by Southgate et al. [20]. Briefly, following removal of the adventitial layer using sterile forceps and scissors, the vein was longitudinally opened. The lumenal surface of the vein was scraped gently with a rubber policeman to remove the endothelium. Using fine forceps, thin segments of the medial layer were stripped from the vein. The stripped medial segments were then cut into approximately 1 mm^2 pieces using a tissue chopper and then transferred into a sterile tube containing wash media (DMEM medium supplemented with 2 mmol/L of L-glutamine (Invitrogen), 50 μg/mL of penicillin (Life Technologies, Paisley, UK) and 50 μg/L of streptomycin (Life Technologies)). Following two washes in the wash media, the segments were resuspended in a small volume of SMC complete medium (SMC growth medium 2 (PromoCell, Heidelberg, Germany) supplemented with 15% foetal calf serum (FCS) (PAA laboratories, Yeovil, UK), 2 mmol/L of L-glutamine (Invitrogen), 50 μg/mL of penicillin (Life Technologies) and 50 μg/L of streptomycin (Life Technologies)) and were then spread onto the base of 25 cm^2 tissue culture flasks. Following overnight culture at 37 °C in a humidified atmosphere containing 5% CO_2 (*v*/*v*), 5 mL of SMC complete medium was carefully added to each flask. The media were changed every 3–5 days and after 2–3 weeks, the outgrowth of VSMCs was passaged into 75 cm^2 flasks and grown to confluence. All experiments were performed using VSMCs between passages 3 and 6. The VSMCs were plated in SMC complete medium, quiesced in medium (DMEM medium supplemented with 2 mmol/L of L-glutamine (Invitrogen), 50 μg/mL of penicillin (Life Technologies) and 50 μg/L of streptomycin (Life Technologies)) containing 0.2% FCS for 48 h. The plated cells were either placed in fresh 0.2% FCS medium (control) or stimulated with 0.2% medium supplemented with PDGF (PDGF-BB, 20 ng/mL) or with IL-1α (10 ng/mL) or a combination of PDGF and IL-1α for 48 h before RNA isolation.

2.4. RNA Extraction and Purification

For the extraction of RNAs from VSMCs, the lysis of the plated cells was performed with 700 μL of QIAzol lysis reagent (Qiagen). The purification of total RNA was achieved using miRNEasy Mini Kit (Qiagen) according to the manufacturer's instructions. The RNA integrity and concentration were determined using the NanoDrop ND-1000 spectrophotometer.

2.5. TaqMan Low Density Array (TLDA) Experiment

miRNA profiling was performed with the Human MicroRNA Array v2.0 Card A and B which is enabled to quantify 667 human miRNAs catalogued in the miRBase v10. For three independent patients (Pt 1, Pt 2 and Pt 3), 90 ng of total RNA was reverse transcribed using the Megaplex RT set pool A and B and then preamplified following the manufacturer's instructions (Applied Biosystems, Carlsbad, CA, USA). The cDNA that was produced was loaded on the Human MicroRNA Array cards and run on the 7900HT Fast Real Time PCR system (Applied Biosystems) using SDS2.3 software. Data analysis was performed using DataAssist v2.0 Software (Applied Biosystems). The miRNA expression data were normalized using the RNU48 housekeeper expression because of their low Ct variance.

The TLDA experiments for the pig vein graft model and the pig stent model are described in detail in ref [17] and [18], respectively.

2.6. Global Analysis

The literature and databases such as the miRBase Sequence Database (Release 21) permit the determination of the guide and the passenger strands for a thousand stem loops from the data obtained by the TLDA experiments. The level of expression (relative expression, %) was classified into three groups: high (Ct < 23), medium (Ct between 23 and 28) and low (Ct > 28) expressions. The dysregulated miRNAs are represented by a relative quotient (RQ) higher than two or less than 0.5 in expression level.

2.7. Gene Expression Quantitative RealTime-PCR (qRT-PCR)

The cDNA for gene expression analysis was synthesized from 400 ng of RNA using TaqMan Reverse Transcription Reagents (Applied Biosystems). The reverse transcription (RT) reactions were carried out in a volume of 20 μL following the manufacturer's instructions and run in a thermocycler under the conditions: 25 °C for 10 min, 48 °C for 30 min and 95 °C for 5 min, then held at 4 °C.

The TaqMan real-time PCR reactions were performed in triplicate comprising 5 μL of TaqMan Universal Master Mix II, no UNG (Applied Biosystems), 0.5 μL of 20× gene-specific primer and a probe mix (TaqMan gene expression assays, Cat number # 4331182 (CDKN1B, Assay ID: Hs01597588_m1; UBC, Assay ID: Hs00824723_m1)), 1.2 μL of the reverse transcription product and 3.3 μL of nuclease free water. RT-PCR was carried out by using a 7900HT Fast Real Time PCR system (Applied Biosystems) under the conditions: 50 °C for 2 min, 95 °C for 10 min, and then 40 cycles of 95 °C for 15 s, 60 °C for 60 s, then held at 4 °C. The gene expression level was normalized to UBC (Ubiquitin C) and the fold-changes were calculated using the ΔΔCT method.

2.8. miRNA Expression Quantitative RealTime-PCR (qRT-PCR)

All the miRNA specific probes (TaqMan microRNA assays) are commercial and available from Life Technologies: Cat number # 4427975 (hsa-miR-204, Assay ID: 000508; hsa-miR-218, Assay ID: 000521; hsa-miR-1275, Assay ID: 002840; hsa-miR-625*, Assay ID: 002432; hsa-miR-222*, Assay ID: 002097).

The RT reactions were performed using the TaqMan miRNA Reverse Transcription Kit (Applied Biosystems) and miRNA-specific RT stem-loop primers (Applied Biosystems). The RT reactions were carried out in a volume of 7.5 μL following the manufacturer's instructions and run in a thermocycler under the conditions: 16 °C for 30 min, 42 °C for 45 min and 85 °C for 5 min, then held at 4 °C.

The TaqMan real-time PCR reactions were performed in triplicate comprising 5 μL TaqMan Universal Master Mix II, no UNG (Applied Biosystems), 0.5 μL of 20 × miRNA-specific primer and a probe mix of the TaqMan miRNA Assay Kit (Applied Biosystems), 0.7 μL of the RT product and 3.8 μL of nuclease free water. RT-PCR was carried out by using a 7900HT Fast Real Time PCR system (Applied Biosystems) under the conditions: 95 °C for 10 min, then 40 cycles of 95 °C for 15 s, 60 °C for 60 s, then held at 4 °C. The expression level of miRNA was normalized to RNU48 and the fold-changes were calculated using the ΔΔCT method. Only the cells from patients harboring a low variation concerning the baseline expression of miRNAs have been used for the miRNA expression analysis (Figure S1).

2.9. Statistical Analysis

The statistical analysis was performed according to figure legends. The statement N = 3 means that three experiments were performed, three independent times, all in triplicate. Data are expressed as mean ± SEM. Comparisons between the two groups were analyzed using the 2-tailed Student's *t*-test. The one-way ANOVA with Tukey's post hoc multiple comparison test, via Graph Pad Prism version 5.0 (GraphPad Software, San Diego, CA, USA), was used for comparisons among three or more groups. A statistical difference was considered as $p < 0.05$. Heat maps were generated using the R program (version 3.2.4) (R Foundation for Statistical Computing, Vienna, Austria).

3. Results

3.1. Global Analysis of miRNA in VSMCs in Response to Cytokine and Growth Factor Stimulation and in Vascular Injury In Vivo

Human VSMCs were treated with PDGF, interleukin 1α (IL-1α) or a combination of both to mimic a number of the pathologic signals induced by vascular injury. As expected, PDGF induced a significant increase in VSMC migration (Figure S2A), and the pro-proliferative phenotype was confirmed by a reduction in the expression of the cyclin-dependent kinase inhibitor 1B (CDKN1B) (Figure S2B). Following confirmation of the phenotypic alternation in VSMCs, we performed a detailed and unbiased analysis of both the guide and passenger miRNAs expression in VSMCs (Table S1). Indeed, using TaqMan Low Density Arrays (TLDA) we were able to quantify a large number of miRNAs simultaneously in each individual sample. The relative expression in each condition was compared to the unstimulated quiested VSMCs and expressed in fold change. Globally, our data highlighted a low variation between the samples. Bioinformatics tools including the miRBase (Realease 21) (http://www.mirbase.org/) have been used to discriminate "guide" versus "passenger" strands. We validated five miRNAs from our array by subsequent assays (Figure S3) and found miRNA markers in our global analysis such as miR-146a, an inflammation-associated miRNA [21], which showed overexpression following exposure to IL-1α. From our array, the level of expression was classified into three groups following pre-amplification according to cycle threshold (Ct) values (low [Ct > 28], medium [Ct between 23 and 28] and high [Ct < 23] expression) for each miRNA. The global analysis revealed that the miRNA guide strands are more abundantly expressed than the passenger strands (Figure 1A) in all the conditions tested (Figure S4A). A similar pattern was identified following the computational analysis of the array data from the porcine models of vein graft failure [17] and in stent restenosis [18] (Figure S4B and Figure S4C, respectively, Figure 1B and Figure S5). A total of one hundred hairpins were analyzed from the data generated by TLDA. Consistently across the three models, we found that the hairpin of miR-10b, miR-193b, miR-199a, miR-214, miR-30a, miR-335 and miR-99b contained medium or highly expressed passenger strands. Furthermore, a direct comparison of the relative expression of miRNA strands obtained from their corresponding hairpins confirmed that the guide strand was more abundantly expressed than its corresponding passenger strand in more than 60% of cases, a pattern that was consistent across all three data sets (Figure 1C). In human VSMCs, the guide strand appeared to be the most predominantly expressed variant of the hairpin, although there are exceptions where the passenger strands were more abundantly expressed than their corresponding guide strand such as miR-625 and miR-30a (heat map, Figure 1A). The same type of exception has been found across the three models for miR-199a, miR-30a, and miR-335.

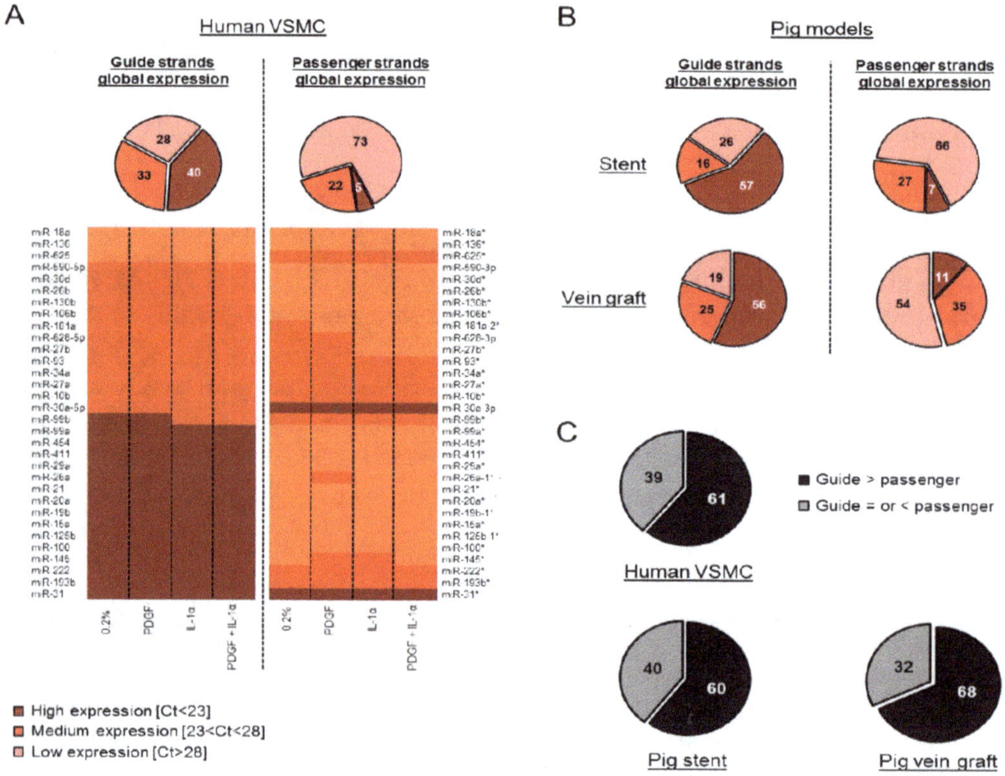

Figure 1. Low passenger strands expression compared to the guide strands in VSMCs and in vivo. One hundred hairpins were analyzed by TaqMan Low Density Array (TLDA) in human VSMCs (average of $n = 3$), the porcine models of vein graft failure and in stent restenosis (average of $n = 6$ for both the porcine models, except for the 7 days DES (drug eluting stent) condition, where $n = 5$). The level of expression (relative expression, %) was classified into three groups: high (Ct < 23), medium (Ct between 23 and 28) and low (Ct > 28) expressions. These groups are represented as a % in each pie chart. (**A**) Each pie chart represents the global expression of the guide and passenger strands in human VSMCs. The heat map illustrates the relative abundance of the number of miRNA hairpins that were consistently expressed across three independent patients and represents the level of expression of the guide versus passenger strands. (**B**) Each pie chart represents the global expression of the guide and passenger strands in the porcine model of in stent restenosis and vein graft failure. (**C**) Each pie chart represents the expression of the guide versus passenger strands within each independent hairpin.

3.2. Passenger Strands Are Dysregulated More Frequently than Guide Strands in Vascular Cells and Vascular Injury In Vivo

Despite their generally lower levels of expression compared to the guide strands, we then quantified the dysregulation of the passenger strands in vascular cells in vitro and in injury models in vivo. Interestingly, in all three settings, we found a substantial and predominant dysregulation of the passenger strands in response to pathological stimuli (Figure 2). The relative number of dysregulated miRNAs included 39% and 32% of the passenger strands and only 15% and 10% of the guide strands in human VSMCs treated with PDGF and IL-1α, respectively (Figure 2A). Furthermore, all together, 20% of the dysregulation concerned the two strands in a hairpin. In vivo, 36% and 43% of the passenger strands were dysregulated compared to 25% and 29% for the guide strands at 7 days following stenting by BMS (bare-metal stent) and DES (drug eluting stent), respectively (Figure 2B). This dysregulation was still at 32% and 39% for the passenger strands compared to 19% and 23% for the guide strands at

28 days following stenting by BMS and DES, respectively (Figure S6A). In our vein graft failure model, we found 59% of the passenger strands dysregulated compared to 40% the guide strands at 7 days following engraftment (Figure 2B). This dysregulation reached 68% for the passenger strands and 33% for the guide strands at 28 days (Figure S6B). Additionally, all together, we found 47% and 31% of the dysregulation concerned the two strands in a hairpin in the pig vein graft model and in the pig stent model, respectively.

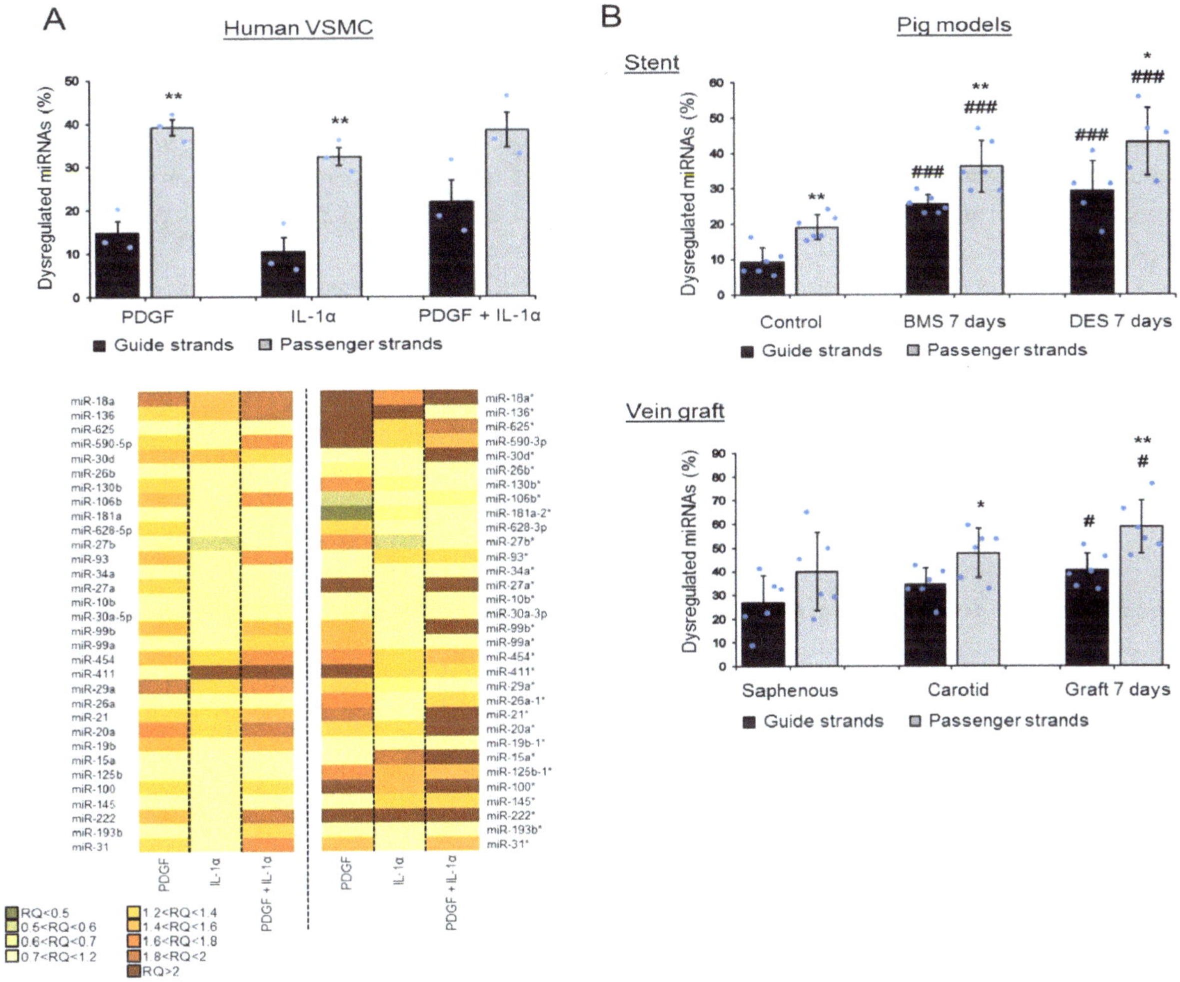

Figure 2. High passenger strand dysregulation compared to the guide strand dysregulation in VSMCs and in vivo. The graphs show the percentage of dysregulated miRNAs for each form in the three independent models. (**A**) The relative dysregulation of each strand of the miRNA hairpin was quantified after stimulation by platelet-derived growth factor (PDGF) and IL-1α in human VSMCs (the statistics have been made between the passenger and guide strands within each condition; unpaired t-test; ** $p < 0.01$ for passenger vs guide strands). The heat map illustrates the relative dysregulation (in relative quotient (RQ)) of a number of miRNA hairpins after stimulation by PDGF and IL-1α across three independent patient samples (average of $n = 3$). (**B**) The dysregulation of the guide and passenger strands was quantified at 7 days following stenting (BMS (bare-metal stent) or DES (drug eluting stent)) or at 7 days following vein grafting in vivo ($n = 6$ for both the porcine models, except for the 7 days in the DES condition where $n = 5$) (the statistics have been made between the passenger and guide strands within each condition; unpaired t-test; * $p < 0.05$ and ** $p < 0.01$; and vs. the corresponding control (unstented control animals or saphenous), # $p < 0.05$ and ### $p < 0.001$).

4. Discussion

VSMCs are typically quiescent and contractile under normal physiological conditions. However, following vascular injury, the release of cytokines and growth factors initiates a transcriptional cascade resulting in the initiation of a process called "phenotypic switching" and this leads the VSMCs to harbor a synthetic, pro-proliferative and pro-migratory state. Since miRNAs regulate mRNA expression and thereby many fundamental biological processes, they are intensely studied as candidates for diagnostic and prognostic biomarkers in a number of diseases. Indeed, miRNAs have been shown to play important roles in many cellular processes such as cell differentiation, proliferation and migration. We decided to conduct a detailed analysis in order to understand the microRNAome through TLDA in the pre-clinical porcine models of vascular injury and human VSMCs. It is important to note that we utilized human arrays in the porcine samples to try and identify a miRNA which would extrapolate to the pathology in the clinical setting, so it is possible that a proportion of miRNAs may be under-represented due to the sequence variation or chromosomal locations between pig and human. Our global analysis shows the dysregulation of a number of miRNAs already described in the literature as phenotypic regulators of VSMCs. For example, we found that miR-222, previously reported to be involved in vascular pathologies associated with excessive rates of SMC proliferation and migration [22], was substantially up-regulated in response to pathological stimuli. Indeed, in our arrays, miR-222 was overexpressed upon PDGF stimulation in VSMCs (Figure 2 and Figure S3) as well as following DES implantation and vein grafting in the porcine models (Figure S6). Furthermore, we found that miR-18a was upregulated following vein grafting and stenting in pig models (Figure S6). This finding is in accordance with the literature, in which miR-18a has been found to be upregulated after rat carotid balloon injury and to promote VSMC differentiation [23]. Furthermore, miR-145 was downregulated following vein grafting and during the early stage after stenting in our porcine models (Figure S6). MiR-145 is extensively studied in vascular biology for its ability to control vascular neointimal lesion formation. Indeed, the restoration of miR-145 in balloon-injured arteries inhibits neointimal growth in rat [24]. Moreover, our global analysis gave a global landscape of the miRNA guide strand as well as passenger strand expression. Indeed, recent studies suggest that these passenger strands can mediate strand-specific roles based on their distinct seed sequences and targets, highlighting their importance [13,14]. We decided to study the behavior of miRNA depending on the strand form after exposure to pathological stimuli via our detailed analysis of the miRNA transcriptome through TLDA. Our results confirmed a higher global expression of the guide strands compared to passenger strands. Few exceptions with a higher expression of the passenger strands of the miRNA:miRNA * duplex have been noted, and this suggests the possibility that they can mediate functional effects on mRNA targets. However, our study is the first to show that a greater proportion of passenger strand miRNAs was significantly altered after exposure to the pathological mediators of vascular remodeling, PDGF and/or IL-1α in vitro and post-stenting or grafting in vivo. Some studies reported a preferred arm switched between samples and tissues [25,26]. In our study, after vascular injury, the selection of the mature miRNA strand via an "arm switching" mechanism could be responsible for the greater proportion of the dysregulated passenger strands that was obtained. It is widely accepted that vascular injury is responsible for the "phenotypic switching" induction in VSMCs but our study suggests that vascular injury could also be implicated in the aforementioned miRNA "arm switching" phenomenon.

In conclusion, we demonstrated via a global analysis that the miRNA passenger strands are frequently dysregulated in acute vascular injury. Our detailed and unbiased analysis emphasized an interesting landscape of miRNAs in vascular biology where the passenger strands contribute more broadly than previously thought and may be attractive targets for disease intervention.

Supplementary Materials:
expression, Figure S2: Validation of VSMC stimulation, Figure S3: Validation of TLDA arrays expression profiles by qRT-PCR, Figure S4: Higher expression of the guide strands compared to the passenger strands in VSMCs and in vivo, Figure S5: Low passenger strands expression compared to the guide strands in stent and vein graft

porcine models, Figure S6: High passenger strands dysregulation compared to the guide strands 28 days following stenting or grafting in porcine models, Table S1: miRNAs expression analyzed by TLDA in human VSMCs.

Author Contributions: Conceptualization, K.P., R.A.M. and A.H.B.; methodology, K.P., L.A.D., K.W. and R.A.M.; investigation and formal analysis, K.P.; writing, K.P. and A.H.B.; funding acquisition, A.H.B.

Acknowledgments: We thank Gregor Aitchison, Elaine Friel and Nicola Britton for technical assistance.

References

1. Wu, L.; Belasco, J.G. Let me count the ways: Mechanisms of gene regulation by miRNAs and siRNAs. *Mol. Cell* **2008**, *29*, 1–7. [CrossRef] [PubMed]
2. Eulalio, A.; Huntzinger, E.; Izaurralde, E. Getting to the root of miRNA-mediated gene silencing. *Cell* **2008**, *132*, 9–14. [CrossRef]
3. Filipowicz, W.; Bhattacharyya, S.N.; Sonenberg, N. Mechanisms of post-transcriptional regulation by microRNAs: Are the answers in sight? *Nat. Rev. Genet.* **2008**, *9*, 102–114. [CrossRef] [PubMed]
4. Bartel, D.P. MicroRNAs: Genomics, biogenesis, mechanism, and function. *Cell* **2004**, *116*, 281–297. [CrossRef]
5. Guo, L.; Zhang, H.; Zhao, Y.; Yang, S.; Chen, F. Selected isomiR expression profiles via arm switching? *Gene* **2014**, *533*, 149–155. [CrossRef] [PubMed]
6. Winter, J.; Diederichs, S. Argonaute-3 activates the let-7a passenger strand microRNA. *RNA Biol.* **2013**, *10*, 1631–1643. [CrossRef]
7. Owens, G.K.; Kumar, M.S.; Wamhoff, B.R. Molecular regulation of vascular smooth muscle cell differentiation in development and disease. *Physiol. Rev.* **2004**, *84*, 767–801. [CrossRef]
8. Dzau, V.J.; Braun-Dullaeus, R.C.; Sedding, D.G. Vascular proliferation and atherosclerosis: New perspectives and therapeutic strategies. *Nat. Med.* **2002**, *8*, 1249–1256. [CrossRef]
9. Ferns, G.A.; Raines, E.W.; Sprugel, K.H.; Motani, A.S.; Reidy, M.A.; Ross, R. Inhibition of neointimal smooth muscle accumulation after angioplasty by an antibody to PDGF. *Science* **1991**, *253*, 1129–1132. [CrossRef]
10. Yang, J.-S.; Phillips, M.D.; Betel, D.; Mu, P.; Ventura, A.; Siepel, A.C.; Chen, K.C.; Lai, E.C. Widespread regulatory activity of vertebrate microRNA* species. *RNA Soc.* **2011**, *17*, 312–326. [CrossRef]
11. Okamura, K.; Phillips, M.D.; Tyler, D.M.; Duan, H.; Chou, Y.; Lai, E.C. The regulatory activity of microRNA* species has substantial influence on microRNA and 3′ UTR evolution. *Nat. Struct. Mol. Biol.* **2008**, *15*, 354–363. [CrossRef]
12. Griffiths-Jones, S.; Hui, J.H.L.; Marco, A.; Ronshaugen, M. microRNA evolution by arm switching. *EMBO Rep.* **2011**, *12*, 172–177. [CrossRef]
13. Poissonnier, L.; Villain, G.; Soncin, F.; Mattot, V. miR126-5p repression of ALCAM and SetD5 in endothelial cells regulates leucocyte adhesion and transmigration. *Cardiovasc. Res.* **2014**, *102*, 436–447. [CrossRef] [PubMed]
14. Zhu, S.; Deng, S.; Ma, Q.; Zhang, T.; Jia, C.; Zhuo, D.; Yang, F.; Wei, J.; Wang, L.; Dykxhoorn, D.M.; et al. microRNA-10A* and microRNA-21 modulate endothelial progenitor cell senescence via suppressing high-mobility group A2. *Circ. Res.* **2013**, *112*, 152–164. [CrossRef] [PubMed]
15. Chang, K.-W.; Kao, S.-Y.; Wu, Y.-H.; Tsai, M.-M.; Tu, H.-F.; Liu, C.-J.; Lui, M.-T.; Lin, S.-C. Passenger strand miRNA miR-31* regulates the phenotypes of oral cancer cells by targeting RhoA. *Oral Oncol.* **2013**, *49*, 27–33. [CrossRef] [PubMed]
16. Yang, X.; Du, W.W.; Li, H.; Liu, F.; Khorshidi, A.; Rutnam, Z.J.; Yang, B.B. Both mature miR-17-5p and passenger strand miR-17-3p target TIMP3 and induce prostate tumor growth and invasion. *Nucleic Acids Res.* **2013**, *41*, 9688–9704. [CrossRef] [PubMed]
17. McDonald, R.A.; White, K.M.; Wu, J.; Cooley, B.C.; Robertson, K.E.; Halliday, C.A.; McClure, J.D.; Francis, S.; Lu, R.; Kennedy, S.; et al. miRNA-21 is dysregulated in response to vein grafting in multiple models and genetic ablation in mice attenuates neointima formation. *Eur. Heart J.* **2013**, *34*, 1636–1643. [CrossRef]
18. McDonald, R.A.; Halliday, C.A.; Miller, A.M.; Diver, L.A.; Dakin, R.S.; Montgomery, J.; McBride, M.W.; Kennedy, S.; McClure, J.D.; Robertson, K.E.; et al. Reducing In-Stent Restenosis: Therapeutic Manipulation of miRNA in Vascular Remodeling and Inflammation. *J. Am. Coll. Cardiol.* **2015**, *65*, 2314–2327. [CrossRef]

19. Angelini, G.D.; Bryan, A.J.; Williams, H.M.; Morgan, R.; Newby, A.C. Distention promotes platelet and leukocyte adhesion and reduces short-term patency in pig arteriovenous bypass grafts. *J. Thorac. Cardiovasc. Surg.* **1990**, *99*, 433–439. [PubMed]
20. Southgate, K.M.; Davies, M.; Booth, R.F.; Newby, A.C. Involvement of extracellular-matrix-degrading metalloproteinases in rabbit aortic smooth-muscle cell proliferation. *Biochem. J.* **1992**, *288*, 93–99. [CrossRef]
21. Raitoharju, E.; Lyytikäinen, L.-P.; Levula, M.; Oksala, N.; Mennander, A.; Tarkka, M.; Klopp, N.; Illig, T.; Kähönen, M.; Karhunen, P.J.; et al. miR-21, miR-210, miR-34a, and miR-146a/b are up-regulated in human atherosclerotic plaques in the Tampere Vascular Study. *Atherosclerosis* **2011**, *219*, 211–217. [CrossRef] [PubMed]
22. Liu, X.; Cheng, Y.; Yang, J.; Xu, L.; Zhang, C. Cell-specific effects of miR-221/222 in vessels: Molecular mechanism and therapeutic application. *J. Mol. Cell. Cardiol.* **2012**, *52*, 245–255. [CrossRef]
23. Kee, H.J.; Kim, G.R.; Cho, S.-N.; Kwon, J.-S.; Ahn, Y.; Kook, H.; Jeong, M.H. miR-18a-5p MicroRNA Increases Vascular Smooth Muscle Cell Differentiation by Downregulating Syndecan4. *Korean Circ. J.* **2014**, *44*, 255–263. [CrossRef]
24. Cheng, Y.; Liu, X.; Yang, J.; Lin, Y.; Xu, D.-Z.; Lu, Q.; Deitch, E.A.; Huo, Y.; Delphin, E.S.; Zhang, C. MicroRNA-145, a novel smooth muscle cell phenotypic marker and modulator, controls vascular neointimal lesion formation. *Circ. Res.* **2009**, *105*, 158–166. [CrossRef]
25. Chiang, H.R.; Schoenfeld, L.W.; Ruby, J.G.; Auyeung, V.C.; Spies, N.; Baek, D.; Johnston, W.K.; Russ, C.; Luo, S.; Babiarz, J.E.; et al. Mammalian microRNAs: Experimental evaluation of novel and previously annotated genes. *Genes Dev.* **2010**, *24*, 992–1009. [CrossRef] [PubMed]
26. Ro, S.; Park, C.; Young, D.; Sanders, K.M.; Yan, W. Tissue-dependent paired expression of miRNAs. *Nucleic Acids Res.* **2007**, *35*, 5944–5953. [CrossRef] [PubMed]

14

Genome-Wide Investigation of the Role of MicroRNAs in Desiccation Tolerance in the Resurrection Grass *Tripogon loliiformis*

Isaac Njaci [1], Brett Williams [1], Claudia Castillo-González [2], Martin B. Dickman [3], Xiuren Zhang [2] and Sagadevan Mundree [1,*]

1 Centre for Tropical Crops and Biocommodities, Queensland University of Technology, Brisbane, QLD 4000, Australia; injaci@gmail.com (I.N.); b.williams@qut.edu.au (B.W.)
2 Department of Biochemistry and Biophysics, Institute for Plant Genomics and Biotechnology, Texas A&M University, College Station, TX 77843, USA; castillo.cm@tamu.edu (C.C.-G.); xiuren.zhang@tamu.edu (X.Z.)
3 Department of Plant Pathology and Microbiology, Institute for Plant Genomics and Biotechnology, Texas A&M University, College Station, TX 77843, USA; mbdickman@tamu.edu
* Correspondence: sagadevan.mundree@qut.edu.au;

Abstract: Drought causes approximately two-thirds of crop and yield loss worldwide. To sustain future generations, there is a need to develop robust crops with enhanced water use efficiency. Resurrection plants are naturally resilient and tolerate up to 95% water loss with the ability to revive upon watering. Stress is genetically encoded and resilient species may garner tolerance by tightly regulating the expression of stress-related genes. MicroRNAs (miRNAs) post-transcriptionally regulate development and other stress response processes in eukaryotes. However, their role in resurrection plant desiccation tolerance is poorly understood. In this study, small RNA sequencing and miRNA expression profiling was conducted using *Tripogon loliiformis* plants subjected to extreme water deficit conditions. Differentially expressed miRNA profiles, target mRNAs, and their regulatory processes were elucidated. Gene ontology enrichment analysis revealed that development, stress response, and regulation of programmed cell death biological processes; Oxidoreductase and hydrolyase molecular activities; and SPL, MYB, and WRKY transcription factors were targeted by miRNAs during dehydration stress, indicating the indispensable regulatory role of miRNAs in desiccation tolerance. This study provides insights into the molecular mechanisms of desiccation tolerance in the resurrection plant *T. loliiformis*. This information will be useful in devising strategies for crop improvement on enhanced drought tolerance and water use efficiency.

Keywords: microRNAs; dehydration; desiccation; resurrection plants; *Tripogon loliiformis*; post-transcriptional gene silencing; miRNAs

---

1. Introduction

The majority of higher plants are sensitive to dehydration and lose viability upon the loss of 41–70% of their total water content [1]. However, some plants are well adapted to adverse environments and implement adaptation mechanisms that mitigate the effects of water loss [2]. Although the adaptation mechanisms are effective under diverse environments, enabling plant survival, severe water loss still leads to the death of the plant. Resurrection plants represent a small but diverse group of angiosperms that exhibit unique tolerance mechanisms to cope with desiccation [3]. This unique group of plants can tolerate desiccation in their vegetative tissues for prolonged periods and rapidly recover their metabolic activity within 48 h of watering with minimal or non-existent tissue damage [3,4].

To achieve this phenomenal tolerance, resurrection plants utilize a repertoire of strategies that include the rapid shutdown of photosynthesis; leaf and cell wall structural adjustment; the stabilization of subcellular milieu through the accumulation of sugars, late embryogenesis abundant proteins (LEA), heat shock proteins, and other compatible solutes; and the induction of extensive antioxidant and ROS scavenging systems during desiccation [5–8].

Tripogon loliiformis is a small tufted diploid (2n = 20) annual to short-lived perennial C4 resurrection grass. It is endemic to a wide range of habitats in Australia and "resurrects" from its desiccated state within 72 h [9,10]. It's short life cycle, ploidy level, and ease of propagation makes it a suitable experimental model for desiccation tolerance studies. *T. loliiformis* physiological responses to desiccation are characterized by structural, physiological, and biochemical changes that include leaf folding, cell wall folding and vacuole fragmentation, the early shutdown of photosynthesis, the retention of chlorophyll (homoiochlorophyllous) [9,11], increased anthocyanin accumulation, and the accumulation of sucrose and trehalose [12]. Response to stress is genetically encoded and transcriptional and post-transcriptional reprogramming are important elements in stress tolerance strategies [13–16]. Previous studies have linked miRNAs with many cellular processes [17,18].

MiRNAs are a class of 20–22 nt endogenous non-coding small RNAs that play important roles in the regulation of gene expression at the transcriptional and post-transcriptional level in animals, plants [19,20], and unicellular organisms [21]. In plants, miRNAs are involved in many metabolic and biological processes, where they play crucial regulatory roles in growth and development [22], phytohormone signaling [23], and adaptive responses to abiotic and biotic stress [17,24,25]. In Arabidopsis, using mannitol as a stress-inducing agent, miR396, miR168, miR167, and miR171 were found to be drought-responsive [26]. In rice, 30 differentially expressed miRNAs were identified under drought stress, while in tobacco, miR395 and miR169 were found to be sensitive to drought stress [27]. Using a combination of high throughput sequencing and microarray technology in a genome-wide study of *Populus euphratica*, Li et al. [28] identified 131 differentially expressed miRNAs under drought stress.

Although studies on miRNAs have been conducted in drought sensitive and tolerant plants including wheat [29], sorghum [30], switch grass [31], brachypodium [32], *Arabidopsis* [25], rice [33], and maize [34], their role in stress tolerance in desiccation tolerant plants has not been investigated. While drought sensitive and tolerant plants employ tolerance mechanisms to withstand mild to moderate dehydration stress, it is postulated that resurrection plants utilize a repertoire of unique desiccation tolerance mechanisms to overcome and adapt to extreme conditions. In this study, *T. loliiformis* plants subjected to desiccation stress at strategic dehydration, desiccation, and rehydration stages were analyzed through high throughput sequencing for the identification and enrichment of conserved miRNAs and elucidation of their potential roles in desiccation tolerance. Future crop production is threatened by the effects of global climate change, erratic weather patterns, and population growth. There is an urgent need for the development of climate resilient and water use efficient crops. Understanding the unique desiccation tolerance mechanisms harbored by resurrection plants presents a great potential for the development of robust crops.

2. Results

2.1. Analysis of Small RNAs in Shoot and Root of Tripogon loliiformis

The unique desiccation tolerance capabilities of resurrection plants suggests underlying tolerance mechanisms and effective regulation of transcription and translation. Studies have shown that miRNAs regulate gene expression at the transcriptional and post-transcriptional level. To investigate the role of miRNAs in desiccation tolerance, small RNA libraries from shoots and roots of hydrated, dehydrating (60% and 40% Relative Water Content (RWC)), desiccated (<10% RWC), and rehydrated plants were sequenced and the identification and expression profiling of miRNAs were conducted. High-throughput sequencing generated 142 million small RNAs reads in the range of 13–16 million reads per library, with an average of 1–3 million unique reads (Figure 1A). The raw reads were pre-processed to remove adapter sequences, low quality reads, and <15 nt reads, resulting in 104 (73%) million clean reads that were used for downstream analysis. Consistent with previous studies, the majority of the small RNAs were in the range of 21–24 nt, with 24 nt having the most reads, followed by 21 nt, across all libraries (Figure 1B).

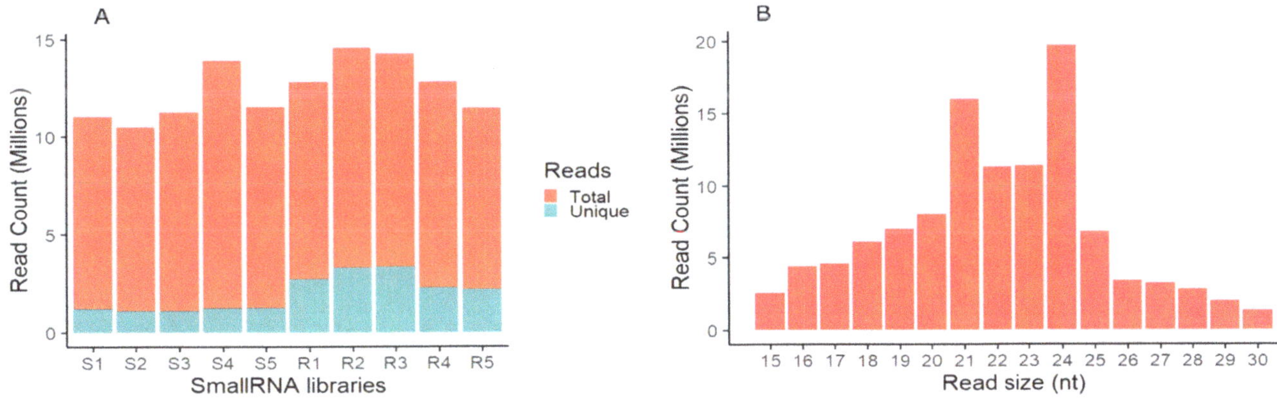

Figure 1. *Tripogon loliiformis* small RNA reads distribution. (**A**) Pre-processed reads distribution in shoot and root small RNA libraries; (**B**) overall reads distribution by size. S1–S5, R1–R5 represent shoot and root libraries at Hydrated (1), 60% RWC (2), 40% RWC (3), <10% RWC (4), and 48 h after rehydration (5), respectively.

2.2. *Evolutionary Conservation of MiRNAs in Tripogon loliiformis*

Many miRNAs are evolutionary conserved in species within the same kingdom and miRNA genes in one species may exist as orthologs or homologs in other species, presenting a powerful strategy to identify new miRNAs through a homology search [35]. To identify conserved miRNAs in *T. loliiformis*, the small RNA sequences were mapped against known plant miRNAs in the miRBase 21.0 database [36] and a total of 265 unique conserved miRNAs comprising 668 family members and isoforms from 60 MIR families were identified (Supplemental Table S1). Many of the identified miRNAs were family members and isoforms of the nine most evolutionary conserved miRNAs in plants [37]. The majority of the identified miRNAs were from closely related monocot species including *Oryza sativa*, *Sorghum bicolor*, *Zea mays*, *Branchypodium distachyon*, *Hordeum vulgare*, *Triticum aestivum*, and *Aegilops tauschii*, as well as from the model species *Arabidopsis thaliana* (Supplemental Figure S1).

2.3. *Spatiotemporal Expression of MiRNAs in Tripogon loliiformis Tissues*

Previous studies suggest that resurrection plants are genetically primed to respond to dehydration, even at a hydrated state, through the constitutive expression of stress response mechanisms [38,39]. Do resurrection plants gain their tolerance by post-transcriptionally regulating transcription more tightly than sensitive plants? Differential expression analysis identified 183 conserved miRNAs that differentially accumulated in shoots and roots during dehydration (Supplemental Table S2). A higher number of miRNAs were up-regulated in the shoots compared to the roots (Figure 2A). In the shoots at 60% RWC, 47 miRNAs were up-regulated and 29 down-regulated. At 40% RWC, 37 miRNAs were up-regulated while 40 were down-regulated. Surprisingly, miRNA accumulation was observed in desiccated tissue at <10% RWC, where 41 miRNAs showed an increased abundance, while 40 displayed a decreased accumulation. In rehydrating shoots, 51 miRNAs were up-regulated while 22 were down-regulated. In contrast, a reduced accumulation of miRNAs was observed in the roots where a higher number showed downregulation as dehydration ensued. In summary, 26 miRNAs were up-regulated and 54 down-regulated at 60% RWC, 24 up-regulated and 50 down-regulated at

40% RWC, and 30 up- and 54 down-regulated at <10% RWC, while 23 and 10 miRNAs were up- and down-regulated respectively on rehydration (Figure 2A, Supplemental Figure S2).

Tissue specific expression was observed where some miRNAs exhibited disparate expression between shoots and roots (Figure 2B). For example, miR399b, miR399j, miR167a, and miR393 h:i:j:k were up-regulated, while miR164c:h, miR169r:a, miR528a:b, and miR160d were down-regulated in the shoots, but were not expressed in the roots. The miRNAs miR444f, miR160e, and miR6300 were up-regulated in the shoots and down-regulated in the roots (Figure 2A), while other miRNAs showed a similar expression trend.

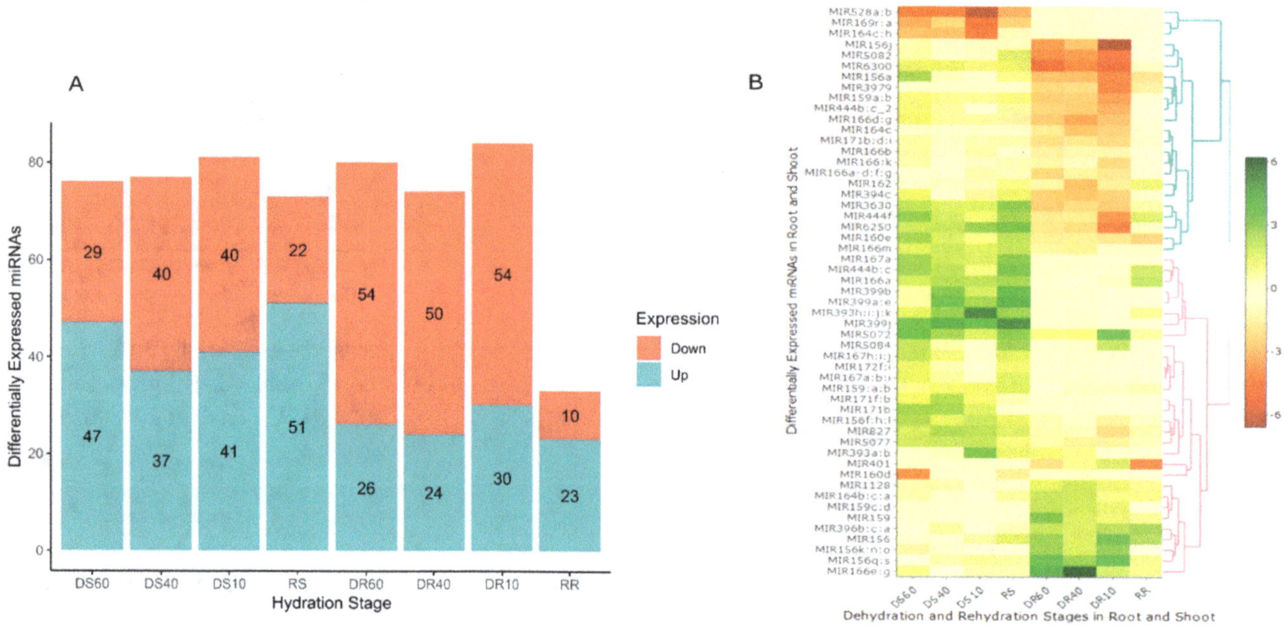

Figure 2. Differentially expressed miRNAs in shoots and roots under dehydration stress. (**A**) Bar graph showing disparate expression between shoot and roots. (**B**) Heatmap showing differentially expressed genes between shoots and roots at the different dehydration, desiccation, and rehydration stages. DS: Dehydrated Shoot, DR: Dehydrated Root, RS: Rehydrated Shoot, RR: Rehydrated Root.

2.4. *Tripogon loliiformis* Stress-Associated MiRNAs Targets

MicroRNAs exert their post-transcriptional gene silencing role through base complementarity pairing with their cognate mRNA transcripts, leading to cleavage or translation repression. A large number of gene targets were predicted from *T. loliiformis* contigs annotated from RNAseq data generated in the same experiment [40], *S. italica* coding sequences, and *A. thaliana* transcripts preloaded in the psRNATarget genomic library for target prediction. A total of 1236 unique contigs targeted by conserved miRNAs were predicted in *T. loliiformis* (Supplemental Table S3). The majority of the predicted targets were orthologs of known conserved miRNAs targets involved in a broad range of biological processes including metabolism, response to abiotic stress, post-transcriptional gene silencing, regulation of development, and gene expression. The targets included transcription factors (TFs), protein kinases, transporters, chaperones, antioxidants, and carbohydrate metabolism associated genes. The transcription factors and other genes targeted by the miRNAs were interpreted to be differentially expressed based on the accumulation of their cognate miRNAs. The TFs associated with down-regulated miRNAs included members of Auxin response factor (ARF 5, 13, 22) targeted by miR162a, miR160a, and miR164b; *Squamosa* promoter binding-like protein (SPL 17, 18, 19) targeted by miR157d and miR156c; MYB (miR396, miR1128, miR159a), MADs-box (miR157a, 164b, miR1128), NAC (miR390), bHLH (miR408), and WRKY 38, 39, 70 targeted by miR167a, miR395b, and miR390. The TFs associated with up-regulated miRNAs included GATA 12 (miR396a),

Scarecrow (miR171c), WRKY 4, 14 (miR5139, miR2916), ethylene-responsive transcription factor (ERF1) (miR5021), AP2/EREBP (miR166), bZIP (miR5021), and the nuclear factor Y (NFY A, C) as targets of miR169b and miR167a, respectively. However, it is worth noting that some members of the same transcription factor family, for example, WRKY, SPL, and MYB, showed disparate expression (Table S3). The predicted chaperones and heat shock proteins and factors targeted by miRNAs included HSP70, HSP90, and DNAJ. The antioxidants ascorbate peroxidase 4-like, peroxidase, and glutathione-s-transferase were among the targets for down-regulated miRNAs. A number of carbohydrate metabolism associated gene targets with a low accumulation of their cognate miRNAs included sucrose synthase (miR195a), Galactinol-sucrose galactosyltransferase (miR1128), fructokinase-4-like (miR319), hexokinase (miR319g:l, miR319a-d:f:h), and sucrose-phosphate synthase (miR167c) (Table S3). The accumulation of sucrose during dehydration and decline in glucose and fructose levels has been previously reported in a number of desiccation tolerant plants, including *T. loliiformis* [12,41].

The majority of the miRNAs targeting protein kinases were down-regulated during dehydration, indicating the active involvement of kinases in stress response signaling. The predicted kinases included class members of calcium-dependent protein kinase 1, 2, 3 targeted by miR157, miR169p, and miR159a, and CBL-interacting protein kinase as a target of miR408e. The expression of some kinases such as cysteine-rich receptor-like protein kinase (miR156h), MAP kinase (miR166h), serine/threonine-protein kinase (miR529), and phosphatidylinositol 4-kinase (miR408) was enhanced, while family members of some kinases displayed disparate expression. Transport associated members of ABC transporters, aquaporin NIP and H^+ antiporter, were targeted by down-regulated miRNAs, while the K^+ antiporter and aquaporin TIP and SIP were targets of up-regulated miRNAs during dehydration stress (Table S3).

2.5. *Functional Roles of MiRNA Targets*

MiRNAs control various cellular physiological, biochemical, and molecular processes. To investigate the potential functions and biological relevance of the predicted miRNA target genes, Gene ontology (GO) and enrichment analysis was conducted using Blast2GO [42] and GO terms viewed on Cytoscape [43]. The over-represented GO terms for the up-regulated miRNA gene targets included development associated processes such as cell development, growth and regulation of cell morphogenesis, metabolic and DNA catabolic processes, and regulation of programmed cell death. Other processes were related to vesicle fusion, membrane fusion, and protein transport and localization (Supplemental Figure S3). Over-represented metabolic activities included oxidoreductase, hydrolyase, nuclease, and ligase. The biological processes associated with down-regulated miRNA target genes included stress responses, defense and innate immunity, lipid and fatty acids metabolism, membrane transport, gene expression regulation, and post-transcriptional gene silencing (Supplemental Figure S4).

3. Discussion

MiRNAs research has been extensively conducted and their critical role in many biological processes enumerated [17,24,25,44]. In plants, most miRNA studies have focused on drought sensitive and tolerant species and no studies on desiccation responsive miRNAs in desiccation tolerant plants have been reported. Drought sensitive and tolerant plants employ stress tolerance mechanisms geared towards the restoration of homeostasis through water retention, albeit at different efficacies. Although resurrection plants utilize similar mechanisms at mild and moderate dehydration stress, they are primed for desiccation and utilize additional desiccation tolerance mechanisms to limit cellular damage to repairable levels through maintenance of the cellular structure and physiological integrity [2]. Based on the different response strategies, it is plausible that resurrection plants employ distinct regulatory processes at the post-transcriptional level that may not be observed in other vascular plant species. In this study, we performed a genome-wide profiling of miRNAs and their expression and targets in the desiccation tolerant *T. loliiformis*. High-throughput sequencing of small RNA libraries confirmed the evolutionary conservation of miRNAs through the identification of conserved miRNAs in *T. loliiformis*.

The high number of miRNAs identified suggests the critical role of miRNAs in gene expression regulation and their indispensable position in desiccation tolerance strategies in resurrection plants. The elevated accumulation of miRNAs in desiccated tissues suggested the existence of enhanced regulatory activities as dehydration ensued. However, the presence of miRNAs at a desiccated state could imply the repackaging and storage of miRNAs as the tissues dehydrated below 40% RWC, evidenced by high numbers during rehydration (Figure 2). The high number of miRNAs expressed during rehydration could be attributed to the regulation of cellular processes for protection against rehydration associated damage observed in resurrection plants [45].

Recent transcriptomics studies in resurrection plants indicated transcripts expression at extreme dehydration and attributed signal transduction proteins and retroelements to the instrumental role of gene silencing during desiccation [13,14,46]. In *T. loliiformis*, a similar trend was observed, where miRNAs expression was recorded at a desiccated state. Tissue specific expression was observed where miRNAs were induced in shoots and their expression was suppressed in roots (Figure 2). The expression disparity pointed to an inherent stress adaptation strategy where miRNAs act as master modulators of processes associated with energy metabolism, growth, and development in the shoots, while redirecting resources to promote stress tolerance and enhance protective mechanisms, some of which could be targeted to the roots. In Arabidopsis miR393, miR390/159, and miR159, TIR1, TCP/MYB, and SBP-LIKE, respectively, were induced under biotic and abiotic stress to suppress development-related processes for morphological adaptation to stress [17]. In resurrection plants, transcriptional reprogramming redirecting resources from growth processes towards cellular protection has been previously reported [13].

The majority of plant miRNAs target transcription factors that bind to conserved *cis*-acting promoter elements to affect the gene expression response, particularly those induced by abiotic stress [47]. Among the predicted transcription factors associated with down-regulated miRNAs during desiccation in *T. loliiformis* were nuclear factor Y NF-YA, NF-YB, and NF-YC targeted by miR169m, miR528, and miR167a, respectively. NF-Y is a CCAAT-DNA binding transcription factor that imparts significant tolerance to drought and facilitates increased yields in corn [48]. Members of the squamosa promoter-binding-like proteins (SPL) involved in the temporal regulation of shoot development and phase transition were differentially expressed, as previously reported [49]. The SPL transcription factors SPL17 (miR156a/c) and SPL18 (miR157) were cognates for down-regulated miRNAs and SPL6 (miR166), SPL12 (miR157), and SPL15 (miR5059) were predicted to be down-regulated. Under water deficit conditions, SPL are usually down-regulated as growth and developmental-associated processes including shoot development, vegetative phase transition, flowering, and leaf polarity are normally shutdown [17]. The observed differential expression implies that some members of SPL have stress tolerance-related functions in *T. loliiformis*, as recently reported in wheat [50]. The predicted members of the WRKY transcription factors modulate many processes in plants, including the response to abiotic and biotic stress [51]. The expression of WRKY 4, 38, 39, and 70 showed enhanced expression under extreme dehydration, pointing to their abiotic stress regulatory roles. In Arabidopsis, WRKY 70 was implicated in the positive regulation of defence, negative regulation of senescence, and modulation of osmotic stress tolerance through the regulation of stomatal aperture [52,53]. In the resurrection plant *Haberlea rhodopensis*, WRKY transcription factors were induced during desiccation [13]. The observed delayed senescence and early photosynthesis shutdown in *T. loliiformis* could be attributed to the regulation of WRKY 70 by miR390. Recent studies have associated NAC TFs with an unfolded protein response (UPR) in the ER stress response [54], while the regulation of GAMYB and ARF by miR159 and miR160 has been reported [49].

Protein synthesis and folding machinery in the endoplasmic reticulum (ER) can be compromised under unfavorable conditions, resulting in ER stress that activates the UPR [55,56]. The ER resident molecular chaperones play a critical role in the folding of newly synthesized proteins, maintenance of proteome integrity, and protein homeostasis [57]. In *T. loliiformis*, DNAJ and calnexin chaperones were predicted to be targeted by miR408 and miR5021, respectively. Calnexin (Cnx), an integral membrane protein, coordinates the processing of newly synthesized N-linked glycoproteins. DNAJ, on the other hand, binds directly on the luminal binding protein BiP, an HSP70 molecular chaperone that interacts with the newly synthesized polypeptides [58]. The observed down-regulation of miR408 and miR5021 would enhance the accumulation of DNAJ and calnexin chaperones to counter the ER stress during dehydration stress in *T. loliiformis*. Heterologous expression of *Oryza sativa* calnexin in tobacco conferred dehydration tolerance under mannitol stress [59].

Carbohydrate metabolism is a key process observed in resurrection plants during desiccation [7,60]. The accumulation of sucrose, oligosaccharides, and compatible solutes such as Late Embryogenesis Abundant (LEA) proteins and small heat shock proteins during desiccation leads to cellular stabilization through cytosolic vitrification [61]. Desiccation-induced accumulation of non-reducing trehalose sugar in *T. loliiformis* was recently associated with the induction of the cytoprotective autophagy pathways [12]. Comparative metabolic analysis of sucrose accumulation between desiccation tolerant and sensitive *Eragrostis nindensis* confirmed the crucial role of sucrose in desiccation tolerance, as sucrose only accumulated in the leaves of tolerant species [62]. Sucrose accumulation correlated with the expression of carbohydrate metabolic enzymes during desiccation, as previously reported in the resurrection plants *Craterostigma plantagineum* and *Haberlea rhodopensis* [13]. In *T. loliiformis*, the observed down-regulation of miRNAs targeting sucrose synthesis genes could lead to the enhanced expression of transcripts encoding sucrose synthase, sucrose 6-phosphate synthase, sucrose transporter, and galactinol synthase.

The lack of cellular damage and severe oxidative stress suggests elaborate protective mechanisms involving energy metabolism, growth, and development programs that regulate the induction of stress response mechanisms to provide a cushion against water loss. Differential miRNA expression analysis indicated distinct tissue specific accumulation patterns between the shoots and roots under desiccation stress. The majority of miRNAs were up-regulated in the shoots and down-regulated in the roots. For example, miR528a/b predicted to target LEA proteins and Calmodulin-like proteins (CML21) was down-regulated in shoots and had no expression in the roots. The accumulation of LEA proteins as a stress response mechanism has been previously reported. LEA proteins were reported to accumulate in shoot and scutellar, but not in root tissue of desiccation tolerant wheat seedling [63]. A variety of stress responses are mediated by Ca^{2+} signaling. For example, in rice, the multi-stress responsive gene2 (OsMSR2) was induced by multiple abiotic stress stimuli, and its overexpression in Arabidopsis enhanced tolerance to drought and salinity [64]. Chlorophyll biosynthesis is controlled by the scarecrow SCL27 gene through the regulation of photochlorophyllide oxidoreductase [65]. In *T. loliiformis*, miR171b predicted to target SCL27 was up-regulated in shoots and had zero expression in the roots. Disparate miRNA expression has been previously reported in other plants under diverse stress conditions [17], but not under desiccation tolerance. The observed spatiotemporal miRNAs expression in *T. loliiformis* suggests a unique adaptation strategy where extensive regulatory processes take place in the shoots, including early photosynthesis shutdown, induction of antioxidant systems, stress signaling, and resources mobilized and redirected to the roots to enhance survival.

4. Materials and Methods

4.1. Plant Materials and Stress Treatments

Tripogon loliiformis seeds from a single plant originally collected from Charleville in South western Queensland, Australia, were germinated in 65 mm plastic pots containing 50% native red soil and seed sowing potting mix in a growth chamber at 27 °C and for a 16 h photoperiod. The *Tripogon loliiformis* biological cycle completes in 12–14 weeks after germination. The plants are in their vegetative stage between two to six weeks and reproductive stage from the seventh to eighth week [66]. For this study, the dehydration experiment was conducted at the sixth week when the plants were in the vegetative stage. Twenty four hours prior to dehydration, fifteen pots containing multiple plants were well-watered to saturation. Three replicate samples from the hydrated plants were randomly collected. Since *T. loliiformis* is a small plant of approximately 5 cm in height, the entire aerial part (shoots) and roots, except the corm, were sampled per plant. The remaining plants were dehydrated by withholding water until they were air dry and their RWC dropped below 10% after six days [9]. During dehydration, triplicate shoot and root samples were collected at 60%, 40%, and <10% RWC. The collected samples were snap frozen in liquid nitrogen and stored at −80 °C until RNA extraction. Desiccated plants were watered and rehydrated samples collected after 48 h. The percentage RWC was determined using the leaf tissues and calculated according to Barrs, Weatherley [67]. *T. loliiformis* leaves were weighed upon sampling to get the fresh weight (FW). The leaves were placed in petri dishes containing water in a 4 °C fridge for 4–6 h. The leaves were blotted dry with a paper towel and weighed to get their turgid weight (TW). The leaves were dried overnight in a vacuum oven at 80 °C. The following day, the samples were cooled at room temperature in an aspirator to avoid error due to condensation, after which the dry weight (DW) was weighed. The percentage RWC of the leaf tissue was calculated using the formula

$$\text{RWC (\%)} = ((\text{Fresh Weight} - \text{Dry Weight})/(\text{Turgid Weight} - \text{Dry Weight})) \times 100.$$

4.2. Total RNA Extraction and High Throughput Sequencing

Total RNA was extracted from shoot and root tissues harvested from hydrated, dehydrating (60%, 40% RWC), desiccated (<10% RWC), and rehydrated plants using the Trizol Reagent (Invitrogen) according to the manufacturer's instructions with modifications. Briefly, 50 mg of sample was ground in liquid nitrogen and transferred into a microfuge tube containing 1 mL of Trizol reagent, mixed thoroughly, and incubated for 5 min at room temperature. A total of 200 μL of chloroform was added, mixed by vortex, and centrifuged for 15 min at 4 °C, 14,000 rpm. Then, 350 μL of the top aqueous phase was homogenized in a Qiagen QIAshredder spin column and centrifuged at maximum speed for 2 min, and the flow-through was transferred into a new 2 mL collection tube for ethanol precipitation by adding 0.5 times 96% ethanol. SmallRNAs enriched total RNA bidding was done by transferring the mixture into a miRNeasy™ minElute™ spin column. Total RNA quality and purity were checked with the NanoDrop® 2000 spectrophotometer (Thermo Scientific, Wilmington, DE, USA) and the integrity and quality were verified using a Bioanalyser (Agilent technologies, Santa Clara, CA, USA) (Supplemental Figure S5). Five triplicate shoot and root small RNAs libraries were prepared using the TruSeq™ Small RNA Sample Preparation protocol from Illumina® and 50 bp single-end reads sequenced at Texas A&M AgriLife Genomics and Bioinformatics service, USA, using an Illumina HiSeq 2500 Sequencer (Illumina Inc., San Diego, CA, USA). The reads have been deposited in the Sequence Read Archive (SRA) at NCBI, Accession number SRP113187 (https://www.ncbi.nlm.nih.gov/Traces/study/?acc=SRP113187).

4.3. Small RNAs Data Analysis and MiRNAs Identification

Small RNA reads were pre-processed for quality and trimmed for removal of the 3′ adapter sequences using CLC genomics workbench 6.52 [68]. The pre-processed reads were mapped against the plant miRNA sequences in the miRBase release 21 [36] using CLC genomics workbench 6.52 [68]. Due to the absence of *T. loliiformis* genomic information at the time of this study, annotated miRNA sequences from 71 plant species in miRBase [36] were used as proxy references against which the miRNAs were annotated. A mismatch allowance of two was used in the conserved miRNA search. The 5′ mature sequences in the size range of 20–22 nt were filtered from the resultant hits as potential conserved miRNAs.

4.4. MiRNAs Expression Analysis

The expression profiles of identified miRNAs were determined using CLC genomics Workbench. The raw small RNA reads were mapped against the identified miRNAs to determine miRNA expression values in the shoot and root tissues across respective dehydration and rehydration states. To determine the differential expression of miRNAs throughout dehydration, desiccation, and rehydration, the expression values were enumerated and normalized against total read counts reported as reads per million (RPM). Using the hydrated stage as the control, all hydration stages were compared to identify the miRNAs that were constitutively expressed, induced, or suppressed. MiRNAs with a fold change of ± 2 and a Bonferroni and FDR corrected p-value ≤ 0.05 were considered significant.

4.5. MiRNAs Targets and Their Functional Enrichment

Gene targets of the conserved miRNAs were predicted using the plant small RNA Target Analysis Server, psRNATarget [69], with the following parameters; maximum expectation (0–5):3, length of complementarity (15–30):19, and maximum energy to un-pair the target site (UPE, 0–100):25. The *Tripogon loliiformis* de novo assembled transcriptome dataset [40] and mRNA sequences from the closely related *S. italica*, as well as Arabidopsis, were used for the targets prediction. To identify the functional categories of the differentially expressed miRNA targets (Table 1), *T. loliiformis* transcriptome functional mapping encompassing all the plant metabolic pathways and enzyme functions was developed using the MapMan Mercator tool and visualization of predicted miRNA targets against the mapping was conducted using MapMan [70]. For Gene Ontology (GO) enrichment analysis, a BLAST search of *T. loliiformis* contigs against NCBI non-redundant (nr) Plantae sequences was conducted. The BLAST output was analyzed for GO enrichment with the Fischer exact test and a p-value of 0.05 to identify over-represented GO terms using Blast2GO [42]. The results were visualized using the Cytoscape BiNGO plugin [43].

Table 1. MapMan functional categories of potential targets of differentially expressed conserved miRNAs during dehydration in *Tripogon loliiformis*.

Name	Contig Number	Accession Number	Target Description
ABIOTIC STRESS RESPONSE			
MIR408	Contig_1263	XP_004951273.1	BAG family molecular chaperone regulator 6
MIR172j	Contig_4654	EMS50802.1	DnaJ homolog subfamily C member 7
MIR3630	Contig_195	CAA47948.2	heat shock protein 70
MIR6250	Contig_29066	XP_003578630.1	hydrophobic protein OSR8-like
MIR166d	Contig_10495	XP_003557818.1	probable methyltransferase PMT2-like
CARBOHYDRATE METABOLISM			
MIR319a	Contig_91996	XP_003568782.1	hexokinase-7-like
MIR172h	Contig_914	XP_004984086.1	sucrose synthase 2-like
MIR167c	Contig_12310	XP_004965756.1	sucrose-phosphate synthase 3-like
MIR394c	Contig_1	XP_003566746.1	callose synthase 7-like
MIR166b	Contig_1239	XP_004984579.1	galactinol synthase 2-like
MIR167e	Contig_3855	XP_004958079.1	xylose isomerase-like
DEVELOPMENT ASSOCIATED			
MIR5021	Contig_16313	XP_004966491.1	senescence-associated protein DIN1-like
MIR172b	Contig_2082	XP_004962343.1	serine/threonine-protein kinase TOR-like
MIR528	Contig_16524	XP_004951509.1	serine-threonine kinase receptor protein
MIR156c	Contig_10275	XP_004957197.1	squamosa promoter-binding-like protein 17-like
MIR156	Contig_10275	XP_004957197.1	squamosa promoter-binding-like protein 17-like
HORMONE METABOLISM			
MIR160d	Contig_39233	EMT29346.1	1-aminocyclopropane-1-carboxylate synthase 7
MIR167h	Contig_16664	XP_004961973.1	ABSCISIC ACID-INSENSITIVE 5-like protein
MIR164a	Contig_14909	XP_004985140.1	Allene oxide synthase 2-like
MIR172b	Contig_975	XP_004956401.1	auxin transport protein BIG-like
MIR171e	Contig_10826	XP_004976258.1	IAA-amino acid hydrolase ILR1-like 5-like

Table 1. *Cont.*

Name	Contig Number	Accession Number	Target Description
DNA REPAIR/SYNTHESIS/CHROMATIN STRUCTURE			
MIR166h	Contig_3556	XP_003576514.1	DNA mismatch repair protein Msh6-1-like
MIR167a	Contig_548	XP_004975516.1	DNA repair helicase XPB1-like
MIR172a	Contig_49068	NP_001151792.1	DNA binding protein
MIR319g:l	Contig_5793	NP_001065884.1	HUA enhancer 2
MIR399a:e	Contig_961	XP_004961835.1	probable histone H2A.4-like

5. Conclusions

Molecular studies have highlighted significant differences between resurrection plants and other dehydration sensitive species at the gene expression level. Some of the observed differences could be attributed to the post-transcriptional regulatory role of miRNAs. In this study, the results obtained suggest that *T. loliiformis*, a desiccation tolerant grass, elicits significant stress responses at the post-transcriptional level. *T. loliiformis* survival under extreme conditions could be attributed to efficient regulatory processes. Our findings in this study will enhance the understanding of *T. loliiformis* miRNA-mediated regulatory mechanisms and provide a foundation for further investigation of the gene pool of resurrection plants that could be holding many answers to agricultural crops.

Supplementary Materials:
68/s1, Figure S1: Annotation of *Tripogon loliiformis* conserved miRNAs using miRBase database, Figure S2: Cross-comparison Venn diagram showing the number of differentially expressed genes between shoots and roots, Figure S3: Gene Ontology of overrepresented GO terms associated with the targets of down-regulated miRNAs, Figure S4: Gene Ontology of overrepresented GO terms associated with the targets of up-regulated miRNAs, Figure S5: Representative Electrophoresis assay image of total RNA used for high-throughput sequencing. RNA with RNA Integrity Number (RIN) above 7.5 was used for sequencing library preparation, Table S1: List of conserved miRNAs identified in *Tripogon loliiformis*, Table S2: List of differential expressed conserved miRNAs in shoots and roots during dehydration, Table S3: List of predicted miRNAs targets associated with *T. loliiformis* contigs.

Author Contributions: Conceived and designed the experiments: I.N., B.W., C.C.-G., S.M., M.B.D., and X.Z. Performed the experiments: I.N., B.W., and C.C.-G. Analyzed the data: I.N., B.W., S.M., M.B.D., and X.Z. Wrote the paper: I.N., B.W., and S.M.

Acknowledgments: This work was supported by an Australia Awards - Africa PhD scholarship (I.N.), QUT Vice Chancellor's Research Fellowship (B.W.), and a professional capacity development grant (S.M.). Computational resources and services used in this work were provided by the HPC and Research Support Group, QUT, Brisbane, Australia. Library preparation and Illumina sequencing was performed by Genomics and Bioinformatics services at Texas A&M University. The authors also thank Peraj Karbaschi for assistance with the plant propagation.

References

1. Höfler, K.; Migsch, H.; Rottenburg, W. Über die Austrocknungresistenz landwirtschaftlicher Kulturpflanzen. *Forschungsdienst* **1941**, *12*, 50–61.
2. Rascio, N.; Rocca, N.L. Resurrection Plants: The Puzzle of Surviving Extreme Vegetative Desiccation. *Crit. Rev. Plant Sci.* **2005**, *24*, 209–225. [CrossRef]
3. Scott, P. Resurrection Plants and the Secrets of Eternal Leaf. *Ann. Bot.* **2000**, *85*, 159–166. [CrossRef]
4. Bernacchia, G.; Salamini, F.; Bartels, D. Molecular characterization of the rehydration process in the resurrection plant Craterostigma plantagineum. *Plant Physiol.* **1996**, *111*, 1043–1050. [CrossRef] [PubMed]
5. Farrant, J.M. A comparison of mechanisms of desicassion tolerance among 3 angiosperms resurrection plant species. *Plant Ecol.* **2000**, *151*, 29–39. [CrossRef]
6. Moore, J.P.; Farrant, J.M. A Systems-Based Molecular Biology Analysis of Resurrection Plants for Crop and Forage Improvement in Arid Environments. In *Improving Crop Resistance to Abiotic Stress*; Wiley-VCH Verlag GmbH & Co. KGaA: Weinheim, Germany, 2012; pp. 399–418. [CrossRef]

7. Peters, S.; Mundree, S.G.; Thomson, J.A.; Farrant, J.M.; Keller, F. Protection mechanisms in the resurrection plant Xerophyta viscosa (Baker): Both sucrose and raffinose family oligosaccharides (RFOs) accumulate in leaves in response to water deficit. *J. Exp. Bot.* **2007**, *58*, 1947–1956. [CrossRef] [PubMed]
8. Asami, P.; Mundree, S.; Williams, B. Saving for a rainy day: Control of energy needs in resurrection plants. *Plant Sci.* **2018**, *271*, 62–66. [CrossRef] [PubMed]
9. Karbaschi, M.R.; Williams, B.; Taji, A.; Mundree, S.G. Tripogon loliiformis elicits a rapid physiological and structural response to dehydration for desiccation tolerance. *Funct. Plant Biol.* **2016**. [CrossRef]
10. Orchard, E.; Wilson, A. Flora of Australia. In *Poaceae Australian Biological Resources Study*; K Mallett, T.M., Ed.; CSIRO: Canberra, Australia, 2005.
11. Gaff, D.; Pate, J.S. *The Biology of Resurrection Plants*; Pate, J.S., McComb, A.J., Eds.; The Biology of Australian Plants, University Western Australia Press: Nedlands, Australia, 1981; pp. 114–146.
12. Williams, B.; Njaci, I.; Moghaddam, L.; Long, H.; Dickman, M.B.; Zhang, X.; Mundree, S. Trehalose Accumulation Triggers Autophagy during Plant Desiccation. *PLoS Genet.* **2015**, *11*, e1005705. [CrossRef] [PubMed]
13. Gechev, T.S.; Benina, M.; Obata, T.; Tohge, T.; Sujeeth, N.; Minkov, I.; Hille, J.; Temanni, M.-R.; Marriott, A.S.; Bergström, E. Molecular mechanisms of desiccation tolerance in the resurrection glacial relic Haberlea rhodopensis. *Cell. Mol. Life Sci.* **2013**, *70*, 689–709. [CrossRef] [PubMed]
14. Rodriguez, M.C.S.; Edsgard, D.; Hussain, S.S.; Alquezar, D.; Rasmussen, M.; Gilbert, T.; Nielsen, B.H.; Bartels, D.; Mundy, J. Transcriptomes of the desiccation-tolerant resurrection plant Craterostigma plantagineum. *Plant J.* **2010**, *63*, 212–228. [CrossRef] [PubMed]
15. Zhou, J.; Wang, X.; Jiao, Y.; Qin, Y.; Liu, X.; He, K.; Chen, C.; Ma, L.; Wang, J.; Xiong, L. Global genome expression analysis of rice in response to drought and high-salinity stresses in shoot, flag leaf, and panicle. *Plant Mol. Biol.* **2007**, *63*, 591–608. [CrossRef] [PubMed]
16. Zhou, L.; Liu, Y.; Liu, Z.; Kong, D.; Duan, M.; Luo, L. Genome-wide identification and analysis of drought-responsive microRNAs in Oryza sativa. *J. Exp. Bot.* **2010**, *61*, 4157–4168. [CrossRef] [PubMed]
17. Khraiwesh, B.; Zhu, J.K.; Zhu, J. Role of miRNAs and siRNAs in biotic and abiotic stress responses of plants. *Biochim. Biophys. Acta* **2012**, *1819*, 137–148. [CrossRef] [PubMed]
18. Sunkar, R. MicroRNAs with macro-effects on plant stress responses. *Semin. Cell Dev. Biol.* **2010**, *21*, 805–811. [CrossRef] [PubMed]
19. Lagos-Quintana, M.; Rauhut, R.; Lendeckel, W.; Tuschl, T. Identification of novel genes coding for small expressed RNAs. *Science* **2001**, *294*, 853–858. [CrossRef] [PubMed]
20. Park, W.; Li, J.; Song, R.; Messing, J.; Chen, X. CARPEL FACTORY, a Dicer Homolog, and HEN1, a Novel Protein, Act in microRNA Metabolism in Arabidopsis thaliana. *Curr. Biol.* **2002**, *12*, 1484–1495. [CrossRef]
21. Zhao, T.; Li, G.; Mi, S.; Li, S.; Hannon, G.J.; Wang, X.-J.; Qi, Y. A complex system of small RNAs in the unicellular green alga Chlamydomonas reinhardtii. *Genes Dev.* **2007**, *21*, 1190–1203. [CrossRef] [PubMed]
22. Yang, T.W.; Xue, L.G.; An, L.Z. Functional diversity of miRNA in plants. *Plant Sci.* **2007**, *172*, 423–432. [CrossRef]
23. Liu, Q.; Chen, Y.Q. Insights into the mechanism of plant development: Interactions of miRNAs pathway with phytohormone response. *Biochem. Biophys. Res. Commun.* **2009**, *384*, 1–5. [CrossRef] [PubMed]
24. Shukla, L.I.; Chinnusamy, V.; Sunkar, R. The role of microRNAs and other endogenous small RNAs in plant stress responses. *Biochim. Biophys. Acta Gene Regul. Mech.* **2008**, *1779*, 743–748. [CrossRef] [PubMed]
25. Sunkar, R.; Zhu, J.K. Novel and stress-regulated microRNAs and other small RNAs from Arabidopsis. *Plant Cell* **2004**, *16*, 2001–2019. [CrossRef] [PubMed]
26. Liu, H.H.; Tian, X.; Li, Y.J.; Wu, C.A.; Zheng, C.C. Microarray-based analysis of stress-regulated microRNAs in Arabidopsis thaliana. *RNA* **2008**, *14*, 836–843. [CrossRef] [PubMed]
27. Frazier, T.P.; Sun, G.; Burklew, C.E.; Zhang, B. Salt and Drought stresses induce the aberrant expression of microRNA genes in tobacco. *Mol. Biotechnol.* **2011**, *49*, 159–165. [CrossRef] [PubMed]
28. Li, B.; Qin, Y.; Duan, H.; Yin, W.; Xia, X. Genome-wide characterization of new and drought stress responsive microRNAs in Populus euphratica. *J. Exp. Bot.* **2011**, *62*, 3765–3779. [CrossRef] [PubMed]
29. Sun, F.; Guo, G.; Du, J.; Guo, W.; Peng, H.; Ni, Z.; Sun, Q.; Yao, Y. Whole-genome discovery of miRNAs and their targets in wheat (*Triticum aestivum* L.). *BMC Plant Biol.* **2014**, *14*, 142. [CrossRef] [PubMed]
30. Zhang, L.; Zheng, Y.; Jagadeeswaran, G.; Li, Y.; Gowdu, K.; Sunkar, R. Identification and temporal expression analysis of conserved and novel microRNAs in Sorghum. *Genomics* **2011**, *98*, 460–468. [CrossRef] [PubMed]

31. Matts, J.; Jagadeeswaran, G.; Roe, B.A.; Sunkar, R. Identification of microRNAs and their targets in switchgrass, a model biofuel plant species. *J. Plant Physiol.* **2010**, *167*, 896–904. [CrossRef] [PubMed]
32. Budak, H.; Akpinar, A. Dehydration stress-responsive miRNA in Brachypodium distachyon: Evident by genome-wide screening of microRNAs expression. *OMICS* **2011**, *15*, 791–799. [CrossRef] [PubMed]
33. Zhao, B.; Liang, R.; Ge, L.; Li, W.; Xiao, H.; Lin, H.; Ruan, K.; Jin, Y. Identification of drought-induced microRNAs in rice. *Biochem. Biophys. Res. Commun.* **2007**, *354*, 585–590. [CrossRef] [PubMed]
34. Ding, D.; Zhang, L.; Wang, H.; Liu, Z.; Zhang, Z.; Zheng, Y. Differential expression of miRNAs in response to salt stress in maize roots. *Ann. Bot.* **2009**, *103*, 29–38. [CrossRef] [PubMed]
35. Zhang, B.; Pan, X.; Cannon, C.H.; Cobb, G.P.; Anderson, T.A. Conservation and divergence of plant microRNA genes. *Plant J.* **2006**, *46*, 243–259. [CrossRef] [PubMed]
36. Griffiths-Jones, S.; Grocock, R.J.; Van Dongen, S.; Bateman, A.; Enright, A.J. miRBase: MicroRNA sequences, targets and gene nomenclature. *Nucleic Acids Res.* **2006**, *34*, D140–D144. [CrossRef] [PubMed]
37. Luo, Y.; Guo, Z.; Li, L. Evolutionary conservation of microRNA regulatory programs in plant flower development. *Dev. Biol.* **2013**, *380*, 133–144. [CrossRef] [PubMed]
38. Oliver, M.J.; Guo, L.; Alexander, D.C.; Ryals, J.A.; Wone, B.W.; Cushman, J.C. A sister group contrast using untargeted global metabolomic analysis delineates the biochemical regulation underlying desiccation tolerance in Sporobolus stapfianus. *Plant Cell Online* **2011**, *23*, 1231–1248. [CrossRef] [PubMed]
39. Gechev, T.; Dinakar, C.; Benina, M.; Toneva, V.; Bartels, D. Molecular mechanisms of desiccation tolerance in resurrection plants. *Cell. Mol. Life Sci.* **2012**, *69*, 3175–3186. [CrossRef] [PubMed]
40. Sra, N. Bioproject: Transcriptome Analysis of Tripogon Loliiformis throughout Dehydration, Desiccation and Rehydration. Available online: https://www.ncbi.nlm.nih.gov/bioproject/?term=PRJNA288839 (accessed on 22 April 2018).
41. Whittaker, A.; Bochicchio, A.; Vazzana, C.; Lindsey, G.; Farrant, J. Changes in leaf hexokinase activity and metabolite levels in response to drying in the desiccation-tolerant species Sporobolus stapfianus and Xerophyta viscosa. *J. Exp. Bot.* **2001**, *52*, 961–969. [CrossRef] [PubMed]
42. Conesa, A.; Götz, S. Blast2GO: A Comprehensive Suite for Functional Analysis in Plant Genomics. *Int. J. Plant Genom.* **2008**, *2008*, 619832. [CrossRef] [PubMed]
43. Maere, S.; Heymans, K.; Kuiper, M. BiNGO: A Cytoscape plugin to assess overrepresentation of gene ontology categories in biological networks. *Bioinformatics* **2005**, *21*, 3448–3449. [CrossRef] [PubMed]
44. Mallory, A.C.; Vaucheret, H. Functions of microRNAs and related small RNAs in plants. *Nat. Genet.* **2006**, *38*, S31–S36. [CrossRef] [PubMed]
45. Alpert, P.; Oliver, M.J. 1 Drying Without Dying. In *Desiccation and Survival in Plants: Drying without Dying*; CABI: Wallingford, UK, 2002; p. 3.
46. Farrant, J.; Cooper, K.; Hilgart, A.; Abdalla, K.; Bentley, J.; Thomson, J.; Dace, H.W.; Peton, N.; Mundree, S.; Rafudeen, M. A molecular physiological review of vegetative desiccation tolerance in the resurrection plant Xerophyta viscosa (Baker). *Planta* **2015**, *242*, 407–426. [CrossRef] [PubMed]
47. Zhu, J.K. Salt and drought stress signal transduction in plants. *Annu. Rev. Plant Biol.* **2002**, *53*, 247–273. [CrossRef] [PubMed]
48. Nelson, D.E.; Repetti, P.P.; Adams, T.R.; Creelman, R.A.; Wu, J.; Warner, D.C.; Anstrom, D.C.; Bensen, R.J.; Castiglioni, P.P.; Donnarummo, M.G. Plant nuclear factor Y (NF-Y) B subunits confer drought tolerance and lead to improved corn yields on water-limited acres. *Proc. Natl. Acad. Sci. USA* **2007**, *104*, 16450–16455. [CrossRef] [PubMed]
49. Ferdous, J.; Hussain, S.S.; Shi, B.J. Role of microRNAs in plant drought tolerance. *Plant Biotechnol. J.* **2015**, *13*, 293–305. [CrossRef] [PubMed]
50. Wang, B.; Geng, S.; Wang, D.; Feng, N.; Zhang, D.; Wu, L.; Hao, C.; Zhang, X.; Li, A.; Mao, L. Characterization of Squamosa Promoter Binding Protein-LIKE genes in wheat. *J. Plant Biol.* **2015**, *58*, 220–229. [CrossRef]
51. Wang, C.; Deng, P.; Chen, L.; Wang, X.; Ma, H.; Hu, W.; Yao, N.; Feng, Y.; Chai, R.; Yang, G.; et al. A Wheat WRKY Transcription Factor TaWRKY10 Confers Tolerance to Multiple Abiotic Stresses in Transgenic Tobacco. *PLoS ONE* **2013**, *8*, e65120. [CrossRef] [PubMed]
52. Li, J.; Besseau, S.; Törönen, P.; Sipari, N.; Kollist, H.; Holm, L.; Palva, E.T. Defense-related transcription factors WRKY70 and WRKY54 modulate osmotic stress tolerance by regulating stomatal aperture in Arabidopsis. *New Phytol.* **2013**, *200*, 457–472. [CrossRef] [PubMed]

53. Besseau, S.; Li, J.; Palva, E.T. WRKY54 and WRKY70 co-operate as negative regulators of leaf senescence in Arabidopsis thaliana. *J. Exp. Bot.* **2012**, *63*, 2667–2679. [CrossRef] [PubMed]
54. Ruberti, C.; Kim, S.-J.; Stefano, G.; Brandizzi, F. Unfolded protein response in plants: One master, many questions. *Curr. Opin. Plant Biol.* **2015**, *27*, 59–66. [CrossRef] [PubMed]
55. Howell, S.H. Endoplasmic reticulum stress responses in plants. *Annu. Rev. Plant Biol.* **2013**, *64*, 477–499. [CrossRef] [PubMed]
56. Williams, B.; Verchot, J.; Dickman, M. When Supply Does Not Meet Demand-ER Stress and Plant Programmed Cell Death. *Front. Plant Sci.* **2014**, *5*, 211. [CrossRef] [PubMed]
57. Kim, Y.E.; Hipp, M.S.; Bracher, A.; Hayer-Hartl, M.; Ulrich Hartl, F. Molecular chaperone functions in protein folding and proteostasis. *Annu. Rev. Biochem.* **2013**, *82*, 323–355. [CrossRef] [PubMed]
58. Gupta, D.; Tuteja, N. Chaperones and foldases in endoplasmic reticulum stress signaling in plants. *Plant Signal. Behav.* **2011**, *6*, 232–236. [CrossRef] [PubMed]
59. Sarwat, M.; Naqvi, A.R. Heterologous expression of rice calnexin (OsCNX) confers drought tolerance in Nicotiana tabacum. *Mol. Biol. Rep.* **2013**, *40*, 5451–5464. [CrossRef] [PubMed]
60. Ghasempour, H.; Gaff, D.; Williams, R.; Gianello, R. Contents of sugars in leaves of drying desiccation tolerant flowering plants, particularly grasses. *Plant Growth Regul.* **1998**, *24*, 185–191. [CrossRef]
61. Farrant, J.M.; Moore, J.P. Programming desiccation-tolerance: From plants to seeds to resurrection plants. *Curr. Opin. Plant Biol.* **2011**, *14*, 340–345. [CrossRef] [PubMed]
62. Illing, N.; Denby, K.J.; Collett, H.; Shen, A.; Farrant, J.M. The signature of seeds in resurrection plants: A molecular and physiological comparison of desiccation tolerance in seeds and vegetative tissues. *Integr. Comp. Biol.* **2005**, *45*, 771–787. [CrossRef] [PubMed]
63. Ried, J.L.; Walker-Simmons, M. Group 3 late embryogenesis abundant proteins in desiccation-tolerant seedlings of wheat (*Triticum aestivum* L.). *Plant Physiol.* **1993**, *102*, 125–131. [CrossRef] [PubMed]
64. Bender, K.W.; Snedden, W.A. Calmodulin-Related Proteins Step Out from the Shadow of Their Namesake. *Plant Physiol.* **2013**, *163*, 486–495. [CrossRef] [PubMed]
65. Ma, Z.; Hu, X.; Cai, W.; Huang, W.; Zhou, X.; Luo, Q.; Yang, H.; Wang, J.; Huang, J. Arabidopsis miR171-Targeted Scarecrow-Like Proteins Bind to GT cis-Elements and Mediate Gibberellin-Regulated Chlorophyll Biosynthesis under Light Conditions. *PLOS Genet.* **2014**, *10*, e1004519. [CrossRef] [PubMed]
66. Le, T.T.T. Molecular and Functional Characterisation of an Osmotin Gene from the Resurrection Plant Tripogon Loliiformis. Ph.D. Thesis, Queensland University of Technology, Brisbane, Australia, 2018. Available online: https://eprints.qut.edu.au/115835 (accessed on 30 August 2018).
67. Barrs, H.; Weatherley, P. A re-examination of the relative turgidity technique for estimating water deficits in leaves. *Aust. J. Biol. Sci.* **1962**, *15*, 413–428. [CrossRef]
68. Qiagen. CLC Genomics Workbench. Available online: https://www.qiagenbioinformatics.com/products/clc-genomics-workbench/ (accessed on 22 April 2018).
69. Dai, X.; Zhao, P.X. psRNATarget: A plant small RNA target analysis server. *Nucleic Acids Res.* **2011**, *39*, W155–W159. [CrossRef] [PubMed]
70. Lohse, M.; Nagel, A.; Herter, T.; May, P.; Schroda, M.; Zrenner, R.; Tohge, T.; Fernie, A.R.; Stitt, M.; Usadel, B. Mercator: A fast and simple web server for genome scale functional annotation of plant sequence data. *Plant Cell Environ.* **2014**, *37*, 1250–1258. [CrossRef] [PubMed]

Permissions

All chapters in this book were first published by MDPI; hereby published with permission under the Creative Commons Attribution License or equivalent. Every chapter published in this book has been scrutinized by our experts. Their significance has been extensively debated. The topics covered herein carry significant findings which will fuel the growth of the discipline. They may even be implemented as practical applications or may be referred to as a beginning point for another development.

The contributors of this book come from diverse backgrounds, making this book a truly international effort. This book will bring forth new frontiers with its revolutionizing research information and detailed analysis of the nascent developments around the world.

We would like to thank all the contributing authors for lending their expertise to make the book truly unique. They have played a crucial role in the development of this book. Without their invaluable contributions this book wouldn't have been possible. They have made vital efforts to compile up to date information on the varied aspects of this subject to make this book a valuable addition to the collection of many professionals and students.

This book was conceptualized with the vision of imparting up-to-date information and advanced data in this field. To ensure the same, a matchless editorial board was set up. Every individual on the board went through rigorous rounds of assessment to prove their worth. After which they invested a large part of their time researching and compiling the most relevant data for our readers.

The editorial board has been involved in producing this book since its inception. They have spent rigorous hours researching and exploring the diverse topics which have resulted in the successful publishing of this book. They have passed on their knowledge of decades through this book. To expedite this challenging task, the publisher supported the team at every step. A small team of assistant editors was also appointed to further simplify the editing procedure and attain best results for the readers.

Apart from the editorial board, the designing team has also invested a significant amount of their time in understanding the subject and creating the most relevant covers. They scrutinized every image to scout for the most suitable representation of the subject and create an appropriate cover for the book.

The publishing team has been an ardent support to the editorial, designing and production team. Their endless efforts to recruit the best for this project, has resulted in the accomplishment of this book. They are a veteran in the field of academics and their pool of knowledge is as vast as their experience in printing. Their expertise and guidance has proved useful at every step. Their uncompromising quality standards have made this book an exceptional effort. Their encouragement from time to time has been an inspiration for everyone.

The publisher and the editorial board hope that this book will prove to be a valuable piece of knowledge for researchers, students, practitioners and scholars across the globe.

List of Contributors

Michaela Dostalova Merkerova, Hana Remesova, Zdenek Krejcik, Nikoleta Loudova, Andrea Hrustincova, Katarina Szikszai, Jaroslav Cermak and Monika Belickova
Institute of Hematology and Blood Transfusion, U Nemocnice 1, 128 20 Prague 2, Czech Republic

Anna Jonasova
General University Hospital, 128 08 Prague, Czech Republic

Ceren Eyileten, Dagmara Mirowska-Guzel, Aleksandra Soplinska and Marek Postula
Department of Experimental and Clinical Pharmacology, Medical University of Warsaw, Center for Preclinical Research and Technology CEPT, 02-097 Warsaw, Poland

Zofia Wicik
Rheumatology Division, Hospital das Clinicas HCFMUSP, Universidade de Sao Paulo, Sao Paulo, SP 01246-903, Brazil

Salvatore De Rosa
Division of Cardiology, Department of Medical and Surgical Sciences, "Magna Graecia" University, 88100 Catanzaro, Italy

Ciro Indolfi
Division of Cardiology, Department of Medical and Surgical Sciences, "Magna Graecia" University, 88100 Catanzaro, Italy
URT-CNR, Department of Medicine, Consiglio Nazionale delle Ricerche of IFC, Viale Europa S/N, 88100 Catanzaro, Italy

Iwona Jastrzebska-Kurkowska and Anna Czlonkowska
2nd Department of Neurology, Institute of Psychiatry and Neurology, 02-957Warsaw, Poland

Zhanhui Zhang and Jihua Tang
State Key Laboratory of Wheat and Maize Crop Science, Henan Agricultural University, Zhengzhou 450002, China

Sachin Teotia
State Key Laboratory of Wheat and Maize Crop Science, Henan Agricultural University, Zhengzhou 450002, China
Department of Biological Sciences, Michigan Technological University, Houghton, MI 49931, USA
Department of Biotechnology, Sharda University, Greater Noida 201306, India

Guiliang Tang
State Key Laboratory of Wheat and Maize Crop Science, Henan Agricultural University, Zhengzhou 450002, China
Department of Biological Sciences, Michigan Technological University, Houghton, MI 49931, USA

Teng Sun and Ji-Min Cao
Key Laboratory of Cellular Physiology, Ministry of Education, Department of Physiology, Shanxi Medical University, Taiyuan 030001, China

Meng-Yang Li and Pei-Feng Li
Institute for Translational Medicine, Qingdao University, Qingdao 266021, China

Tereza Brachtlova and Victor W. van Beusechem
Amsterdam UMC, Vrije Universiteit Amsterdam, Medical Oncology, Cancer Center Amsterdam, De Boelelaan 1117, 1007 MB Amsterdam, The Netherlands

Nardos Tesfaye Woldemariam and Rune Andreassen
Department of Life Sciences and Health, Faculty of Health Sciences, OsloMet–Oslo Metropolitan University, Pilestredet 50, N-0130 Oslo, Norway

Oleg Agafonov
Bioinformatics Core Facility, Department of Core Facilities, Institute of Cancer Research, Radium Hospital, Oslo University Hospital, 0379 Oslo, Norway

Bjørn Høyheim
Department of Basic Sciences and Aquatic Medicine, Faculty of Veterinary Medicine, Norwegian University of Life Sciences, 0454 Oslo, Norway

Ross D. Houston
Division of Genetics and Genomics, The Roslin Institute and Royal (Dick) School of Veterinary Studies, University of Edinburgh, Midlothian EH25 9RG, UK

John B. Taggart
Institute of Aquaculture, University of Stirling, Stirling, FK9 4LA, UK

Y.-H. Taguchi
Department of Physics, Chuo University, Tokyo 112-8551, Japan

Joseph L. Pegler, Jackson M. J. Oultram, Christopher P. L. Grof and Andrew L. Eamens
Centre for Plant Science, School of Environmental and Life Sciences, Faculty of Science, University of Newcastle, Callaghan 2308, Australia

Biao Chen, Jiao Yu, Lijin Guo, Zhijun Wang, Xiaolan Chen, Haiping Xu and Qinghua Nie
Department of Animal Genetics, Breeding and Reproduction, College of Animal Science, South China Agricultural University, Guangzhou 510642, China
National-Local Joint Engineering Research Center for Livestock Breeding, Guangzhou 510642, China
Guangdong Provincial Key Lab of Agro-Animal Genomics and Molecular Breeding, and Key Laboratory of Chicken Genetics, Breeding and Reproduction, Ministry of Agriculture, Guangzhou 510642, China

Mary Shannon Byers
Department of Biological Sciences, College of Life and Physical Sciences, Tennessee State University, Nashville, TN 37209, USA

Muhammad Shahbaz and Marinus Pilon
Biology Department, Colorado State University, Fort Collins, CO 80523-1878, USA

Hsiuying Wang
Institute of Statistics, National Chiao Tung University, Hsinchu 30010, Taiwan

Érika Frydrych Capelari
Programa de Pós-graduação em Biologia Celular e Molecular (PPGBCM), Centro de Biotecnologia, Universidade Federal do Rio Grande do Sul, Porto Alegre 91501-970, Brazil
Programa de Pós-graduação emGenética e Biologia Molecular (PPGBM), Universidade Federal do Rio Grande do Sul, Porto Alegre 91501-970, Brazil

Guilherme Cordenonsi da Fonseca and Frank Guzman
Programa de Pós-graduação em Biologia Celular e Molecular (PPGBCM), Centro de Biotecnologia, Universidade Federal do Rio Grande do Sul, Porto Alegre 91501-970, Brazil

Rogerio Margis
Programa de Pós-graduação em Biologia Celular e Molecular (PPGBCM), Centro de Biotecnologia, Universidade Federal do Rio Grande do Sul, Porto Alegre 91501-970, Brazil
Programa de Pós-graduação emGenética e Biologia Molecular (PPGBM), Universidade Federal do Rio Grande do Sul, Porto Alegre 91501-970, Brazil
Centro de Biotecnologia, Laboratório de Genomas e Populações de Plantas (LGPP), Universidade Federal do Rio Grande do Sul, Av. Bento Gonçalves, 9500—Laboratório 206 Prédio 43422, Porto Alegre 91501-970, Brazil

Karine Pinel, Louise A. Diver, Katie White, Robert A. McDonald and Andrew H. Baker
Institute of Cardiovascular and Medical Sciences, BHF Glasgow Cardiovascular Research Centre, University of Glasgow, Glasgow G12 8TA, UK

Isaac Njaci, Brett Williams and Sagadevan Mundree
Centre for Tropical Crops and Biocommodities, Queensland University of Technology, Brisbane, QLD 4000, Australia

Claudia Castillo-González and Xiuren Zhang
Centre for Tropical Crops and Biocommodities, Queensland University of Technology, Brisbane, QLD 4000, Australia
Department of Biochemistry and Biophysics, Institute for Plant Genomics and Biotechnology, Texas A&M University, College Station, TX 77843, USA

Martin B. Dickman
Department of Plant Pathology and Microbiology, Institute for Plant Genomics and Biotechnology, Texas A&M University, College Station, TX 77843, USA

Index

www.ingramcontent.com/pod-product-compliance
Lightning Source LLC
LaVergne TN
LVHW080949210926
848485LV00004B/811

* 9 7 8 1 6 3 9 8 7 7 2 2 5 *